Lecture Notes in Electrical Engineering

Volume 188

For further volumes:
http://www.springer.com/series/7818

R. Malathi · J. Krishnan
Editors

Recent Advancements in System Modelling Applications

Proceedings of National Systems Conference 2012

Editors
R. Malathi
J. Krishnan
Department of Electronics and
Instrumentation Engineering
Faculty of Engineering and Technology
Annamalai University
Annamalai Nagar
Tamil Nadu
India

ISSN 1876-1100 ISSN 1876-1119 (electronic)
ISBN 978-81-322-1720-6 ISBN 978-81-322-1035-1 (eBook)
DOI 10.1007/978-81-322-1035-1
Springer New Delhi Heidelberg New York Dordrecht London

Softcover re-print of the Hardcover 1st edition 2013

Printed on acid-free paper

Springer is part of Springer Science+Business Media (www.springer.com)

Preface

This series in Electrical Engineering is an abridged form of selected papers originally submitted to the 36th National Systems Conference on Recent Advancements in System Modelling Applications (NSC 2012), held between 6 and 8 December 2012 at Annamalai University, Annamalai Nagar, Tamil Nadu, India. The conference is part of the Silver Jubilee celebrations of the Department of Electronics and Instrumentation Engineering attached to the Faculty of Engineering and Technology, Annamalai University.

This conference is an annual event of the System Society of India (SSI) held at national level to epitomise the latest technological advances in system modelling. The objective of the conference is to bring together researchers and developers from academia and industry working in the areas of engineering sciences, life sciences, agricultural sciences and management sciences, with a special focus on system modelling.

A systems approach to problem solving requires an ingenious perspective to deduce the important system variables and establish the relationships between them. The key to understanding thereafter evolves from modelling the system, and realising how the variables change over time as the problem parameters are changed. The system dynamics that originate as a function of the variations in parameters serve to evolve measures to archive the desired response.

The goal of these lecture notes is to describe the central concepts in system modelling and offer an insight into their applications. The findings in the domain of computer science include areas of image and speech signal processing, intelligent optimisation techniques and communication networks. The research progress in the discipline of electrical science is pivoted towards power electronics and drives. The problems that surface in computational biology and its modelling techniques are addressed in the biological science. A host of methodologies suggested tackling issues that pertain to immunology beehives the paramedical science.

The contributions peer reviewed by the programme committee members/technical committee reviewers are included in the lecture series. The reviews echoed a focus centred on originality, quality and relevance to the theme of the conference.

A number of parallel sessions with a view to accommodate the wide scope of the central theme and provide an interactive platform for the participants to share and debate their research findings highlighted the proceedings.

The editors express their thanks to the authors for submitting their research papers to the conference, the program committee/technical review committee and the committee members for their dedicated efforts. The editors profoundly thank the authorities of the Annamalai University for providing the infrastructural support to this conference.

R. Malathi
J. Krishnan

About the Editors

Dr. R. Malathi is a Professor in the Department of Electronics and Instrumentation Engineering, attached to the Faculty of Engineering and Technology at Annamalai University, Annamalai Nagar. She received her Ph.D. in Computational Bioengineering from the Indian Institute of Technology, Chennai. Her career spans 20 years of experience in teaching and 6 years in research and she has to her credit more than 25 research publications in peer reviewed International journals and conferences. She is a member of the Board of Studies and the Board of Exams in various institutions. Her experience has brought out a number of M.E./M.Tech. theses in Instrumentation Engineering and is currently supervising two Ph.D. scholars in the area of Bioengineering. She holds the distinction of carving out societal findings through AICTE/UGC sponsored projects. She is a member/Life member of various National and International Society such as Biophysical Society, ISTE, American Heart Association, ISI, etc. She carries with her the distinction of bagging the Best Thesis Award from the Indian Institute of Technology, Madras and the Best Paper Award in International conferences. Her research and teaching interests include Bioengineering, Soft computing techniques and control and Instrumentation.

Dr. J. Krishnan acquired his doctoral degree in the field of Computational Bioengineering from the Indian Institute of Technology, Chennai. His research pursuit has brought to light more than 30 research articles in peer reviewed International journals and conferences. He has followed it up through a number of DST/AICTE sponsored projects to his credit. His innovative ideas have surfaced in a number of post graduate dissertations in the Control and Instrumentation domain and he has six research scholars under him. He is a member of various national and international societies such as Biophysical Society, ISTE, American Heart Association and ISI among others. He carries with him 21 years of teaching experience and 5 years of research experience. He is currently a professor in the Department of Electronics and Instrumentation Engineering, attached to the

Faculty of Engineering and Technology at Annamalai University, Annamalai Nagar. His research and teaching interests include Computational Bioengineering, Control system and Modelling of dynamical systems.

Contents

Chapter 1
Smart Grid Environment with Effective Storage and Computational Facilities

T. Rajeev and S. Ashok

Abstract Smart grid technology provides good support for power generation from consumer premises using solar/wind. The difficulty associated with operation of such grids are, lack of an operating platform for the coordinated operation of large number of distributed sources and requirement of huge computational and storage facilities. The immense potential of cloud computing technology can be utilized to address these issues. Also the resources in various substations can be shared to reduce the cost of operation. The cloud computing architecture provides a user friendly environment for the reliable operation of smart grids and it supports various smart grids applications. The proposed architecture for data storage has been realized in the open stack cloud environment.

Keywords Cloud computing · Infrastructure as a service · Storage infrastructure (swift) · Middleware · Distributed generation

1.1 Introduction

A cloud computing environment for smart grid operation enables better integration of renewable forms of energy, better consumer interaction and involvement in power management, promoting power generation from consumer premises using renewable sources like solar/wind. The new strategic decision from Govt. of India plans to supply solar panels to consumers on subsidized rates. The idea is to

---

T. Rajeev (✉)
Department of Electrical Engineering, College of Engineering, Trivandrum, India
e-mail: mail2rajeevt@gmail.com

S. Ashok
Department of Electrical Engineering, NIT, Calicut, India

R. Malathi and J. Krishnan (eds.), *Recent Advancements in System Modelling Applications*, Lecture Notes in Electrical Engineering 188,
DOI: 10.1007/978-81-322-1035-1_1, © Springer India 2013

promote power generation from premises. The new deregulation policies permit power generation by different entities. More advanced smart grid management is needed to tackle the situation, in which large number of distributed resources with complicated control functionalities [6]. Cloud computing is evolving as a key computing platform for sharing resources that include infrastructures software, applications, and business processes. Virtualization is a core technology for enabling cloud resource sharing [9], is extremely well suited to a dynamic cloud infrastructure because it provides important advantages in sharing, manageability, and isolation. Multiple users and applications can share physical resources without affecting one another.

Cloud computing allows the operational entities to create elastic environments that expand and contract based on the workload and target performance parameters. Cloud services are generally classified into three categories (i) infrastructure as a service (IaaS) (ii) Platform as a service (PaaS) and (iii) Software as a service (SaaS). IaaS is the delivery of computer infrastructure (server, storage and network), and associated software (OS virtualization technology) as a service. The consumer uses fundamental computing resources such as processing power, storage, networking components or middleware. Software as a service features a complete application offered as a service on demand. A single instance of the software runs on the cloud and services multiple end users or client organizations. All the applications that run on the cloud and provide a direct service to the customers located in the SaaS layer. The application developers can either use the PaaS layer to develop and run their applications.

An effective and efficient cloud computing architecture for enabling application deployment environment offers smooth application sharing over the internet and it provides resource sharing. The application is available in the internet for common users on pay on go basis. Wenjun Zhang [11] address the serious problems in cloud application development such as complex architecture, web services application programming interfaces, cost, poor user interface etc., and found that these problems arises from inappropriate functionality segmentation between client-side and server-side. That is, web services, application logic and transaction logic are overly concentrated on the server side, while on the client side computing power has not been fully utilized. The author had demonstrated 2-Tier cloud-architecture (2TCAR), which contains rich client tier based on rich internet application(RIA) and server-side cloud tier based on Simple data base storage cloud. Rich client tier is maximized to implement almost all functionalities of cloud application and transaction logic. Functionality in server-side cloud tier is simplified only to implement data storage and query, communication between two tiers is also simplified. The paper also discusses about the design of simple data base cloud in server-side cloud tier, and communication between two tiers.

Network-based cloud computing is rapidly expanding as an alternative to conventional office-based computing. As cloud computing becomes more widespread, the energy consumption of the network and computing resources that underpin the cloud will grow. In [1] authors presented an analysis of energy consumption in cloud computing. The analysis considered both public and private clouds, and includes energy consumption in switching and transmission as well as data processing and data storage. They showed that energy consumption in transport and switching can be a significant percentage of total energy consumption in cloud computing. Cloud computing can enable more energy-efficient use of computing power, especially when the computing task are of low intensity or infrequent. Under such circumstances cloud computing consume more energy than conventional computing where each user performs all computing on their own personal computer.

1.2 Cloud Computing in Smart Grid

The distribution system was initially designed to operate with minimum real time intervention except for responding to fault, allowing routine maintenance and network modifications [2]. Many distribution systems have utilized advanced digital devices such as relaying systems, controls, remote monitoring and communication technologies. With the emergence of smart grid, the demand of data storage, computing and real time operational necessity are increasing steadily. The existing centralized power system is incapable of handling the large number of resources and its control functionalities. The operation mode of smart micro grids facilitates both islanded mode and grid connected mode of operation [4]. An important responsibility of network managers is maintaining control over the security of the network [10] and data storage. The necessary security and fire wall protection has to be incorporated to the system for applications deployed for public cloud. Cloud computing in large power grid includes cloud data service center, is considered as one of the central options which can integrate current infrastructure resources of the enterprise like hardware, high-performance distributed computing and data platform etc. and provide the required computing and storage need of a smart grid. Leema (2011) proposed a layered model of cloud computing application in smart grid for smart devices. It will make full use of existing resources and shields the heterogeneity, providing unified interface to the up-layer applications, which will have the characteristics of expandability, availability, virtualization, user transparency, service diversity and intelligence. The interaction between cloud computing and smart grid through the algorithms which involve massive data centers has been presented in [5]. The authors focused on one design possibility that can improve load balancing in grid by carefully distributing the service request among data centers in a cloud computing system.

1.3 Proposed Architecture

The entities in a restructured power system have to process a large amount of transactions with the integration of distributed generation into the power grid. The architecture proposed here offers flexible environment that securely permits to integrate smart grid applications. The architecture is capable of accommodating more features. By incorporating suitable data handling and data messaging services, hardware requirement in smart grid environment can be reduced considerably [8]. The architecture considered here consists of different modules as shown in Fig. 1.1. The storage computing and networking is represented in the infrastructure layer. End users access cloud based applications through a web browser or a light weight desktop or mobile application while the business software and data are stored on servers at a remote location [3]. Infrastructure as a service delivers basic storage and compute capabilities as standardized services over the network. Servers, storage systems, switches, routers, and other systems are pooled and made available to handle workloads that range from application components to high-performance computing applications. Operating system requires hypervisor and open stack compute control the hypervisors through API server. There are different hypervisors available. The process for selecting a hypervisor means priority and making decisions based on resource constraints, numerous supported features and required technical specifications. Kernel virtual machine (KVM) is used as hypervisor in the architecture and there is flexibility for selecting multiple hypervisors for different zones.

Open stack is the cloud middleware used in the architecture. More efficient use of the computing capacity is achieved with this, thereby increasing productivity, speedy deployment of different applications, protecting sensitive data while making savings in capital cost. There are three main service families under open stack namely compute infrastructure (nova), storage infrastructure (swift) and imaging service (glance) [10]. Networked on line storage facilities are provided in the architecture, where data is stored in a virtualized pool of storages. The virtualized resources are used by the customers according to the requirement. The resources may span across multiple servers. The storage services may accessed through a web service interface application programming. Swift is an object store to store large number of objects distributed across multiple hardware. Swift has built in redundancy and features like backing up, archiving data/services. It is

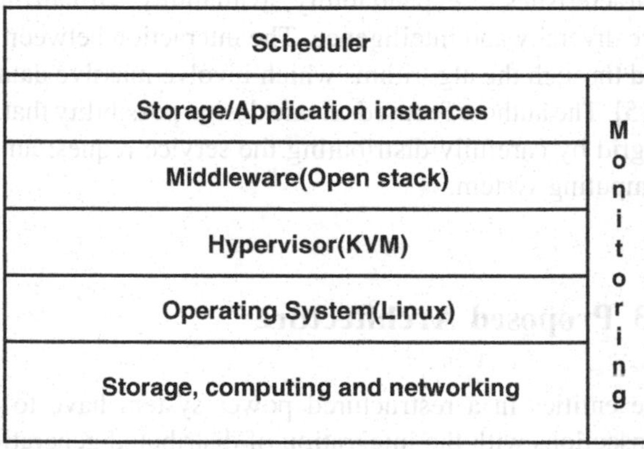

Fig. 1.1 Logical view of cloud test bed for smart grid

Scheduler	
Storage/Application instances	M o n i t o r i n g
Middleware(Open stack)	
Hypervisor(KVM)	
Operating System(Linux)	
Storage, computing and networking	

Table 1.1 Hardware/software requirements

Item	Recommended hardware/software
Cloud controller	HpProliant, 64bit X86, 2048KCache, 12 GB RAM, 2X1TBHDD, 1 GB NIC
Client node	Intel Pentium4 3.20 GHz; 2048 Cache; 8 GB RAM; 1 TB HDD.2X1GBNIC
Operating system	Ubuntu11.10
Hypervisor	KVM
Middleware	Openstack
Monitoring	Web stress tool
Testing	Smart grid applications

scalable to multiple data objects and data. The functions includes storing the machine images, working as an independent data container offering elasticity and flexibility of cloud based storage for web applications. The discovery, storage and retrieval of virtual machine images can be provided through Glance, the open stack Image Service. It includes three components: glance-api, glance-registry and the image store. The glance-api accepts API calls and actual image blobs are placed in the image store. The glance-registry stores and retrieves metadata about images. System has been divided into two sections frontend and backend. They connect to each other through a network, usually the Internet. The hardware/software requirements of the test bed are shown in Table 1.1. Cloud storage is a model of networked online storage where data is stored on virtualized pools of storage. Depending upon the requirement, entities can decide about setting up their own storage or lease storage capacity from the service providers and use it for their storage needs. Cloud storage services have gained popularity in the last years. These services allow users to store data off-site, masking the complexities of the infrastructure supporting the storage service.

To store large amounts of data from thousands of generating sources cloud storage systems build their services over distributed storage infrastructures, more scalable and more reliable than centralized solutions. The architecture proposed here offers dynamic interoperability of heterogeneous storage services. A private cloud computing is intended for the work where as a public cloud environment, the storage grid is a virtualized origin of geographically distributed storage systems. Hence security requirements are high. The consumer uses an application, but does not control the operating system, hardware or network infrastructure on which it is running. Software as a service features a complete application offered as a service on demand. A single instance of the software runs on the cloud and services multiple end users or client organizations.

1.4 Results and Discussions

A large amount of time data from the renewable sources needs to be processed for the real time monitoring and control of the future smart grids. The application has been deployed in the cloud, as per the logical diagram. The cloud computing system has been divided into two sections frontend and backend. They connect to each other through a network, usually the Internet. The front end is the side the computer user or client. The back end is the "cloud" section of the system. The performance of data storage applications in open stack cloud test bed has been realized in the work. The storage applications were deployed as instances in the test bed. A linearly increasing rate of 50/s was applied to test the system performance. The performance guarantee of the cloud environment is needed for attracting service entities interest in this area. The result obtained is as shown in Fig. 1.2 shows that data centre utilization is dynamically adjusted and the resources are allocated on time. In conventional computing environment the error percentage increases with increase in the rate of user request. But the dynamic resource allocation in cloud environment effectively manages the situation and maintains a good success rate for high values user request. Open stack dashboard is configured for monitoring the status of instances, resource allocation, capacities assigned with various applications etc. The advantage of using open stack environment is that it offers the flexibility and freedom to incorporate the progressing developments in the smart grid sector.

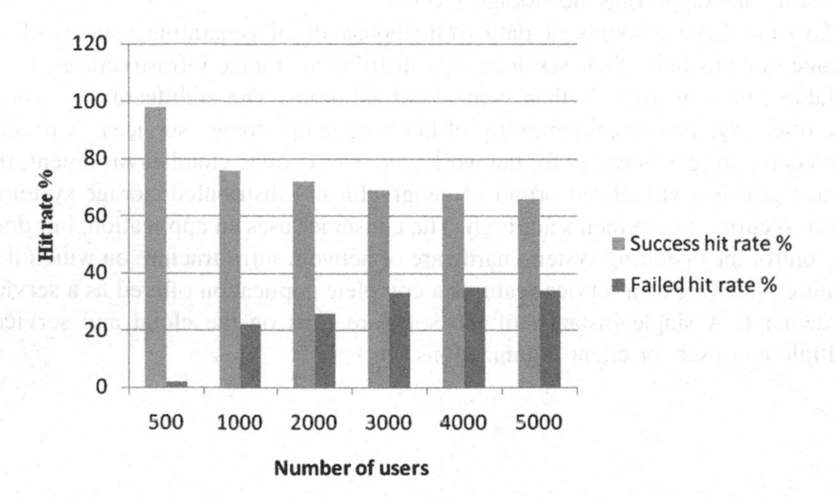

Fig. 1.2 Performance result of test bed

1.5 Conclusions

Technological advances in distributed generation have resulted in smart micro grids that provide a flexible, efficient and environmental friendly power source. The reliable and efficient operation of a distribution system with large number of renewable sources needs to address huge data storage and processing issues associated with integrating many small DG sources. The application model in cloud computing environment coordinates data storage and processing needs. The proposed architecture for application deployment in smart grid environment has been implemented in the test bed. The performance of open stack cloud environment for data storage application has been verified.

References

1. Baliga J, Ayre RWA, Hinton K, Tucker RS (2011) Green cloud computing balancing energy in processing, storage, and transport. Proc IEEE 99(1):149–166
2. Katiraei F, Irvani M, Lehn PW (2005) Microgrid autonomous operation during and subsequent to Islanding process. IEEE Trans Power Deliv 20(2):248–257
3. Kelly T, Zhang A (2006) Predicting performance in distributed enterprise applications. HP Labs Tech Report, HPL-2006-76
4. Krishnamurthy S, Jahns TM, Lasseter RH (2008) The operation of a diesel gensets in a CERTS microgrid. In: IEEE power and energy society general meeting—conversion and delivery of electrical energy in 21st centuary, pp 1–8
5. Mohsenian-Rad A-H, Leo-Garcia A (2010) Cordination of cloud computing and smart power grids. In: Proceedings IEEE international conference on smart grid communications, pp 368–372
6. Moreno-Munoz A (ed) (2007) Power quality mitigation technologies in a distributed environment, 1st edn. Springer
7. OpenStack Beginner's Guide (for Ubuntu—Precise) v3.0 (2012) 7 May 2012
8. Rajeev T, Ashok S (2011) A cloud computing approach for power management of micro grids. In: Proceedings IEEE PES innovative smart grid technologies India(ISGT-India), pp 49–52
9. Wood T, Shenoy P, Venkataramani A, Yousif M (2007) Black-box and grey-box strategies for virtual machine migration. In: Proceedings 4th USENIX symposium network system design implementation, pp 229–242
10. Zhabelova G, Vyatkin V (2012) Multiagent smart grid automation architecture based on IEC 61850/61499 intelligent logical nodes. IEEE Trans Ind Electron 59(5):2351–2362
11. Zhang W (2012) Two tier cloud architecture and applications in electronic health reccord. J Softw 7(4):765–772

Chapter 2
Implementation of Kalman Filter to Monitor the Level Fluctuations in a Dam Using FPGA

K. Shashank, Nitin Ravi, M. Rakshith and J. V. Alamelu

Abstract In this paper we study the design, implementation and evaluate the performance of a Kalman filter using FPGA. It is essential to be familiar with minimum mean square error filtering and state space methods. It is important that the set of equations, their relevance to one another and indeed the overall functionality of the algorithm that defines the Kalman filter require complete understanding. The filter will be implemented with field programmable gate arrays (FPGA), to monitor the level fluctuations for a dam/reservoir.

Keywords Kalman · FPGA · State space method · Mean square error filtering

2.1 Introduction

The Kalman Filter is a means to predict the future behavior of a system based on past behavior. A system's past behavior is, in a way, remembered and used along with measurements to make the predictions of how the system might behave in the future.

K. Shashank (✉) · N. Ravi · M. Rakshith · J. V. Alamelu
Department of Instrumentation Technology, M.S. Ramaiah Institute of Technology, Bangalore, India
e-mail: shashbeckmmm@gmail.com

N. Ravi
e-mail: nitinravi1@gmail.com

M. Rakshith
e-mail: rakshith_swk@yahoo.co.in

J. V. Alamelu
e-mail: jvalamelu@gmail.com

R. Malathi and J. Krishnan (eds.), *Recent Advancements in System Modelling Applications*, Lecture Notes in Electrical Engineering 188,
DOI: 10.1007/978-81-322-1035-1_2, © Springer India 2013

According to the paper published by Kleeman (1995) the reason that mathematical models such as the Kalman Filter are useful to a designer is because virtually all systems are non-deterministic. In other words, few if any systems are devoid of randomness or stochastic behavior. Whether a system contains stochastic processes or the environment that may act upon a system is itself stochastically governed is non-deterministic [1].

A DSP (Digital Signal Processor) processor on the other hand is a normal processor optimized for faster floating point calculations to aid in signal processing without much modification. Preferably FPGA is chosen when aimed to test/simulate. Current DSP's have one two MAC (Multiply Accumulator) units. In our summary of results for Kalman Filtering we draw heavily upon the work of Dan Simon [2].

These units are used sequentially. If one needs more than two MAC's (for example, over 100 tap FIR filter with sample rate of 200 MHZ) then parallel MAC's with single cycle computation is possible to realize only using FPGA's with current trends.

Why FPGA and not ASIC?

- Integrated circuit costs are rising aggressively
- ASIC complexity has lengthened development time
- R&D resources and headcount are decreasing.
- Revenue losses for slow time-to-market are increasing.
- Financial constraints in a poor economy are driving low-cost technologies.

These trends make FPGA's a better alternative than ASIC's for a larger number of higher-volume applications than they have been historically used for, to which the company attributes the growing number of FPGA design starts.

The paper is explained under the following topics:

- Kalman filter.
- Design optimization.
- Application and Outcome of the project.
- Block diagram for implementation.
- Results and discussions.
- Advantages of designing the model.

2.2 Kalman Filter

The Kalman filter equations are a set of mathematical equations that provide an efficient computational means to estimate the state of a process, in a way that minimizes the mean of the squared error [3]. The filter is a very powerful device as it supports the estimation of past, present and future states. It even extends its functionality so it can carry out this procedure when the precise nature of the modelled system is unknown. The system may or may not be subjected to a series of random disturbances, when this occurs it is required to estimate the state variables from noisy observations. The Kalman filter takes inaccurate, incomplete and noisy data combined with environmental disturbances beyond a designer's control and over time develops an optimal estimate of desirable quantities [3]. The Kalman Filter is a means to predict the future behaviour of a system based on past behaviour. A system's past behaviour is, in a way, remembered and used along with measurements to make the predictions of how the system might behave in the future.

The filter estimates its process by using a form of feedback control, as implied in the previous section. The filter will estimate the process state at some time and then obtains its feedback in the form of noisy measurements. These equations fall into the category of either Time update equations or measurement update equations.

$$X_k = AX_{k-1} + Bu_{k-1} + w_{k-1} \tag{2.1}$$

$$z_k = HX_k + v_k \tag{2.2}$$

2.2.1 Time Update Equations

The time update equations are used to predict the current state and covariance matrix, used in time t + 1 to predict the previous state. These equations can be generally seen as predictor equations as they are responsible for projecting forward in time. K is representative of the time step, so the time update equations are basically indicative of K + 1.

2.2.2 Measurement Update Equations

The measurement equations are responsible for feedback and for correcting the errors that have been made in the time update equations [3]. In a sense they are back propagating to get new values for the prior state to improve the guess for the next state. These equations can be seen as corrector equations and the final estimation algorithm resemble that of a predictor corrector algorithm. So by definition measurement equations adjust the projected estimate by an actual measurement at that time.

TIME UPDATE (PREDICT):

$$\hat{x}_{-k} = A\hat{x}_{k-1} + Bu_k \tag{2.3}$$

$$P_{-k} = AP_{k-1}A^T + Q \tag{2.4}$$

MEASUREMENT UPDATE (CORRECT):

$$K_k = P_{-k}H^T\left(HP_{-k}H^T + R\right)^{-1} \tag{2.5}$$

$$\hat{x}_k = \hat{x}_{-k} + K_k(z_k - H\hat{x}_{-k}) \tag{2.6}$$

$$P_k = (I - K_kH)P_{-k} \tag{2.7}$$

P_k Prior Error Convergence
K Kalman Gain
z_k State Measurement
x Posterior State Estimate
R_k Measurement Error Covariance
Q_k Random White Noise
A_k Variable
B_k Variable
μ_k Control Variable
H_k Matrix—valued Function

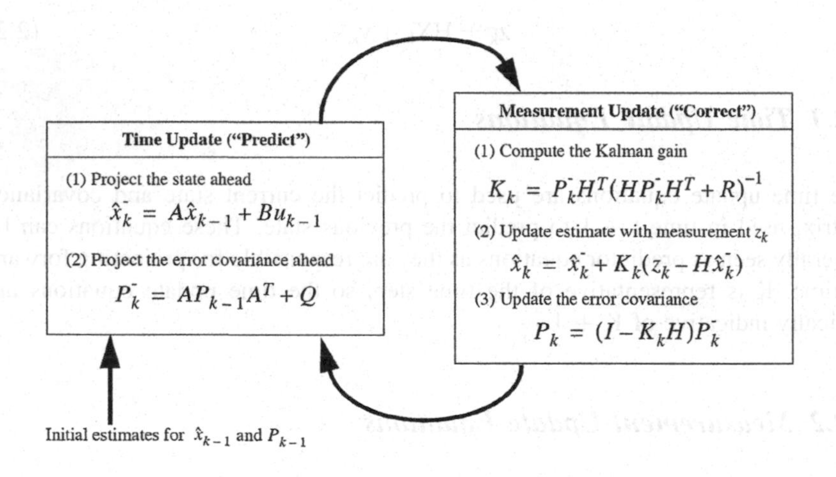

Time Update ("Predict")
(1) Project the state ahead
$\hat{x}_k^- = A\hat{x}_{k-1} + Bu_{k-1}$
(2) Project the error covariance ahead
$P_k^- = AP_{k-1}A^T + Q$
Measurement Update ("Correct")
(1) Compute the Kalman gain
$K_k = P_k^- H^T (HP_k^- H^T + R)^{-1}$
(2) Update estimate with measurement z_k
$\hat{x}_k = \hat{x}_k^- + K_k(z_k - H\hat{x}_k^-)$
(3) Update the error covariance
$P_k = (I - K_k H)P_k^-$
Initial estimates for $\hat{x}_{k-1}$ and P_{k-1}

Fig. 2.1 Operation of discrete Kalman filter

Above set of equations provide a definition of variables in the Kalman filter equations. In the equations, a measurement of the process, Z_k and X_k are previously defined by linear stochastic difference equations. For practical examples, process noise covariance Q and measurement noise covariance R matrices, might change with each time step or measurement. However for the purposes of our project, we have assumed them to be constant values. A is an n by n matrix in the difference equation and relates the state at the previous time step k − 1 to the state at the current time step k, without the presence of process noise. Once again A is assumed to be fixed despite the fact that this would more realistically be susceptible to change with each time step. Matrix B relates the control variable to the state x. Matrix H relates the state to the measurement Z_k (Fig. 2.1) [4].

The primary goal of this design is to maximize speed or throughput or drive through as much as possible. As a secondary goal, minimization of area and power will also be considered (Fig. 2.2).

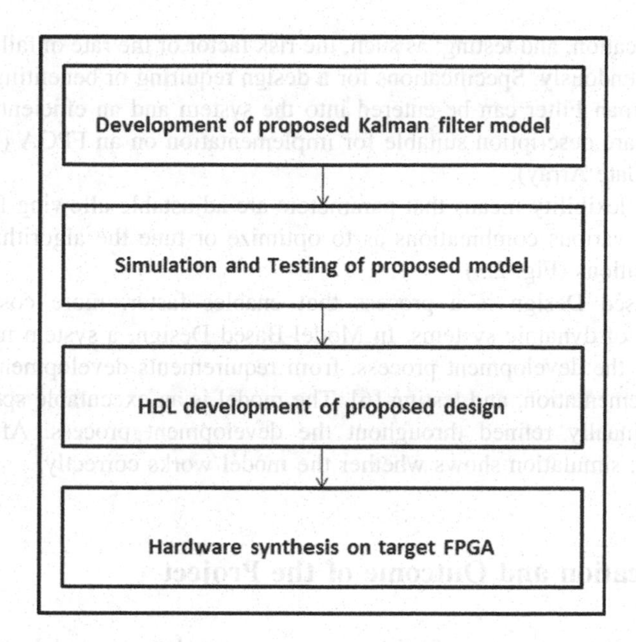

Fig. 2.2 Flow chart of methodology

2.3 Design Optimization

Design optimization can be accomplished in several ways depending on what type of optimization is required [5]. For the Kalman filter described by us, optimization for speed is most critical. Parallelization and pipelining are two methods used to help create a hardware design that fulfils this requirement. These optimizations are studied in order of priority as they relate to our system design. The designed model proposes a system that allows a designer to greatly reduce the time needed for

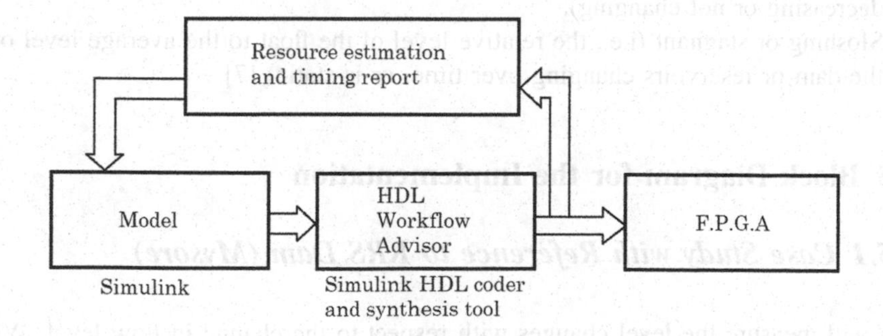

Fig. 2.3 Block diagram to convert MATLAB to Verilog

design, verification, and testing; as such, the risk factor or the rate of failure is also reduced tremendously. Specifications for a design requiring or benefiting from the use of a Kalman Filter can be entered into the system and an efficient and optimized hardware description suitable for implementation on an FPGA (Field Programmable Gate Array).

The code flexibility means that parameters are adjustable allowing for experimentation of various combinations as to optimize or tune the algorithm for different applications (Fig. 2.3).

Model-Based Design is a process that enables faster, more cost-effective development of dynamic systems. In Model-Based Design, a system model is at the centre of the development process, from requirements development, through design, implementation, and testing [6]. The model is an executable specification that is continually refined throughout the development process. After model development, simulation shows whether the model works correctly.

2.4 Application and Outcome of the Project

The Kalman filter removes noise by assuming a pre-defined model of a system. Therefore, the Kalman filter model is meaningful.

The Kalman filter model can be set up in all dams across the world. The outcome of this project is complete automation of dam operations such as the opening and closing of the gates based on the set level values. Leakage or any irregularity in functioning of the dam can be detected by comparing measured value with the estimated level value.

This technique aims at estimating the level of water in the tank, which is unknown. The measurements obtained are from the level of the float. This could be an electronic device, or a simple mechanical device

The water could be:

1. Filling, emptying or static (i.e., the average level of the tank is increasing, decreasing or not changing).
2. Sloshing or stagnant (i.e., the relative level of the float to the average level of the dam or reservoirs changing over time, or is static) [7].

2.5 Block Diagram for the Implementation

2.5.1 Case Study with Reference to KRS Dam (Mysore)

We will measure the level changes with respect to the change in flow level. We will also measure if any loss of water occurs due to leakage. We will present a look up table (LUT) with the parameters of the dam such as inlet capacity, outlet capacity, level (storage capacity) (Fig. 2.4).

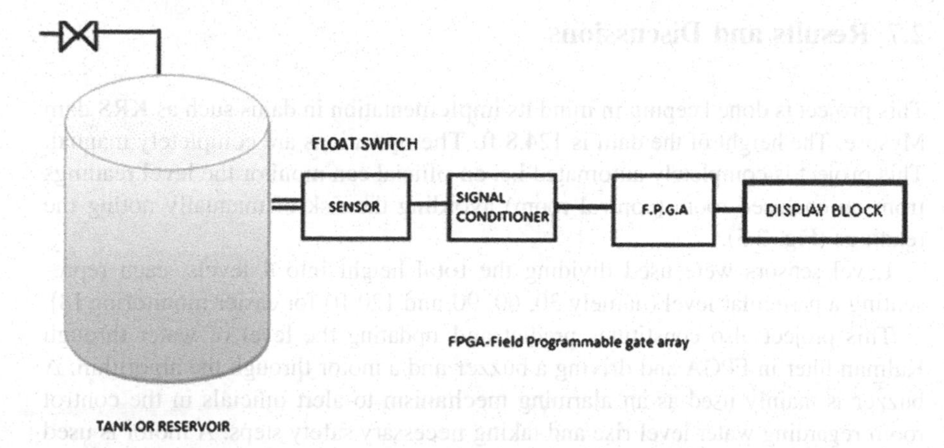

Fig. 2.4 Block diagram of the prototype model

2.6 Project Diagram

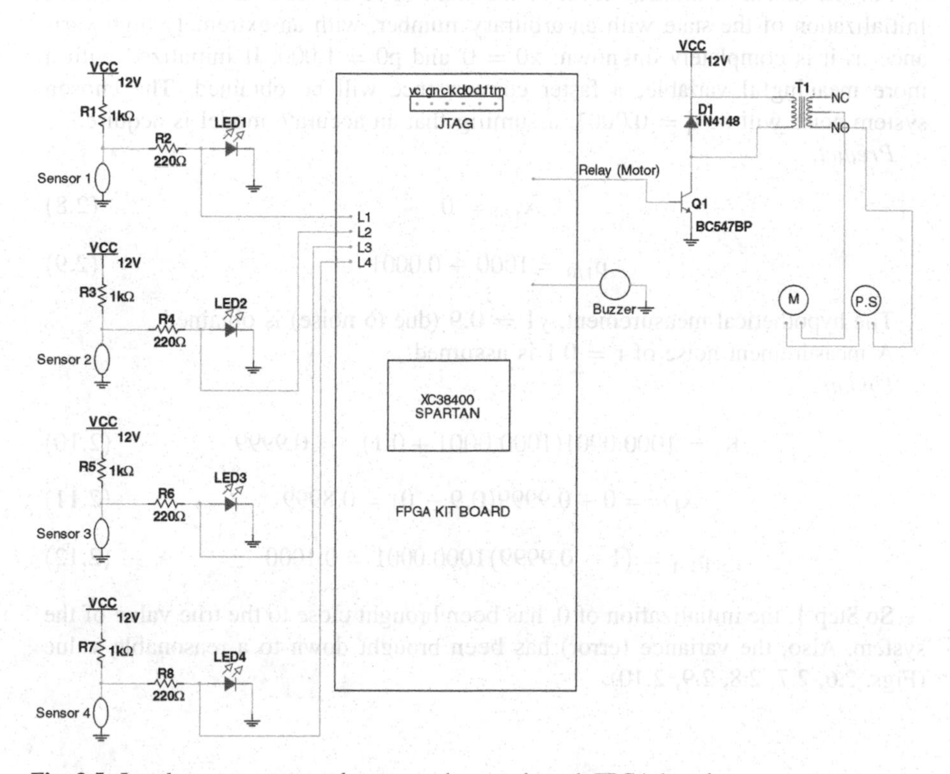

Fig. 2.5 Level sensors connected to motor, buzzer through FPGA board

2.7 Results and Discussions

This project is done keeping in mind its implementation in dams such as KRS dam Mysore. The height of the dam is 124.8 ft. The operations are completely manual. This project is completely automated i.e. an official can monitor the level readings from an enclosed room (control room) avoiding the risk of manually noting the readings (Fig. 2.5).

Level sensors were used dividing the total height into 4 levels, each representing a particular level (namely 30, 60, 90 and 120 ft) for easier monitoring [8].

This project also constitutes predict and updating the level of water through Kalman filter in FPGA and driving a buzzer and a motor through the algorithm. A buzzer is mainly used as an alarming mechanism to alert officials in the control room regarding water level rise and taking necessary safety steps. A motor is used as gate mechanism. When overflow level is reached, gates (motor) will automatically be switched on reducing the possibility of human delay and error in operations. The shortcomings of the present operational mechanism is that the level readings are noted down manually only twice a day (0630 and 1,830 hrs). This type of manual level monitoring is very risky during rainy seasons. This project being automated reduces the element of risk involved.

For the first test, the true level of the dam or reservoir is L = 1 is assumed. Initialization of the state with an arbitrary number, with an extremely high variance as it is completely unknown: x0 = 0 and p0 = 1,000. If initialized with a more meaningful variable, a faster convergence will be obtained. The chosen system noise will be q = 0.0001, assuming that an accurate model is acquired.

Predict:

$$x_{1/0} = 0 \tag{2.8}$$

$$p_{1/0} = 1000 + 0.0001 \tag{2.9}$$

The hypothetical measurement, y1 = 0.9 (due to noise) is obtained.

A measurement noise of r = 0.1 is assumed.

Update:

$$K_1 = 1000.0001(1000.0001 + 0.1)^{-1} = 0.9999 \tag{2.10}$$

$$x_{1/1} = 0 + 0.9999(0.9 - 0) = 0.8999 \tag{2.11}$$

$$p_{1/1} = (1 - 0.9999)1000.0001 = 0.1000 \tag{2.12}$$

So Step 1, the initialization of 0, has been brought close to the true value of the system. Also, the variance (error) has been brought down to a reasonable value (Figs. 2.6, 2.7, 2.8, 2.9, 2.10).

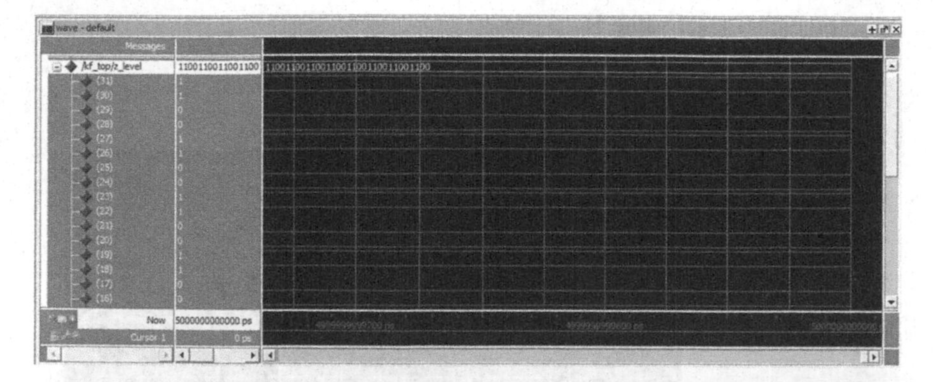

Fig. 2.6 Intial level input [bit 31–16]

Fig. 2.7 Intial level input [bit 15–0]

Here the initial level input is
Z_level = 11001100110011001100110011001100

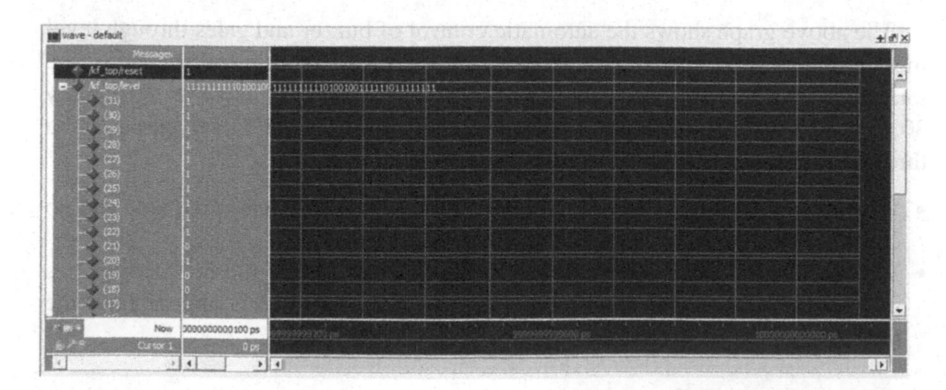

Fig. 2.8 Predicted level value [bit 31–16]

The predicted level value is
level = 11111111101001001111101111111

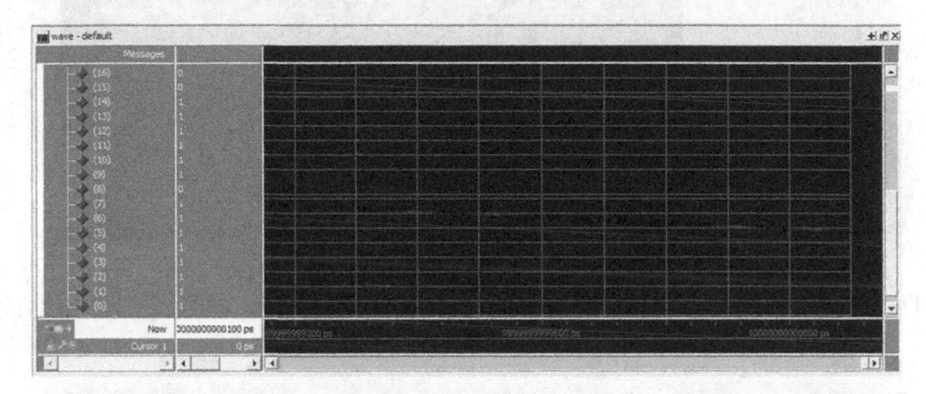

Fig. 2.9 Predicted level value [bit 15–0]

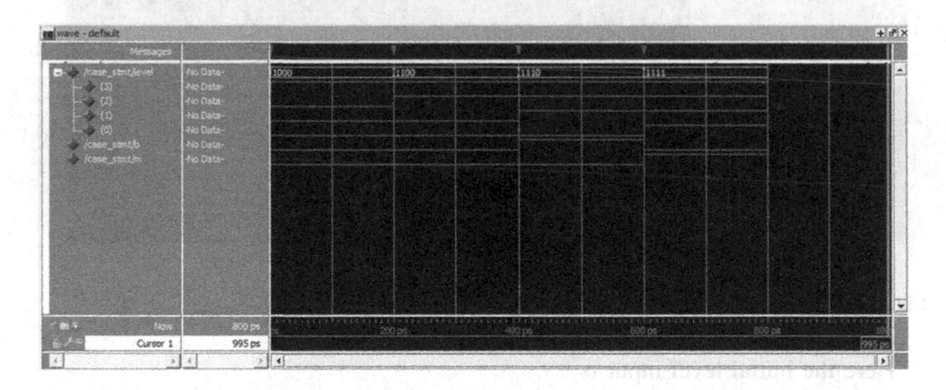

Fig. 2.10 Signals of sensors actuating the motor and buzzer

The above graph shows the automatic control of buzzer and gates through level measurements. Taking into consideration the KRS dam Mysore specifications, the total height of the dam is considered as 125 ft. This is divided into 4 levels namely 30, 60, 90, and 120 ft for operational purpose. However for safety purpose, the threshold level value is considered to be 120 ft.

- When water reaches 30 and 60 ft, only the level measurements are made and no controlling is required.
- When water reaches 90 ft, near to threshold level (120 ft) the buzzer switches on (alarm) alerting officials in control room that the level is approaching the threshold mark.

- When water reaches 120 ft i.e. threshold mark, the gates are automatically opened and required amount of water flows out of the dam. Once the water level falls below 120 ft the dam gates automatically close.
- The dam operations are hence completely automated. dam operations are hence completely automated.

2.8 Advantages of Designing Model in FPGA

1 Cost of the design is less compared to ASIC.
2 Unlike ASIC models, any changes can be implemented without having to make changes in the hardware i.e. it is flexible [9].
3 The model can be implemented into VLSI chip design which can be generated in large numbers making the model implementation easy and cost effective [10].

2.9 Scope for Future Work

This model strives to develop a VLSI chip as the end product and implement them in dams across the world. This work falls under the category of CPS (Cyber Physical Systems).CPS gives equal importance to the link between computational elements and physical elements unlike embedded systems which gives importance only to the computational elements. Development in this area will lead us to live in a more advanced and user friendly 'Cyber Physical Society'.

References

1. Kleeman L (1995) Understanding and applying Kalman filtering. Department of Electrical and Computer Systems Engineering Monash University, Clayton
2. Kleinbauer R (2004) Kalman filtering implementation in MATLAB, study report in the field of study geodesy and geoinformatics. Universtat Stuttgart, Helsinki
3. Pasricha R, Sharma S (2009) An FPGA—based design of fixed-point Kalman filter. DSP J 9(1):1–9
4. Welch G, Bishop G (2001) An introduction to Kalman filtering. TR 95-041
5. Simon D (2001) Kalman filtering. Embed Syst Program
6. Cornell (2008) Kalman filter tank filling, subject MI63
7. Sorensen H (1985) Kalman filtering: theory and applications. IEEE Press, Los Alamitos
8. Chen G, Li G (2005) The FPGA implementation of Kalman filter. University of Science and Technology of China, China
9. Chi-Jui C, Mohanakrishnan S, Evans JB (1994) FPGA implementation of digital filters. University of Kansas, Lawrence, pp 66045–2228
10. Vij V, Mehra R (2011) FPGA based Kalman filter for wireless sensor networks. Lawrence

Chapter 3
Medical Image Watermarking Based on DWT and ICA for Copyright Protection

P. Mangaiyarkarasi and S. Arulselvi

Abstract Digital watermarking has been proposed to increase medical image security, confidentiality and integrity. Medical image watermarking is a special subcategory of image watermarking in the sense that the images have special requirements. Particularly, watermarked medical images should not differ perceptually from their original counterparts, because the clinical reading of the images for diagnosis must not be affected. Hence, in this paper, a robust and imperceptible watermarking technique based on discrete wavelet transform (DWT) is proposed for medical images. For extraction, a blind source separation technique, namely, independent component analysis is attempted. Conventional extraction techniques need embedding parameters such as strength, location and information about watermark or original image, whereas, ICA extracts the watermark without the use of these parameters. The quality of watermarked image and extracted watermark are measured in terms of peak signal to noise ratio (PSNR) and normalized correlation (NC) values respectively. Robustness of the proposed scheme is validated against various image processing attacks. The proposed work is attempted on color images also. A true color image is split into red, green and blue components, where watermark is embedded in each plane individually followed by extraction. Performance of the proposed scheme is compared among the color components, and it is recommended that blue plane is the better choice of embedding watermark for medical color images.

Keywords Medical image watermarking · Discrete wavelet transform · Independent component analysis · Copyright protection

---

P. Mangaiyarkarasi (✉) · S. Arulselvi
Department of Electronics and Instrumentation Engineering, Annamalai University, Chidambaram, TamilNadu, India
e-mail: mangai_me@yahoo.co.in

S. Arulselvi
e-mail: arulselvi_2k3@yahoo.co.in

R. Malathi and J. Krishnan (eds.), *Recent Advancements in System Modelling Applications*, Lecture Notes in Electrical Engineering 188,
DOI: 10.1007/978-81-322-1035-1_3, © Springer India 2013

3.1 Introduction

Digital information management in hospitals, hospital information system (HIS) and its special cases of radiology information system (RIS), picture archiving and communication system (PACS) forms the information infrastructure of modern health care [1]. Recently the advent of multimedia has boosted the potential of telemedicine applications ranging from teleconsulting, telediagnosis etc., to cooperative working session and telesurgery [2]. These advances in information and communication technology provide in fact new ways to score, access and distribute medical data, and introduces new practices for the profession, as well as the patient themselves by accessing to their own medical files [3]. With these benefits, there are concomitant risks for electronic patient records (EPR) and strictly personal documents circulating in open networks, and being accessible, e.g., via Internet. Thus, it is a widely shared point of view that there is an urgent need for network level security measures and protocols in medical information systems.

This paper aims to provide a medical image watermarking technique based on discrete wavelet transform and extraction using independent component analysis. The proposed embedding technique is based on computation of noise visibility function (NVF), where the strength of watermarking is controlled which results in watermarks at texture and edge areas are stronger than flat areas. For extraction, a blind source separation technique namely ICA, is implemented. Multiple ICA algorithms are in existence and among them, FastICA is chosen in this work, since its convergence rate is high [4]. The features of FastICA are quick convergence, easy to implement and suitable for watermark applications [5].

The paper is organized as follows: Sect. 3.2 reviews literatures on watermarking for medical images and Sect. 3.3 elaborates the theory of DWT. Section 3.4 discuss the proposed watermarking approach includes embedding and extraction procedures. Simulation results are presented in Sect. 3.5 and conclusions are drawn in Sect. 3.6.

3.2 Literature Review

Medical image watermarking systems can be broken into three broad categories: robust, fragile, and semi-fragile. This section explains these terms and provides a brief review of existing watermarking systems in each category.

Robust watermarks are designed to resist attempts to remove or destroy the watermark [6]. They are used primarily for copyright protection and content tracking. A number of robust medical image watermarking systems have been developed. For example one system uses a spread spectrum technique to encode copyright and patient information in images [7]. The drawback in this scheme is, it is less robust against attacks and can easily tampered when the embedding locations are identified. Another scheme embeds a watermark in a spiral fashion around the Region of Interest (ROI) of an image [8]. Any image tampering that occurs will severely degrade the image quality. The Gabor transform has also been applied to hide information in medical images, where the image quality is not guaranteed in this type of watermarking [9].

Fragile watermarks are used to determine whether an image has been tampered or modified. The watermark is destroyed if the image is manipulated in the slightest manner [10]. Fragile invertible authentication schemes have been proposed for medical images, whereby a watermark can be removed from a watermarked image [11]. Another medical image watermarking system embeds information in bit planes, which results in watermarked images with very low normalised root mean square errors (NRMSE), indicating that the watermark is practically invisible [12].

Semi-fragile watermarks combine the properties of both robust and fragile watermarks [10]. Like robust methods, they can tolerate some degree of change to the watermarked image and they are also capable of localizing regions of an image that are authentic and those that have been altered. Recently, much emphasis has been placed on semi fragile medical image watermarking. Jagadish et al. investigated interleaving hidden information in the Discrete Cosine Transform (DCT) and the Discrete Wavelet Transform (DWT) domains [12]. One observation that is generally applicable to robust systems is the greater the robustness of the watermark, the lower the image quality [13]. Hence, to maintain the tradeoff between imperceptibility and robustness, the proposed scheme uses the combination of DWT and FastICA.

3.3 Discrete Wavelet Transform

Many literatures have reported watermarking schemes based on DWT. The main advantage of wavelet analysis is that they allow both spatial and frequency resolution during decomposition [14]. Wavelet transform allows the decomposition of the signal in narrow frequency bands while keeping the basis signals space limited. Figure 3.1 shows one level DWT decomposition using low pass and high pass analysis filters as $h(-m)$ and $g(-m)$ respectively. c_j and d_j are the low band and high band output coefficients at level j. If the level of decomposition is increased, the approximate image will be more stable. But the complexity increases and the amount of information that can be embedded will be decreased. As a compromised way, the original image is decomposed into two levels. During decomposition, the original image is decomposed into an approximate image LL1 and three detail images LH1, HL1 and HH1 as shown in Fig. 3.1. For the next level decomposition, the approximate image LL1 is further decomposed to obtain four lower-resolution sub-band images, namely, LL2, LH2, HL2 and HH2. The approximate image holds the most important information of the original image and others contain some high-frequency information such as the edge details [15].

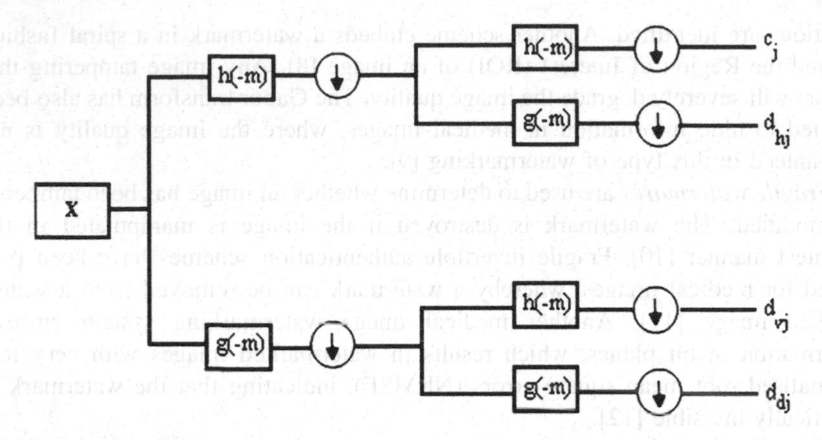

Fig. 3.1 One level decomposition of an image using DWT

DWT analysis is given by

$$c_{j+1}[m,n] = \big(c_j(m,n) * h[-m]\big) \downarrow 2 \tag{3.1}$$

$$d_{j+1}[m,n] = \big(c_j(m,n) * g[-m]\big) \downarrow 2 \tag{3.2}$$

3.4 Proposed Watermarking Scheme

The proposed watermarking scheme comprises of two parts:

Watermark embedding stage and watermark extraction stage. The following section briefly explains the proposed scheme.

3.4.1 Watermark Embedding

The resultant subbands obtained from two level DWT are LL2, LH2, HL2 and HH2. The approximation components at LL2 sub-band are not chosen to embed watermark because they will seriously degrade the image quality. Similarly the diagonal detail coefficients HH2 are also not considered because security is poor when watermark is embedded in that subband. Hence, the watermark can be embedded either in HL2 or LH2 or both. A stochastic model of the cover image is applied to an adaptive watermark by computing NVF with non-stationary Gaussian model. In this case, NVF can be expressed by

$$NVF(i,j) = \frac{1}{1 + \sigma_x^2(i,j)} \tag{3.3}$$

where $\sigma_x^2(i, j)$ denotes variance of the cover image in a window centred on the pixel with coordinates (i, j). By applying NVF, the watermark in texture and edges becomes stronger than in flat areas. The quality of the watermarked image in the proposed scheme is increased by using this NVF mask in the embedding equations. The embedding equations are given below:

$$\begin{aligned} I'LH_2(i,j) &= LH_2(i,j) + E(LH_2)\alpha_1(1 - NVF(i,j))W(i,j) \\ &+ \frac{E(LH_2)}{10}\alpha_2.NVF(i,j).W(i,j) \end{aligned} \tag{3.4}$$

$$\begin{aligned} I'HL_2(i,j) &= HL_2(i,j) + E(HL_2)\alpha_1(1 - NVF(i,j))W(i,j) \\ &+ \frac{E(HL_2)}{10}\alpha_2.NVF(i,j).W(i,j) \end{aligned} \tag{3.5}$$

where $I'LH_2(i,j)$ and $I'HL_2(i,j)$ are watermarked transform coefficients, $E(LH_2)\alpha_1$ and $E(HL_2)\alpha_1$ denote the watermark strengths of texture and $\frac{E(LH_2)}{10}\alpha_2$ and $\frac{E(HL_2)}{10}\alpha_2$denote the watermark strengths of edge regions for LH and HL subbands, respectively. α_1 and α_2 are smoothing factors at the texture regions and flat regions, E denotes the mean and $W(i,j)$ is the watermark.

To obtain the watermarked image, inverse DWT is performed on the watermarked subbands. In order to apply ICA for watermark detection, the embedding process needs to create a private key for extraction. In this work, the logo watermark is embedded into medical images using Eqs. (3.4) and (3.5) and inverse transformed to obtain watermarked image. For color images, the same procedures are implemented into each color planes individually, inverse transformed and combined together to form watermarked images [16].

3.4.2 Watermark Extraction

During extraction, the watermarked image is once again decomposed to two levels using discrete wavelet transform. From the resultant subbands, LH_2 and HL_2 along with key are used to generate input mixture for ICA detector. From the mixtures, ICA extracts the watermark from the watermarked image. Since ICA doesn't require original image for extraction, the proposed extraction is said to be blind [17].

3.4.3 Independent Component Analysis

Independent component analysis is a novel statistical technique that aims at finding linear projections of the data that maximize their mutual independence [5]. ICA has received attention because of its potential applications in signal processing such as in feature extraction, and blind source separation with special emphasis to physiological data analysis and audio signal processing. The goal of ICA is to recover the source signals given only sensor observations that are linear mixtures of independent source signals. It aims at extracting unknown hidden components from multivariate data using only the assumption that the unknown factors are mutually independent [18]. The algorithm starts with the assumption that data vector $X = (X_1, \ldots, X_M)$ can be represented in terms of linear superposition of basis functions, as given by

$$X = A.S \tag{3.6}$$

$$Y = B.X \tag{3.7}$$

where $S = (S_1, \ldots, S_N)$ are unknown source coefficients, A is an $M \times N$ matrix and the columns $a_1, \ldots, a_N$ are called basis functions. Basis functions are constant while coefficients vary with data. The goal of ICA is to find matrix B in (3.7), which results in estimates of coefficient values of Y being statistically as independent as possible over a set of data X. It is noted that the watermark and the original image can be regarded as unknown sources (S_i) and the watermarked image as a mixture (X_i). By creating different mixtures, one can perform ICA to extract the watermark.

In this work, a linear mixture of watermarked image with key is generated as input signal to the ICA and the ICA separates the watermark as the output from the mixtures. The novelty of this detector is that it does not require original image and embedding parameters such as watermark location and strength. Moreover, it is fast in convergence, easy to implement and suitable for watermarking applications. With the help of a private key K to create different mixtures, one can extract successfully the watermark to claim the ownership.

Mixtures are created by the following equations:

$$X_1 = a_{11}I' + a_{12}W + a_{13}K \tag{3.8}$$

$$X_2 = a_{21}I' + a_{22}W + a_{23}K \tag{3.9}$$

$$X_3 = a_{31}I' + a_{32}W + a_{33}K \tag{3.10}$$

where X_1, X_2, X_3 are mixtures, I' is the watermarked image, a is a mixing matrix, W is the encrypted watermark matrix and K is a random key in the embedding process [4].

3.5 Results and Discussions

The proposed watermarking is tested on some test images (Chest, Knee, Head and Finger) of size 256 × 256. The host images are decomposed up to two levels by discrete wavelet transform. A binary logo watermark of size 64 × 64 is embedded in the middle frequency sub-bands LH_2 and HL_2 using Eqs. (3.4) and (3.5). To obtain the watermarked image, inverse DWT is performed on the watermarked coefficients. Figure 3.2 shows input images and their corresponding watermarked images obtained by DWT and IDWT. The visual quality of watermarked images shows high degree of imperceptibility. This can be statistically proved by the PSNR values tabulated in Table 3.1, where all the watermarked images have values greater than 40 dB. This value should vary from image to image because of the effectiveness of NVF.

The robustness of the proposed watermarking scheme is validated against attacks like, Gaussian noise addition, Salt and Pepper noise addition, Blurring, Sharpening, Rotation, JPEG Compression and Cropping. Figure 3.3a shows the watermarked image added with Gaussian noise of zero mean and variance 0.01. The extracted watermark by FastICA is also shown in that image. Figure 3.3b shows the watermarked image added with Salt and Pepper noise with noise density 0.1. Figure 3.3c, d shows the blurred image obtained by average filtering and sharpened image by high pass filtering. Figure 3.3e shows the rotated image by 10 ° and Fig. 3.3f shows the JPEG compressed image. Figure 3.3g shows the watermarked image with cropping, where the pixels of the watermarked image are replaced by zeros. The PSNR values are calculated after the above mentioned attacks and they are tabulated in Table 3.2. From the table, it is inferred that the proposed scheme generates PSNR values higher than 25 dB for the above mentioned attacks. Hence, the robustness of the proposed scheme is proven visually and statistically.

Similarly, the watermark detection using ICA extracts the watermark perfectly from the watermarked image. To justify the extraction, performance measure namely normalized correlation value is calculated between original and extracted watermark. NC value indicates the degree of similarity between two images and it should vary from 0 to 1. Using FastICA, extracted watermarks from the attacked images are also shown in Fig. 3.3a–g and their corresponding normalized correlation values are tabulated in Table 3.2. From the table, it is noted that the proposed scheme can extract watermark from the attacked images with a similarity measure more than 94 %. Hence, the proposed scheme performs better in embedding as well as extraction of medical data. It is suggested that, DWT with FastICA can be well suited for copyright protection and authentication of medical images.

An attempt is made to implement the proposed scheme on color images. Figure 3.4 shows a true color knee image and the original watermark image. First, the color components are separated as red, green and blue as shown in Fig. 3.5. The embedding procedure explained in Sect. 3.4 is implemented on these components individually and inverse discrete wavelet transform is performed to obtain watermarked images. The watermarked images obtained from the three planes are shown in Fig. 3.6. The embedded logo is extracted from these planes individually are also shown in Fig. 3.6 and their performance metrics are tabulated in Table 3.3.

From the table, it is noted that blue component generates a high PSNR value when compared to other two and in extraction also, it is less correlated with color

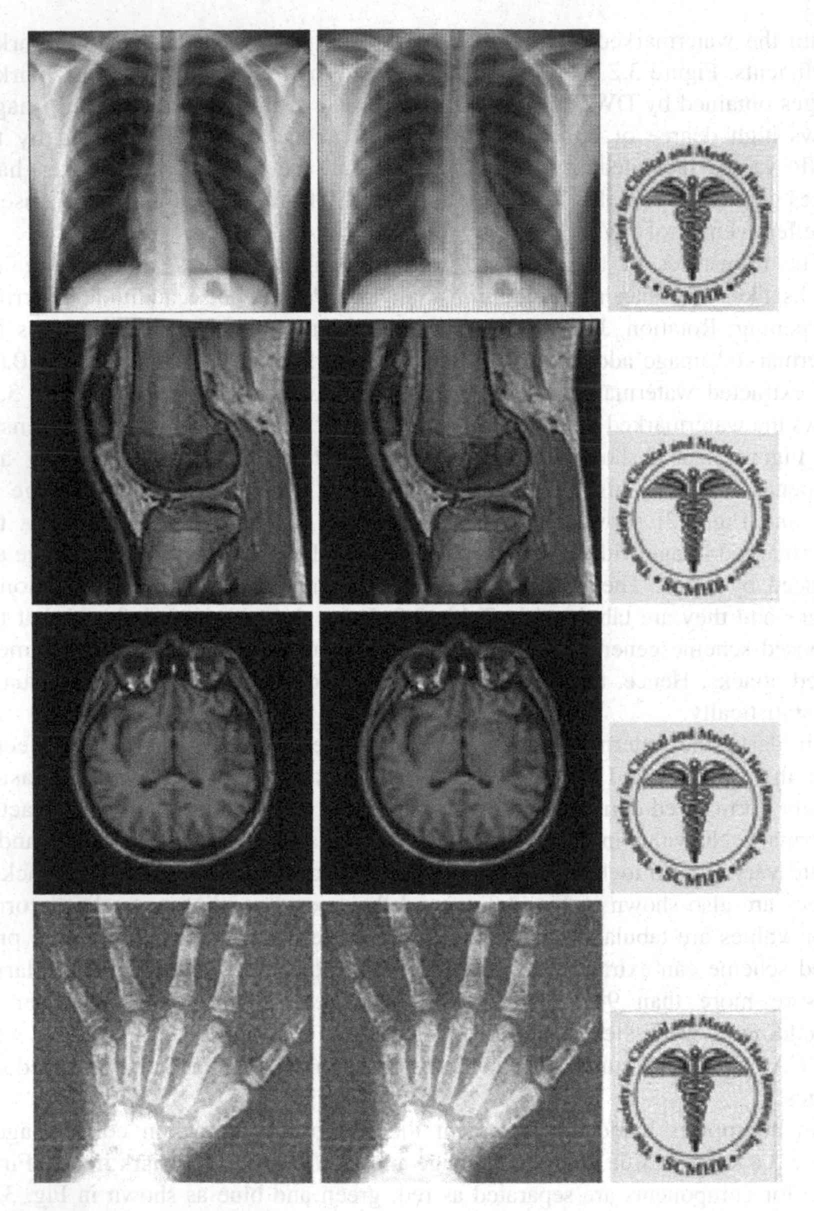

Fig. 3.2 Input images, their corresponding watermarked images and extracted watermarks using DWT and ICA

pixels and thus producing 97 % of degree of similarity. From the results, it is inferred that blue plane retain the energy during embedding and also performs better in extraction than other planes. Hence, it is suggested that embedding watermark in blue component can provide a good imperceptible and robust watermarked images. The robustness of the color image is also validated against

Table 3.1 Performance measures of proposed watermarking scheme for different input images

Images	PSNR (dB)	NC
Chest (X-ray)	42.0331	0.9677
Knee (MRI)	59.2049	0.9696
Head (MRI)	58.1882	0.9702
Finger (X-ray)	61.1611	0.9677

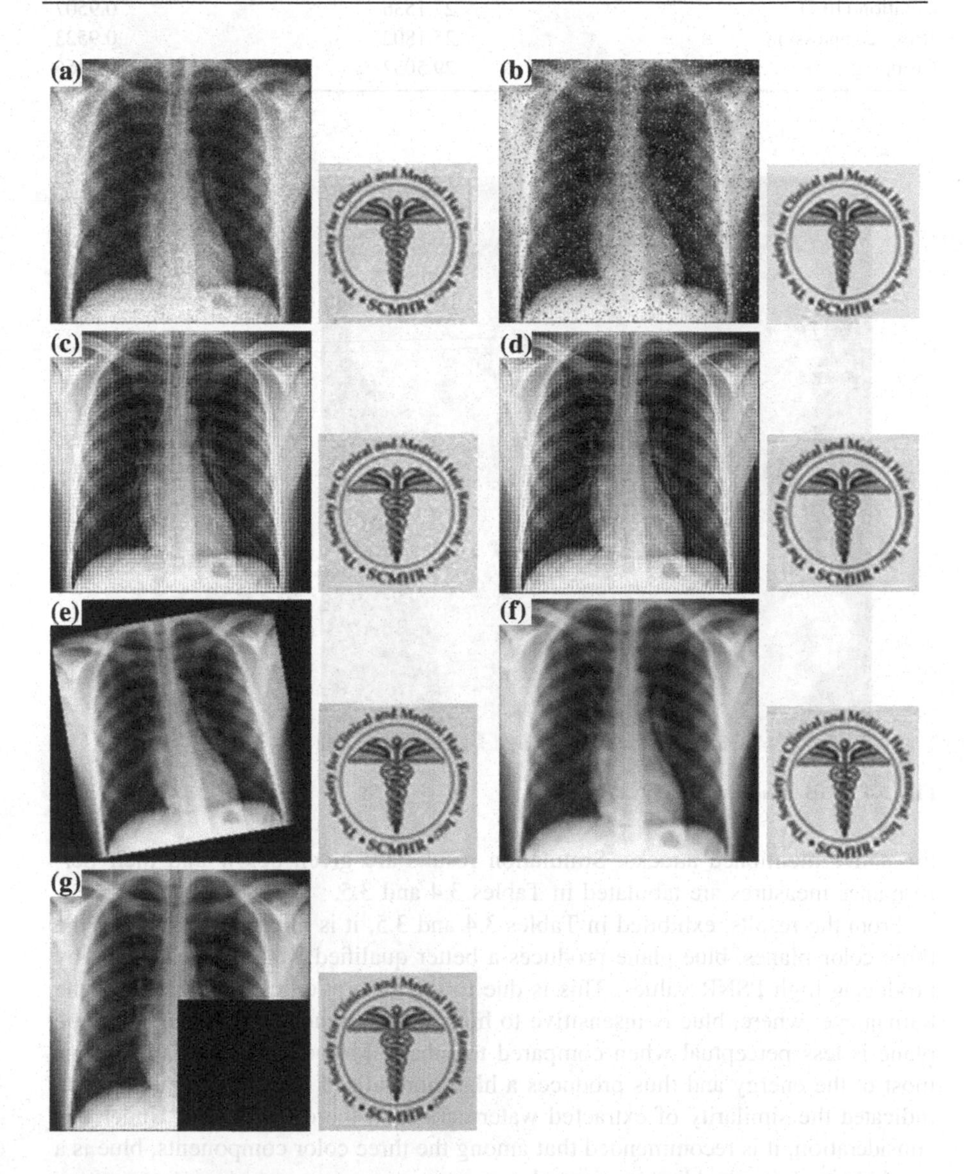

Fig. 3.3 Various attacks on watermarked image of chest (X-ray) and their corresponding extracted watermarks. **a** Gaussian noise added. **b** Salt and pepper noise added. **c** Blurred. **d** Sharpened. **e** Rotated. **f** JPEG compressed. **g** Cropped

Table 3.2 Robustness of the proposed scheme under various attacks

Attacks	PSNR (dB)	NC
Gaussian noise (0.01)	28.9455	0.9456
Salt and pepper noise (0.1)	33.5802	0.9444
Blurring (0.2)	29.9484	0.9472
Sharpening (0.2)	29.8973	0.9438
Rotation (10 °)	27.1536	0.9507
JPEG compression	25.1802	0.9523
Cropping	29.5032	0.9466

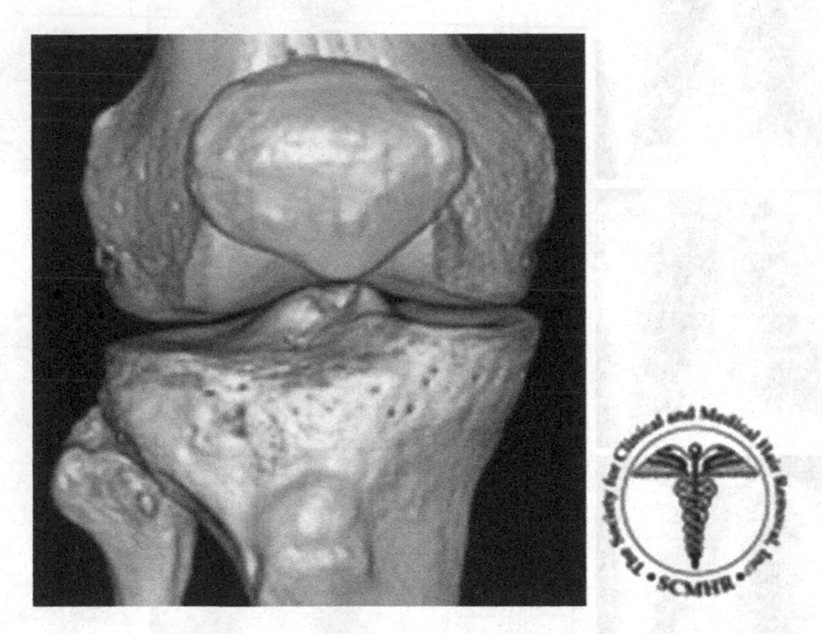

Fig. 3.4 RGB image and logo watermark

the above mentioned attacks. Simulation results are given below and their performance measures are tabulated in Tables 3.4 and 3.5.

From the results, exhibited in Tables 3.4 and 3.5, it is inferred that among the three color planes, blue plane produces a better qualified watermarked image by producing high PSNR values. This is due to the absorption characteristics of the human eye, where, blue is insensitive to human eye. Hence, for human eye, blue plane is less perceptual when compared to others. Moreover, blue plane retains most of the energy and thus produces a high normalized correlation value which indicated the similarity of extracted watermark with the original one. Under this consideration, it is recommended that among the three color components, blue is a good choice of embedding watermark.

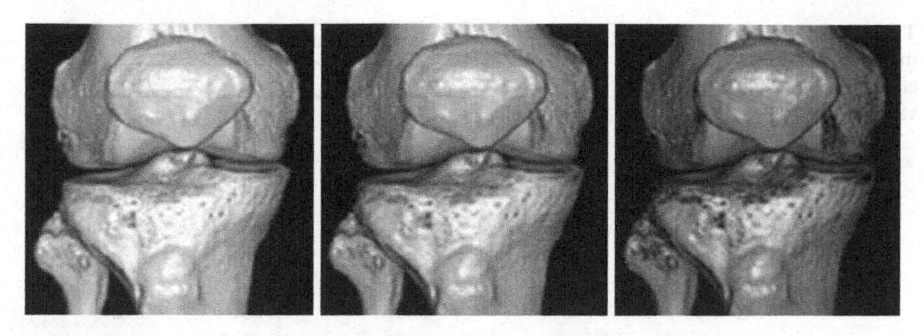

Fig. 3.5 *Red*, *Green* and *Blue* components of the color image

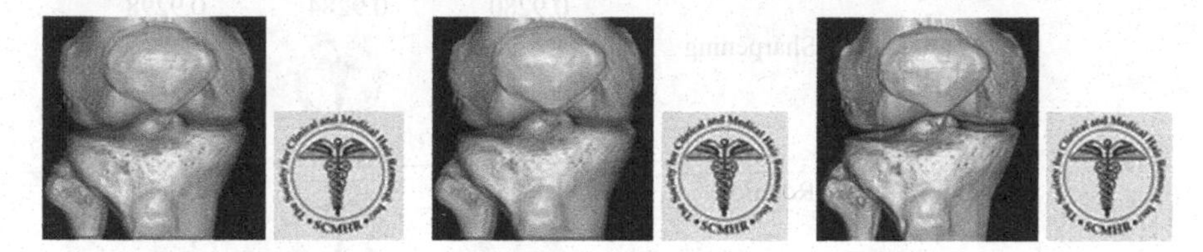

Fig. 3.6 Watermarked images and corresponding extracted watermarks from R, G and B components

Table 3.3 Performance of proposed watemarking scheme for color image

Color	PSNR (dB)	NC
Red	49.6921	0.9688
Green	49.6657	0.9687
Blue	49.9626	0.9703

Table 3.4 Robustness of the proposed scheme for color Image in terms of PSNR values

Attacks	PSNR (dB)		
	R	G	B
Gaussian noise	29.1411	29.1681	29.2977
Salt and pepper noise	29.5325	29.5053	29.7791
Blurring	29.4338	29.4017	29.6000
Sharpening	29.4251	29.3934	29.5911
Rotation	27.5958	27.5953	27.7572
Cropping	28.3711	28.3785	28.5737
JPEG compression	25.9793	26.8452	29.1959

3.6 Conclusion

A robust watermarking algorithm using DWT and extraction by FastICA for medical images is presented in this paper. The proposed algorithm is adaptive by implementing NVF, in which the watermark in texture and edges are stronger in

Table 3.5 Performanance of the extraction scheme for color image in terms of NC values

Attacks	Extracted watermarks and their NC values		
	R	G	B
Gaussian noise	0.9300	0.9301	0.9306
Salt and pepper noise	0.9259	0.9236	0.9279
Blurring	0.9280	0.9284	0.9298
Sharpening	0.9279	0.9284	0.9298
Rotation	0.9313	0.9315	0.9317
Cropping	0.9314	0.9316	0.9317
JPEG compression	0.9313	0.9318	0.9325

flat areas. The results obtained from simulation using MATLAB.7 are presented and tabulated. From the results, it is inferred that imperceptibility and robustness of the proposed scheme is considerably good. The performance of the proposed scheme is evaluated in terms PSNR and NC values. The advantage of using FastICA algorithm for watermark detection is that it does not need information about location and strength of the embedding parameters. For color images also, the proposed scheme performs in a better way when the watermark is embedded in blue plane. Hence the proposed watermarking algorithm is better suitable for copyright protection and data security of clinical records.

References

1. Kolodner S (1999) Filmless radiology, collection health informatics. Springer, NewYork
2. Miaou SG, Hsu CH, Tsai YS, Chao HM (2000) A secure data hiding technique with heterogeneous data-combining capability for electronic patient records. In: Proceedings of

the world congress on medical physics and biomedical engineering, session, electronic healthcare records, IEEE-EMB, Ed., Chicago
3. Macq B, Deweyand F (1999) Trusted Headers for Medical Images. In: DFG VIII-DII watermarking workshop, Erlangen
4. Thirugnanam G, Natarajan M, Mangaiyarkarasi P, Malmurugan N (2009) Comparison of independent component analysis for DWT based digital image watermarking. Int J Adv Res Comput Eng 3(1):165–169
5. Hyvarinen A, Karhunen J, Oja E (2001) Independent component analysis, 1st edn. Wiley, London
6. Lin ET, Podilchuk CI, Delp EJ (2000) Detection of image alterations using semi-fragile watermarks. In: Proceedings of the SPIE international conference on security and watermarking of multimedia contents II, vol 3971. San Jose', pp 152–163
7. Tachibana H Harauchi H, Ikeda T, Iwata Y, Takemura A, Umeda T (2005) Practical use of new watermarking and vpn techniques for medical image communication and archive. RSNA 2002 archive site: http://archive.rsna.org/index.cfm,2002. Accessed 4 Jan 2005
8. Wakatani A (2002) Digital watermarking for ROI medical images by using compressed signature image. In: Annual Hawaii international conference on system sciences, Hawaii, pp 2043–2048
9. Kong X, Feng R (2001) Watermarking medical signals for telemedicine. IEEE Trans Inf Technol Biomed 5(3):195–201
10. Fridrich J, Goljan M, Du R (2001) Invertible authentication. In: Proceedings of SPIE, security and watermarking of multimedia contents III, vol 3971. San Jose, pp 197–208
11. Jagadish N, Bhat PS, Acharya R, Niranjan UC (2004) Simultaneous storage of medical images in the spatial and frequency domain: a comparative study. Biomedical engineering online, 3(1):record
12. Puech W, Rodrigues JM (2004) A new crypto-watermarking method for medical images safe transfer. In Proceedings of the 12th European signal processing conference, Vienna, pp 1481–1484
13. Kutter M, Petitcolas FAP (1999) A fair benchmark for image watermarking systems. In: Proceedings of SPIE security and watermarking of multimedia contents, vol 3657. San Jose, pp 226–239
14. Joo S, Suh Y, Shin JKikuchi H, Cho SJ (2002) A new robust watermark embedding into wavelet DC components. J ETRI 24(5):401–404
15. Mangaiyarkarasi P, Arulselvi S (2012) A robust digital image watermarking technique based on DWT and FastICA. CiiT Int J Digit Image Process 4(2):100–105
16. Mangaiyarkarasi P, Arulselvi S (2012) Robust color image watermarking technique based on DWT and ICA. Int J Comput Appl 44(23):6–12
17. Hien TD, Nakao Z, Chen YW (2004) ICA- based robust logo image watermarking. In: Proceedings of SPIE, security, steganography, and watermarking of multimedia contents
18. Nguyen TV, Chandra patra J (2007) A simple ICA-based digital image watermarking scheme. Sci Direct J Digit Signal Process 18:762–776

Chapter 4
AANN-Based Online Handwritten Tamil Character Recognition

AN. Sigappi and S. Palanivel

Abstract The paper develops an autoassociative neural network (AANN) based online handwritten character recognition method for Tamil scripts. The coordinate positions traced during the pen movement in the digital tablet are preprocessed to remove noise, normalize the characters to a uniform height, and rescale the pen positions to result in uniformly spaced x and y positions. The rescaled x and y positions are individually provided as feature vectors to an AANN classifier and trained using the back propagation algorithm. The network training results in the adjusted weights that minimize the error between the input and output and captures the distribution of feature vectors in the input space effectively to create separate x and y models for 156 Tamil characters. The number of rescaled points is varied in order to experiment with various AANN structures and determine the best recognition rate. The confidence scores individually obtained from the x and y models are concatenated to give rise to the weighted xy model through a weighted sum rule. The experimental results reveal a higher recognition rate of 89.74 % for a weighting factor of 0.3 when applied to the scores of 96 rescaled points. The strength of the proposed approach lies in the computational simplicity and the consistency of the results claim its use in real world applications.

Keywords Online handwritten character recognition · Autoassociative neural networks · Rescaling · Recognition rate

AN. Sigappi (✉) · S. Palanivel
Department of Computer Science and Engineering, Annamalai University,
Annamalainagar, Cuddalore 608002, India
e-mail: aucse_sigappi@yahoo.com

S. Palanivel
e-mail: spal_yughu@yahoo.com

R. Malathi and J. Krishnan (eds.), *Recent Advancements in System Modelling Applications*, Lecture Notes in Electrical Engineering 188,
DOI: 10.1007/978-81-322-1035-1_4,

4.1 Introduction

The emergence of modern input modalities such as stylus or pen used on a digital tablet offer a natural and convenient means for the user to input text in a regional language like Tamil. In spite of the availability of several methods for online recognition of handwritten characters, the problem still continues to receive the attention of researchers in a direction to improve the robustness of the approach so as to accommodate both between class and within class variations [1]. It is a more significant and challenging issue particularly for a language like Tamil because of the large character set and the mixed literacy levels leading to individualistic and varied styles in writing. Tamil is a classical, dravidian language spoken by nearly 80 million people living in the Indian state of Tamilnadu and in various countries all over the world. The Tamil writing script owes its origin to the Brahmi script and the language comprises of 12 vowels, 18 consonants, 216 compound characters formed by combining the consonants with vowels, and a special character called *'ayutham'*. Five consonants from Sanskrit add to the total of 247 along with their compound characters to arrive at 276 letters. There are only four modifier symbols that attach with most of the compound characters to group the complete set of characters into 156 classes of unique symbols [2].

In view of the fact that character recognition techniques require to cope with a high variability of the handwritten letters and the associated intrinsic ambiguity, a host of techniques appear to evolve to address issues in this domain. Though template matching, statistical techniques and structural techniques [1] find their use in this perspective, neural network classifiers owing to their parallel structure and intelligent learning capabilities can serve as a viable option for recognizing online handwritten characters. Autoassociative neural network models (AANNs) open up a new dimension to recognize handwritten characters by capturing the distribution of patterns in the input space [3].

4.2 Related Work

A stroke has been represented as a string of shape features and a test stroke identified by means of a string matching procedure that compares the test stroke with the prestored database of strokes. The method has been developed for Tamil [4] and extended to Malayalam [5] and Telugu characters [6]. The pixel densities calculated for the 64 different zones of the scanned image of a character have been used as the features to train a support vector machine (SVM) based offline handwritten Tamil character recognition system [7].

A micro inertial measurement unit based on micro electro mechanical systems (MEMS) sensors has been built to record the three dimensional accelerations and angular velocities of the motions during handwriting. The unsupervised self organizing map (SOM) has been trained with the discrete cosine transform (DCT) coefficients of the gathered data and found to produce good character recognition results [8]. The Devnagari handwritten character recognition has been compared using 12 different classifiers and four sets of features obtained from curvature and gradient information of the character images. The experimental results have been obtained on 36,172 samples of Devnagari basic characters using a fivefold cross validation strategy [9].

A precision constrained Gaussian model (PCGM) has been proposed with fewer number of tied and untied parameters than the modified quadratic discriminant function (MQDF) models and the maximum discriminant decision rule used for the classification of handwritten Chinese characters. The PCGM-based approach has been reported to strike a good trade-off between recognition accuracy and memory requirement [10]. A template matching method has been used for online Arabic handwriting recognition and stretched to handle complex diacritic attachment variations with the dual-graph approach by branch-and-bound techniques [11]. A prototype based dynamic time warping (DTW) classifier based on agreement and rejection distance thresholds has been evolved to classify handwritten Tamil characters by calculating the distance between an unknown sample and a set of labeled prototypes and the performance of the classifier studied [12].

4.3 Proposed Methodology

Pen-based input appears to evince interest in view of the ease with which characters in regional languages can be input to a computer. It necessitates an online recogniser with distinct stages of data collection, preprocessing and classification to accept and process the information about the pen position. The data are collected from the sequence of points traced by the pen as the character is written. The (x, y) coordinate positions of the pen movements on a digital tablet are sampled at equally spaced time intervals and stored as (x_i, y_i) values.

4.3.1 Preprocessing

The acquired data are preprocessed to reduce the noise, normalize and rescale the pen positions to result in the feature vectors for classification [13]. The noise that creeps in either due to the device characteristics or the writing style is reduced using a Gaussian low pass filter without loss of essential information. Normalization maps all the characters to a uniform height with the same aspect ratio as that of the original character through the use of Eqs. 4.1 and 4.2.

$$x_i^n = \frac{x_i - x_{\min}}{x_{\max} - x_{\min}} \tag{4.1}$$

$$y_i^n = \frac{y_i - y_{min}}{y_{max} - y_{min}} \quad (4.2)$$

The normalized $\{x_i^n\}$, $\{y_i^n\}$ values in [0,1] range are rescaled to obtain a fixed number of evenly spaced points for all characters, irrespective of the manner in which the character is written and the number of strokes in the character. The ratio of the number of points in each stroke of the normalized input character is determined and the desired number of points distributed in the same ratio for the rescaled character as well. The steps to be applied to each stroke of the character during the rescaling process are:

1. Find the length of the stroke, l.

$$l = \sum_{i=1}^{c} \sqrt{(x_{i+1} - x_i)^2 + (y_{i+1} - y_i)^2} \quad (4.3)$$

where c denotes the number of points in the input character.
2. Determine the space between adjacent pair of points in the rescaled character, s_r, as

$$s_r = \frac{l}{r - 1} \quad (4.4)$$

where r is the chosen number of rescaled points for all characters.
3. Calculate the new rescaled points as, s_r, as

$$x_i^r = \begin{cases} x_i & , i = 1 \\ x_{i-1} + s_r & ,\text{otherwise} \end{cases} \quad (4.5)$$

$$y_i^r = \begin{cases} y_i & , i = 1 \\ y_{i-1} + s_r & ,\text{otherwise} \end{cases} \quad (4.6)$$

The computed x_i^r and y_i^r rescaled points form the corresponding feature vectors for the classification task and the Fig. 4.1 displays an input Tamil character, its normalized and rescaled equivalent images.

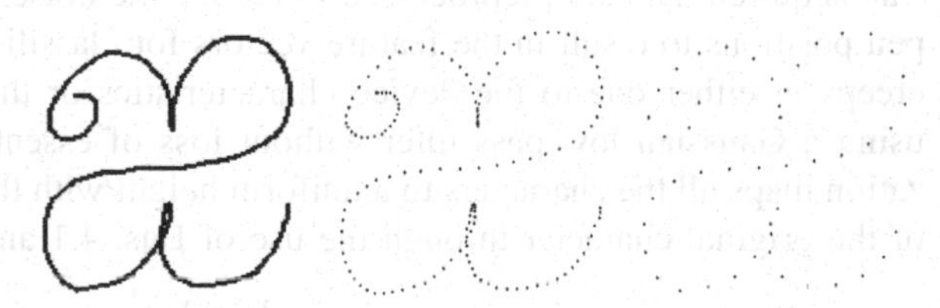

Fig. 4.1 An input character, its normalized and rescaled equivalent

4.3.2 Classification Using AANN

The feed forward autoassociative neural networks (AANNs) consist of five layers whose processing units are connected through weights that can be positive, zero, or negative [14]. The AANN model achieves an identity mapping of the input space whose architecture shown in Fig. 4.2 comprises of one input layer, one output layer and three hidden layers between the input and output layers [15].

The processing units in the first and third hidden layers use nonlinear activation functions whereas the units in the second layer utilize either linear or nonlinear functions. The network performs compression of the input parameters in the first three layers of the network and decompression in the last three layers of the network so as to capture the distribution of the feature vectors.

The AANN is trained with the input feature vectors and the initial random weights are adjusted after each epoch to minimize the error between the output and the input patterns using the back propagation algorithm. The points in the input space influence the shape of the hypersurface that is created through its projection on the compression layer and the distribution capturing capability of the AANN is determined by the constraints in the network structure. The nonlinear output function for the network is chosen as tanh(s), where s is the activation value of the unit and the error e_i for the data point i is governed by,

$$p_i = exp\left(\frac{-e_i}{\alpha}\right) \tag{4.7}$$

where alpha is a constant. The plot of the error seen as a probability surface appears to end up with greater amplitude for a smaller error e_i that relates a closer match of the network for the chosen cluster of input points.

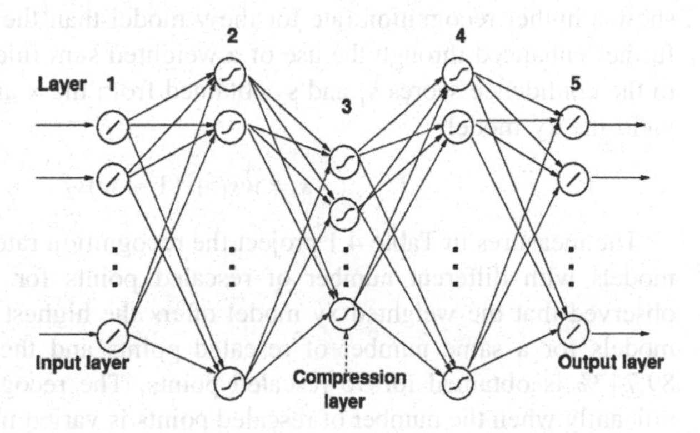

Fig. 4.2 Architecture of AANN

4.4 Experimental Results

The AANN-based online handwritten character recognition technique is investigated using the HP Tamil dataset that includes around 500 isolated samples of 156 unique symbols from the Tamil script. The collection consists of characters written by school children, university graduates and adults from different regions of India. 10 samples of each character written by 20 users constitute the corpus and a leave-one-out approach is used for testing the proposed method. A series of 20 tests are carried out with the samples from 19 writers used for training and the remaining one user for testing.

The rescaled x and y points of each character obtained using the steps explained in Sect. 4.3.1 are individually considered for construction of separate x and y models using the five layer AANN model described in Sect. 4.3.2. The number of strokes and points in each stroke for each character varies from person to person and even for the same person it varies from time to time. The number of units in each layer is appropriately arrived based on the chosen number of rescaled points to give rise to several network structures. For example, the structure assumes the form 16L-32N-8N-32N-16L in a case with 16 rescaled points. It is significant to note that the number of units in the input and output layers are the same, while the hidden layer is designed with the same number of units for the second and fourth layers and a reduced number for the middle layer. The back propagation algorithm is used to train the network for the combined rescaled training samples of each character and the weights computed at the end of 3,500 epochs constitute the model for each character.

Each test sample is matched against the 156 models to result in an array of 156 confidence scores and the model that corresponds to the highest confidence score is considered to be the matched model. Therefore, a stream of trials spread over different rescaled points are conducted for each character with x model and y model and the performance is measured in terms of recognition rate depending on the number of times the character is correctly recognized. The experimental results show a higher recognition rate for the y model than the x model. The exercise is further enhanced through the use of a weighted sum rule given in Eq. 4.8 applied to the confidence scores s_1 and s_2 obtained from the x and y models separately to yield the xy model.

$$s = ws_1 + (1 - w)s_2 \tag{4.8}$$

The measures in Table 4.1 project the recognition rates obtained from the three models with different number of rescaled points for the 156 characters. It is observed that the weighted xy model offers the highest score over the other two models for a same number of rescaled points and the best recognition rate of 89.74 % is obtained for 96 rescaled points. The recognition rate increases significantly when the number of rescaled points is varied from 8 to 48, marginally in the range 48–96, and remains the same thereafter. The performance is relatively

Table 4.1 Recognition rate (in %) for various rescaled points

Number of rescaled points	x model	y model	xy model
8	55.77	62.17	75.64
16	60.89	69.87	79.49
32	67.94	75.00	84.61
48	71.79	77.56	87.82
64	72.43	78.84	88.46
80	72.43	79.49	89.10
96	73.72	80.13	89.74
112	74.35	80.13	89.74
128	74.35	80.13	89.74

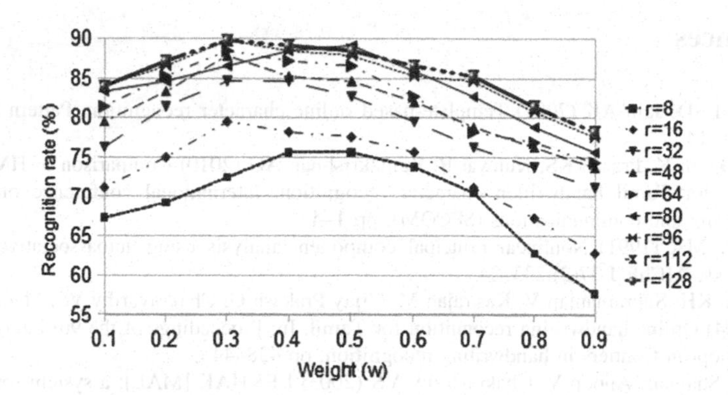

Fig. 4.3 Recognition rates for different weights over various rescaled points (r)

better for y than for x and accordingly xy model yields the best recognition rate when the weight is higher for y.

The investigative study proceeds over a variation in the weights from 0.1 to 0.9 and the results are displayed in Fig. 4.3. It follows from the line chart that the proposed approach offers the optimum performance for a weight corresponding to 0.3 in the xy model.

4.5 Conclusion

AANN-based online handwritten character recognition method applicable for Tamil scripts has been developed to allow user friendly character input using a pen and digital tablet. The methodology has been evolved through preprocessing and classification stages to ensure a noise free and rescaled x and y coordinate positions for the construction of models. The HP Tamil dataset has been availed for both training and testing the models with 156 characters written by different class of people. Three different models have been pronounced based on the rescaled x, y positions and weighted scores of x and y models. The use of a weighted sum rule based method has been found to add strength to enhance the recognition performance. The results have been found to be more or less consistent over a series of trials conducted using the leave-one-out approach. The method has been found to extract the best performance for 96 rescaled points for the xy model with a weight of 0.3. The formulation has been based on simple underlying principles and therefore will turn out to be a suitable tool to interpret handwritten Tamil characters.

Acknowledgments The authors thank the authorities of Annamalai University for providing the necessary facilities to carry out this research work. The authors also acknowledge HP Labs India for making the HP Tamil dataset available for this work.

References

1. Connell SD, Jain AK (2001) Template-based online character recognition. Pattern Recogn 34(1):1–14
2. Shashikiran K, Prasad KS, Kunwar R, Ramakrishnan AG (2010) Comparison of HMM and SDTW for Tamil handwritten character recognition, international conference on signal processing and communications (SPCOM), pp 1–4
3. Kramer MA (1991) Nonlinear principal component analysis using autoassociative neural networks. AIChE J 37(2):233–243
4. Aparna KH, Subramanian V, Kasirajan M, Vijay Prakash G, Chakravarthy VS, Madhavnath S (2004) Online handwriting recognition for Tamil. In: Proceedings of the 9th international workshop on frontiers in handwriting recognition, pp 438–443
5. Gowri Shankar, Anoop V, Chakravarthy VS (2003) LEKHAK [MAL]: a system for online recognition of handwritten Malayalam characters, National conference on communications
6. Srinivas Rao M, Gowrishankar, Chakravarthy VS (2002) Online recognition of handwritten Telugu characters, international conference on universal knowledge and language
7. Shanthi N, Duraiswamy K (2010) A novel SVM-based handwritten Tamil character recognition system. Pattern Anal Appl 13(1):173–180
8. Zhou S, Dong Z, Li WJ, Kwong CP (2008) Handwritten character recognition using MEMS motion sensing technology, IEEE/ASME International conference on advanced mechatronics
9. Pal U, Wakabayashi T, Kimura F (2009) Comparative study of Devnagari handwritten character recognition using different feature and classifiers, international conference on document analysis and recognition (ICDAR), pp 1111–1115
10. Wang Y, Huo Q (2009) Template-based online character recognition. Pattern Recogn 42:3296–3302
11. Sternby J, Morwing J, Andersson J, Friberg C (2009) On-line Arabic handwriting recognition with templates. Pattern Recogn 42:3278–3286
12. Niels R, Vuurpijl L (2005) Dynamic time warping applied to Tamil character recognition, international conference on document analysis and recognition (ICDAR), pp 730–734
13. Deepu V, Sriganesh M, Ramakrishnan AG (2004) Principal component analysis for online handwritten character recogntion, 17th international conference on pattern recognition, pp 327–330
14. Berthold MR, Sudweeks F, Newton S, Coyne R (1998) It makes sense: using an autoassociative neural network to explore typicality in computer mediated discussions. In: Rafaeli S, Sudweeks F, McLaughlin M (eds) Network and netplay: virtual groups on the internet, AAAI/MIT Press, Cambridge, pp 191–219
15. Yegnanarayana B, Kishore SP (2002) AANN: an alternative to GMM for pattern recognition. Neural Netw 15(3):259–269

Chapter 5
Obstacle Detection Techniques for Vision Based Autonomous Navigation Systems

R. Karthikeyan, B. Sheela Rani and K. Renganathan

Abstract The aim behind this paper is to develop, design and implement a autonomous vehicle on real time environment with suitable performance measures. For this purpose, an algorithm using image processing is modeled. The vehicle considered for study is a differentially steered toy car fixed with a web cam. The image data are transferred to the central computer through cables. The vehicle uses image acquisition tools and image processing tools to handle the images and find the obstacle. Two algorithms are being developed based on the above idea along with evaluation of appropriate performance measures.

Keywords Navigation systems · Sum of absolute difference · Correlation · Performance analysis

5.1 Introduction

Due to the advancement of Technology in Science and need for automation there has always been a thrust to improve the capability of autonomous vehicle toward a step forward in driverless techniques, to cater the demands of the end users and

---

R. Karthikeyan (✉) · K. Renganathan
Department of Instrumentation and Control Engineering, Sri Sairam Engineering College, Chennai 600 044, Tamilnadu, India
e-mail: karthikeyan6173@yahoo.com

K. Renganathan
e-mail: renganathan_1980@yahoo.com

B. Sheela Rani
Sathyabama University, Chennai 600 119, Tamilnadu, India
e-mail: kavi_sheela@yahoo.co.in

R. Malathi and J. Krishnan (eds.), *Recent Advancements in System Modelling Applications*, Lecture Notes in Electrical Engineering 188,
DOI: 10.1007/978-81-322-1035-1_5,

their safety. In this regard, vision based systems plays an important role and has been one of the most concentrated areas of research, design and development.

The main idea behind this paper is to design, develop and implement algorithms for achieving autonomous obstacle avoidance in real time environment along with the evaluation of performance measures, with useful study made from robots used in surveillance [1], Human motion detections [2, 3], Algorithms on compressed domain [4], Artificial Intelligence [5], and Image processing [6].

When the obstacle is dynamic i.e., when the obstacle comes in front of the mobile platform from left or right, the proposed algorithm will process on the images acquired real time and give an output to control the vehicle. The program will control the vehicle by sending a signal through the serial port to the micro-controller, from where it is transmitted to the vehicle. When the program is executed, the vehicle starts moving and the images are acquired. When the obstacle is introduced, the vehicle would pause and divert as per the needs.

Here the proposed algorithm processes on consecutive images for controlling the vehicle. As the working is based on a vision based system, the constraints faced while doing it real time were and the resolution of the images acquired by the web camera should be in a good state. If the images are dull and dark, the result of the program will be affected. As a web camera resolution of 352 × 288 pixels are used to acquire images, the background need to be really bright or environment should not be dark. This can be avoided by using high resolution CCD camera which can acquire high definition images. The web camera is wired so the prototype has a limiting distance. This can be avoided by using a wireless CCD camera. There are mainly two techniques to identify the obstacle in front of the mobile platform (1) Sum of Absolute Difference (2) Correlation technique.

The performance of the vehicle is mainly based on and processing of images through the USB web camera. The vehicle uses image acquisition tools and image processing tools to handle the images and find the obstacle. The Image Acquisition Toolbox is a collection of functions that extend the capability of the MATLAB numeric computing environment. The toolbox supports a wide range of image acquisition operations, including acquiring images through many types of image acquisition devices, from professional grade frame grabbers to USB-based Webcams. Viewing a preview of the live video stream triggers acquisitions (which include external hardware triggers), thereby configuring callback functions that execute when certain events occur bringing the image data into the MATLAB workspace. The toolbox also supports a wide range of image processing operations, including Spatial image transformations, morphological operations, neighborhood and block operations, linear filtering and filter design Transforms Image analysis, and enhancement Image registration De blurring Region of interest operations.

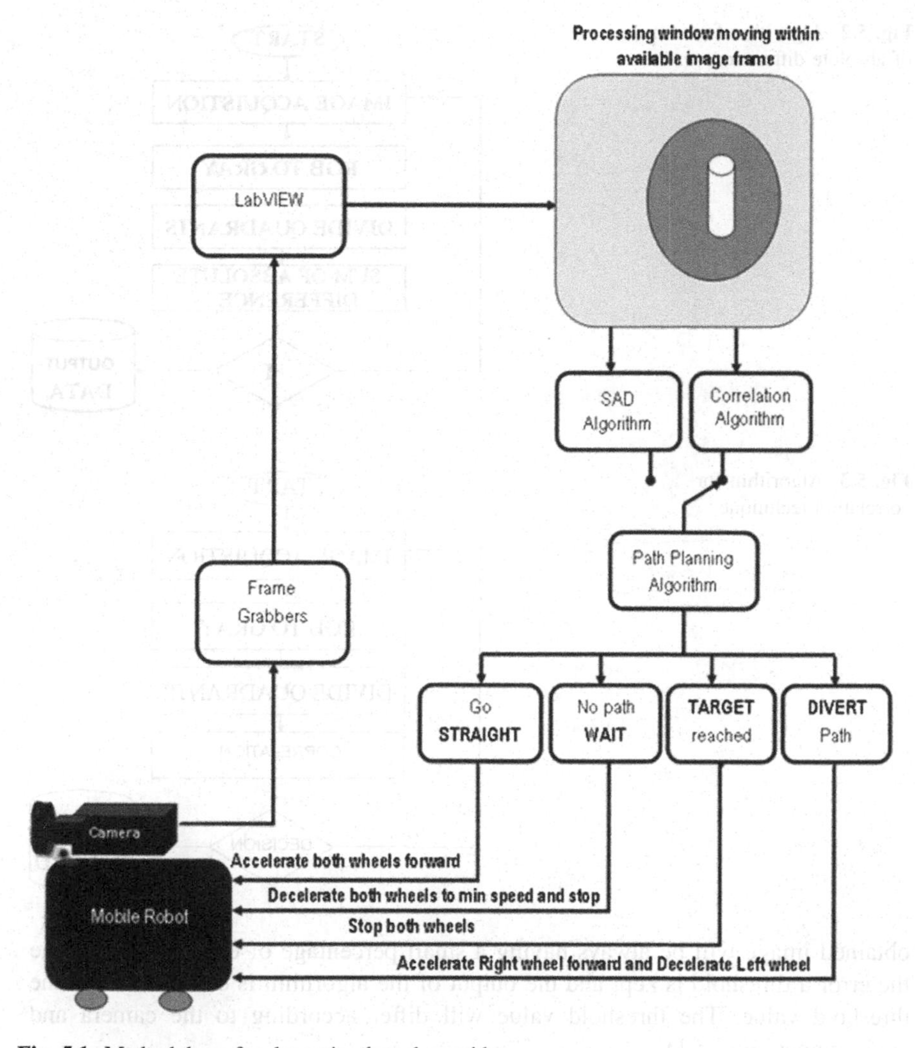

Fig. 5.1 Methodology for dynamic obstacle avoidance

5.2 Methodology

The main objective of the proposed algorithms is to compare the consecutive frames that are obtained from the camera and to verify whether there is any change of parameters. In the Sum of Absolute Difference algorithm (SAD) the consecutive frames are first divided into quadrants for easier process and converted into gray scale image, then the final step is to calculate the sum of absolute difference, which is verified with threshold value. The correlation algorithm has the same method, but in the final step the correlated values are obtained and the minimum value is compared with the threshold value. The purpose of setting a threshold value is, the

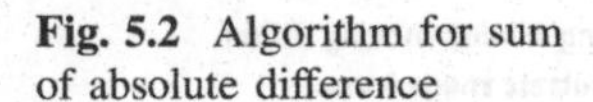

Fig. 5.2 Algorithm for sum of absolute difference

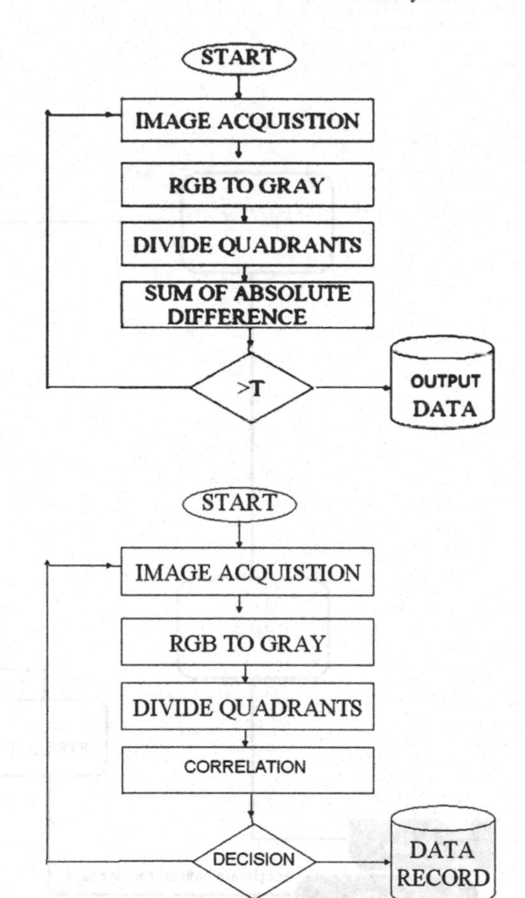

Fig. 5.3 Algorithm for correlation technique

obtained image will be always having a small percentage of error. To overcome the error a threshold is kept and the output of the algorithm is compared with the threshold value. The threshold value will differ according to the camera and environment (Fig. 5.1).

5.3 Sum of Absolute Difference Technique

This technique computes absolute difference of two images. It subtracts each element in array Y from the corresponding element in array X and returns the absolute difference in the corresponding element of the output array Z. X and Y are real, non sparse, numeric or logical arrays with the same class and size. Z has the same class and size as X and Y. If X and Y are integer arrays, elements in the output that exceeds the range of the integer type are truncated (Fig. 5.2).

Here the above flow chart explains about the process involved in sum of absolute difference technique. This method involves a formula based on which the

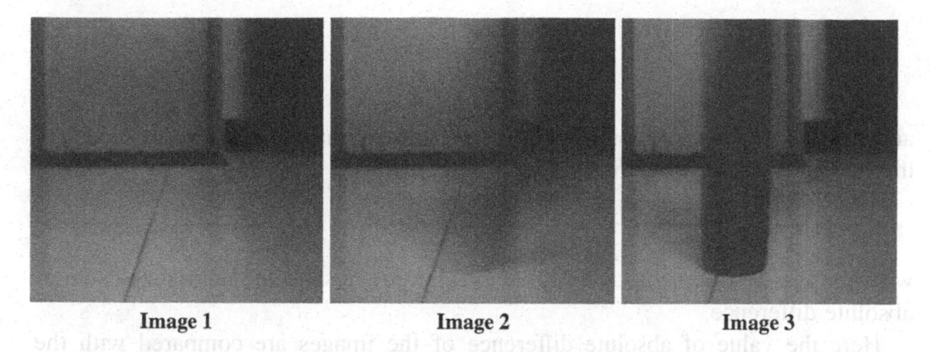

Fig. 5.4 Real time images of background, image 1, image 2, image 3

absolute differences between the two images are calculated. The formula is as given below:

$$D(t) = \frac{1}{N}\sum \left|I(t_i) - I(t_j)\right|, \tag{5.1}$$

where N is the number of pixels used as scaling factor, $I(t_i)$ is the image I at time t, $I(t_j)$is the image at time j and D(t) is the normalized sum of absolute difference for that time.

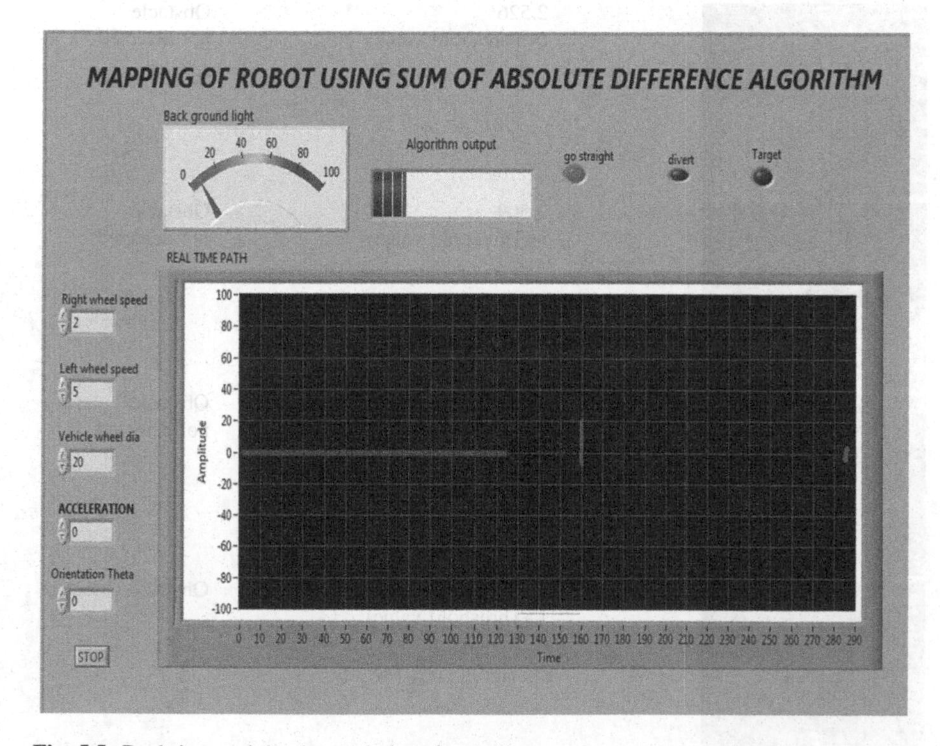

Fig. 5.5 Real time and the detected obstacle position

In ideal case when there is no motion

$$I(t_i) = I(t_j), \tag{5.2}$$

and D(t) = 0. However, noise is always present in images and a better model of the images in the absence of motion is given as

$$I(t_i) = I(t_j) + n(p), \tag{5.3}$$

where n(p) is a noise signal. The value D(t) here represents the normalized sum of absolute difference.

Here the value of absolute difference of the images are compared with the threshold value which will be fixed according to the environment and the camera used for the operation.

Table 5.1 SAD and output values for a threshold of 4,500

Image	Sad value	Output
	2,518 (<Threshold value)	Obstacle not detected
	2,526 (<Threshold value)	Obstacle not detected
	3,509 (<Threshold value)	Obstacle not detected
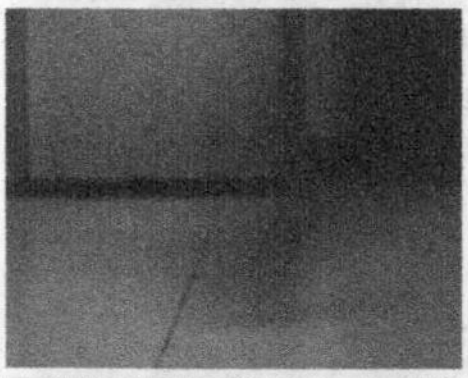	3,982 (<Threshold value)	Obstacle not detected
	6,590 (>Threshold value)	Obstacle detected

5.4 Correlation Technique

This technique computes the two-dimensional correlation coefficient between two matrices. Correlation method processes gray images only, so the acquired RGB image should be converted to a grayscale image (Fig. 5.3).

The formula used is here is given by

$$r = \frac{\sum_m \sum_n (A_{mn} - \bar{A})(B_{mn} - \bar{B})}{\sqrt{\left(\sum_m \sum_n (A_{mn} - \bar{A})^2\right)\left(\sum_m \sum_n (B_{mn} - \bar{B})^2\right)}} \tag{5.4}$$

where $\bar{A}$ = mean2(A), and $\bar{B}$ = mean2(B).

Here the value 'r' is compared with the threshold value which is fixed according to the environment and the camera used for the operation. This method is implemented to correlate consecutive images to find any obstacle on the path. The value may vary from 0 to 1, if the consecutive images are almost the same the value of 'r' will be almost 1 or a little less than 1. So if the consecutive images does not have any relation or an obstacle comes in its way, then the value of 'r' will be closer to 0. The threshold value must be such a way that it must detect the

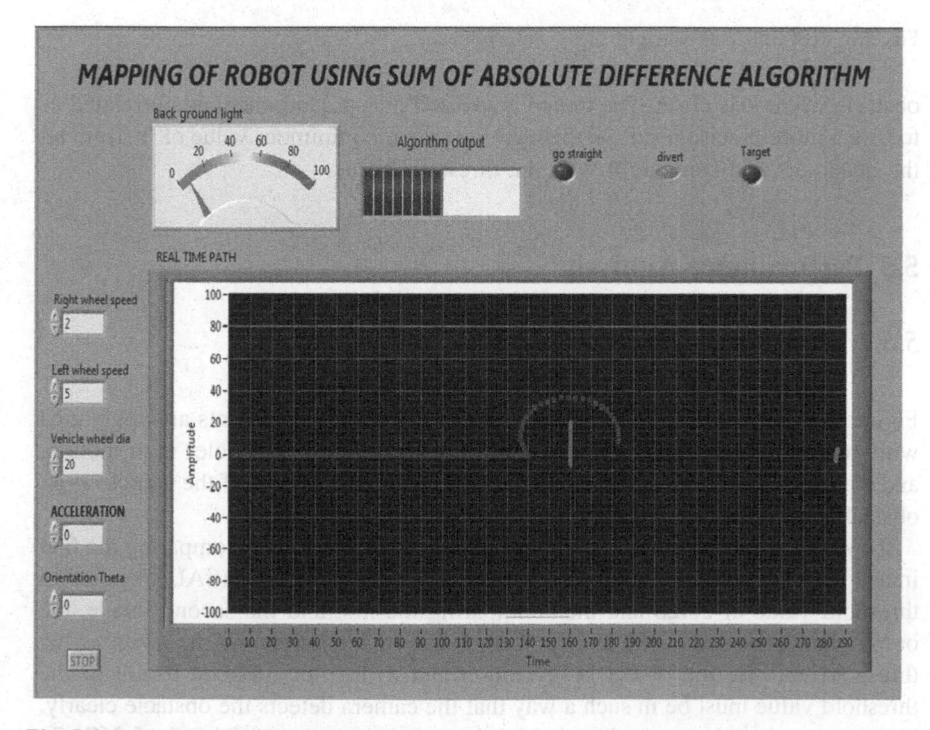

Fig. 5.6 Real time path of avoiding the obstacle

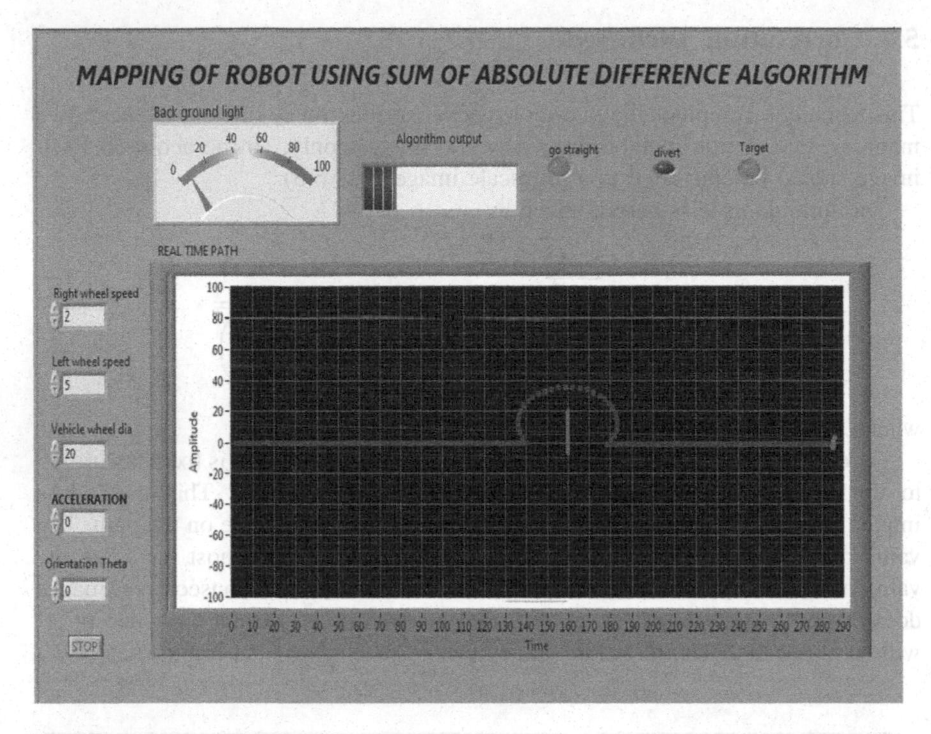

Fig. 5.7 Real time path when of successful avoidance

obstacle when it is close. The image is divided into 4 quadrants and correlated as to give a more precise result. In this way we take the minimum value of 'r' from all the quadrants and compare it with the threshold value.

5.5 Performance Analysis

5.5.1 Evaluation of Threshold

Figure 5.4 shows three different images. The first Image represents an empty area with no obstacle, the second image represents the object (obstacle) is in motion, and the third image represents that the object has come in front of the vehicle as an obstacle. Here the three images are consecutive images.

Now as per the environment, the threshold values are set. By comparing the first image with the previous image the Sum of Absolute Difference (SAD) resulted in threshold value of 2,428 and then comparing the first and the second image, the output value resulted 3,480 as it is in motion. This value cannot be taken as the threshold because the object is in motion and it has not surfaced itself. So the threshold value must be in such a way that the camera detects the obstacle clearly. Now comparing the second and third image the output of SAD is 5,267. So

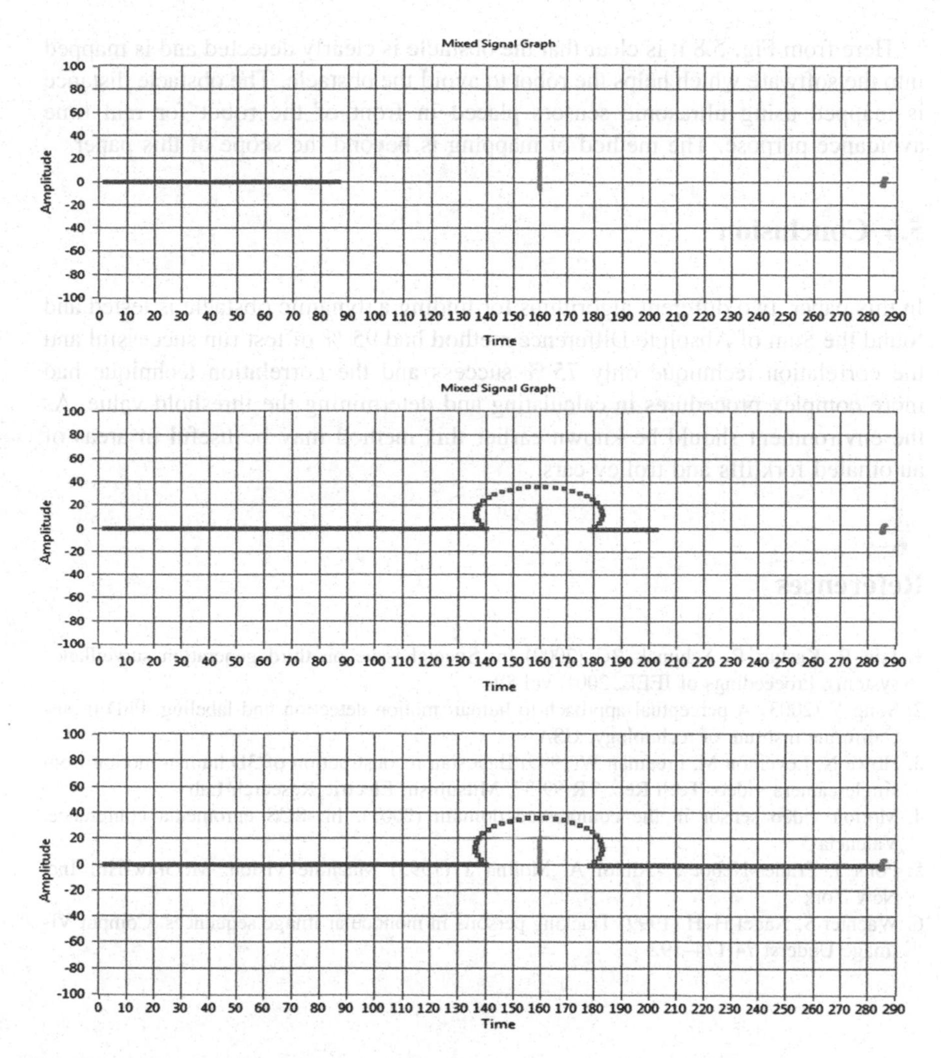

Fig. 5.8 Various real time path in a single trial run

threshold value is at 4,500 to make a condition such that the SAD value should be greater than threshold value so as to detect the object. Hence in this way it is most suitable for the preferable environments (Fig. 5.5 and Table 5.1).

5.5.2 Results

As the output value depends on the environment, the value may vary accordingly, but generally it is found that when there is a object in front of the vehicle the SAD value will go beyond 4,500. So the robot car was successfully designed and the obstacle is avoided (Figs. 5.6 and 5.7).

Here from Fig. 5.8 it is clear that the obstacle is clearly detected and is mapped into the software which helps the robot to avoid the obstacle. The obstacle distance is mapped using ultrasonic sensors placed in front of the robot for real time avoidance purpose. The method of mapping is beyond the scope of this paper.

5.6 Conclusion

In this paper, two different algorithms for finding a dynamic obstacle is tested and found the Sum of Absolute Difference method had 95 % of test run successful and the correlation technique only 75 % success and the correlation technique had more complex procedures in calculating and determining the threshold value. As the environment should be known earlier this method may be useful in areas of automated forklifts and trolley cars.

References

1. Jain R, Kasturi R, Schunck BG (2009) In: Special issue on third generation surveillance systems, Proceedings of IEEE, 2001, vol 89
2. Song Y (2003) A perceptual approach to human motion detection and labeling. PhD thesis, California institute of technology, USA
3. Howe N, Leventon M, Freeman W(1999) Bayesian reconstruction of 3D human motion from single-camera video. Tech Rep. TR-99-37, Mitsubishi Electric Research Lab
4. Motion video sensor in the compressed domain (2001). In: SCS euromedia conference, Valencia
5. Pons J, Prades-Nebot J, Albiol A, Molina J (1995) Machine vision. McGraw-Hill Inc, New York
6. Wachter S, Nagel H–H (1999) Tracking persons in monocular image sequences. Comput Vis Image Underst 74:174–192

Chapter 6
Multilevel Renyi's Entropy Threshold Selection Based on Bacterial Foraging Algorithm

P. D. Sathya and V. P. Sakthivel

Abstract A novel stochastic optimization approach to solve multilevel thresholding problem in image segmentation using bacterial foraging (BF) technique is presented. The BF algorithm is based on the foraging behavior of *E. Coli* bacteria which is present in the human intestine. The proposed BF algorithm is used to maximize Renyi's entropy function. The utility of the proposed algorithm is aptly demonstrated by considering several benchmark test images and the results are compared with those obtained from particle swarm optimization (PSO) and genetic algorithm (GA) based methods. The experimental results show that the proposed algorithm could demonstrate enhanced performance in comparison with PSO and GA in terms of solution quality and stability. The computation speed is accelerated and the quality improved through the use of this strategy.

Keywords Multilevel thresholding · Image segmentation · Renyi's entropy · Bacterial foraging algorithm

6.1 Introduction

Image thresholding which extracts object from the background in an input image is one of the most common applications in image analysis. Thresholding is a necessary step for image preprocessing in automatic recognition of machine printed or a hand-written text, in shape recognition of objects, and in image enhancement.

P. D. Sathya (✉) · V. P. Sakthivel
Department of Electrical Engineering, Annamalai University, Chidambaram 608002, India
e-mail: pd.sathya@yahoo.in

V. P. Sakthivel
e-mail: vp.sakthivel@yahoo.com

R. Malathi and J. Krishnan (eds.), *Recent Advancements in System Modelling Applications*, Lecture Notes in Electrical Engineering 188,
DOI: 10.1007/978-81-322-1035-1_6, © Springer India 2013

Among the image thresholding methods, bi-level thresholding separates the pixels of an image into two regions i.e. the object and the background. One region contains pixels with gray values smaller than the threshold value and the other contains pixels with gray values larger than the threshold value. Further, if the pixels of an image are divided into more than two regions, it is called multilevel thresholding. In general, the threshold is located at the obvious and deep valley of the histogram. However, when the valley is not so obvious, it is very difficult to determine the threshold.

During the past decade, many research studies have been devoted to the problem of selecting the appropriate threshold value. A thorough survey of a variety of thresholding techniques has been presented by Sahoo et al. [1]. Among those techniques, global, histogram based algorithms [2] have been widely used to determine the threshold, and they can be classified as parametric and non-parametric approaches.

In the parametric approaches [3], the gray level distribution of each class is assumed to have a probability density function and has been assumed to be a Gaussian distribution. The attempts have been focused to find an estimate of the parameters of the distribution that offers the best fit the given histogram data in the least squares sense. The result has been typically a nonlinear optimization problem and focused to be computationally expensive and time-consuming solution.

In the non-parametric approaches, the effort has been to determine the thresholds that separate the gray-level regions of an image in an optimum manner according to some discriminate criteria such as the between-class variance [4], entropy [5], and cross entropy [6]. The non-parametric approaches have been computationally efficient and simple to implement, compared to the parametric approaches.

In bi-level thresholding the existing non-parametric methods are robust and computationally fast for time-critical applications. However, the computational complexity of those methods has been focused to exponentially increase and the selected thresholds generally become less credible as the number of classes to be separated increases. In multilevel image thresholding, the pixels have been classified into many classes, not just foreground and background. A host of methods have been proposed for multilevel thresholding [7–10].

In [8], the Otsu's function has been modified by a fast recursive algorithm along with a look-up-table for multilevel thresholding. A fast thresholding computation using Otsu's function has been presented [9]. Another fast multilevel thresholding technique has been proposed by Yin [10].

Various deterministic methods have been applied to solve multilevel thresholding problem in image segmentation. Several techniques using genetic algorithms (GAs) have also been proposed to solve the multilevel thresholding problem [11, 12]. The particle swarm optimization (PSO) has been applied to the multilevel thresholding for image segmentation [13].

In this paper, the BF algorithm is employed to solve the multilevel thresholding problem in image segmentation. The algorithm is based on the foraging (methods for locating, handling and ingesting food) behavior of *E. Coli* bacteria present in the human intestine. It is successfully used to solve various kinds of engineering problems [14–16]. The proposed BF algorithm is compared with the PSO and GA methods over six benchmark images with respect to the following performance measures: Solution quality, convergence speed and Peak to Signal Noise Ration (PSNR) measure. It is shown that the BF algorithm offers superior performance than the PSO and GA.

6.2 Problem Formulation with Renyi's Entropy

Global threshold selection methods usually use the gray-level histogram of the image. The optimal thresholds are determined by optimizing some criterion function obtained from the gray-level distribution of the image. A new thresholding technique using Renyi's entropy utilize two probability distributions (object and background) which are derived from the original gray-level distribution of an image.

Let there be L gray levels in a given image and these gray levels are in the range {0, 1, 2… (L−1)}. Then one can define $Pi = h(i)/N$, $(0 \leq i \leq (L-1))$ where h (i) denotes number of pixels for the corresponding gray-level L and N denotes total number of pixels in the image which is equal to $\sum_{i=0}^{L-1} h(i)$.

Renyi's bi-level thresholding can be described as follows:

$$f(t) = \text{argmax}\left[H_0^\alpha(t) + H_1^\alpha(t)\right] \tag{6.1}$$

and

$$H_0^\alpha(t) = \frac{1}{1-\alpha} \ln \sum_{i=0}^{t-1} \left(\frac{P_i}{P^A}\right)^\alpha, P^A = \sum_{i=0}^{t-1} P_i$$

$$H_1^\alpha(t) = \frac{1}{1-\alpha} n \sum_{i=t}^{L-1} \left(\frac{P_i}{P^B}\right)^\alpha, P^B = \sum_{i=t}^{L-1} P_i.$$

where α is a positive parameter.

This Renyi's entropy criterion method can also be extended to multilevel thresholding and it is described as follows:

$$f(t) = \text{argmax}[H_0^\alpha(t) + H_1^\alpha(t) + H_2^\alpha(t) + \ldots + H_m^\alpha(t)] \tag{6.2}$$

Where

$$H_0^\alpha(t) = \frac{1}{1-\alpha} \ln \sum_{i=0}^{t_1-1} \left(\frac{P_i}{P^A}\right)^\alpha, P^A = \sum_{i=0}^{t_1-1} P_i$$

$$H_1^{\alpha}(t) = \frac{1}{1-\alpha} \ln \sum_{i=t_1}^{t_2-1} \left(\frac{P_i}{P^B}\right)^{\alpha}, P^B = \sum_{i=t_1}^{t_2-1} P_i$$

$$H_2^{\alpha}(t) = \frac{1}{1-\alpha} \ln \sum_{i=t_2}^{t_3-1} \left(\frac{P_i}{P^C}\right)^{\alpha}, P^C = \sum_{i=t_2}^{t_3-1} P_i \ldots$$

$$H_m^{\alpha}(t) = \frac{1}{1-\alpha} \ln \sum_{i=t_m}^{L-1} \left(\frac{P_i}{P^m}\right)^{\alpha}, P^m = \sum_{i=t_m}^{L-1} P_i.$$

6.3 Bacterial Foraging Optimization Algorithm

BF algorithm is a newly introduced evolutionary optimization algorithm that mimics the foraging behavior of Escherichia coli (commonly referred to as *E. coli*) bacteria. The BF algorithm appears to be first introduced by Passino [14]. There are successful applications of BF algorithm in image processing, such as Image Watermarking [16, 17], Image Enhancement [18], Image Circle Detection [19] and Filtering [20].

The BF models the movement of *E. coli* bacterium moves using a pattern of two types of movements: tumbling and swimming. Tumbling refers to a random change in the direction of movement, and swimming refers to moving in a straight line in a given direction. A bacterium in a neutral medium alternates between tumbling and swimming movements.

Suppose it is desired to search for the position X in an N-dimensional space. Let X_i be the initial position of bacterium i in the search space, $i = 1, 2, \ldots, S$, where S is the number of bacteria. In biological bacteria populations, S can be as high as 10^9 and N is three. Let F (X_i) represent an objective function. Let F $(X_i) < 0$, F $(X_i) = 0$, and F $(X_i) > 0$ represent the bacterium at location X_i in nutrient rich, neutral, and noxious environments, respectively. Chemotaxis is a foraging behavior that captures the process of optimization, where bacteria to climb up the nutrient concentration gradient (i.e., bacteria try to achieve positions having lower values of F (X_i) and avoid being at positions X_i, where F $(X_i) \geq 0$).

The bacterium i at position X_i takes a chemotactic step j with the step size C(i) and evaluates itself for objective function $F(X_i)$ at each step. If at position $X_i(j+1)$, the objective value F is better than at position $X_i(j)$, then another step of same size C(i) in the same direction will be taken again, if that step resulted in a position with a better value than at the previous step. It is referred to as a swimming step. Swimming is continued until for a maximum number of steps N_s. After N_c chemotactic steps, a reproduction steps is taken in which the population is sorted in ascending order of the objective function value F and least healthy bacteria are replaced by the copies of the healthier bacteria. After N_{re} reproduction steps, an elimination-dispersal step is taken. Here, a bacterium is eliminated and a new bacterium is created at a random location in the search space with probability ρ_{ed}. The optimization stops after N_{ed} elimination-dispersal steps.

Bacteria create swarms by means of cell-to-cell signaling via an attractant and a repellant. Cell-to-cell attraction for bacterium i is represented with $F_{cc}(X_g, X_i)$, i = 1, 2,..., S. this is defined as follows:

$$F_{cc}(X_g, X_i) = \sum_{i=1}^{S}\left[d_{attract}\exp\left(-\omega_{attract}\sum_{n=1}^{N}(X_g - X_i)^2\right)\right] + \sum_{i=1}^{S}\left[h_{repellant}\exp\left(-\omega_{repellant}\sum_{n=1}^{N}(X_g - X_i)^2\right)\right]$$

The cell-to-cell signaling F_{cc} ( ) helps cells to move toward other cells, but not very close to them. In BF algorithm, the maximum number of objective function evaluations is S. N_c. N_s. N_{re}. N_{ed}. A general biologically inspired thumb-of-rule for choosing the parameters of BF is: $N_c > N_{re} > N_{ed}$. The detailed pseudo-code for BF algorithm is given in Algorithm 1.

Algorithm 3 Pseudo-code for BF algorithm

1: Initialize S, N, N_c, N_s, N_{re}, N_{ed}, ρ_{ed}, $d_{attract}$, $h_{repellant}$, $\omega_{attract}$, $\omega_{repellant}$, X_{min} and X_{max}
2: Initialize X_i randomly for i = 1, 2,.., S
3: Initialize C (i) for i = 1, 2,..., S
4: Set the loops counters j, k and l to 0
5: **//Elimination-Dispersal loop:**
6: **while** $l \le N_{ed}$ **do**
7: $l = l + 1$
8: **//Reproduction loop:**
9: **while** $k \le N_{re}$ **do**
10: $k = k + 1$
11: **//Chemotaxis loop:**
12: **while** $j \le N_c$ **do**
13: $j = j + 1$
14: **for** each bacterium i = 1, 2,..., S **do**
15: Compute F(i, j, k, l)
16: Let F (i, j, k, l) = F(i, j, k, l) + $F_{cc}(X_g, X_i)$
17: Let F_{last} = F (i, j, k, l)
18: **//Tumble:**
19: Generate a N-dimensional random vector Δ_m (i), i = 1, 2,..., N on [−1, 1]
20: **//Move:**
21: Let $X_i(j+1, k, l) = X_i(j, k, l) + C(i)\frac{\Delta(i)}{\sqrt{\Delta^T(i)\Delta(i)}}$

22: Compute $F(i, j + 1, k, l)$ with $X_i(j + 1, k, l)$
23: **//Swim:**
24: Let $m = 0$
25: **while** $m < N_s$ **do**
26: Let $m = m + 1$
27: **if** $F(i, j + 1, k, l) < F_{last}$ **then**
28: Let $F_{last} = F(i, j + 1, k, l)$
29: Let $X_i(j + 1, k, l) = X_i(j + 1, k, l) + C(i)\frac{\Delta(i)}{\sqrt{\Delta^T(i)\Delta(i)}}$
30: Use this $X_i(j + 1, k, l)$ to compute new $F(i, j + 1, k, l)$
31: **else**
32: $m = N_s$
33: **end if**
34: **end while**
35: **end for**
36: **end while**
37: Compute for each bacterium i, for given k and l $F^i_{health} = \sum_{j=1}^{N_c+1} F(i, j, k, l)$
38: Eliminate S_r fraction of bacteria with highest F_{health} and split the other bacteria into two at their locations.
39: **end while**
40: For each bacterium, with probability ρ_{ed} eliminate the bacterium and create a new one at a random position.
41: **end while**

6.4 Experimental Results and Evaluation

The performance of the proposed algorithm is examined using several test images, the results obtained are compared with the results of well known heuristic algorithms such as PSO and GA. The tested algorithms belong to the population-based thresholding algorithm. The proposed algorithm is implemented with a core2duo 2 GHz personal computer in MATLAB language. The test images with their corresponding histograms are shown in Fig. 6.1. Parameters used for the proposed algorithm are summarized in Appendix.

Table 6.1 shows the optimal thresholds and the corresponding objective values obtained by all the algorithms using Renyi's objective function. Higher value of objective function indicates the better results in the segmentation. The proposed approach offers an improved objective function value over the PSO and GA methods as seen from Table 6.1, clearly explaining its ability to locate better solutions than other methods.

Fig. 6.1 Test images **a** Lena. **b** Pepper. **c** Baboon. **d** Hunter. **e** Airplane. **f** Butterfly

The quality of the thresholded images can be evaluated by PSNR measure. The higher value of PSNR means the quality of the thresholded image is better. Obviously, PSNR can be used as a criterion for optimal thresholding. It can be observed from Table 6.2 that the results of the proposed method have higher PSNR than the other two methods.

The Table 6.3 summarizes the standard deviation values of three different approaches. Through 50 trials, the proposed approach yielded smaller standard deviations of objective function values.

The Table 6.4 illustrates the CPU time taken to find the optimal threshold values of all the three algorithms. It is seen from the table that the time requirement of the proposed BF method is less and either comparable or better than the other mentioned methods. And also, the CPU time increases with the number of thresholds. So as a whole, it can be said that the BF method is efficient than previously mentioned methods.

The results of 5-level thresholding in Table 6.1 are illustrated in Figs. 6.2, 6.3 and 6.4. Figure 6.4 shows that the quality of the thresholded images is better by Renyi based BF algorithm than the other two algorithms.

All over, the proposed Renyi's entropy based BF algorithm provides better efficiency, PSNR value and stability. And also, the proposed method converges faster than the other two algorithms.

Table 6.1 Comparison of optimal threshold values and their objective values obtained by MCE based evolutionary algorithms

Test images	m	Optimal threshold values			Objective values		
		BF	PSO	GA	BF	PSO	GA
Lena	2	99,166	99,166	99,166	12.3897	12.3897	12.3897
	3	67,122,178	93,137,193	88,168,201	15.3020	15.2254	14.9657
	4	63,94,126,174	75,114,144,178	85,129,149,179	17.6047	17.4717	17.2935
	5	73,104,132,164,194	82,114,147,174,201	87,116,156,177,204	20.6058	20.4798	20.3066
Pepper	2	80,150	80,150	80,150	12.5809	12.5809	12.5809
	3	72,120,161	74,128,196	88,149,204	15.5931	15.4238	15.2091
	4	82,115,155,197	70,110,145,179	76,106,147,185	17.6938	17.5800	17.3779
	5	53,91,132,167,197	75,117,145,172,200	79,116,150,181,220	21.0780	20.8113	20.6959
Baboon	2	77,143	77,143	77,143	12.2914	12.2914	12.2914
	3	60,118,158	59,128,155	63,87,171	15.2906	15.1011	14.9583
	4	70,107,142,171	73,110,155,191	84,120,160,189	17.3988	17.2551	17.0455
	5	42,88,118,146,175	54,94,128,163,193	69,102,131,161,191	20.7277	20.6136	20.4514
Hunter	2	91,178	91,178	91,178	12.4967	12.4967	12.4967
	3	51,131,184	85,128,171	61,107,201	15.6301	15.4790	15.3496
	4	45,96,144,184	58,100,132,187	62,104,139,193	17.3064	17.1923	17.0108
	5	55,103,150,182,220	57,93,128,185,208	82,123,150,182,220	21.3092	21.1116	20.9974
Airplane	2	77,170	77,170	77,170	12.2407	12.2407	12.2407
	3	75,114,185	87,121,176	77,118,198	15.3752	15.2436	15.0324
	4	73,114,146,184	73,112,153,193	61,121,153,195	18.2087	18.0972	17.8989
	5	66,99,135,166,191	77,111,144,170,194	67,104,140,183,212	20.7815	20.5993	20.3574
Butterfly	2	96,144	96,144	96,144	10.6317	10.6317	10.6317
	3	69,111,151	99,120,171	66,127,170	12.8903	12.7196	12.5904
	4	84,108,138,170	85,107,138,170	68,114,142,177	15.1106	15.0799	14.8957
	5	71,96,121,144,174	63,99,122,143,165	80,113,137,153,179	17.1092	17.0044	16.8461

Table 6.2 The standard deviation value of four multilevel thresholding methods

Test images	m	Standard deviation		
		BF	PSO	GA
Lena	2	0.0000	0.0000	0.0000
	3	0.0263	0.0674	0.0864
	4	0.0481	0.0966	0.1415
	5	0.0612	0.1145	0.1750
Pepper	2	0.0000	0.0000	0.0000
	3	0.0311	0.0948	0.1493
	4	0.0312	0.1425	0.2071
	5	0.0427	0.1980	0.2672
Baboon	2	0.0000	0.0000	0.0000
	3	0.0220	0.0631	0.0984
	4	0.0226	0.0956	0.1255
	5	0.0557	0.1067	0.3191
Hunter	2	0.0000	0.0000	0.0000
	3	0.0131	0.0444	0.0964
	4	0.0360	0.1056	0.1390
	5	0.0595	0.1200	0.1511
Airplane	2	0.0000	0.0000	0.0000
	3	0.0313	0.0604	0.0775
	4	0.0471	0.0622	0.1079
	5	0.6111	0.0888	0.1854
Butterfly	2	0.0000	0.0000	0.0000
	3	0.0145	0.0519	0.0604
	4	0.0353	0.0972	0.1016
	5	0.0773	0.1086	0.1247

Table 6.3 The PSNR measure by four multilevel thresholding methods

Test images	m	PSNR (db)		
		BF	PSO	GA
Lena	2	15.2436	15.2436	15.2436
	3	17.5192	17.3691	16.4330
	4	18.0722	17.6204	17.2404
	5	21.0964	20.5180	20.0175
Pepper	2	12.7412	12.7412	12.7412
	3	16.4483	15.8646	15.7222
	4	18.9852	17.8329	17.4007
	5	20.3446	20.2480	19.9707
Baboon	2	12.6285	12.6285	12.6285
	3	15.7325	14.9961	14.2945
	4	19.8436	18.6163	18.1994
	5	21.8513	21.4614	20.5676

(continued)

Table 6.3 (continued)

Test images	m	PSNR (db)		
		BF	PSO	GA
Hunter	2	12.7530	12.7530	12.7530
	3	16.2716	15.3867	15.2920
	4	18.2297	17.2325	16.8773
	5	19.5559	18.5961	18.0875
Airplane	2	13.7735	13.7735	13.7735
	3	15.0386	14.5306	14.0115
	4	16.1041	15.4407	15.4115
	5	18.7440	17.6627	17.0447
Butterfly	2	14.2756	14.2756	14.2756
	3	15.6140	15.2239	14.7754
	4	18.7420	17.5003	17.0536
	5	19.8837	18.9507	18.4744

Table 6.4 The CPU time taken by various multilevel thresholding methods

Test images	m	CPU time (Seconds)		
		BF	PSO	GA
Lenna	2	4.0938	4.4313	4.9219
	3	4.2500	4.6878	4.9531
	4	4.5313	4.9844	5.2031
	5	4.6875	5.3594	5.8438
Pepper	2	4.1250	4.4688	4.8906
	3	4.2031	4.9063	5.1563
	4	4.3281	5.0001	5.5313
	5	4.7500	5.3124	5.5594
Baboon	2	3.7656	3.9844	4.0313
	3	3.9327	4.2969	4.6875
	4	4.1250	4.5313	4.9375
	5	4.3594	4.7500	5.0781
Hunter	2	4.2813	4.4063	4.9531
	3	4.4750	4.8594	5.2513
	4	4.5016	5.0156	5.4688
	5	4.8343	5.6094	6.2500
Airplane	2	4.1719	4.5781	4.9692
	3	4.2031	4.4500	5.2188
	4	4.3275	4.8594	5.3594
	5	4.5938	5.0938	4.9991
Butterfly	2	4.4825	4.7344	4.9991
	3	4.4375	4.9219	5.1563
	4	4.6250	5.2656	5.5313
	5	4.8125	5.4063	5.8906

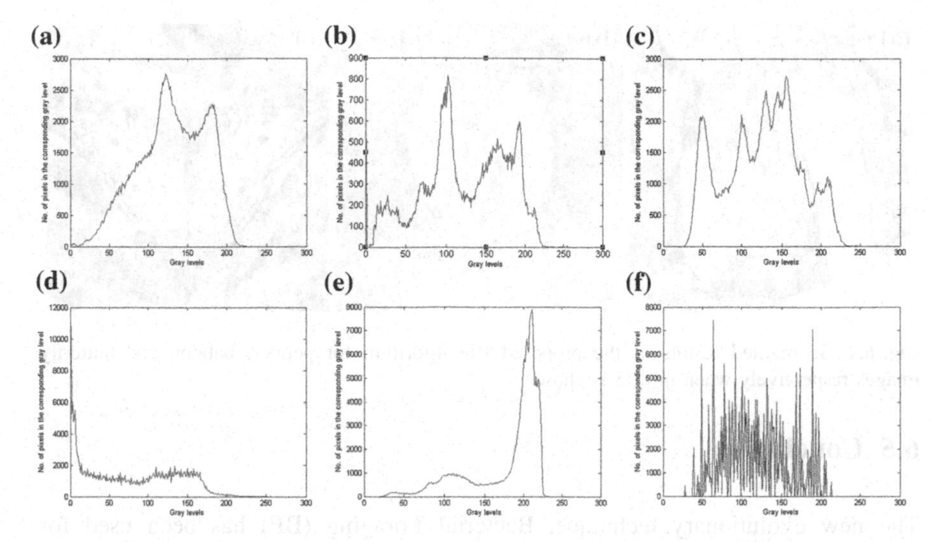

Fig. 6.2 Histogram of test images **a** Lena. **b** Pepper. **c** Baboon. **d** Hunter. **e** Airplane **f** Butterfly

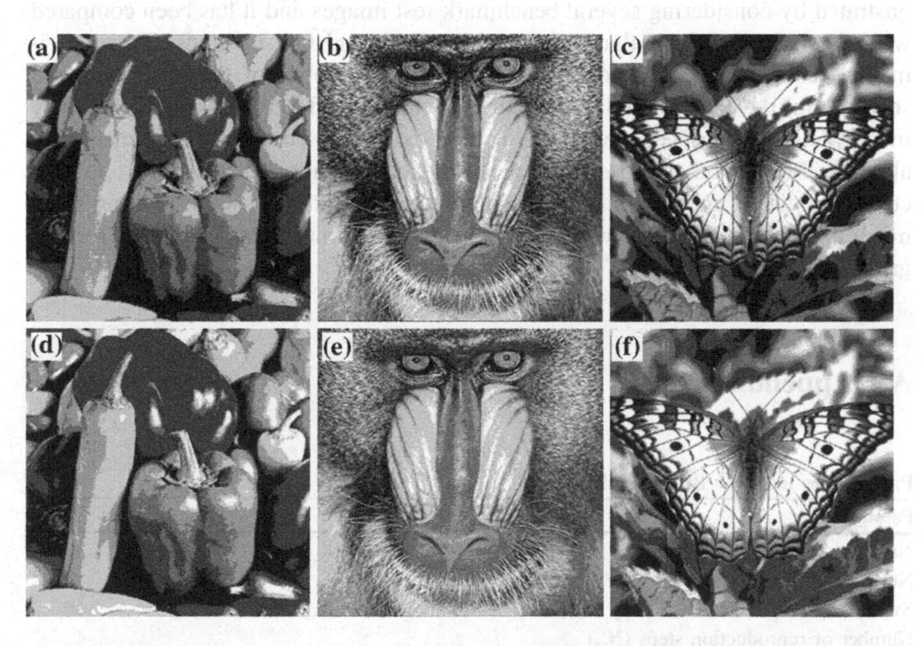

Fig. 6.3 Segmented results of PSO algorithm for pepper, baboon and butterfly images respectively when m = 5 is chosen

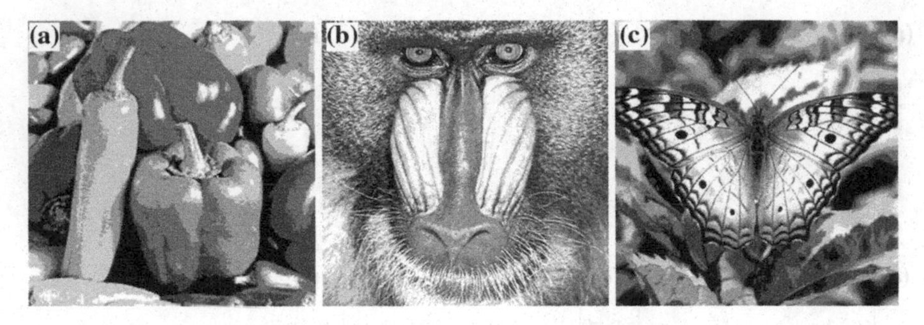

Fig. 6.4 Segmented results of the proposed BF algorithm for pepper, baboon and butterfly images respectively when m = 5 is chosen

6.5 Conclusion

The new evolutionary technique, Bacterial Foraging (BF) has been used for solving multilevel thresholding problem, with an endeavor to maximize the Renyi's entropy function. The utility of the proposed algorithm has been demonstrated by considering several benchmark test images and it has been compared with other evolutionary algorithms such as PSO and GA methods. The experimental results have been found to confirm the potential of the BF method in solving multilevel thresholding problem and show its effectiveness and superiority over PSO and GA. The scope of the method extends to cover proposed method is also suitable for other types of images, and can be applied to a wide class of computer vision applications, such as character recognition, watermarking technique and segmentation of wide variety of medical images and will thus foray a greater role in this domain.

A.1 Appendix

Parameters used for BF algorithm

Parameter	Value
Number of bacterium (s)	20
Number of chemotatic steps (N_c)	10
Swimming length (N_s)	10
Number of reproduction steps (N_{re})	4
Number of elimination of dispersal events (N_{ed})	2
Depth of attractant ($d_{attract}$)	0.1
Width of attract ($\omega_{attract}$)	0.2
Height of repellent ($h_{repellent}$)	0.1
Width of repellent ($\omega_{repellent}$)	10
Probability of elimination and dispersal (P_{ed})	0.02

References

1. Sahoo PK, Soltani S, Wong AKC (1988) A survey of thresholding techniques. Comput Vis Graph Image Process 41(2):233–260
2. Glasbey CA (1993) An analysis of histogram based thresholding algorithms. CVGIP: Graph Models Image Process 55:532–537
3. Weszka JS (1979) A survey of threshold selection techniques. Comput Vis Graph Image Process 7:259–265
4. Otsu N (1979) A threshold selection method from gray-level histograms. IEEE Trans Syst Man and Cybern, SMC 9(1):62–66
5. Kapur JN, Sahoo PK, Wong AKC (1985) A new method for gray-level picture thresholding using the entropy of the histogram. Comput Vis Graph Image Process 29:273–285
6. Li CH, Lee CK (1993) Minimum cross entropy thresholding. Pattern Recogn 26(4):617–625
7. Yin PY, Chen LH (1997) A fast iterative scheme for multilevel thresholding methods. Signal Process 60:305–313
8. Liao PST, Chen S, Chung PC (2001) A fast algorithm for multilevel thresholding. J Inf Sci Eng 17:713–727
9. Lin KC (2003) Fast image thresholding by finding zero(s) of the first derivative of between class variance. Mach Vis Appl 13:254–262
10. Yin P-Y, Chen L-H (1997) A fast iterative scheme for multilevel thresholding methods. Signal Proces 60(3):305–313
11. Yin PY (1999) A fast scheme for optimal thresholding using genetic algorithms. Signal Proces 72:85–95
12. Lai CC, Tseng DC (2004) A hybrid approach using Gaussian smoothing and genetic algorithm for multilevel thresholding. Int J Hybrid Intell Syst 1(3):143–152
13. Maitra M, Chatterjee A (2008) A hybrid cooperative-comprehensive learning based PSO algorithm for image segmentation using multilevel thresholding. Expert Syst Appl 34:1341–1350
14. Passino KM (2002) Biomimicry of bacterial foraging for distributed optimization and control. IEEE Trans Control Syst Mag 22(3):52–67
15. Huang H-C, Chen Y-H, Lin G-Y (2009) Fuzzy-based bacterial foraging for watermarking applications. In: International conference on hybrid intelligent systems, Shenyang, pp 214–217
16. Huang H-C, Chen Y-H, Abraham Ajith (2009) Optimized watermarking using swarm-based bacterial foraging. J Inf Hiding Multimedia Signal Process 1(1):51–58
17. Hanmandlu M, Verma OP, Kumar NK, Kulkarni M (2009) A Novel optimal fuzzy system for color image enhancement using bacterial foraging. IEEE Trans Instrum Meas 58(2):2867–2879
18. Dasgupta S, Biswas A, Das S, Abraham A (2008) Automatic circle detection on images with an adaptive bacterial foraging algorithm. In: International conference on genetic and evolutionary computation, Atlanta, USA, pp 1695–1696
19. Bakwad KM, Pattnaik SS, Sohi BS, Devi S, Panigrahi PK, Sastry Gollapudi VRS (2009) Bacterial foraging optimization technique cascaded with adaptive filter to enhance peak signal to noise ratio from single image. IETE J Res 55(4):173–179
20. Das TK, Venayagamoorthy GK, Aliyu UO (2008) Bio-inspired algorithms for the design of multiple optimal power system stabilizers: SPPSO and BFA. IEEE Trans Ind Appl 44(5):1445–1457

Chapter 7
Re-Routing Strategy for Wireless Virtual Private Networks with CDMA Nodes

C. Mahalakshmi and M. Ramaswamy

Abstract This paper attempts to develop a pioneering mechanism to re-route the data through an alternate path on the occurrence of an exigency in a Wireless Virtual Private Network (WVPN). The fact that Code Division Multiple Access (CDMA) technology facilitates large scale data transmission and is relatively free from interference augur the choice of CDMA nodes to examine the feasibility of a re-routing concept. It is designed to assuage a restricted usage of bandwidth to serve the elaborate needs of the growing traffic. The proposed Cluster based Adhoc On-demand Distance Vector (CAODV) methodology allows the Cluster Head (CH) to perceive a viable swap of the packets in the next possible minimum bandwidth path. The Network Simulator-2 (NS-2) simulation results are compared in terms of its performance metrics with that obtained in a normal environment to highlight the applicability of the projected strategy in the utility world.

Keywords Bandwidth utilization · CAODV · CDMA nodes · WVPN

7.1 Introduction

The Wireless networks are thought of to serve as a media where ever it may not be possible to construct the requisite platform. The sporadic spurt in technological innovations paves the way for higher echelons of frame work to suit the increasing

C. Mahalakshmi (✉) · M. Ramaswamy
Electrical Engineering Department, Annamalai University, Annamalai Nagar, Chidambaram, India
e-mail: maha_c2008@yahoo.com

M. Ramaswamy
e-mail: aupowerstaff@gmail.com

R. Malathi and J. Krishnan (eds.), *Recent Advancements in System Modelling Applications*, Lecture Notes in Electrical Engineering 188,
DOI: 10.1007/978-81-322-1035-1_7, © Springer India 2013

traffic. It is equivocal to explore routing patterns to elicit the transfer of data in the vicinity of specific nature of source and destination points.

A Wireless Virtual Private Network (WVPN) envisages transfer of data through remote access in a dedicated environment [1]. The basic philosophy in a VPN is to acquire the required bandwidth and accord a safe passage of data, unaffected by the existing traffic sent through the physical network. It is an onerous responsibility from the view of the service provider to guarantee the bandwidth needs, while ensuring the transition is accomplished with minimum resources [2].

The elaborate presence of CDMA nodes in the emerging networks creates a focus to explore its routing methodology along with the VPN nodes. The CDMA enabled technology however preludes a standard for multiple access and enjoy a facility to transmit and/or receive data over multiple codes. It permits multiple nodes to function in the same bandwidth simultaneously through the use of a unique spreading code. Though it ushers in the need of a larger bandwidth for transfer of data, hybrid approaches appear to evolve and provide a spatial reuse of the medium.

Owing to the fact that the traffic is ever increasing and the infrastructure is not growing at the same pace, the network appears to face the threat of insecurity. It is becoming unreliable in the sense it pre empts the possibility of loss of data, an increase in the delay and inability to handle traffic congestion. The need for high speed data transfer further adds to the dimension and it becomes inexplicable to foresee alternate paths for imminent transfer of data.

A non co-operative game power control model has been proposed based on the Code Division Multiple Access (CDMA) model of Wireless Sensor Network (WSN) and a distributed power control algorithm developed [3]. The simulation results have been found to display that the algorithm effectively reduces the total transmitting power of nodes, saves energy and prolongs the network lifetime efficiently. An energy-efficient cross layer protocol for providing application specific reservations in WSN called the "Unified Clustering and Communication Protocol" (UCCP) has been presented [4]. The approach has been motivated by providing an integrated solution for self-organization and end-to-end communication in WSN. The results have been found to demonstrate that it provides an energy-efficient and scalable solution to meet the application specific QoS demands in resource constrained sensor nodes. A distributed, randomized clustering algorithm, to organize the sensors in a WSN into clusters has been explained. It has been extended to generate a hierarchy of Cluster Heads (CHs) and display that the energy savings increase with the number of levels in the hierarchy [5]. An optimal clustering algorithm that balances each cluster and minimizes the total distance between sensor and master nodes has been suggested [6]. While balancing has been found to evenly distribute the load on all master nodes, minimizing the total distance helps in reducing the communication overhead and hence the energy dissipation. A new clustering method has been proposed for increasing the lifetime of the network. The sensors with a high-energy for managing the cluster head have been distributed to decrease their responsibilities in network and the performance evaluated through simulation [7]. An energy efficient heterogeneous clustered scheme for WSN based on weighted election probabilities of each node to become a cluster head according to the residual energy in each node has been detailed. The simulation results have been found to demonstrate that its effectiveness in prolonging the network lifetime [8]. Energy efficient multilevel clustering schemes have been designed for WSNs, constituted of extremely energy constrained nodes. The network has been found to require a multi-level clustering protocol that enables far-off nodes to communicate with the base station [9]. A Cluster Based Hierarchical Wireless Sensor Network Architecture has been outlined to facilitate more than one application sharing the whole or a part of a WSN, where each application may have its own network, processing requirements and protocols [10]. A new error control strategy has been proposed to evaluate the energy level performance of CDMA WSN and the results compared with similar existing methods [11]. The advantages of M-ary orthogonal modulation has been sought to create a WSN framework that can simultaneously monitor multiple source events each of which has multiple states [12].

Though the review echoes the trends only in the field of WSNs it is an ardent vision to perceive a similar effort with VPN in the wireless domain. The continuously increasing traffic confronts difficulties in transmission and may experience constraints in the transmission path necessitating to explore avenues for a mechanism through which an uninterrupted data transfer can be envisaged.

7.2 Problem Description

The basic focus is to carve out a Cluster based Adhoc on-demand Distance Vector (CAODV) strategy that serves to retrieve data congestion in a WVPN. It is developed so as to access an alternate path and ensure the continuous transfer of data between the desired source and destination. The scheme permits the highest energy node to be chosen as the initial CH and migrate to the next alternate viable path that acquires for itself a minimum possible bandwidth to perpetuate the flow of traffic the moment a contingency arises. The algorithm is articulated to arrive at the best values for the performance metrics for the path that uses the minimum bandwidth in addition to fostering continuity in data transmission.

7.3 System Modelling

A network configured with a judicious mix of CDMA and VPN nodes is shown in Fig. 7.1. The spatially distributed fifty nodes arrange themselves in the form of clusters to enable data transmission through three paths that connect the chosen source and destination. It corners the use of a network with CDMA nodes even though they are costly on account of the forging need for high speed transmission.

Fig. 7.1 WVPN topology

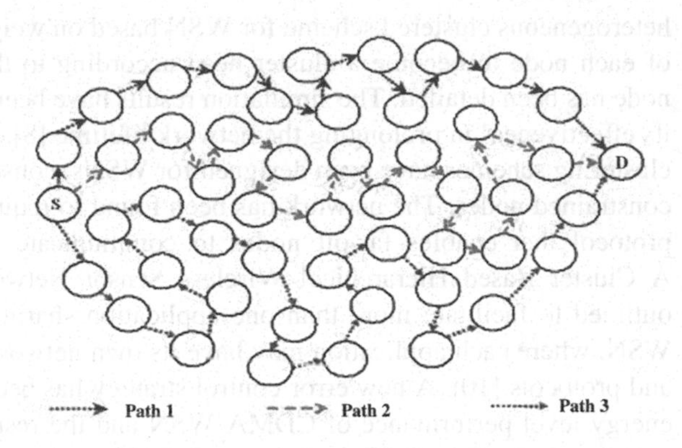

The scheme evolves to enable the flow of traffic with minimum required bandwidth over a specified time frame.

7.4 Proposed Scheme

The proposed approach involves the use of a CAODV methodology to establish an alternate path on the occurrence of an exigency to continue the transfer of data between the preferred source and destinations. It attempts to discover new routes with minimum bandwidth using an expanding ring search procedure availing the service of large neighbourhoods [13]. The focus of the re-routing mechanism is to ensure the minimum usage of bandwidth and extract acceptable values of performance metrics in terms of smaller delay, lower packet loss and energy consumption and higher number of packets received and Packet Delivery Ratio (PDR).

The Adhoc On-demand Distance Vector (AODV) routing minimizes the number of broadcasts by creating routes on-demand. The source broadcasts a route request packet and the neighbours in turn broadcast the packet to their neighbours till it reaches an intermediate node that offers recent route information about the destination or till it reaches the destination. The route request packet uses sequence numbers to ensure that the routes are loop free and ensures that if the intermediate nodes reply to route requests, they offer the latest information only. It records in its tables the route request and uses this information to construct the reverse path for the route reply packet [14].

The reactive routing exhibits considerable bandwidth and overhead advantages over proactive approaches. It facilitates quick adaptation to dynamic link conditions, less processing time, low memory overheads, and smaller network utilization. The AODV is susceptible to security threats and any malicious intention may compromise its overall performance. It is inevitable to explore other options, retaining the advantages of the use of AODV.

Clustering is essentially an exercise that offers scalability to a large number of nodes in the sense it brings out the need for balancing the load and enables efficient resource utilization [15–18]. It can be extremely effective in one-to-many, many-to-one, one-to-any, or one-to-all (broadcast) communication and be of use for data fusion.

The essential operation in clustering is to select a set of CHs among the nodes in the network, and cluster the rest of the nodes with these heads. The CHs serve to co-ordinate among the nodes within their clusters and reduce communication interference. Periodic re-clustering can select nodes with higher residual energy and prolong the network lifetime by reducing the number of nodes contending for channel access, summarizing network state information and updates at the CHs using intra-cluster coordination. It facilitates routing with an overlay among cluster heads, which has a relatively small network diameter.

The strategy revolves around with a view to annihilate the drawbacks of AODV routing and employ a Cluster based AODV to channelize the undeterred data transfer. It is constituted of a set up phase involving the formation of clusters. The nodes broadcast their ID number in this phase over a specific time frame in a pre defined range of the cluster. The chosen CHs evolve a schedule to progressively continue the choice of CH in a way that serves to increase the lifetime of the network.

The execution phase orients to maintain a track of the entry in the table to identify the node from which the first request generated. The CAODV routing pattern uses symmetric links to permit the nodes to traverse the forward route in order that the route reply packet reverses back to the source. It is built to inherit the ability to follow this procedure till the moving source acquires the ability to reinitiate the route discovery as and when needed. If one of the intermediate nodes move then the moved nodes neighbour realizes the link failure and sends notification to its upstream neighbours and so on till it reaches the source from where an alternate path can be established. The implementation of CAODV algorithm is explained through the flow chart depicted in Fig. 7.2.

7.5 Simulation Results

A wireless topology comprising of VPN source and destination nodes (1 & 9) and the remaining are treated to be CDMA nodes. The strategy is created to envision transfer of data through three different paths between a chosen source and destination in a space of 1,000 × 1,000 m. The procedure initiates a sequential methodology by which it facilitates data to flow in one path at a time over specific time frames. The stream continues to extradite the smooth passage of the packets and land at the best performance indices in the path which utilizes the minimum bandwidth.

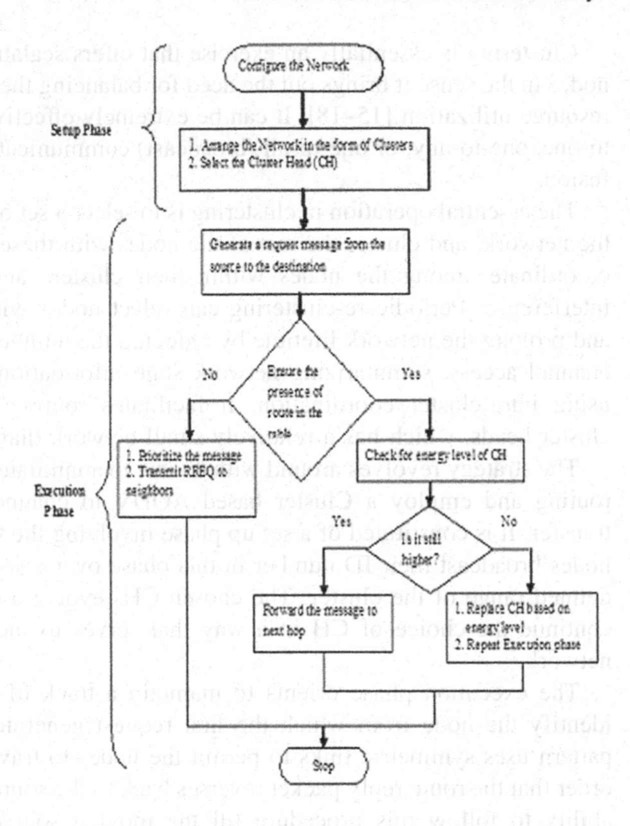

Fig. 7.2 CAODV algorithm

The algorithm includes a facility to acquire higher resources as and when traffic congestion arises. It alternatively constitutes a sequence to drift down the size of the packets and allows to follow an alternate path evolved using the cluster based mechanism only if both ventures are not successful. The re-routing scheme is tailored to epitomize the usage of bandwidth that is strictly required to ensure the continuity of the flow of data.

The bar charts depicted through Figs. 7.3, 7.4, 7.5, and 7.6 elaborate the performance of the strategy in the process of delivery for varying packet sizes in the

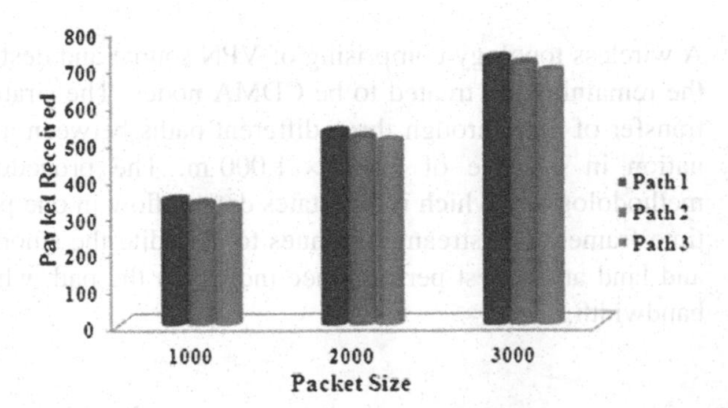

Fig. 7.3 Packets received versus packet size

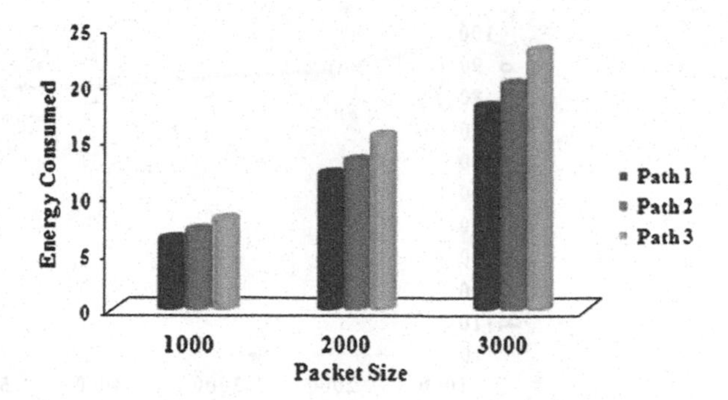

Fig. 7.4 Energy consumed versus packet size

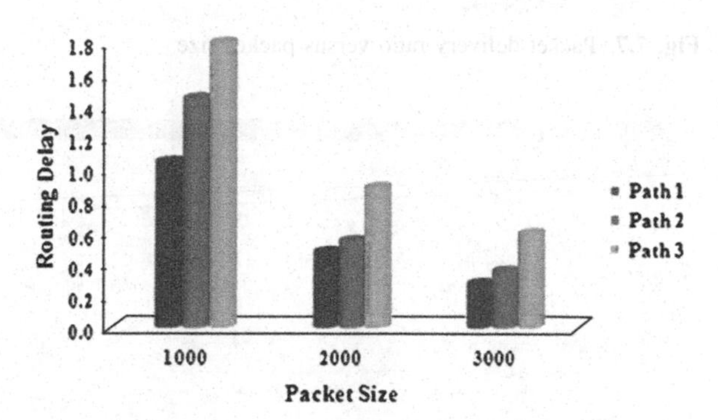

Fig. 7.5 Routing delay versus packet size

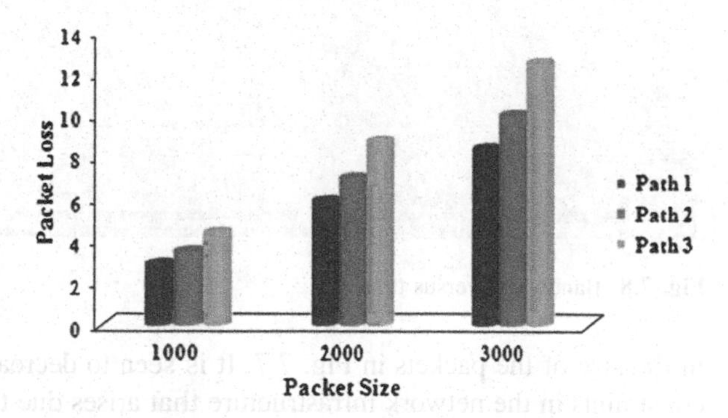

Fig. 7.6 Packet loss versus packet size

three paths through predefined time intervals. It is observed that the path which uses the minimum bandwidth extracts the best performance in view of the fact that it enjoys a smaller delay, lower packet loss and energy consumption and higher number of packets received thus emphasizing the merits of the projected mechanism.

The network Packet Delivery Ratio (PDR) computed as a ratio of the packets received to the packets sent across a specific time period is related over an increase

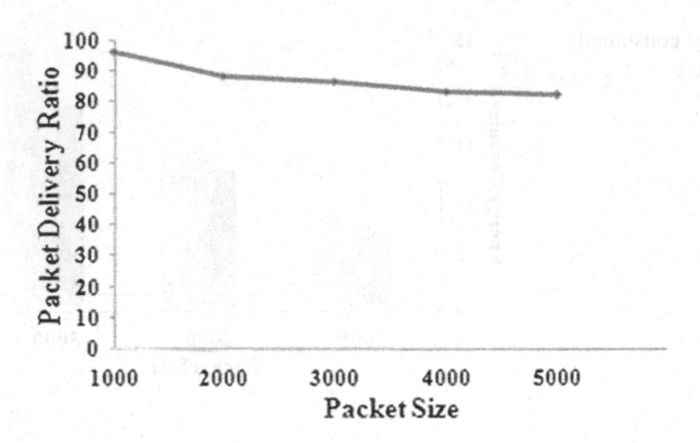

Fig. 7.7 Packet delivery ratio versus packet size

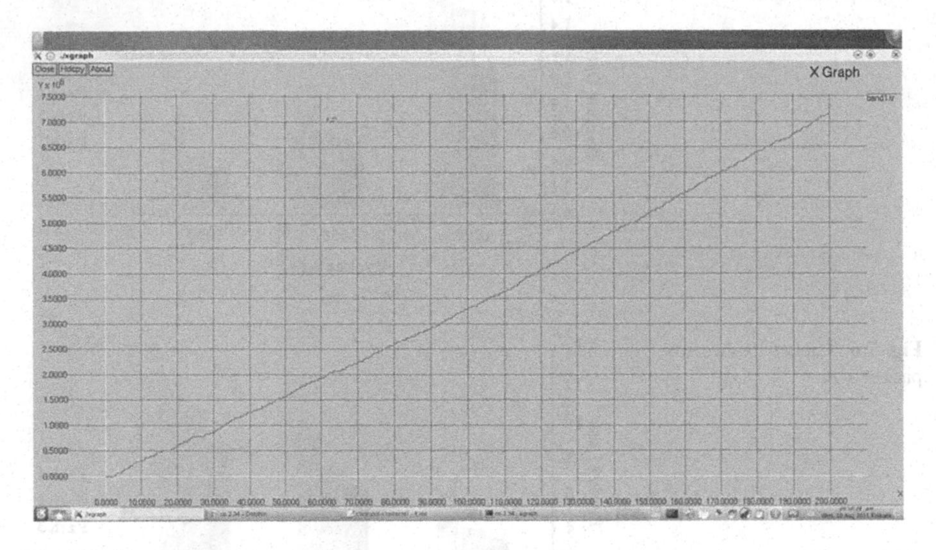

Fig. 7.8 Bandwidth versus time

in the size of the packets in Fig. 7.7. It is seen to decrease gradually owing to the constraints in the network infrastructure that arises due to the growing traffic load.

A traffic congestion is assumed to evolve in that path and augurs a re-routing procedure with a view to comply the safe transfer of the packets. The NS-2 graphs for the different indices during the period of congestion are displayed in Figs. 7.8, 7.9, 7.10, 7.11 and 7.12.

The ability of the proposed approach to handle congestion in the minimum bandwidth path is effectively explained through a distinct improvement in the metrics that are accrued after the retrieval is effected. It is interesting to observe from the entries in Table 7.1 that the re-routing procedure tailors to acclaim an enhanced performance with a similar bandwidth requirement in the newly created path.

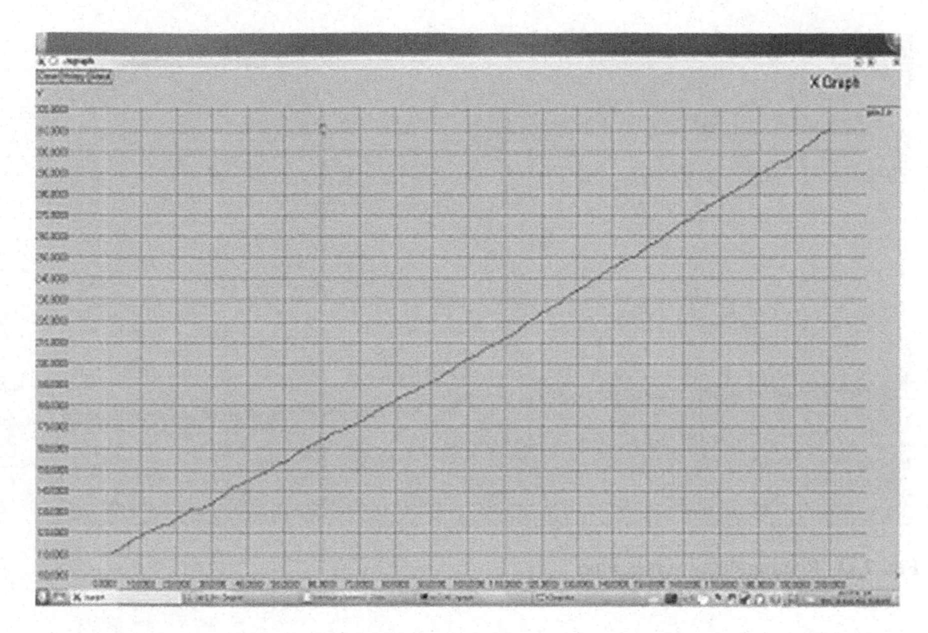

Fig. 7.9 Packets received versus time

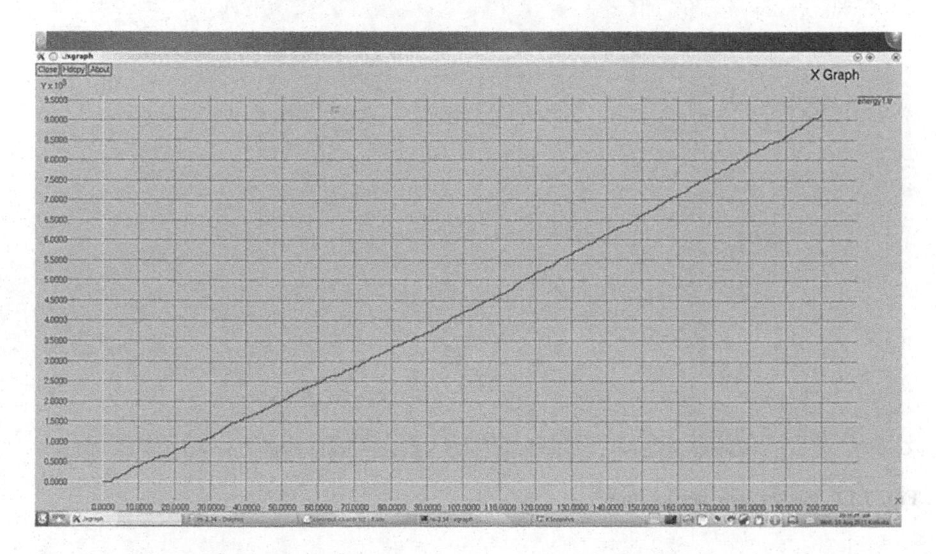

Fig. 7.10 Energy consumed versus time

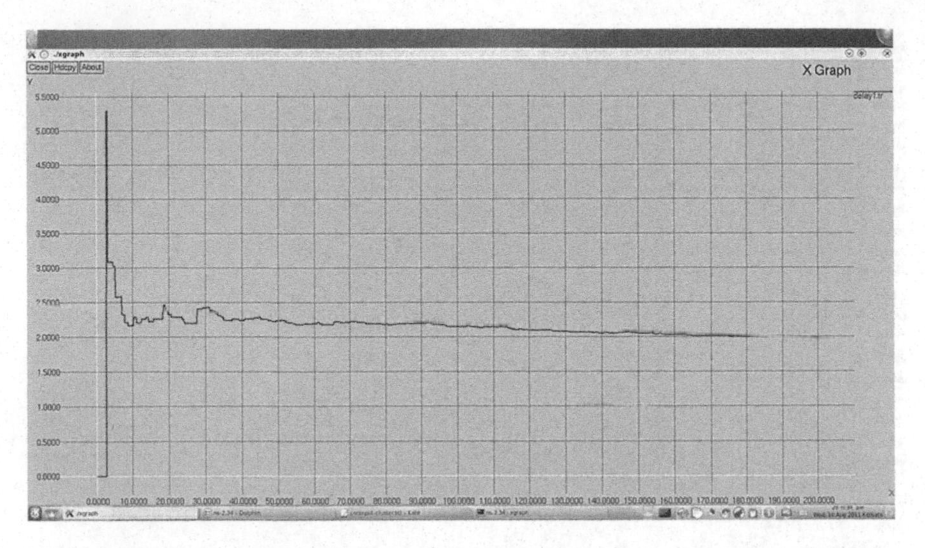

Fig. 7.11 Routing delay versus time

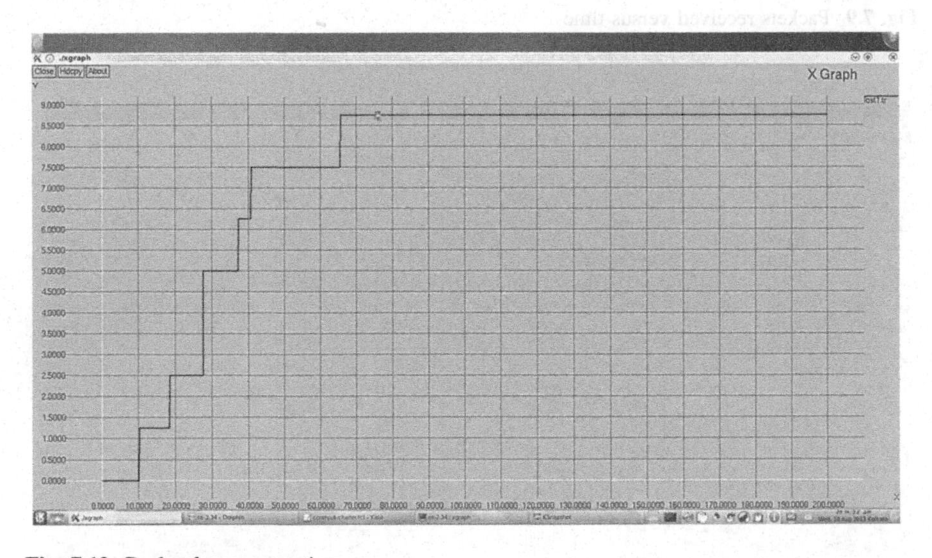

Fig. 7.12 Packet loss versus time

Table 7.1 Performance comparison

Constraint	Bandwidth*10^6	Packets received	Energy consumed*10^3	Routing delay	Packet loss
Before congestion	3.7	340	6.25	1.05	3
During congestion	7.2	310	9.1	5.4	9
After congestion	3.6	390	4.7	0.75	2

7.6 Conclusion

A re-routing strategy suitable for VPN in a wireless environment has been designed to demonstrate its capability to cater to traffic congestion. The performance has been investigated through NS-2 simulation and the extracted indices serve to validate the projected philosophy. The simulated graphs have been obtained during and after the retrieval of exigency to illustrate the ability of the scheme to garner the desired performance. The algorithm has been developed to ensure a continuous data transfer and will go a long way in exploring new dimensions for re-routing mechanisms and facilitate to venture better perspectives in the field of mobile communications.

Acknowledgments The authors thank the authorities of Annamalai University for providing the necessary facilities in order to accomplish this piece of work.

References

1. Munasinghe GKS, Shahrestani SA (2005) Wireless VPNs: an evaluation of QoS metrics and measures. In: Proceedings of the international conference on mobile business (ICMB'05), pp 616–622
2. Juttner A, Szabo I, Szentesi A, (2003) On bandwidth efficiency of the hose resource management model in virtual private networks. In IEEE Proceedings of the INFOCOM'03, pp 386–395
3. Gengzhong Z, Sanyang L, Xiaogang Q (2010) A power control algorithm based on non-cooperative game for wireless cdma sensor networks. Int J Digital Content Technol its Appl 4(3):137–145
4. Aslam N, Phillips W, Robertson W (2008) A unified clustering and communication protocol for wireless sensor networks. Int J Comput Sci 35(3):249–258
5. Bandyopadhyay S, Coyle EJ (2003) An energy efficient hierarchical clustering algorithm for wireless sensor networks. In: Proceedings of IEEE INFOCOM, pp 713–1723
6. Ghiasi S, Srivastava A, Yang X, Sarrafzadeh M (2002) Optimal energy aware clustering in sensor networks. Sensors 2:258–269
7. Babaie S, Zade AK, Hosseinalipour A (2010) New clustering method to decrease probability of failure nodes and increasing the lifetime in WSNs. Int J Comput Sci Inf Secur 7(2):73–76
8. Kumar D, Aseri TC, Patel RB (2009) EEHC: energy efficient heterogeneous clustered scheme for wireless sensor networks. J Comput Commun 32:662–667
9. Soni S, Chand N (2010) Energy efficient multi-level clustering to prolong the lifetime of wireless sensor networks. J Comput 2(5):158–165
10. Yadav P, Yadav N, Varma S (2007) Cluster based hierarchical wireless sensor networks (chwsn) and time synchronization in CHWSN. In: International symposium on communications and information technologies, pp 1149–1154
11. Datta U, Bharath KD, Amit Kumar B (2011) Performance of a hybrid ARQ scheme in CDMA wireless sensor network. Int J Energy, Inf Commun 2(3):59–74
12. Sim EH, Yang L-L (2012) DS-CDMA with M-ary orthogonal modulation for wireless sensor networks simultaneously monitoring multiple events. In: IEEE 75th vehicular technology conference: vtc2011-spring, 6–9 MAY, pp 1–5
13. Liu J-S, Richard Lin C-H (2005) Energy-efficiency clustering protocol in wireless sensor networks. Ad Hoc Networks 3:371–388

14. Hoda MN, Singh U, Milind, Kumar R (2008) Simulation of AODV, DSR and DSDV routing protocols for mobile ad-hoc networks. In: Proceedings of the second national conference, pp 1–4
15. Younis O, Fahmy S (2004) Distributed clustering in ad-hoc sensor networks: a hybrid, energy-efficient approach. In: Proceedings of IEEE INFOCOM, pp 629–640
16. Lin CR, Gerla M (1997) Adaptive clustering for mobile wireless networks. IEEE J Sel Areas Commun 1(7):1265–1275
17. Banerjee S, Khuller S (2001) A clustering scheme for hierarchical control in multi-hop wireless networks. In: Proceedings of twentieth annual joint conference of the IEEE Computer and Communications Societies, pp 1028–1037
18. Estrin D, Govindan R, Heidemann J, Kumar S (1999) Next century challenges: scalable coordination in sensor networks. In: Proceedings of the fifth ACM/IEEE international conference on mobile computing and networking (MOBICOM), pp 263–270

Chapter 8
Stopping Power of Proton Beam in Water Phantom: A Simulational Study

Sirisha Sathiraju Naga lakshmi and Sonali Bhatnagar

Abstract Geant4 is a Monte-Carlo simulation tool-kit developed for the virtual study of high energy physics. Hadrontherapy is an open source code, available in this toolkit for application in radiation therapy developed by the MC-INFN group. The aim of this work is to study the passive transport beam line, which is installed at Laboratory Nazionali del Sud (INFN) in Catania, Italy. The theoretical models implemented in the Geant4 code are studied. The physical interpretation of dose distribution curves for proton beam in the calculation of stopping power of proton beam in the calculation of stopping power of proton beam in water medium at different energies are discussed. In the end we study the spectra of secondary particles produced in the interactions for 60 MeV proton beam which is relevant for the study of Radioactive Biological Efficiency (RBE) and Spread Out in Bragg Peak (SOBP).

Keywords Monte-carlo simulation · Hadrontherapy · MC-INFN · Geant4 · Water phantom

8.1 Introduction

Hadrontherapy is a different technique of radiotherapy that makes use of fast non-elementary particles i.e. protons, neutrons and light nuclei in curing tumors. These particles stop at a certain depth in a substance and deposit a large portion of its

---

S. Sathiraju Naga lakshmi · S. Bhatnagar (✉)
Department of Physics and Computer Science, Dayalbagh Educational Institute,
Agra 282110, India
e-mail: deisonali.bhatnagar@gmail.com

S. Sathiraju Naga lakshmi
e-mail: sirisha.dei@gmail.com

R. Malathi and J. Krishnan (eds.), *Recent Advancements in System Modelling Applications*, Lecture Notes in Electrical Engineering 188,
DOI: 10.1007/978-81-322-1035-1_8, © Springer India 2013

energy near the end of its range. The deposited energy forms a narrow dose peak known as the Bragg Peak [1]. This physical characteristic of dose deposition sterilizes the cells in target volume while minimizing damage to healthy tissues along the radiation path.

The energy range required for radio therapeutic use is typically 60–230 MeV [2] for protons and 120–400 MeV/nucleon [2] for carbon ions. These beam energies are accelerated by use of cyclotrons, synchrotrons or linacs. But the choice of accelerators mainly depends on dose delivery system namely passive and active beam lines. For a passive spreading of beam, a large amount of mechanical energy is necessary and also doesn't allow in exploiting the ballistic properties of the charged hadron beams [3]. In case of active method, it is a small pencil beam used to paint the tumors correspondingly with the help of magnetic forces.

Cyclotron is a compact accelerator that delivers a high density, continuous beam of constant extraction energy where these characteristics will be adjusted by means of a passive system. In cyclotrons, protons are mostly preferable due to existence of required extraction energy for treatment [3], Synchrotron allows the beam at low energy and accelerated to desired energy level in order to cover the complete tumor thickness of about a few mm. It is ideally suited for active beam delivery technique.

In order to perform this therapy practically there are few challenges such as necessity of accurate and real time patient for monitoring. During treatment if the patient moves, causes a serious damage. The machines also needed to be smaller and cheaper, which is currently motivating new developments in the field of accelerator physics. To overcome these challenges, there is need to study a simulation tool for the design of beam irradiation systems and for proposing treatment planning. In recent years Monte-Carlo techniques have been used for this purpose and in clinical applications, the simulation must reproduce dose distributions in three dimensions with the best possible accuracy to ensure the safety of the patient.

This paper gives an overview of the study of the Hadrontherapy code in Geant4 which is developed by the Instituto Nazionale di Fisica Nucleare (INFN) group, Italy [4] where the ongoing experiment is installed at Laboratory Nazionali del Sud (INFN-LNS), in Catania, the first Italian Hadrontherapy facility named (Centro di Adro Terapia ed Applicazioni Nucleari Avanzate) CATANA. Section 8.2 discusses the theoretical models which are implemented in the Geant4 toolkit. Section 8.3 discusses the parameters calculated and their results.

8.2 Theoretical Models

The Monte-Carlo technique is implemented in Hadrontherapy for verification of treatment planning system and to calculate the dosc of low-energy ($\sim$250 MeV) clinical proton beams [5]. But in case of nuclear reactions it is a complex task to implement the Monte-Carlo program because it requires combination of several phenomenological models which account for the complete spectrum of particles, energies and absorbers of a given application.

The Geant4 physics lists are a combination of "processes" and "models" that define the interaction probabilities. All the electromagnetic and nuclear processes induced by the proton beams with energies $\leq$250 MeV [5] in low-Z absorbers were taken into consideration.

In the present work, a physics list was used to study the simulation of passive proton beam line that passes through an irradiation system. This physics list consists of three modules that correspond to (a) electromagnetic, (b) elastic scattering, (c) inelastic scattering of protons, neutrons and heavier ions [5]. These three modules are used in three approaches during simulation. One of the approaches is activating the Reference Physics List by setting the environment variables. It contains all the physics processes necessary to a particle transport. The second approach is activating the builders which are specific of a given model (electromagnetic processes, hadronic etc.). The last approach is activating the local physics lists, these models belong to the ion–ion interactions. To get accurate results, it is strongly recommended to activate the reference physics list. The details and significance of the models using in this toolkit is discussed below.

The electromagnetic interactions in Geant4 is comprised of 'standard' and 'low-energy parameterized' models. Among these, the standard electromagnetic package is an analytical model that derives directly from Quantum Electro Dynamics (QED) [5] calculations and describes the interactions of photons and all charged particles down to 1 keV. Photoelectric effect, Compton scattering, bremstrahlung and e^+, e^- annihilation are the interactions that are considered for standard model. In case of low energy model, Rayleigh scattering model was also included. The energy loss of hadrons is done in standard model using Bethe-Bloch formula <2 MeV [5], whereas above this level (>2 MeV) [5] the hadron ionization is analyzed by the low-energy parameterized model.

The nuclear models results in collection of cross-section data sets to build its own hadronic physics list. In Geant4 there are three cascade models: the Binary cascade model, the Bertini model and low-energy parameterized models are used for simulation of inelastic nuclear scattering in hadronic collision which gives rise to an intra-nuclear cascade [5]. A Pre-equilibrium phase is automatically evoked by the intra-nuclear cascade which comprises of two models. One of these, the pre-compound model is used for implementation of the inelastic scattering, particularly for incoming particle of energies below 100 MeV [5]. It is based on the Griffins semi-classical description of composite nucleus decay. This pre-compound model is invoked by binary cascade in the code. The other is pre-equilibrium model which describes the emission of photons, nucleons and light fragments. It is also invoked by the Bertini cascade, which combines a Quantum Molecular Dynamics (QMD) [5] component and a classical cascade component to study the inelastic scattering of protons, neutrons and light ions between (100 MeV–10 GeV) [5]. The low-energy parameterized model of the nuclear category is used to implement the elastic models.

8.3 Results and Physical Interpretation

8.3.1 Geant4 Hadrontherapy and its Validation

The work is based on study of simulation of a passive proton beam produced from a Superconducting Cyclotron (SC), was set and used for eye treatments. The functions of the elements studied are as follows:

1. Scattering system: spreads the beam geometrically.
2. Collimators: avoid the scattering radiation.
3. Modulator: it modulates the energy of the beam in a wide spectrum.
4. Monitor chambers: these are special transmission ionization chambers used to control the particle flux during the irradiation.
5. Final collimator: it defines the final shape of the beam before reaching the patient.
6. Water phantom: a box f water where the energy deposited and dose is stored.

The geometry setup used in the simulation for the study of proton beam is shown in Fig. 8.1:

Validation of the code is necessary for the study of the dose distribution in proton beam. Here a proton of incident energy 62 MeV is allowed to accelerate through cyclotron. The cubic phantom (a uniform solid with same material) has been divided in 200 slices with 4 cm thickness in depth and positioned orthogonal

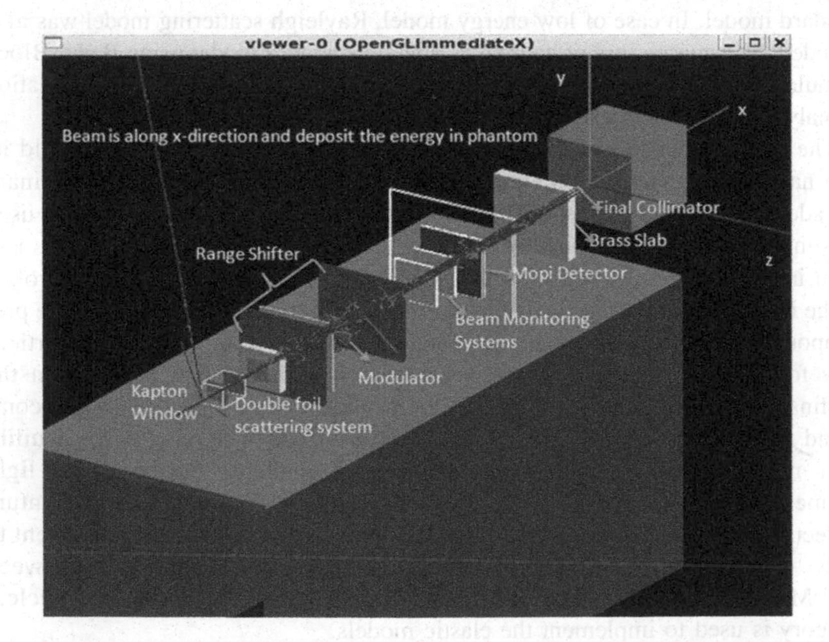

Fig. 8.1 The output of the Geant4 simulation of proton irradiation beam line

to beam direction, inside which total dose is stored. During simulation the transport parameters also should be carefully chosen. So, production cut which determines how many secondary electron and gamma are transported is set 0.01 mm. The lower its value, then larger is the number of secondaries and resulting total computational time. The data obtained by simulation is compared with data acquired at INFN-LNS, Catania for proton beams. Here the experimental data is also available in the toolkit. A satisfactory agreement from Fig. 8.2 has been found between the experimental and simulated curves for the proton beams.

This result demonstrates the accuracy of Hadrontherapy and Geant4 in the dose construction.

8.3.2 Stopping Power

Generally the fast charged particles such as protons ionize the atom or molecule when passes through matter and gradually loses energy in small steps, resulting in maximum energy deposition in Bragg Peak. This can be done by the stopping power, i.e. the average energy loss of the particle per unit path length denoted by $-dE/dx$ and measured in MeV/cm. The stopping power of particle can be calculated using Bethe-Bloch Formula [3]:

$$-\frac{dE}{dx} = \frac{5.08 \times 10^{-31} \times Z^2 \times N}{\beta^2}[F(\beta) - \ln(I)]\frac{\text{MeV}}{\text{cm}} \tag{8.1}$$

where $F(\beta) = \frac{\ln(1.02\times10^6\times\beta^2)}{(1-\beta^2)} - \beta^2$, $N = N_a \times \rho \times \frac{Z}{A}$.

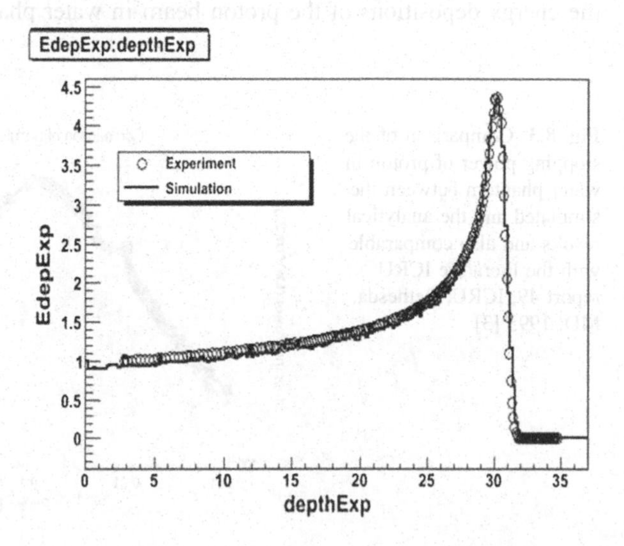

Fig. 8.2 Validation in the simulated and experimental depth dose profiles at 62 MeV proton beam in water phantom

Here E is energy, ρ is density, I is mean excitation potential of target, $\beta = v/c$. For calculating the velocity of proton we use the classical formula of mass energy relation as the rest mass of proton is larger than the kinetic energy

$$K.E = {}^{1}/_{2}\, mv^{2}. \tag{8.2}$$

The stopping power of proton in water phantom is simulated and also analytically done within an energy range (0.001–100000 MeV). Here the comparison between the Geant4 output and the analytical result using Matlab 7.0.1 is shown in the Fig. 8.3.

We can observe the dependence of stopping power on proton energy. Based on an approximate theory i.e. the Thomas–Fermi model of atom [3], Bohr suggested that for higher energies above 100 keV regions, the stopping power decreases as the particle velocity approaches the velocity of light. When the velocity of the particle is comparable with speed of light, the normal spherical field becomes distorted which leads to higher energy loss at higher energies. Stopping power is an important parameter because of its various applications in the study of biological effects, radiation damage dosage rates and energy dissipation at various depths of an absorber. These studies are crucial in designing the detection systems, radiation technology etc.

8.3.3 Bragg Peaks at Different Energies

During the radiation therapeutic use, protons of energy in the range 60–230 MeV [6] is suitable for passive spreading of dose delivered in the tumor region of the patient rather than using a active dose distribution. Thus, energy deposited in the water phantom is studied at different energies. The Fig. 8.4 illustrates the plots of the energy depositions of the proton beam in water phantom using Root 5.28.0.

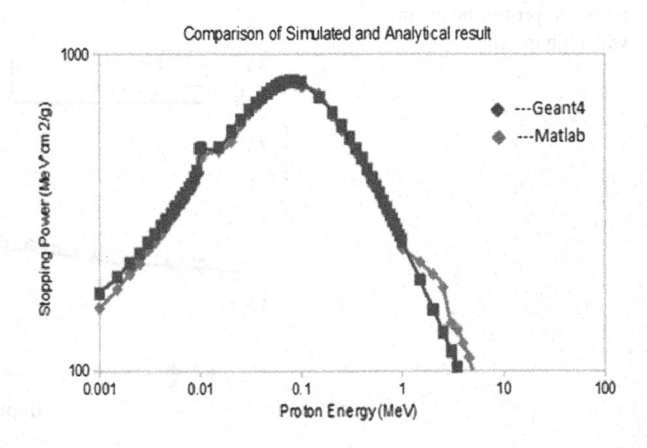

Fig. 8.3 Comparison of the stopping power of proton in water phantom between the simulated and the analytical results and also comparable with the literature ICRU report 49, ICRU, Bethesda, MD, 1993 [3]

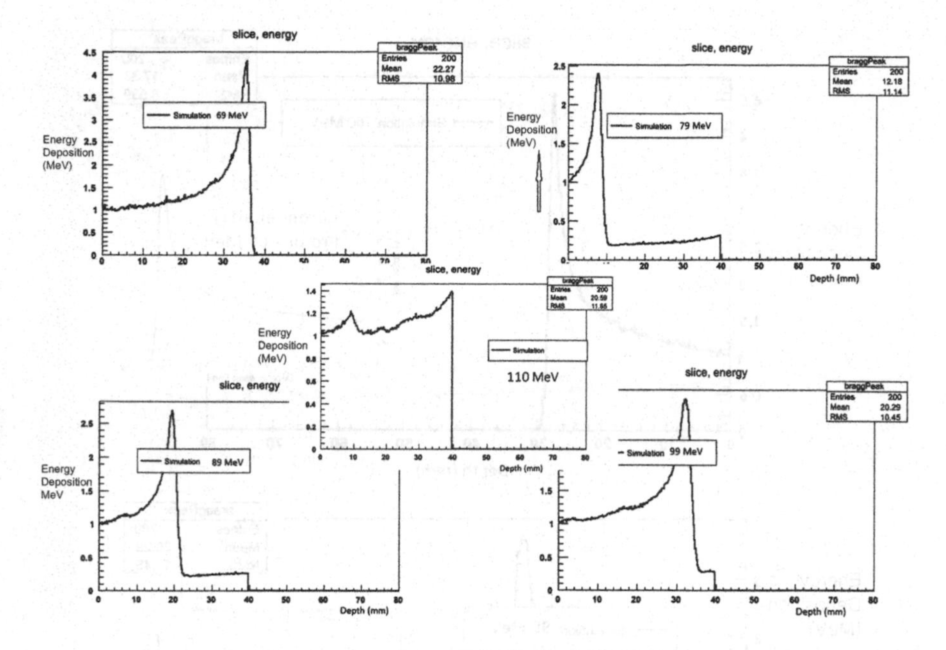

Fig. 8.4 The energy deposition of the proton beam at different energies

As seen in Fig. 8.4, the Bragg peak shifts to lower depths as we move to higher energies. The peak is clearly not visible at 110 MeV. To get this peak we will study the effect of hadronic models at higher inelastic interactions. The amount of Bragg Peak position in the water phantom depth (mm) is illustrated in the Fig. 8.5 which is also compared with the work done by Cirrone et al. [1], where they used a passive proton beam, the Bragg Peak position at 60 MeV is 32 mm depth of water phantom. In the present study the Bragg Peak position is at 29 mm depth of water phantom. Secondly in the paper by L. Grevilliot et al. [7], they used a pencil beam of energy 98.71 MeV which gave the Bragg Peak position at 78 mm depth in water phantom whereas in the present work we calculate a 31 mm depth of water phantom at 99 MeV proton energy. This observed variation in results at same energies with the other works gives us a lead to study the inclusion and effect of hadronic models at higher energies.

To perform all these studies, we followed the approach 2 which was discussed in the Sect. 8.2 i.e. the builders that belong to the respective models (electromagnetic and hadronic) are activated. Table 8.1 gives an overview of the models used in this study.

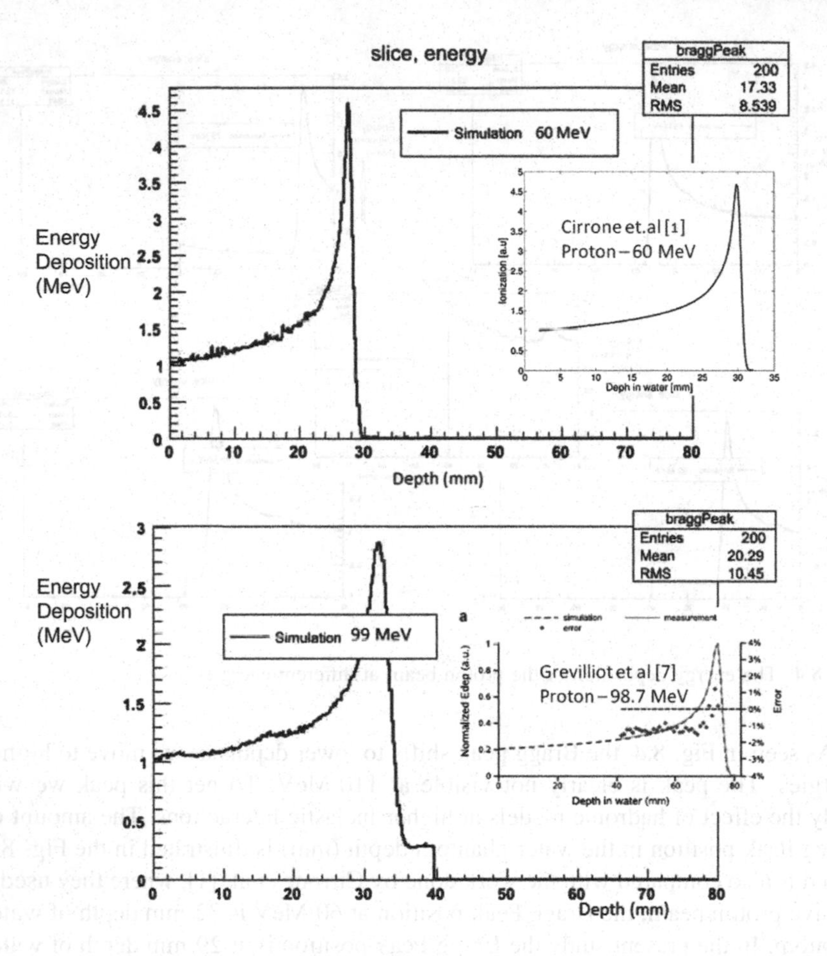

Fig. 8.5 Comparison of the energy deposited in phantom depth with literature (Cirrone et al. [1], Grevilliot et al. [7])

Table 8.1 Theoretical models used

Electromagnetic model	Emstandard_opt3
Hadronic elastic model	Elastic
Hadronic inelastic model	Binary

8.3.4 Significance of Secondary Particles

Using the models, the amount of dose stored by the protons and its secondaries such as secondary protons, alpha, gamma, electron and deuteron are plotted in root as shown in Fig. 8.6.

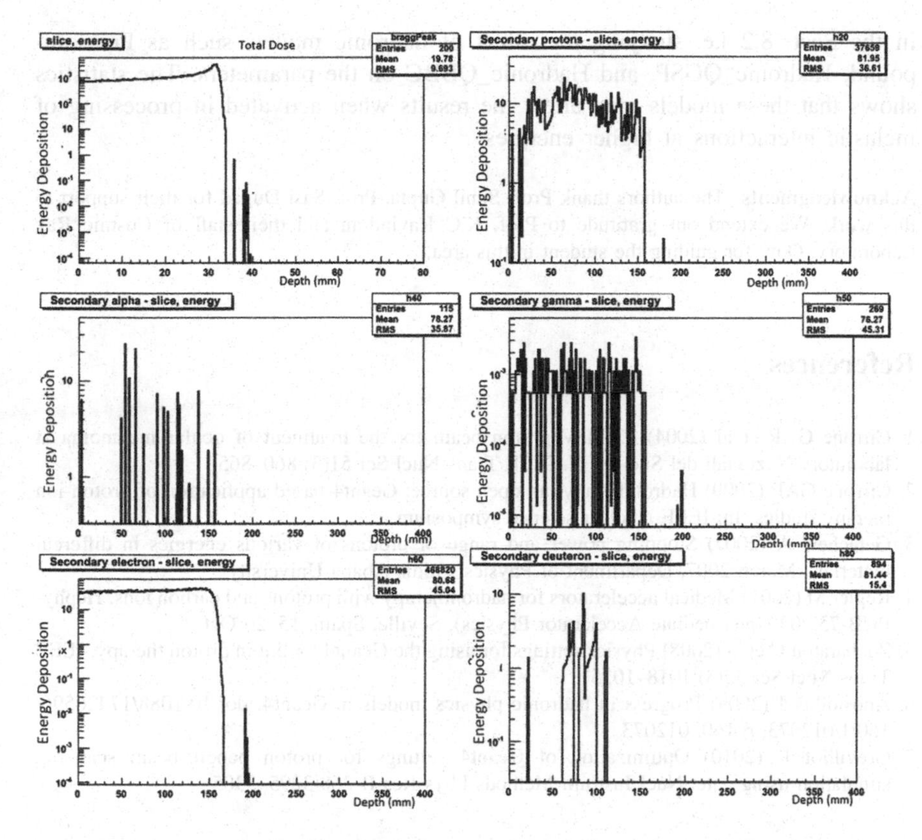

Fig. 8.6 Dose by particle corresponds to the hadronic (binary) inelastic model

Here, secondaries undergo non-elastic interactions and these products have a significant contribution in delivering the total dose to the tumor region and hence, needed to be tracked. Their yield also depends upon the beam delivery system. The individual dose rate and their energy distribution of these secondaries (primary protons, alpha) results in the calculation of Radioactive Biological Efficiency (RBE) parameter. The dose of these secondaries also influences the peak/plateau ratio of an unmodulated beam and the flatness of the Spread Out in Bragg Peak (SOBP).

8.4 Conclusion

The Bragg Peaks of the proton beam at different energies are plotted in Fig. 8.4. The results of the Bragg Peak positions in Fig. 8.5 are comparatively less than peaks in the work of Cirrone et al. [1], Grevilliot et al. [7]. The expansion of this work will be to incorporate the approach 1 and 3 which we have discussed in the Sect. 8.2 i.e. studying the effect of hadronic models such as Pre-compound, Hadronic_QGSP, and Hadronic_QBBC on the parameters. The statistics shows that these models may affect the results when activated in processing of inelastic interactions at higher energies.

Acknowledgments The authors thank Prof. Sunil Gupta, Prof. Sasi Dugad for their support to this work. We extend our gratitude to Prof. K.C Ravindran and their staff of Cosmic Ray Laboratory, Ooty for guiding the student in this area.

References

1. Cirrone GAP et al (2004) 62 MeV proton beam for the treatment of ocular melanoma at laboratory Nazionali del Sud-INFN. IEEE Trans Nucl Sci 51(3):860–865
2. Cirrone GAP (2009) Hadrontherapy: an open source, Geant4 based application of proton-ion therapy studies. In: IEEE Nuclear science symposium
3. Getachew A (2007) Stopping power and range of protons of various energies in different materials, March 2007, Department of Physics, Addis Ababa University
4. Regler M (2001) Medical accelerators for hadrontherapy with protons and carbon ions. Hephy-PUB-757/02 (Intermediate Accelerator Physics), Seville, Spain, 15–26 Oct
5. Zacharatou C et al (2008) Physics settings for using the Geant4 toolkit in proton therapy. IEEE Trans Nucl Sci 55(3):1018–1025
6. Apostolkis J (2009) Progress in hadronic physics models in Geant4. doi:10.1088/17426596/160/1/012073, p 160, 012073
7. Grevilliot L (2010) Optimization of Geant4 settings for proton pencil beam scanning simulation using gate. Nucl Instrum Methods Phys Res B 268:3295–3305

Chapter 9
Decidable Utility Functions Restricted to a System of Fuzzy Relational Equations

Garima Singh, Dhaneshwar Pandey and Antika Thapar

Abstract This work considers a multiobjective optimization problem with max-product fuzzy relational constraints. The utility function based approach is proposed that translates the multidimensional criterion space to single dimensional one. Further, a hybridized genetic algorithm is applied to solve the transformed single objective optimization problem. As the feasible domain has the inherent non-convexity; traditional metaheuristics cannot be applied in their original form. So with a careful study of the feasible domain a specific hybridized genetic algorithm is designed that keeps the new solutions inside the feasible domain and results in a set of solutions offering close approximation of the efficient set and hence, makes the problem decidable.

Keywords Fuzzy relation equation · Multiobjective optimization · Hybridized genetic algorithm

9.1 Introduction

Fuzzy systems have broader utility in different engineering and practical applications particularly in image processing, approximate reasoning, decision-making support systems, control systems, and data analysis. In general, a fuzzy system can be characterized by a set of conditional (IF–THEN) fuzzy propositions, collectively known as the fuzzy rule base, the main functioning tool behind making fuzzy

---

G. Singh (✉) · D. Pandey · A. Thapar
Faculty of Science, Department of Mathematics, Dayalbagh Educational Institute, Agra 282110, India
e-mail: singhgarima.dei@gmail.com

R. Malathi and J. Krishnan (eds.), *Recent Advancements in System Modelling Applications*, Lecture Notes in Electrical Engineering 188,
DOI: 10.1007/978-81-322-1035-1_9,

inferences. Any IF–THEN fuzzy proposition can be represented by a fuzzy relation between the input and the output variable that correspond to its antecedent and consequence respectively. So, fuzzy relations are considered as the basic mathematical tool to build fuzzy models of various real life and hypothetical systems with the help of fuzzy implications and approximate reasoning. An important application of fuzzy relations is fuzzy relation equations (FRE), processing fuzzy information in relational structures especially in knowledge based systems as they play the key role behind the fuzzy reasoning inference system. Fuzzy relational optimization is a branch of fuzzy optimization dealing with the optimization problems with one or more objective functions subjected to fuzzy relational equations based on certain algebraic compositions.

Real-life optimization problems often require multiple objectives to be optimized simultaneously. Such problems are categorized as 'Multiobjective or multicriterion optimization problems (MOOP)'. In MOOP, the rise of several conflicting objectives enhances its complexity, conflicting in the sense that improvement in one objective can cause deterioration in the other. In this situation it is hardly possible to find solutions that optimize all the objectives at the same time. So, the aim is to determine one or more good compromising solutions for the problem.

In multiobjective optimization, several criteria are optimized simultaneously. In this case, instead of a single optimal solution, a set of good alternative solutions generally called as the Pareto optimal set is obtained. All the solutions in this set are incomparable to each other and each solution is considered an acceptable solution for the problem when all objectives are considered at the same time. Although a wide class of methods is available to deal MOOP's but an efficient procedure is always at demand.

Evolutionary algorithms (EA) have emerged as a powerful tool to deal these problems efficiently, as they have some inherent features to solve such kind of problems regardless of much mathematical background required. Moreover, evolutionary algorithms perform well even in multimodal, nonlinear search spaces. Due to this striking characteristic, EA do better than the other optimization methods used to solve MOOP.

The two general approaches that are used to handle MOOP are (1) Utility function approach that scalarizes multiple objectives into single objective by translating multi-criterion space to single criterion space. This translation requires some prior information about the problem from the decision maker in terms of some utility function or predefined preference vector of weights denoting the measures of importance of the different objectives. (2) The second general approach is to determine the entire Pareto optimal front or a representative subset of solutions. Though the method is a smart way to estimate the set of optimal solutions for the problem but is computationally challenging and suffers from inefficiency as the problem size and the number of objectives increase. In addition to it many times the method struggles to obtain the diverse set of solutions. The first approach to solve MOOP is a good option as it is efficient and easy to design but is generally ignored because of its sensitivity to the prior inputs it demands to solve the problem.

We are considering a fuzzy relational multiobjective optimization problem (FRMOOP) with the feasible domain designed by a system of max-product fuzzy relational equation constraints. The proposed approach is based on the application of decidable utility functions for the considered optimization problem. The target is to generate a set of approximate efficient solutions using some decidable utility functions. The method allows the decision maker to choose the best compromise solution from the set of solutions obtained, as decidability is considerably a good approximation of the Pareto front.

The area of multiobjective optimization has been explored by numerous researchers [1–6], as it has always been challenging to deal with numerous objectives having different characteristics together. In area of fuzzy relational optimization it is still in nascent stage. Firstly, Wang [7] studied the problem of multiobjective mathematical programming with multiple linear objective functions subjected to constraints defined by max–min composite fuzzy relational equations.

Later, Loetamonphong et al. [8] studied MOOP with multiple objective functions subjected to max–min fuzzy relational equations. Taking advantage of the special structure of the solution set, they developed a reduction procedure to simplify the problem and proposed a genetic algorithm to find the Pareto optimal solutions. Khorram and Zarei [9] considered a multiple objective optimization model subject to a system of fuzzy relational equations with max-average composition and presented a reduction procedure and then used a modified genetic algorithm to solve the problem. Jiménez et al. [10] considered multiobjective linear programming problems. By using the idea of fuzzy goals for each of the objective functions they proposed a general procedure to obtain a non-dominated solution, which is also fuzzy-efficient. Zhang et al. [11] proposed a method utilizing a max-pro optimum scheme for solving the max–min decision function in a fuzzy optimization environment. Recently, Thapar et al. [12] considered a multiobjective optimization problem subjected to a system of fuzzy relational equations based upon the max-product composition. A well structured non-dominated sorting genetic algorithm was applied to solve the problem.

In most of the existing literature on FRMOOP, max–min composition based FRE has been discussed. A restriction with max–min composition is that it is conservative in nature; it has limitations over the application towards the real world decision problems. It is generally used when a system requires conservative solutions in the sense that the goodness of one value cannot compensate the badness of another value [8]. In this paper we consider the multiobjective optimization problem with fuzzy relational equations composed of max-product composition which is non-conservative and propose a utility function based approach to handle the multiple objectives in the problem. With a careful observation of the feasible domain, an effective hybridized genetic algorithm is proposed without solving the system of FRE.

The paper is organized as follows: Section 9.1 presents the background and motivation behind the problem. In Sect. 9.2 and 9.3 considered problem and the utility function approach have been described respectively. Section 9.4 presents the design of the genetic algorithm used to solve the problem. In the fifth section the results obtained with the computational experiments are presented. A concluding remark is given in the last section.

9.2 The Problem

Let $A = [a_{ij}]$, $0 \leq a_{ij} \leq 1$, be a $m \times n$ dimensional fuzzy matrix and $b = [b_1, b_2, \ldots, b_n]$, $0 \leq b_j \leq 1$, be a n-dimensional vector, then the system of fuzzy relational equations is defined by A and b as follows:

$$x \circ A = b \tag{9.1}$$

where $\circ$ denotes max $- t$ composition of x and A; t denotes a compositional operator from product algebra over residuated lattice $L = \langle [0,1], \wedge, \vee, t, \Theta_t, 0, 1 \rangle$. It is intended to find a solution vector $x = [x_1, x_2, \ldots, x_m]$, with $0 \leq x_i \leq 1$, such that

$$\max_{i=1}^{m} (x_i \cdot a_{ij}) = b_j, \; \forall j = 1, 2, \ldots, n \tag{9.2}$$

Each input of (9.2) may require certain amount of resources which can be considered cost, and a decision maker may wish to achieve certain objectives-a situation which is usually the case for the real-world applications. Let $I = \{1, 2, \ldots, m\}$ and $J = \{1, 2, \ldots, n\}$ be the index sets. We are interested in the following multiobjective optimization model with max-product fuzzy relational equations as constraints:

$$\begin{aligned} &\text{Min}\, \{z_1 = f_1(x), z_2 = f_2(x), \ldots, z_s = f_s(x)\} \\ &\quad \text{s.t.}\ \max_{i \in I} (x_i \cdot a_{ij}) = b_j, \; \forall j \in J \end{aligned} \tag{9.3}$$

where $f_k(x)$ is a linear or nonlinear objective function, $k \in K = \{1, 2, \ldots, s\}$.

Let $X(A,b) = \{x \in [0,\ 1]^m \,|x \circ A = b\}$ be the solution set of fuzzy relation equation (9.2). For any x^1, $x^2 \in X$, we say $x^1 \leq x^2$ if and only if $x_i^1 \leq x_i^2$, $\forall\, i \in I$. Therefore, $\leq$ forms a partial ordering relation on X and $(X, \leq)$ becomes a lattice. Equations in (9.2) form a system of latticized polynomial equations. $\widehat{x} \in X(A,b)$ is the maximum solution of (9.2), if $x \leq \widehat{x}$, $\forall\, x \in X(A,b)$. Similarly, $\breve{x} \in X(A,b)$ is a minimal solution, if $x \leq \breve{x}$ implies $x = \breve{x}$, $\forall\, x \in X(A,b)$. According to [13], if $X(A,b) \neq \phi$, then it is, in general, a non-convex set which can be completely determined by unique maximum solution $\widehat{x}$ and several minimal solutions $\breve{x}$. The maximum solution of system in (9.2) can be computed explicitly using the residual implicator (pseudo complement) by assigning

$$\widehat{x} = A \Theta_t b = \left[\min_{j \in J} (a_{ij} \Theta_t b_j) \right]_{i \in I} \tag{9.4}$$

where $a_{ij}\Theta_t b_j = \sup\{x_i \in [0,1]\, |x_i \cdot a_{ij} \le b_j\}$. For the product t-norm, the operator is given as:

$$a_{ij}\Theta_t b_j = \begin{cases} 1 & \text{if } a_{ij} \le b_j \\ b_j/a_{ij} & \text{if } a_{ij} > b_j \end{cases} \quad (9.5)$$

If $\check{X}(A, b)$ denotes the set of all minimal solutions, then the complete solution set of fuzzy relation equations (9.2) can be formed as follows:

$$X(A,b) = \bigcup_{\check{x} \in \check{X}(A,b)} \left\{x \in [0,1]^m | \check{x} \le x \le \hat{x}\right\}$$

Definition 1 For each $x \in X(A, b)$, we define $z^x = (z_1^x, z_2^x, \ldots, z_s^x)$ to be its criterion vector where $z_k^x = f_k(x), \quad \forall k \in K$.

Let us define objective space $Z = \{z^x = f(x) | x \in X(A, b)\}$ as the image of the decision space under the mapping $f : X \to R^s$ where R^s is the s-dimensional Euclidean space. The image of a solution under this mapping in the objective space is known as the criterion vector.

Definition 2 A point $x' \in X(A, b)$ is an efficient or a Pareto optimal solution to the problem (9.3) iff there does not exist any $x \in X(A, b)$ such that $f_k(x) \le f_k(x')$, $\forall\, k \in K$, and $f_k(x) < f_k(x')$ for at least one $k \in K$. Otherwise, x' is an inefficient solution.

Definition 3 For any two criterion vectors, z^1, z^2 we say that z^1 dominates z^2 iff $z^1 \le z^2$ and $z^1 \ne z^2$ i.e. $z_k^1 \le z_k^2$, $\forall\, k \in K$ and $z_k^1 < z_k^2$ for at least one k.

Definition 4 $z' \in Z$ is said to be non-dominated iff there does not exist any $z \in Z$ that dominates z'. Otherwise, z' is a dominated criterion vector. The set of all efficient solutions is called as the efficient set and the image of the efficient set in objective space is the non-dominated set.

Definition 5 The criterion vector z^* composed of the least attainable objective function values in the problem domain is called the ideal point i.e., $\forall\, k \in K$

$$z_k^* = \{\bar{z}_k \,|\, \bar{z}_k = \min(z_k(x)),\ x \in X\}$$

In general, the idea of ideal point is impractical and it corresponds to a non-existent solution. But it plays an important role in searching the Pareto optimal solutions in numerous methods used to solve MOOP [1].

Definition 6 An objective vector z^{**} formed with components slightly less than that of the ideal objective vectors is known as the utopian objective vector i.e., $z_k^{**} = z_k^* - \varepsilon_k$ with $\varepsilon_k > 0\ \forall\, k \in K$ [1].

9.3 Utility Function Approach for FRMOOP

The utility function approach is always considered a simple technique to solve the multi-criterion optimization problems. A utility function $U : R^s \to R$ is a mathematical representation of decision maker's preferences mapping criterion vectors into the real line giving a value of utility for decision maker. Using the utility function, the problem defined in (9.3) transforms to the following mathematical programming problem:

$$\text{Minimize } U(z_1(x), z_2(x), \ldots, z_s(x))$$
$$\text{s.t. } x \in X(A, b)$$

Different utility functions have been studied in the literature in this field [1–3, 6]. The most general utility functions that are used in literature are weighted metric utility function and weighted linear utility function. Based on these utility functions the two transformed optimization problems can be described as follows:

9.3.1 Weighted Linear Utility Function

This method linearly combines different objectives to a scalarized single objective as follows:

$$\text{Min } Z(x) = \sum_{k=1}^{s} \lambda_{k} f_k(x)$$
$$\text{s.t. } x \in X(A, b)$$

where $f_k(x)$ is the kth objective function with the weight vector $\lambda = (\lambda_1, \lambda_2, \ldots, \lambda_s)$, meeting the conditions $\lambda_k > 0$, $\sum_{k=1}^{s} \lambda_k = 1$; for $k \in K$. i.e., the aggregated function to be optimized is the strict convex combination of all the objectives under consideration. Generally, the weights represent the relative importance of the objectives. A detailed analysis of this method is presented in works [1, 2]. This method works well as far as convex multiobjective optimization problems are concerned but lacks sometimes in finding certain Pareto optimal solutions in case of non-convex MOOP.

9.3.2 Weighted Metric Utility Function

This method presents another approach of combining multiple objectives via the distance minimization of the particular solution from the reference point z^* based on different metrics. In this method, for a non-negative weight vector the utility function using the distance measure based on p-metric is considered as follows:

$$\text{Min } Z_p(x) = \left(\sum_{k=1}^{s} \lambda_k |f_k(x) - z_k^*|^p \right)^{1/p}$$
$$\text{s.t. } x \in X(A, b)$$

where $p \in (1, \infty)$ is the distance metric. When $p = 1$, the resulting problem is equivalent to the weighted sum approach. For $p = 2$, the Euclidean distance is minimized. When $p = \infty$, the metric is also known as the Tchebycheff metric, the transformed utility function has a special name as Weighted Tchebycheff function given as:

$$Z_\infty(z, z^{**}, \lambda) = \max_{k=1}^{s} \{ \lambda_k | (f_k(x) - z_k^{**}) | \}$$

where $\lambda = (\lambda_1, \lambda_2, \ldots, \lambda_s)\ \forall \lambda_k \geq 0$, and z^{**} is the reference point. When the Tchebycheff metric is used, each and every Pareto optimal solution of the MOOP can be found if the utopian point is used as the reference point. For each efficient solution there exists a weighted Tchebycheff scalarizing function such that it is optima of the considered MOOP [1, 3].

9.4 Materials and Methods

Once the original problem has been transformed to the single utility function based optimization problem, a hybridized genetic algorithm is applied to solve the modified optimization problem. Hybridized genetic algorithm combines the elements of genetic algorithm with the idea of local search. The combination results in an improved searching tool to explore new and more diversified solutions of the considered problem. The design of the proposed algorithm can be summarized in following sections:

9.4.1 Initialization

The feasible domain for the considered problem has a special structure. As the feasible region is designed by a system of fuzzy relational equations, it might be the case that some of the variables assume some specific value. The values of such variables need to be fixed for the sake of solvability of the system. To identify such variables we define sets $I_j = \left\{ i \in I | \widehat{x}_i \cdot a_{ij} = b_j \right\},\ \forall j \in J.$

Definition 7 If I_j is singleton, say $I_j = \{i'\}_{i' \in I}$ for some $j \in J$ then $x_{i'}^* = \widehat{x}_{i'}$, $i' \in I$, then variable $x_{i'}$ is fixed as it is the only variable to satisfy the jth equation. Such variables assume fixed values in all solutions.

Once fixed variables have been detected their value is fixed as $x_i = \widehat{x}_i = \breve{x}_i$. Then an initial population of fixed size is created with the fixed variables assuming the value $\widehat{x}_i$ and the variables that are not fixed assuming a random value in the range $\left[0\ \widehat{x}_i\right]$ in all the solutions. Now the feasibility of generated solutions is examined. The following algorithm has been used to maintain the feasibility of the solutions:

Algorithm 1 *For maintaining feasibility of solutions*

- Choose a violated constraint j. Let $D_j = \{i \in I |\ a_{ij} \geq b_j\}$
- Randomly choose an element $k \in D_j$. For $a_{kj} > b_j$ or $\widehat{x}_k = b_j / a_{kj}$, set $x_k = b_j / a_{kj}$. Otherwise, assign a random number between $\left[b_j / a_{kj}, \widehat{x}_k\right]$ and x_k.
- Check the feasibility of the new solution. If the solution is still infeasible, then go to Step 1 and repeat the process. Otherwise, stop.

9.4.2 Selection

Once the initial population of fixed size has been generated, selection of good individuals is the next step to apply the recombination operators. The selection of good individuals at this step is performed via the rank selection scheme. The individuals are evaluated using the aggregated objective function defined in Sect. 9.3. A normalized weight vector giving a new utility function is randomly selected for each run. The probability of selection for each individual is calculated as follows:

$$p(x^i) = \frac{\max_{i=1}^{N}(z(x^i)) - z(x^i)}{\sum_{i=1}^{N} z(x^i) - \max_{i=1}^{N}(z(x^i))}, \quad i = 1, 2, \ldots, N$$

where $z(x^i)$ denotes the fitness of the *ith* individual and N be the population size. Using the selection scheme a pre-specified number of parent solutions are selected at the end. Selected individuals undergo the recombination process so as to create a new population.

9.4.3 Crossover

Owing to the nature of solution space of the problem the conventional real coded crossover techniques are not feasible. So, a specific crossover scheme is designed that generates feasible individuals at the end. The algorithm used for crossover can be described in the following steps:

Algorithm 2 *Crossover*

Get the matrices A, b and find the maximum solution $\widehat{x}$ by (9.4) and set parameters $0 \leq \alpha \leq 1$, $\beta \geq 1$, $0 \leq \zeta \leq 1$, $0 \leq \delta \leq 1$.

Randomly select two individuals x_1, x_2 from the selected population.

For $i = 1, 2$

Generate a random number $\varepsilon \in [0, 1]$

If $(\varepsilon \geq \zeta)$

$$x_i = \beta x_i - (\beta - 1)\widehat{x}$$

Else

$$x_i = \alpha x_i + (1 - \alpha)\widehat{x}$$

Generate a random number $\varepsilon_2 \in [0, 1]$

If $(\varepsilon_2 \geq \delta)$

Go to evaluation procedure

Else

$$x_1^{next} \leftarrow x_1$$

$$x_1 = \alpha x_1 + (1 - \alpha)x_2$$

If $x_1 \circ A = b$

Go to evaluation procedure

Else

$$x_1 \leftarrow x_1^{next}$$

$$x_1 = \beta x_1 - (\beta - 1)x_2$$

If $x_1 \circ A \neq b$

$$x_1 \leftarrow x_1^{next}$$

If $x_1 \circ A \neq b$

Make x_1 feasible using Algorithm 1
Go to evaluation procedure.

End

The repeated linear combinations of individuals draw the generated individuals inside the feasible space. Here, α, β are small numbers close to 1 respectively and are generally kept small. For our problem we are taking $\alpha = 0.99$, $\beta = 1.0085$, $\zeta = 0.012$, $\delta = 0.99$.

9.4.4 Mutation

Mutation randomly perturbs a candidate solution by exploiting the search space with a hope to create a better solution in the problem domain. We adopt the following mutation procedure to solve our problem:

Algorithm 3 Mutation

- Get the matrices A, b and find the maximum solution $\widehat{x}$ by (9.4) and set the mutation probability $\theta = 0.1$.
- Generate $r_i \in [0, 1]$ for each bit of every individual in the crossed population.
- For $\forall i \in I$ if $r_i \leq \theta$, randomly assign x_i a number from $\left[0, \widehat{x}_i\right]$.
- For the modified $x = (x_1, x_2, \ldots, x_m)$ check feasibility $x \circ A = b$.
  If $x \circ A = b$, go to the evaluation procedure else make the solution feasible via Algorithm 1.

9.4.5 Local Search Scheme

The population formed after the recombination operators undergo the local search operation. The local search is probabilistically applied to good solutions that avoid the unfruitful exploitation of the search space leading to bad new generated solutions. The tournament selection with tour size 3 is applied to select the particular solution to undergo local search procedure based upon the utility function considered with the current weight vector for that particular run. When local search is applied to a solution a random new point is selected from the neighborhood of the current point. We consider the neighborhood $[-0.05, 0.05]$ of radius 0.05. If the new solution provides a better value of the current utility function for that run; the selected solution is replaced by the current solution. Otherwise, some other neighbor is selected and tested against the current solution. To avoid the unnecessary search the number of solutions examined in neighborhood of a solution is pre-fixed. The method terminates if no further improvement is possible until the number of successive fails to examine the improved solution ends up. If no improved solution is obtained the initial solution is kept as such in the population. This procedure is repeated for all the solutions in the population that undergo the local search procedure. This kind of local search scheme was used in work [4] for a multiobjective combinatorial optimization problem.

9.5 Results and Discussions

In this section, implementation of the proposed hybridized genetic algorithm for multiple linear and nonlinear objective functions is discussed. We consider multiobjective linear and nonlinear optimization problems and a system of fuzzy relation equations with max-product composition to investigate the nature of the solutions obtained using the proposed procedure.

Example Consider a four dimensional problem with fuzzy matrices A and b as follows:

$$A = \begin{bmatrix} 0.5042 & 0.0569 & 0.3641 & 0.2527 \\ 0.9398 & 0.6578 & 0.0359 & 0.1663 \\ 0.4979 & 0.3937 & 0.5715 & 0.9849 \\ 0.7182 & 0.0330 & 0.9476 & 0.1271 \end{bmatrix},$$

$$b = [0.6120 \quad 0.4284 \quad 0.8075 \quad 0.1083]$$

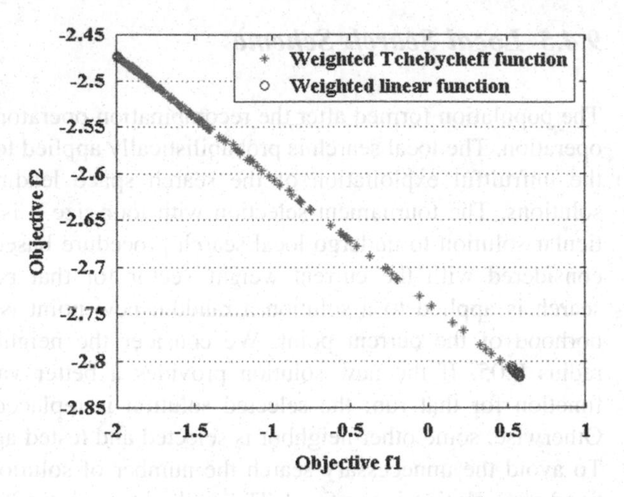

Fig. 9.1 Approximation of non-dominated set for Case I

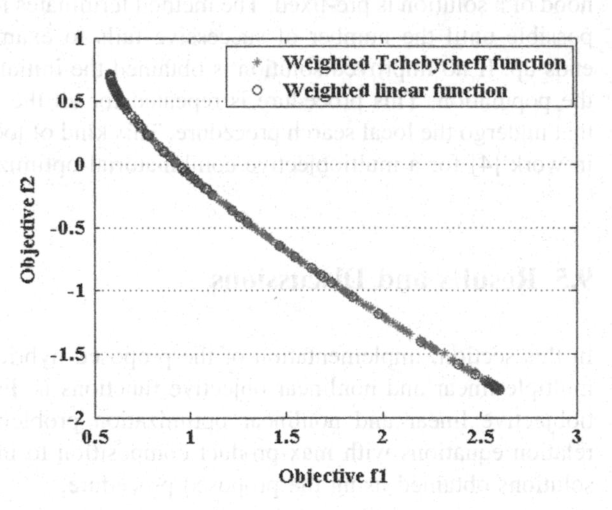

Fig. 9.2 Approximation of non-dominated set for Case II

The maximum solution obtained using (9.4) come out to be $\hat{x} = [0.4286 \quad 0.6512 \quad 0.1100 \quad 0.8521]$. For this particular problem, the values of x_2 and x_4 of all solution vectors have to be fixed at 0.6512 and 0.8521 respectively. Now the problem reduces to two dimensions. Therefore, we can focus on values of x_1 and x_3. The test results for some multiple linear and nonlinear optimization problems in Case I and Case II with this system of fuzzy relational equations as constraints are discussed below.

$$\text{Case I : Min} \begin{bmatrix} f_1(x) = -0.6x_1 + 0.5x_2 + 0.1x_3 + 0.3x_4, \\ f_2(x) = 0.8x_1 - 0.4x_2 + 0.2x_3 - 0.3x_4. \end{bmatrix}$$

$$\text{Case II : Min} \begin{bmatrix} f_1(x) = 10(x_1 - 0.45)^2 + 10(x_3 - 0.35)^2, \\ f_2(x) = -6(x_1 - 0.7)^2 + 10(x_3 - 0.45)^2. \end{bmatrix}$$

The generated local optima obtained using the 200 randomly selected weighted Tchebycheff functions and weighted linear functions for case I and II are shown in Figs. 9.1 and 9.2 respectively. Each approximate solution is the optima of the corresponding utility function for that run. It is clear that the weighted Tchebycheff functions result in more diversified solutions in comparison of weighted linear sum functions.

9.6 Conclusions

A fuzzy relational multiobjective optimization problem has been discussed. The suggested utility function based approach generates a good approximation of the efficient set with least computational effort required. The single objective hybridized genetic algorithm used effectively decides and results in good approximations of the Pareto solutions in cases with both linear and nonlinear functions. Two kinds of utility functions have been considered; out of which weighted Tchebycheff functions perform exceptionally well even when the shape of non-dominated front is complicated in nature. The proposed algorithm offers an effective way out to handle FRMOOP even when the dimension of criterion space is large and do well even for the large scale problems.

Acknowledgments Authors are thankful to referees for their valuable suggestions.

References

1. Deb K (2001) Multi-objective optimization using evolutionary algorithms for solving multi-objective problems. Wiley, New York
2. Marler RT, Arora JS (2010) The weighted sum method for multi-objective optimization: new insights. Struct Multidisc Optim 41:853–862. doi:10.1007/s00158-009-0460-7
3. Steuer RE (1986) Multiple criteria optimization: theory, computation, and application. Wiley, New York
4. Ishibuchi H, Yoshida T (2002) Hybrid evolutionary multi-objective optimization algorithms. In: Proceedings of the 2nd international conference on hybrid intelligent systems. Santiago, Chile, pp 163–172
5. Coello CAC (2000) An updated survey of GA-based multiobjective optimization techniques. ACM Comput Surv 32(2):109–143
6. Marler RT, Arora JS (2004) Survey of multi-objective optimization methods for engineering. Struct Multidisc Optim 26:369–395
7. Wang WF (1995) A multiobjective mathematical programming problem with fuzzy relation equation constraints. J Multi-Criteria Dec Anal 4:23–35

8. Loetamonphong J, Fang SC, Young RE (2002) Multi-objective optimization problems with fuzzy relation equation constraints. Fuzzy Sets Syst 127:141–164
9. Khorram E, Zarei H (2009) Multi-objective optimization problems with fuzzy relation equation constraints regarding max-average composition. Math Comput Modell 49:856–867
10. Jiménez F, Cadenas JM, Sánchez G, Gómez-Skarmeta AF, Verdegay JL (2006) Multiobjective evolutionary computation and fuzzy optimization. Int J Approximate Reasoning 43:59–75
11. Zhang BX, Chung BTF, Lee ET (1997) A successive approximation method for solving multiobjective fuzzy optimization problems. Kybernetes 26:392–406
12. Thapar A, Pandey D, Gaur SK (2011) Satisficing solutions of multi-objective fuzzy optimization problems using genetic algorithm. Appl Soft Comput 12:2178–2187
13. Klir GJ, Yuan B (1995) Fuzzy sets and fuzzy logic: theory and applications. Prentice Hall, Upper Saddle River, NJ

Chapter 10
Fuzzy Model for Optimal Operation of a Tank Irrigation System

N. Manikumari and A. Murugappan

Abstract Water is a basic human need. Three fourth of our earth is surrounded by water. Water is stored in lakes and large tanks for human purposes like drinking, bathing, washing and irrigation. Effective use and operation of this storage is of prime importance. Veeranam tank is the largest tank irrigation system in Tamilnadu which commands two taluks of Kattumannarkoil and Chidambaram comprising of a large number of villages and hamlets which sums upto 120. In addition it has recently supplemented Chennai urban water supply. Optimal operation of the tank is thus need of the hour. Abundant literature is available on Artificial Intelligence techniques and their successful application for prediction, simulation, identification, classification and optimization in the field of water resources management. But literature on operation of tanks is meager. Tanks have been operated on adhoc basis and experience, with not much scientific logic. This paper addresses the need for optimal operation of Veeranam tank irrigation system using the concept of a fuzzy rule-based system for tank operation. The rule-base was built on the basis of the expert's knowledge and heuristic data. MATLAB Fuzzy logic toolbox was used for simulation in this study. Model was evaluated using various metrics like RMSE, MAE and AI.

Keywords Tank Irrigation · Fuzzy sets · Optimal operation · Fuzzy Rule · Root mean square error · Mean absolute error and agreement index

---

N. Manikumari (✉) · A. Murugappan
Department of Civil Engineering, Annamalai University,
Annamalai Nagar, Chidambaram, India
e-mail: aumani@sify.com

A. Murugappan
e-mail: profam@sify.com

R. Malathi and J. Krishnan (eds.), *Recent Advancements in System Modelling Applications*, Lecture Notes in Electrical Engineering 188,
DOI: 10.1007/978-81-322-1035-1_10, © Springer India 2013

10.1 Introduction

Water is a prime natural resource, a basic human need and a precious national asset. It is a scarce resource which is to be planned, developed, conserved and managed, and on an integrated and environmentally sound basis, keeping in view the socio-economic aspects and needs of mankind. Tanks are considered to be the most economical water storage systems and require less expenditure for construction and, also require small area to be submerged for impounding water [1]. It is vital to operate a tank in an optimal way to manage water during the failure of monsoons and floods. Several researchers [2–6] applied the concept of a fuzzy rule-based system for reservoir operation. The rule-base was built on the basis of the expert's knowledge. The MATLAB Fuzzy logic toolbox was used for the simulation in most studies. It was also reported that fuzzy based models could effectively handle data with uncertainty, vagueness and imprecision as well.

10.2 Data Base for Study

Table 10.1

10.3 Database Collected

The various measured data used in the study were:

Table 10.1 Hydraulic particulars of Veeranam tank

Full tank level (F.T.L.)	47.50 ft (14.470 m)
Maximum water level (M.W.L.)	50.00 ft (15.240 m)
Tank bund level (T.B.L.)	54.00 ft (16.460 m)
Catchment	165 miles^2 (or) 422 km^2
Capacity at F.T.L.	1465 Mcft (or) 41.46 M cu.m
Waterspread at F.T.L.	15 miles^2 (or) 38.40 km^2
Maximum width of tank	5.63 km
Circumference	40.225 km
Length of main bund	15.30 km
Width of main bund	8 m (average)
Length of foreshore bund	35.32 km
Number of sluices	34 (28 in the main bund and 6 in the foreshore bund)
Sill level of lowest sluice	Periyamadagu, +27.67 ft (8.44 m)
Sill level of highest sluice	Anaikkal, +39.81 ft (12.13 m)
Sill level of highest sluice of foreshore	Kudigadu, +41.00 ft (12.50 m)
Tank reference sluice	Radha +31.90 ft (11.92 m)

- Daily data on maximum temperature, minimum temperature, maximum humidity, minimum humidity, actual bright sunshine hours, wind speed and Daily rainfall observed at IMD observatory, Annamalainagar for computing Daily Reference crop ET and Effective Rainfall.
- Daily actual inflow into the tank from Lower Coleroon through the Vadavar feeder canal.
- Daily storage in the tank.
- Daily release from the tank.

Then crop water requirement was arrived by computing Reference Crop Evapotranspiration adopting the standard FAO Penman–Monteith method. The performance of the tank irrigation system was assessed by conducting a water balance study taking the command of a few major canal systems in the different seasons taking a period of 21 successive years. A fuzzy rule based model for optimal operation of the tank was developed.

10.4 Phases of Model Development

The fuzzy logic model was constructed using Matlab fuzzy tool box of the recent version of Matlab software.

- Fuzzification of inputs
- Formulation of the fuzzy rule set
- Application of the fuzzy operator
- Implication of rules
- Defuzzification

The methodology used for developing the fuzzy rule based model for the tank irrigation system is shown in Fig 10.1. This show the sequential steps in the modeling procedure, which was adopted for each of the successive week models starting from week 1 to week 25 of each of the two group of seasons August having eleven seasons and September, six seasons. The input data matrix incorporated every crop season in the Fuzzy Rule Based (FRB) model. The three inputs namely weekly inflow, initial storage at the beginning of the week and computed weekly demand were fuzzified using Mamdani Inference system. Weekly release was the output. Trapezoidal membership functions for two end fuzzy sets and Triangular membership functions for three intermediate fuzzy sets were adopted for each of the three inputs and one output. The fuzzy rule set was formulated based on expert knowledge.

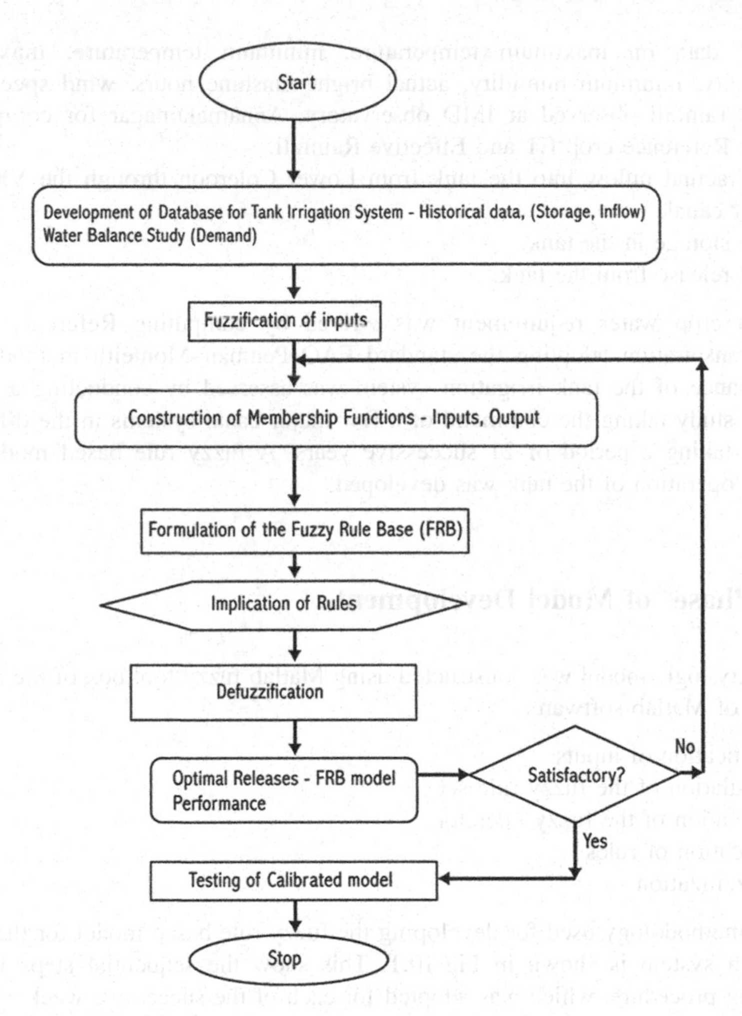

Fig. 10.1 Methodology flow chart

10.5 Application of Fuzzy Operators

After fuzzifying inputs and creating the rule, it was found that antecedent of a given rule was having more than one part, due to which fuzzy operator is applied to obtain one number, that represents the result of the antecedent for that rule [8]

The input to the fuzzy operator is two or more membership values from fuzzified input variables. In fuzzy logical operations, any number of well-defined methods can fill in for the *AND* operation or the *OR* operation. In the fuzzy logic tool box, two built in *AND* methods are supported: *min* (minimum) and *prod* (product). Two built in *OR* methods are also supported: *max* (maximum) and *probor* (the probabilistic *OR* method). But in this study, the *AND* operator was only employed. For example, if a set of input variables, in this case, three actual values as inputs, one for weekly inflow (which has membership in categories, very low, low, medium, high and very high), another for initial storage and a third input for weekly irrigation demand (both with same 5 membership categories), this may give the rules in form:

- If inflow is "very low" *AND* storage is "very low", *AND* demand is "very low", then release is "very low".
- If inflow is "low" *AND* storage is "low", *AND* demand is "low" then release is "low".
- If inflow is "low" *AND* storage is "high", *AND* demand is "medium", then release is "medium".
- If inflow is "medium" *AND* storage is "low", *AND* demand is "medium", then release is "medium".
- If inflow is "medium" *AND* storage is "medium", *AND* demand is "medium", then release is "medium".
- If inflow is "medium" *AND* storage is "high", *AND* demand is "high", then release is "high".
- If inflow is "high" *AND* storage is "low", *AND* demand is "medium", then release is "medium".
- If inflow is "high" *AND* storage is "medium", *AND* demand is "high", then release is "high".
- If inflow is "high" *AND* storage is "high", *AND* demand is "high", then release is "high".
- If inflow is "very high" *AND* storage is "high", *AND* demand is "very high", then release is "very high".

Actually the rule base constructed consisted of 92 feasible rules of the combination of three inputs and one output, simulating tank operation for meeting irrigation demand. The above is a sample of feasible combination of inputs that arise in the operation of the tank system.

10.6 Defuzzification

Defuzzification is a method of converting a fuzzy quantity into a crisp quantity. The result obtained from implication was in the form of a fuzzy set. For real application, this has to be defuzzified.

The "centroid" evaluation method is used for defuzzifying the output into crisp numbers. This method returns the center of area under the membership curve and directly computes the real valued output as a normalized combination of membership values.

Table 10.2 Fuzzy model variables For week 25 of August group

Season	Inflow (M cu.m)	Storage (M cu.m)	Demand (M cu.m)	Historical release (M cu.m)	Model release (M cu.m)
1991–1992	5.05	0.00	3.74	1.59	2.09
1995–1996	12.54	25.91	4.96	3.89	3.96
1996–1997	4.69	26.90	5.44	0.93	4.27
1997–1998	31.15	8.13	3.33	6.59	1.90
1998–1999	2.97	22.23	4.56	3.40	3.45
1999–2000	7.00	25.49	3.77	2.03	2.69
2000–2001	5.51	25.91	3.22	1.10	1.75
2001–2002	6.38	28.03	3.50	1.20	2.18
2005–2006	0.73	41.48	4.55	0.17	3.45
2006–2007	21.00	14.61	3.96	5.03	2.78
2007–2008	11.34	31.84	4.55	0.82	3.45

10.7 Model Development

Fuzzy rule based models were developed for the 25 weeks of the crop season, when the tank was operated effectively for irrigation for two of the release groups, August group and September group. July commencement release seasons in the data period taken for study was not modeled as there were just three seasons only which were found insufficient for modeling and calibration of the fuzzy models. The sample seasons of each group were divided into two, one a data set for calibrating the model and another a testing set to test the model after calibration. Table 10.2 shows the typical variables for the 25th Week of August group.

10.8 Model Application

The range of values for the fuzzy sets was fixed by analyzing historical flow data pertaining to inflow and storage, as well as release. The input weekly irrigation demand was the result of the irrigation water balance study carried out for the major canals of the tank as suggested by [7, 8].

Five Membership Functions for each variable were constructed assigning linguistic variables: Very low, Low, Medium, High and Very high by suitably partitioning them into fuzzy sets.

The ranges for these five membership functions for each variable were chosen appropriately based on the values of each group and their variation in practice during the seasons pertaining to each of the two season groups.

The results of fuzzy modeling of the tank were arrived for the week models starting from week 1 to week 25 of the crop seasons commencing in August. Out of the eleven seasons of August group, nine seasons, except 1999–2000 and 2006–2007 were taken for calibrating the model and the remaining two seasons were kept aside

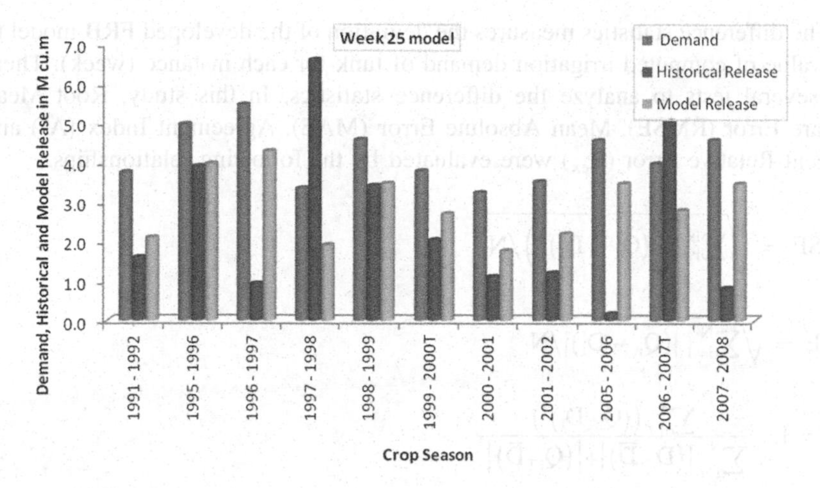

Fig. 10.2 Demand, historical and model releases for August group

for testing the model after calibration. The inputs, weekly inflow and initial storage were arrived from daily flow records of the tank for each season. The demand is the computed irrigation demand arrived from water balance study done in the canal commands from which it was computed for the tank as a whole using a weighting factor. Historical release refers to the weekly actual release from the tank which is the sum of daily releases through the canals also obtained from tank records. This data set of historical releases was used to construct Membership Functions for the output, which the fuzzy model is to evolve. They were an aid to construct tank operation rules also.

The FRB model releases were then compared with the irrigation demand of the corresponding week, being the objective which the tank irrigation system had to meet. This was the criteria based on which the model performance was evaluated. The model after initial runs was tuned iteratively until the model release for each input set matches the irrigation demand as close as possible. This was done by a trial and error process.

The test seasons were then employed to validate the model. Similar procedure was repeated for all the 25 weeks of the crop season of the August group.

Figure 10.2 shows the typical model parameters like demand, historical release and model releases for Week 25 of August group.

10.9 Conclusion

Several statistical routines could be used to analyze relationship between the simulated fuzzy model release and weekly irrigation demand. The model fit and difference of releases evolved by the FRB model is of concern for evaluation of model performance.

The difference statistics measures the deviation of the developed FRB model to the value of computed irrigation demand of tank for each instance (week). There are several tests to analyze the difference statistics. In this study, Root Mean Square Error (RMSE), Mean Absolute Error (MAE), Agreement Index (AI) and Percent Relative Error (R_{ex}) were evaluated by the following relationships.

$$RMSE = \sqrt{\sum_{i=1}^{N}\left((Q_i - D_i)^2\right)/N}$$

$$MAE = \sqrt{\sum_{i=1}^{N}|(Q_i - D_i)|/N}$$

$$A_X = 1 - \frac{\sum_{i=1}^{N}\left((Q_i - D_i)^2\right)}{\sum_{i=1}^{N}\left|(D_i - \overline{D})\right| + \left|(Q_i + \overline{D})\right|^2}$$

$$R_{ex} = \left(\frac{\sum_{i=1}^{N}\left((Q_i - D_i)^2\right)}{\sum_{i=1}^{N}(Q_i)^2}\right) \times 100$$

where N = Number of observations (model releases).
Q = Weekly model release in M. cu.m. and
D = Weekly Irrigation demand in M.cu.m.

From Table 10.3, Mean of Agreement Index 0.69 means the average model efficiency works about to be 69 %. It was seen that except weeks 9, 17 and 18, the remaining models were found to perform satisfactorily.

The three weeks depicted extreme flow situations which the model has not incorporated. Season 2001–2002 exhibited 60.5 % savings in water by way of reduction of surplus. Season 2005–2006 models met the demand fully and resulted in a small surplus of 5 %. Seasons 2006–2007 and 2007–2008 could bring down the deficit from 37 % to 1 % and from 5 % to 0.22 % respectively.

The proposed FRB model is found suitable for simulating tank operation and can serve as an efficient tool for derivation of operational rules of the irrigation tank wherein the operator can input the values of initial storage, inflow and demand to get the suitable release.

The performance of the rules when validated with data demonstrated that these rules could evolve model releases that closely match the irrigation demand during most of the weeks of crop seasons.

It is also confirmed that the calibration of fuzzy rule base with the historical data and fine tuning of the same has improved the performance of the irrigation tank substantially.

Considering the model weekly deficits during calibration and testing phases, the FRB model can be confidently used to decide the release pattern of the tank.

Table 10.3 Statistical parameters of August release FRB models

Week No.	RMSE (M cu.m)	MAE (M cu.m)	AI	Rex
1	1.95	1.21	0.82	23.30
2	2.03	1.70	**0.93**	9.21
3	1.49	1.26	**0.97**	5.35
4	2.75	2.02	0.87	17.30
5	4.29	3.39	0.72	33.97
6	3.95	3.09	0.75	39.93
7	5.01	4.16	0.77	31.44
8	5.00	4.08	0.80	31.74
9	6.36	5.64	0.49	88.55
10	5.58	4.71	0.65	58.71
11	4.70	3.78	0.65	44.28
12	2.87	2.42	0.87	15.92
13	1.84	1.59	**0.95**	5.94
14	3.82	3.03	0.65	41.32
15	2.30	1.66	0.83	21.75
16	2.76	2.29	0.79	27.29
17	21.42	11.86	0.24	81.99
18	49.89	19.08	0.04	96.81
19	4.25	3.93	0.63	46.29
20	4.77	4.08	0.65	49.43
21	4.12	3.18	0.69	44.49
22	4.56	3.65	0.68	41.94
23	4.41	4.02	0.54	57.88
24	3.18	2.71	0.64	44.34
25	2.41	1.80	0.56	68.23
MEAN	6.23	4.01	0.69	41.10

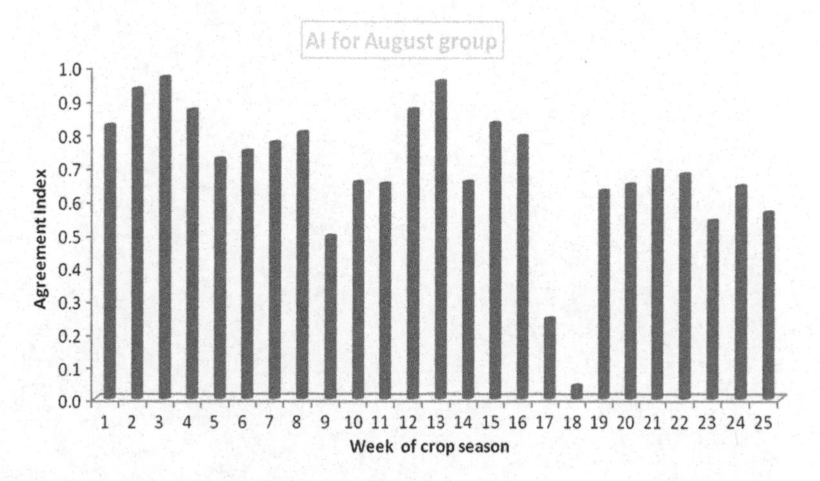

Fig. 10.3 AI of FRB models for August group

In the above Fig. 10.3, Agreement Index—a metric for the performance of the FRB model is shown for all the 25 weeks in the August weekly model group.

This FRB performs well except for the 18th week, which is a flood instance. The advantage of linguistic representation of optimal operating schemes enables fuzzy models to help tank operators to feel comfortable in understanding and implementation.

References

1. Arumugam N, Mohan S (1997) Integrated decision support system for tank irrigation system operation. J Water Resour Plan Manag 123(5):266–273 ASCE
2. Anjaneya Prasad M (2006) Development of operational rules for serial and parallel reservoir systems using fuzzy logic approach. Ph.d. Thesis submitted to Indian Institute of Technology, Chennai
3. Bardossy A, Duckstein L (1995) Fuzzy rule- based modeling with applications with geophysical, biological and Engineering systems. CRC Press. Boca Raton
4. Panigrahi DP, Mujumdar PP (2000) Reservoir operation modelling with fuzzy logic. J Water Resour Manag 14:89–109 Kluwer
5. Mehta R, Jain SK, Kumar V (2003) Fuzzy technique for reservoir operation—effect of membership functions with different number of categories. Hydrol J 28(3–4):17–33
6. Murugappan A (2001) Fuzzy optimization model for tank irrigation system operation. Ph.d. Thesis, Annamalai University, Annamalainagar
7. Shrestha BP, Duckstien L, Stakiv EZ (1996) Fuzzy rule-based modeling of reservoir operation. J Water Resour Plann Manag, ASCE 122(4):262–269
8. Suharyanto Xu C, Goulter IC (1996) Reservoir operating rules with fuzzy programming. In: Russell SO. Campbell PF Discussion, (Paper 9026) 122(3):165–170

Chapter 11
Deconvolving the Productivity of Salespeople via Constrained Quadratic Programming

Gautam K. Bhat and Kush R. Varshney

Abstract With the present market trend, businesses and organisations with large salesforces are experiencing much turnover among their sellers. Movement of salespeople from one company to another is a continual process as long as there is market demand. In the traditional sense, a salesperson's productivity is directly proportional to the revenue that he or she brings to the company. Importantly, the senior leaders in organisations are interested in knowing the variations in sales productivity as a result of hiring and attrition in the salesforce. In this paper we focus our attention on the characterisation of sales productivity based on four categories. When an existing salesperson leaves, what is the sales productivity over time if replaced by a new hire from a university, an experienced new hire, or a transfer from another division in the company? In addition if an organisation ventures into acquisition, what is the anticipated sales productivity from this? We model the sales productivity of new hires as a linear time-invariant system and estimate productivity profiles with a least-squares deconvolution formulation. By applying business constraints on productivity profiles for regularisation, we are left with a constrained quadratic program to solve. We demonstrate the estimation technique on real-world sales data from a global enterprise, finding productivity profiles under the four different cases listed above.

Keywords Business applications · Deconvolution · Estimation theory · Least-squares · Salesforce analytics

G. K. Bhat (✉)
IBM Global Technology Services, Ramapuram, Chennai, India
e-mail: gautambh@in.ibm.com

K. R. Varshney
IBM Thomas J. Watson Research Center, Yorktown Heights, New York, USA
e-mail: krvarshn@us.ibm.com

R. Malathi and J. Krishnan (eds.), *Recent Advancements in System Modelling Applications*, Lecture Notes in Electrical Engineering 188,
DOI: 10.1007/978-81-322-1035-1_11, © Springer India 2013

11.1 Introduction

Systems theory, broadly construed, is concerned with modelling, analysing, and optimising a set of interacting components that form an integrated whole. Those components could be mechanical machines, electrical or electronic elements, or even human beings [1]. In this paper, the particular system that we focus on is a salesforce whose interacting components are individual salespeople. Such study of sellers falls under the scope of business analytics, specifically salesforce analytics [2–4].

When new salespeople join an enterprise, whether hired directly from a university, as a result of a merger or acquisition, through an internal transfer from a different division, or hired with experience from the industry at-large, it takes time for them to get acclimated to the organisation, learn about the product and service offerings, and build contacts. Thus sellers have a period of little to no productivity followed by a ramping up period until they have reached the productivity of sellers that have been with the enterprise for a longer duration. The amount of revenue earned is the traditional measure of productivity for salespeople.

The problem that we study herein is characterising the productivity of newly-hired salespeople, specifically determining productivity profiles as a function of time after hiring (for different types of new hires). Such characterisations are important for planning purposes because the head count of sellers is not a true indication of the productivity of the salesforce; if a target productivity is desired at a particular time in the future, then hiring decisions must be made in the present based on the productivity profiles [5]. Brooks' Law, which states that "adding manpower to a late software project makes it later" [6] applies equally well to sales productivity because of ramp-up time [7]: if there is a shortage of sellers today, then hiring more sellers today will not improve productivity because they will have no initial productivity, and may in fact transiently decrease productivity of the overall salesforce as they are integrated into the enterprise.

In studying novel systems, Willsky asks [8], "How can we extend existing mathematical methodologies? How can we use existing methodologies in the context of a specific physical problem to obtain a tractable formulation which addresses the issues of interest in the more ill-defined physical problem?" The novel system study here is that of sales productivity characterisation. In the context of this specific 'physical' problem, the tractable existing framework within which we pursue our methodology is that of convolutive discrete-time linear time-invariant (LTI) systems [9]. We seek to identify the system that transforms head count at various post-hiring times to sales productivity or revenue. Specifically, we use a least-squares formulation of deconvolution to do so [10, 11], which leads to a quadratic programming optimisation.

The extension of the basic least-squares identification that we present here further models the system through business constraints for purposes of regularisation. We include non-negativity constraints, which often arise in statistical learning optimisations [12]. Additional business constraints lead us to additional mathematical constraints related to monotonicity and smoothness of the seller productivity profiles as well as a saturation level of the profiles. Thus overall, we optimise a constrained quadratic program to characterise the productivity of new sellers as a function of the time since they were hired by the enterprise.

Constrained quadratic programming for time profile estimation also arises in application domains other than salesforce analytics. For example, there is a constrained quadratic programming formulation in [13] to estimate transcriptional profiles of clinical blood samples. Also, there is a similar formulation in [14] to characterise event profiles from functional magnetic resonance imaging data sources. Although there have been theoretical, mathematical models of salesforces in the marketing literature previously [15], systems-theoretic thinking and analysis has not been applied to the problem of new seller productivity. In our previous work [7], we did not adopt a linear time-invariant system and deconvolution perspective to the problem.

This paper discusses the specification, identification, and estimation of sales productivity of salespeople. This involves measurements on a set of criteria at different levels of skill and experience. The numbers of measurements are taken over a long period of time to get as accurate results as possible. In optimising the sales productivity result, we have considered all the requisite constraints that are applicable to the salesforce. Optimisation techniques need to be performed on a huge data set and over a period of time. Once we have the result, that is not the end of the world; it needs to be validated and verified from a logical perspective. Several constraints are taken into account so that a definite logical result is derived. In the context of optimising sales productivity result, it is imperative to say that the result must be non-negative. A salesperson irrespective of being a fresh graduate, experienced, or so on, will only generate revenues, hence the optimised result must always be non-negative. The deconvolution method assumes that: (1) productivity of the fresh sales graduate may not be steep in the initial phase of his career, hence exceptional skills are not taken into consideration here, and (2) same two salespeople may or may not contribute to the same level of sales productivity. Again, recent studies have shown that the workforce productivity may vary from trial to trial.

Here we have divided the paper into several smaller parts. We begin with the system model to give a background of what is being done and a high level understanding of the need to go for deconvolution. Followed by this we provide the algorithm associated with the derivation followed by experiments and results section with details on every step performed and analysed. Finally we have the discussion section ending with conclusion.

11.2 System Model

In this section, we first describe LTI systems and how sales revenue fits into that framework. Then we describe the system model of sales productivity that we employ.

11.2.1 Linear Time-Invariant Systems

LTI systems produce same amount of output for the similar amount of input passed. In other words, if x_1 amount of input is fed to the system, it will produce y_1 amount of output. If x_2 amount of input is applied, then, y_2 amount of output is produced. The following pictorial representation will make it clearer.

$$\xrightarrow{x_1} \boxed{\text{LTI System}} \xrightarrow{y_1} \tag{11.1}$$

The above diagram can be interpreted such that for an input x_1, the output generated is y_1.

$$\xrightarrow{x_2} \boxed{\text{LTI System}} \xrightarrow{y_2} \tag{11.2}$$

For an input x_2, the output generated is say y_2.

Similarly, if the above two inputs are added, the output will also be summed up as shown.

$$\xrightarrow{x_1+x_2} \boxed{\text{LTI System}} \xrightarrow{y_1+y_2} \tag{11.3}$$

The above analogy applies to sales productivity in that if we double the number of sellers, the revenue will double.

Along similar lines, time-invariant systems always produce the same output irrespective of when the input is sent or applied to the system. This means that if we hire someone in August 2010 or in April 2012 or in January 2020, if 7 months have elapsed since hiring, his or her productivity will be the same. This can be visualised through a diagram below.

$$\xrightarrow{x[n]} \boxed{\text{LTI System}} \xrightarrow{y[n]} \tag{11.4}$$

In the above diagram, input x is applied at time n which generates output y for the same time n.

$$x[n-k] \longrightarrow \boxed{\text{LTI System}} \longrightarrow y[n-k] \tag{11.5}$$

Input x applied at time $n - k$ still generates the same output y at time $n - k$.

The transformation of inputs to outputs by LTI systems is described by the convolution operation. The output is the convolution of the input with the unit pulse response of the system. Convolution can be expressed in matrix–vector form by taking either the input signal or the unit pulse response signal of the system as a vector with the other used to construct a convolution matrix.

11.2.2 Sales Productivity Model

As discussed in the previous section, we can model sales productivity as an LTI system taking counts of sellers as input and revenue as output. The unit pulse response of the system is then the productivity profile. Let us first fix the notation so that the input signal $x[k]$ represents the number of sellers that were hired k months ago. Similarly, let $h[k]$ represent the unit pulse response of the system and also the productivity profile so that $h[-k]$ is the productivity of a seller that was hired k months ago. The output y is the total revenue produced by all sellers in the salesforce. Assuming that the system is causal with a finite impulse response of length m, the convolution leading to the output is.

$$y[n] = \sum_{k=0}^{m} x[n-k]\, h[k] \tag{11.6}$$

Here, n represents different months in which we observe the system. We can also write this as the matrix–vector equation

$$\mathbf{y} = \mathbf{X}\mathbf{h} \tag{11.7}$$

where $\mathbf{X}$ is the convolution matrix formed from x.

Since we are dealing with four different classes of new sellers, we in fact have four different productivity profiles h and four different counts of new sellers, but that can be represented similarly to (11.2):

$$\mathbf{y} = [\mathbf{X}_1 \quad \mathbf{X}_2 \quad \mathbf{X}_3 \quad \mathbf{X}_4] \begin{bmatrix} \mathbf{h}_1 \\ \mathbf{h}_2 \\ \mathbf{h}_3 \\ \mathbf{h}_4 \end{bmatrix} \tag{11.8}$$

where the $\mathbf{X}_i$, $i = 1, \ldots, 4$, matrices are again convolution matrices. In the sequel, to keep the notation simple, we refer to (11.2) as the representation of the system.

We measure the revenues for the entire salesforce y and we also measure the head counts of new sellers each month. The productivity profiles are to be estimated.

11.3 Estimating Productivity Profiles Using Quadratic Programming

Since the revenues $\mathbf{y}$ and the head counts $\mathbf{X}$ are measured, our task is to estimate the productivity profile $\mathbf{h}$. We take a least-squares approach by minimising the ℓ_2 norm between the measured revenue and the estimated output of the LTI system:

$$\|\mathbf{y} - \mathbf{X}\mathbf{h}\|_2^2. \tag{11.9}$$

This can be written as a quadratic program with the following objective function to find the productivity profile solution,

$$\min_{\mathbf{h}} \frac{1}{2}\mathbf{h}^T\mathbf{X}^T\mathbf{X}\mathbf{h} - \mathbf{h}^T\mathbf{X}^T\mathbf{y}. \tag{11.10}$$

where $\mathbf{h}$ is the sales productivity vector.

One of the simplest forms of regularisers is given by the sum of squares of the unit pulse response vector elements: $\frac{1}{2}\mathbf{h}^T\mathbf{h}$ [11]. With the least-squares objective and this regulariser, the solution is the Moore–Penrose pseudoinverse

$$\mathbf{h} = \left(\mathbf{X}^T\mathbf{X}\right)^{-1}\mathbf{X}^T\mathbf{y}. \tag{11.11}$$

However, in our sales analytics problem, we have further business constraints to motivate additional regularisation. The constrained quadratic program that we use to find $\mathbf{h}$ is the following:

$$\begin{aligned} \min_{\mathbf{h}} \quad & \tfrac{1}{2}\mathbf{h}^T\mathbf{X}^T\mathbf{X}\mathbf{h} - \mathbf{h}^T\mathbf{X}^T\mathbf{y} \\ \text{s.t.} \quad & \mathbf{h}_{lb} \leq \mathbf{h} \leq \mathbf{h}_{ub} \\ & \mathbf{A}_{in}\mathbf{h} \leq \mathbf{A}_{ub}. \end{aligned} \tag{11.12}$$

We have four constraints motivated by the business application. First, productivity profiles are non-negative because sellers cannot produce negative revenue; if they sell nothing, their productivity is zero. Second, we assume that the productivity profiles are monotonically non-decreasing because over time, the sellers gain experience, knowledge, and contacts, so their productivity does not get worse over time. Third, we assume that the productivity does not rapidly jump from time step to time step, so we constrain increases in the productivity profile to not exceed a certain value. Last, we impose a maximum productivity L, which is a saturation level and the productivity of sellers that have been with the enterprise for more than m months.

Specifically, for the non-negativity constraint, we set $\mathbf{h}_{lb}$ to be a length m vector of all zeroes. For the saturation constraint, we set $\mathbf{h}_{ub}$ to be a length m vector with all entries equal to a parameter L. The other constraints we include are encoded through the matrix $\mathbf{A}_{in}$ and the vector $\mathbf{A}_{ub}$. It is straightforward to encode that successive values of $\mathbf{h}$ be monotonically non-decreasing and also that successive values of $\mathbf{h}$ not increase by more than another parameter value that we set. $\mathbf{A}_{in}$ has blocks that are Toeplitz matrices compose of positive and negative ones. Half of $\mathbf{A}_{ub}$ is all zeroes and the other half is equal to the parameter value indicating the increase limit per time in $\mathbf{h}$.

Having derived the optimisation problem (11.7), we find the optimised value of the sales productivity profiles $\mathbf{h}$ using the `qp` function of Octave. To construct the $\mathbf{X}$ matrix, we utilise the `convmtx` function of Octave. To begin with, we adopted the most simplified solution by assuming that there are no constraints in the problem statement. No constraint scenario is realised by considering empty vectors and matrices on all parameters in the above function.

First, we calculated the matrix with minimal constraints as below:

$\mathbf{h} = \mathtt{qp}(\mathtt{zeros}(m,1),\ \mathbf{X}^T\mathbf{X},\ -\mathbf{X}^T\mathbf{y}^T)$.

To get non-negative values we define a few more variables:

$\mathbf{h} = \mathtt{qp}(\mathtt{zeros}(m,1),\ \mathbf{X}^T\mathbf{X},\ -\mathbf{X}^T\mathbf{y}^T,\ [],\ [],\ \mathtt{zeros}(m,1),\ [])$.

Finally, the full constrained quadratic program is solved as

$\mathbf{h} = \mathtt{qp}(\mathtt{zeros}(m,1),\ \mathbf{X}^T\mathbf{X},\ -\mathbf{X}^T\mathbf{y}^T,\ [],\ [],\ \mathtt{zeros}(m,1),\ L{*}\mathtt{ones}(m,1),\ [],\ \mathbf{A}_{in},\ \mathbf{A}_{\mathrm{ub}})$,

where we use the Octave function `toeplitz` to construct $\mathbf{A}_{in}$.

11.4 Empirical Results

We collected data from one of the business units of International Business Machines (IBM). The data corresponds to one of the sales organisations from a few recent years. Hiring information comes from human resources (HR), and revenue details come from the finance section.

Figure 11.1 shows the total head count of sellers in the organisation during the period of examination. Figure 11.2 shows head count dynamics for the different classes of new sellers. The values shown indicate the number of sellers of each category that joined the organisation during the month. Figure 11.3 displays the solutions for the productivity profiles that we obtained, i.e., it plots time duration in months and revenue generated by the sales person for 4 different cases. Figure 11.4 shows the actual total revenue of the organisation and the revenue reconstructed using the actual head counts and the learned productivity profiles $\mathbf{h}$.

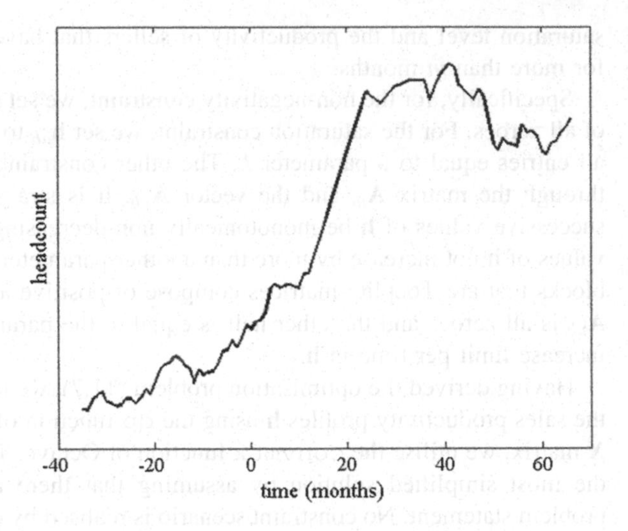

Fig. 11.1 Total head count data of salespeople during the period of examination

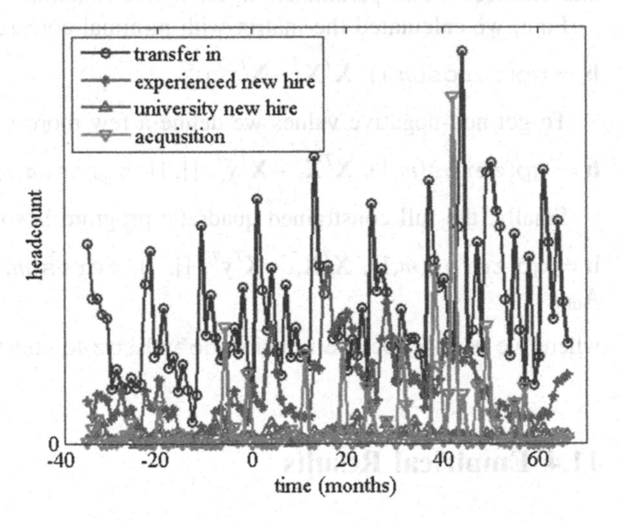

Fig. 11.2 Head count dynamics for different classes of sellers during the period of examination

Case 1 shown in blue colour in Fig. 11.3 indicates the plot for a salesperson who has been transferred from one unit to the other. As shown in the figure, this salesperson is expected to bring in good revenue over a short period of time and remains constant thereafter. Transfers in have 1 month of no productivity and do not reach the steady state productivity L.

Case 2 shown in red colour indicates the performance of an experienced salesperson who joins the organisation. The productivity of this category of salespeople is instantly high as they do not take much time to settle and perform. The vertical scale has values labelled between 0 and 1. This is so because we have normalised the value from 0 to 1 to maintain confidentiality of the sales data of the organisation.

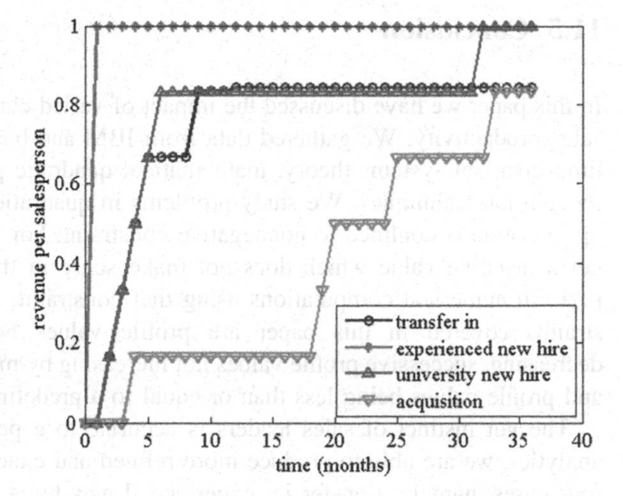

Fig. 11.3 Productivity profiles for four different cases estimated from head count data and revenue generated by sellers for given time duration

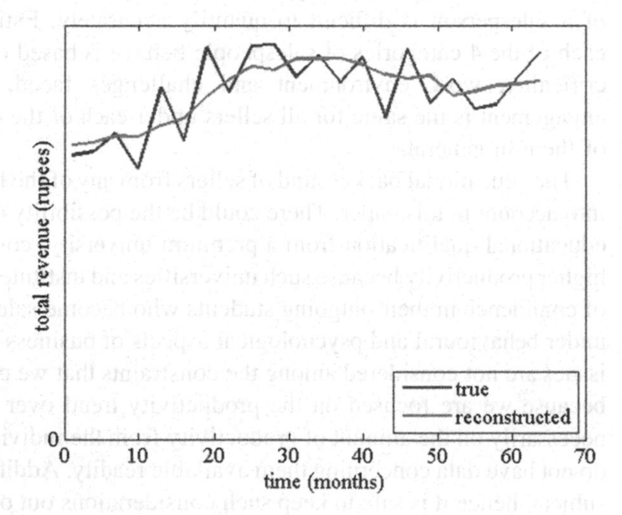

Fig. 11.4 Revenue fit using revenue reconstructed from the actual head counts and the estimated productivity profiles

Case 3 relates to new hires from colleges and universities who do not already have prior work experience. At this point all that they possess is theoretical knowledge requiring practical exposure. They, like the transfers in have 1 month of no productivity, but do reach the saturation value L. Their progression is similar to transfers in.

Case 4 indicates the nature of involvement of the salesforce as a result of acquisition. In this competitive world, acquisitions are not rare and the implications of this on an organisation's sales productivity are extremely important to understand. Accurate forecasting of sales productivity has a direct impact on organisations' income and revenue generation. Those that join due to acquisitions have several months of no productivity at the beginning and ramp up very slowly. They do not reach the steady-state productivity L like the transfers in.

11.5 Conclusion

In this paper we have discussed the impact of varied classes of salespeople on the sales productivity. We gathered data from IBM and based our analysis on linear time-invariant systems theory, mathematical quadratic programming, and pattern recognition techniques. We study problems in quadratic programming where the optimisation is confined to nonnegative constraint. For these problems, we might get a negative value which does not make sense in the sales world and so we perform numerical computations using that constraint. The remaining three constraints covered in this paper are profile values being monotonically non-decreasing, successive profile values not increasing by more than a certain amount, and profile values being less than or equal to a predefined upper limit value.

The gut instinct of sales leaders is accurate to a point, but through business analytics, we are able to produce more refined and exact productivity profiles for four cases, namely, transfer in, experienced new hires, university new hires and those sellers that join the organisation as a result of acquisition. The productivity of a salesperson is difficult to quantify accurately. Estimating the trend of how each of the 4 categories of salespeople behave is based on the assumption that the criticality, work environment and challenges faced, and complexity of the engagement is the same for all sellers under each of the categories and also for all of them in general.

The educational background of sellers from any of the four categories is not taken into account in this paper. There could be the possibility of the common notion that educational qualification from a premium university, college or institute results in higher productivity because such universities and institutes may infuse great amount of confidence in their outgoing students who become salespeople. These issues fall under behavioural and psychological aspects of business analytics. However, these issues are not considered among the constraints that we chose for calculations here because we are focused on the productivity trend over a period of time and not necessarily on the amount of productivity from the individual salespeople; also, we do not have data concerning them available readily. Additionally, this is a debatable subject; hence it is safe to keep such considerations out of the scope of this paper.

References

1. Sekhon HS, Varshney RK, Kumar P, Singh JM, Prasad R (eds) (2005) Glimpses of systems theory and novel applications: felicitation volume in honour of Prof. R. K. Varshney. Navin Press, Aligarh
2. Davenport TH, Harris J, Shapiro J (2010) Competing on talent analytics. Harvard Bus Rev 88:52–58
3. Varshney KR, Mojsilović A (2011) Business analytics based on financial time series. IEEE Signal Process Mag 28(5):83–93

4. Baier M, Carballo JE, Chang AJ, Lu Y, Mojsilović A, Richard MJ, Singh M, Squillante MS, Varshney KR (2012) Sales-force performance analytics and optimisation. IBM J Res Dev 56(6):8.1–8.10
5. Singh M, Bhattacharjya D, Deleris L, Katz-Rogozhnikov D, Squillante M, Ray B, Mojsilović A, Barrera C, Kakrania D, Richard J, Saha A, Fu J (2011) The growth and performance diagnostics initiative: a multidimensional framework for sales performance analysis and management. Serv Sci 3(1):82–98
6. Brooks FP Jr (1995) The mythical man-month: essays on software engineering. Addison-Wesley, Boston
7. Varshney KR, Singh M, Sharma M, Mojsilović A (2011) Estimating post-event seller productivity profiles in dynamic sales organisations. In: Proceedings of the IEEE international conferrence on data mining workshops, Canada, pp 1191–1198
8. Willsky AS (1982) Some solutions, some problems, and some questions. IEEE Contol Syst Mag 2(3):4–16
9. Oppenheim AV, Willsky AS, Nawab SH (1997) Signals and Systems. Prentice Hall, Upper Saddle River
10. Bishop CM (2006) Pattern recognition and machine learning. Springer, New York
11. Theodoridis S, Koutroumbas K (2009) Pattern recognition. Academic Press, Burlington
12. Saul LK, Sha F, Lee DD (2003) Statistical signal processing with nonnegativity constraints. In: Proceedings of the European conference speech communication technology, Switzerland, pp 1001–1004
13. Gong T, Hartmann N, Kohane IS, Brinkmann V, Staedtler F, Letzkus M, Bongiovanni S, Szustakowski JD (2011) Optimal deconvolution of transcriptional profiling data using quadratic programming with application to complex clinical blood samples. PLoS ONE 6(11):1–11
14. Lu Y, Jiang T, Zang Y (2005) Single-trial variable model for event-related fMRI data analysis. IEEE Trans Med Imag 24(2):236–245
15. Moorthy KS (1993) Theoretical modeling in marketing. J Mark 57(2):92–106

Chapter 12
An Integrated Community Economic System: Gateway to a Sustainable New World Order

B. Aashiq and Prem Sewak Sudhish

Abstract The consequences of recent stagnation in economy are being experienced globally, with different industry and business segments, and their employees, bearing the brunt of this recession by varying degrees. Several employees have lost their livelihoods—a penalty of the slump they may not even have been directly responsible for. The present economic crisis calls for a systems approach that is not only sustainable but also absorbs the transients caused by global economic events, ensuring that the basic human necessities of all individuals are catered to at all times. The concept of a holistic community living is explored and is compared with the prevalent self-centered culture that breeds volatility and insecurity. It has been shown that integrating individuals together into local community systems is not only economically viable and stable but also leads to other benefits that ensure an improved lifestyle in a supportive environment that is free from petty strife. The practicability of the proposed systems model is supported by examples and proved through a case study on Dayalbagh, verily a Utopia on earth.

Keywords Economy · Organizational behavior · Community living · Sustainability

B. Aashiq
Dayalbagh Educational Institute, Dayalbagh, Agra, India
e-mail: aashiqb2291@gmail.com

P. S. Sudhish (✉)
Department of Physics and Computer Science, Dayalbagh Educational Institute, Dayalbagh, Agra, India
e-mail: pss@alumni.stanford.edu

R. Malathi and J. Krishnan (eds.), *Recent Advancements in System Modelling Applications*, Lecture Notes in Electrical Engineering 188,
DOI: 10.1007/978-81-322-1035-1_12, © Springer India 2013

12.1 Introduction

An organizational system whether small or big is considered successful only with the support and sustenance of its employees, who form the life-blood of the system. It is, therefore, the primary responsibility of any organization to recharge this blood frequently to achieve greater heights. Employee satisfaction, employee loyalty, employee involvement, employee motivation, employee flexibility and employees' qualitative productivity are some of the vital requirements for any organization.

There is a significant challenge that organizations are often required to address—ensuring that employees not only continue to exhibit but also to nurture and mature the desirable attributes mentioned above. A question that often arises is whether cash incentives are an adequate solution towards employee compensation or if there are other mechanisms that present more promising outcomes.

While the general perception is that cash incentives are the most sought after, certain studies have shown that cash is a poor motivator due to its lack of trophy value and shelf life, higher tax liability, not being a constant visual stimulus and it being redeemable. A study of 6,500 employees at Ralph Head Association Ltd., USA indicated that the corporation could obtain greater motivation and improved performance from $600 merchandise incentive awards, as opposed to $1000 in cash or debit card awards. The merchandise group outperformed the cash group counterparts by 70–80 %. Another lacuna with the cash incentive program is that it is difficult to do away with once it has been introduced. Especially in low pay environments, the end of an incentive program is generally perceived as the compensation benefit being reduced. In fact, no reward will be able to inspire employees enough if their basic needs are not met or if they get little satisfaction from progress at work.

In this paper, we explore commune systems based on the systems way of life and compare them with traditional organizations. In communes, the employees live together to share their common interests, property, possessions, resources, work and income, ensuring that the basic needs of all individuals are satisfied and this has resulted in enormously enhanced employee satisfaction.

12.2 System of Communes

A commune is a community established by people who decide to live together voluntarily to share their common needs and requirements. The primary reasons or objectives for the formation of communes may vary from one region to another depending upon a number of factors. Communes could be formed due to the matching of political ideologies, common beliefs or values, for the purpose of employment, economical living, etc.

12.2.1 Principles of Communes

Communes are often established along some principles and ideologies that govern the members. A study of the various communes shows that almost all of them have three core principles in common—equality, cooperation and harmony with nature. The communes are generally egalitarian, rejecting the idea of hierarchy or graduations of social status as being necessary to social order; they are organized at a human scale, with groups of manageable sizes as the scale of society is very large and are anti–bureaucratic.

12.2.2 Popular Commune Systems

12.2.2.1 Acorn Community

Acorn Community, established in 1993, is located in Virginia, USA and is a member of the Federation of Egalitarian communities. It was formed when Twin Oaks Community was oversized. Having survived financial crunch and interpersonal ideologies mismatch in its earlier days, its current business is on a rise. The Acorn community functions by running the mail order seed business called the Southern Exposure Seed Exchange that includes open pollination, heirloom and traditional varieties. It comprises of 75 acres of land with 24 members and 3 interns. The income generated is shared proportionately amongst the members. The community also holds regular group meetings and members have a work quota of 42 hours a week with all kinds of work being given equal importance.

Although structured in areas such as membership and work, policies are kept to a minimum, preferring freedom to prevail. The commune has a strong sense of environmental awareness and hence works towards its protection. The procedures at the commune afford due importance to the divergent cultures, attitudes and requirements. The work done by the members ranges from office and accounting to childcare, farming, cooking, cleaning and extends to preparations for communal parties etc. A regular member is usually expected to have more free time. Four weeks are allotted in a year for holidays and an additional holiday can be accumulated by working overtime. The labor system is highly flexible as the members simply do what they believe in.

12.2.2.2 East Wind Community

The mission of the East Wind Community is to achieve peace, social justice, economic and social democracy. Located in Missouri Ozarks, USA and founded in 1973, it is a democratic and secular community where all the communes' assets are held by the members in common. It runs several businesses on its 1,045 acres of land. All the decisions are made democratically and all positions in the commune are elected. Food, shelter, clothing, medical care, education and monthly stipend are provided to each member. The sources of income for this community include East Wind Nutbutters, a nut butter business that grosses $500,000 annually. The community also produces roasted peanuts, peanut butter, cashew butter, almond butter and tahini. The members are supposed to work a certain number of hours and income from businesses is distributed amongst the community.

A membership team formed within the commune selects the prospective members who are then invited for a 3 week visitor period, at the end of which they might become provisional members. Following a 1 year provisional membership, full members have the opportunity to call for a vote on any new member who is either failing to fulfill his responsibility or is too disruptive. There is an associate status for people who would not want to make full time commitment.

Community conferences and labor exchanges between the communities are organized and the members set aside a small amount from the income for individual or general medical purposes. Common spaces have been created and activities are organized for recreational purposes. The people live in dorm style buildings with central dining hall, laundry and shower house. Some houses have also been built for visitors. East Wind is also a part of the Federation of Egalitarian communities, which includes several other communes.

12.2.2.3 Amana Colonies

Amana colonies have the striking characteristic that for 80 years they maintained a full-fledged self-sufficient local economy, depending very little on the outside world and the industrializing US economy. The level of physical comfort, housing, possessions, education and socio–cultural amenities were comparable to an average middle class American office worker, factory worker or tradesman of the time. Comprising of seven villages, Amana colonies were a group of settlements in Iowa, USA. Calling themselves the Ebenezer Society or the community of true inspirations, they changed their place to Iowa city from New York state near Buffalo in 1856. The commune was able to achieve a flourishing lifestyle and independence with the aid of specialized handicrafts and farming occupations which were brought from Germany. The master craftsmen passed on their skills, techniques and knowledge of artisans, iron and copper smiths, wood wrights, weavers, shoemakers, cheese-makers etc. to the next generation and this continued from one generation to another. In order to lead a sustainable pedestrian community life, they used hand, horse, wind and water power. Amanians made their own furniture, clothes, blankets, dishes, utensils, children's toys, tools, candles, candy etc. till 1935. Their products are known for their beauty, durability and individuality. Amana is a major tourist attraction spot today, mainly for its restaurants and craft–shops. The Amana colonies have been listed as a national historic landmark of USA since 1965.

12.2.3 Other Commune Traditions

12.2.3.1 Germany

Germany has a long history of intentional communes since centuries. The communes here are usually formed on the basis of political ideologies. Living and working together, having a communal economy, that is, common finances and common property including land, buildings, means of production, having communal decision making—usually by consensus, trying to reduce hierarchy and hierarchical structures, having communalization of housework, childcare and other communal tasks, ensuring equality between men and women and maintaining low ecological footprints through sharing and saving resources are some of the common characteristics of communes in Germany.

12.2.3.2 Israel

Israel is popular for maintaining urban communes. The *Kibbutzim* in Israel are perfect examples of officially organized communes. There are several urban communes called urban Kibbutzim. They are smaller and more anarchist. They emphasize on social change, education and local involvement in the cities they live.

12.2.3.3 Venezuela

The communes in Venezuela have their own productive gardens that grow their own vegetables for self supply. The communes also take administrative and financial decisions independently, including those for budgeting and the use of funding. Venezuela currently has more than 200 communes.

Even though communal living was at its peak in the 1960s and the 1970s, it has made a comeback today. More and more people are becoming dissatisfied with isolated, energy-intensive separate households, and want to live more lightly on the earth. Living in communities has numerous advantages both for individuals and the wider environment.

12.3 Case Study: Dayalbagh—An Indian Utopia

The Dayalbagh community located on the outskirts of Agra was established about a century ago and is one of the oldest, successful and innovative independent intentional societies in the world. The employees at this organization have voluntarily adopted a scheme that greatly subsidizes the cost of healthy living and this is preferred over cash incentives.

For a society to excel and sustain, the economic and social requirements of its members should be well taken care of along with the basic necessities like food, clothing, shelter, employment and health facilities. Accordingly, the members of the community are provided with housing, water, electricity, food grains and dairy products at highly subsidized rates. The community owns agricultural farms with a water canal for irrigation, dairy, pristine houses built in natural surroundings. Essential services and support departments, such as the water and electricity supply department cater to the needs of the members.

The colony is laid out in an open garden setting. The land where the colony was established once consisted of sand dunes. For more than 60 years residents of the colony—men and women, young and old—have worked with quiet dedication in a vast program for reclamation of land, resulting in a lush green 1,200—acre (4.9 km^2) farm where food-grains, oil-seeds, fodder and vegetables are grown.

Nobody owns any property in Dayalbagh individually. The land, the houses and institutions all belong to the community as a whole. People also live and work as a community. For example, the residents share various responsibilities like cleaning up the colony and arranging night security. The colony has its own water supply, electricity distribution and civic services. The colony's dairy provides for the requirements of milk, and a community kitchen supplies food free to pilgrims. The residents can also obtain meals on nominal charge from there and free themselves from household chores.

The community also maintains a hospital with a maternity ward. Facilities exist for ophthalmic and dental treatment, ultrasound, ECG and pathological testing. All consultations and treatment are free for everyone. The majority of patients are from outside the colony. There are also homeopathic and ayurvedic dispensaries.

Small-scale industries known as the Model Industries were established in 1916 to provide employment and a source of livelihood to the persons residing in the colony and they did some pioneering work in the country. The industries have now been decentralized and cottage scale production of goods of daily necessity is taking place in units set up by members across the globe in colonies set up as miniature Dayalbagh.

The members of the community, irrespective of their qualifications or primary occupation, work in the agricultural fields—contributing actively to the local production of food grains and also instilling the virtue of dignity of labor and a sense of equality.

The children are provided with free milk (from the community dairy) and free school uniform. To inculcate the qualities of brotherhood, team spirit, cooperation and to make them aware of the varied national and international culture, heritage, languages and to keep them connected to their roots various centers such the School of Art and Culture, School of Languages, Children Recreation Centre and Training Centre for Music conduct regular classes for students.

The community runs a nursery and play center, primary schools, secondary schools for boys and girls, day boarding school and intermediate colleges that are also open to other children residing in the neighborhood. The community also has a secular University that is well known for its system of education in undergraduate and postgraduate programs in all major disciplines. The system of education aims to bring about physical, intellectual, emotional and ethical integration of an individual. Quality education is provided at nominal fees. Large open grounds are also available for the children and youth to play and regenerate their energy.

For financial security and women's empowerment, the community also runs two banks—one of which is solely run for and by the female members of the community.

The communication, sanitation and transportation facilities are all environment friendly. The workplace is close to the employees' residence; hence the popular mode of transport is the cheap, pollution free and eco–friendly bicycle. The wastage disposal system is fast and efficient.

12.3.1 Socio-Economic Aspects

The literacy of all members and their families is complete. The attainment of a broad-based education has made the thought process of the members modern and flexible. With most household activities in the community, including cooking being organized centrally, the women too, are able to participate in various activities like dress designing, interior decoration, tailoring, working in community kitchen, working in fields, teaching in schools etc. Thus, they form an integral and equal part of the commune. All the members of the community are involved in some primary occupation and no one sits idle. The participants in this system all have a definite objective and they work hard towards the achievement of these goals.

Marriages in this community are extremely simple and in groups with an exemplary sense of equality. The members consciously work towards avoiding ostentatious expenditure and fanfare. The community voluntarily decides on the ceiling on the expenditure to be incurred on these occasions, as also on the number of guests being invited. Dowry and other gifts that pose themselves as a major problem in the Indian society are completely banned.

The members believe in harmonizing with the natural economy and adhere to the principle of "Waste Nothing" that includes food, clothing, thoughts, time and energy. The green lifestyle with an extremely low carbon footprint is ensured by the community's choice for locally grown vegetarian food and a voluntary upper limit on electricity consumption by individual households.

The community has no discrimination on the basis of caste, creed, color, race, earnings, ethnicity, national origin or religion. Ceremonies or rites are virtually non-existent, as is the concept of economic classes. An administrator and an attendant have comparable lifestyles and are treated alike. The people live in harmony sharing their needs and requirements.

12.3.2 Benefits of an Integrated Community Economy

The employees are motivated to work harder every day and strive for their personal optimal performance as they are always free from the worries of their basic needs or their position in the socio-economic infrastructure. The employees are left with more time to work for the organization as their parallel responsibilities are being shared by the community. The loyalty and involvement of the members is high and intellectual freedom entitles everyone to express their thoughts and be involved in decision-making. Employee attrition is non-existent as all their basic requirements along with job security are ensured. The rate of absenteeism is extremely low with a high punctuality as the residences of all the employees are in close proximity of their workplace. A positive and healthy work environment is created by the employees not only sharing their office space but also their needs, wants, demands and the freedom to design ways to attain them. The quality of work also gets a positive influence with the existence of personal touch and belongingness. Not only does the organization achieve employee satisfaction but is also left with adequate capital for expansion.

The integrated community economy also extends the benefit of mass-scale. The group economy not only affords a reduction in cost per unit but also an enhanced bargaining power. As the children grow up in this environment, they are closely aware of the benefits this system is reaping and hence are usually desirous of being a part of this organization. Thus, the burden or expenditure towards hunting for qualified, motivated and loyal employees is a complete saving. The consequences of employee dissatisfaction such as strikes, lockouts, shut downs, damage to property, deliberate work slowdown etc. that are common in small and large organizations nationwide are unheard of in this community.

12.4 Conclusion

Unlike the ancient eastern philosophies that preach "unworldliness", that is, leaving worldly pleasures and renouncing the world in the pursuit of truth or the western materialistic view of "worldliness", that is, seeking materialistic pleasure and fulfillment through self-centered activities, the integrated systems approach of community living instills a sense of "better worldliness". This novel and admirable concept emphasizes on a lifestyle in which the members fulfill their responsibilities cooperatively to the best of their abilities, ensuring not only the fulfillment of their genuine needs but also catering to the needs of their fellow beings in a spirit of service.

During the past few decades, community structures that replicate the Dayalbagh model have been established successfully in various parts of the country, not being limited by the geography or the social structure of the surroundings. These establishments are located far and wide, including communities from deep tribal hinterlands of Rajaborari in Madhya Pradesh and interior rural expanses of Murar in Bihar and Melathiruvengadanathapuram in Tamil Nadu to modern urban centers such as New Delhi and Hyderabad. Efforts are also underway to establish similar communes overseas.

Dayalbagh stands out as a model community for other organizations to follow, not only for peace and prosperity but also for evolving a new world order based on the golden principles of equality, selfless service, dignity of labor and for bringing about harmony through "Fatherhood of God and Brotherhood of man".

References

1. Angus C, Converse PE, Willard Rodgers L (1976) The quality of American life. Russell Sage Foundation, New York
2. Bowling A (1991) Measuring health. Open University Press, Buckingham
3. Robinson DN (1999) Aristotle's psychology. Columbia University Press, New York
4. Russell B (1930) The conquest of happiness. George Allen and Unwin, New York
5. Davis K, Golden HH (1954) Urbanization and development in pre-industrial areas. Econ Dev Cult Change 3(1):6–29
6. Shin DC, Johnson DM (1978) Avowed happiness as an overall assessment of the quality of life. Soc Indic Res 5(1–4):475–492
7. Acorn Community, Egalitarian Intentional Community in Virginia, Feminist Anarchist Communist. http://www.acorncommunity.org/
8. East Wind Community. http://www.eastwind.org/
9. Amana Colonies in Iowa: The Handcrafted Escape. http://amanacolonies.com/
10. "Commune", Wikipedia, The Free Encyclopedia. http://en.wikipedia.org/w/index.php?title=Commune&oldid=494879903
11. Dayalbagh http://www.dayalbagh.org.in/

Chapter 13
Evaluation of New COPWM Techniques for Three Phase Seven Level Diode Clamped Z-Source Inverter

V. Arun, B. Shanthi and S. P. Natarajan

Abstract This paper presents the comparison of the different Carrier Overlapping Pulse Width Modulation (COPWM) techniques for three phase seven level Z-source diode clamped inverter. Due to switch combination redundancies, there are certain degrees of freedom to generate the multi level AC output voltage. This work presents the use of CFD combination. The Z-source based DCMLI is triggered by the different COPWM techniques having sinusoidal reference and triangular carriers. It is observed that the COPWM-3 technique provides reduced harmonics at its output voltage. The effectiveness of the PWM techniques developed using CFD are demonstrated by simulation using MATLAB/SIMULINK.

Keywords CFD · CO · DCMLI · FF · CF · PWM · THD · Vrms

V. Arun (✉)
Department of EEE, Arunai Engineering College, Thiruvannamalai, India
e-mail: varunpse@yahoo.com

B. Shanthi
Centralised Instrumentation and Service Laboratory, Annamalai University, Chidambaram, India
e-mail: au_shan@yahoo.com

S. P. Natarajan
Department of EIE, Annamalai University, Chidambaram, Tamilnadu, India
e-mail: spn_annamalai@rediffmail.com

R. Malathi and J. Krishnan (eds.), *Recent Advancements in System Modelling Applications*, Lecture Notes in Electrical Engineering 188,
DOI: 10.1007/978-81-322-1035-1_13, © Springer India 2013

13.1 Introduction

Multilevel inverters are a viable solution to increase the power with a relatively low stress on the components and with simple control systems. Multilevel inverter presents several other advantages. Multilevel inverter generates better output waveforms with a lower dv/dt than the standard inverter. Then, multilevel inverter can increase the power quality due to the great number of levels of the output voltage: in this way, the AC side filter can be reduced, decreasing its costs and losses. Furthermore multilevel inverter can operate with a lower switching frequency than conventional inverter, so the electromagnetic emissions they generate are weaker, making less severe to comply with the standards. Multilevel inverter can be directly connected to high voltage sources without using transformers; this means a reduction of implementation and costs. Gajanayake et al. [1] developed the closed-loop controller for a Z-source inverter. Peng et al. [2] developed Z-source inverter system and control for general-purpose motor drives. Kanimozhi and Senthil Kumar [3] proposed cascaded Z-source multilevel inverters for uninterruptible power supply application. Kandasamy and Manojchakkaravarthi [4] introduced novel cascaded inverter with Z-source. Bakar et al. [5] described the various PWM techniques. Yousuf et al. [6] introduced multi carrier PWM technique for five level inverter. Loh et al. [7] introduced three-level Z-source NPC inverter. Loh et al. [8] proposed three level NPC inverter with two unique Z-source impedance networks to boost the inverter three-level output waveform. Loh et al. [9] developed two three level cascaded Z-source inverter with step up and step down output voltage levels. Chandrudu et al. [10] developed Z-source inverter for induction motor drive. Ali and Kamaraj [11] introduced double carrier pulse width modulation technique. Rajakaruna and Jayawickrama [12] developed all possible steady state analysis and also design the symmetrical impedance network. Sasikumar et al. [13] described the simulation and harmonic analysis of impedance source inverter (ZSI) fed stand alone wind energy conversion system. Dehghan et al. [14] proposed Z-Source inverter with two ac outputs and two DC inputs. Shanthi and Natarajan [15] proposed carrier overlapping PWM methods for five level flying capacitor inverter. Uthayakumar et al. [16] proposed carrier overlapping PWM techniques for seven level asymmetrical multilevel inverter. This paper presents a three phase diode clamped Z-source inverter topology for investigation with COPWM-1, COPWM-2 and COPWM-3 using sinusoidal reference switching techniques. Simulations were performed using MATLAB/SIMULINK. Harmonics analysis and evaluation of different performance measures for various modulation indices have been carried out and presented.

13.2 Z-Source Seven Level Inverter

Figure 13.1 shows the two-port network that consists of an inductors (L1, L2) and capacitors (C1, C2) and connected in X shape is employed to provide an impedance source (Z-source) coupling the inverter to the dc source. The Z-source

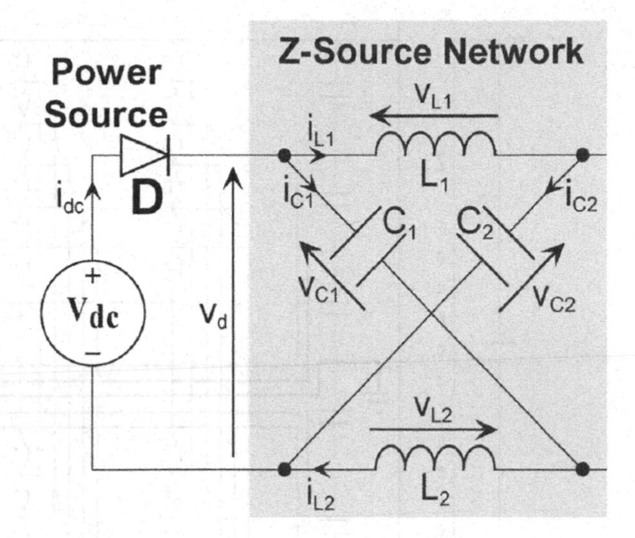

Fig. 13.1 Impedance network

multilevel inverter utilizes shoot-through state to boost the input DC voltage of inverter switches when both switches in the same phase leg are on. The impedance Source inverters are having lower costs, reliable, less complexity and higher efficiency.

Figure 13.2 shows the seven level Z-source diode clamped inverter, Which delivers the staircase output voltage using several levels of DC voltages developed by input DC capacitors. If m is the number of level, then the number of capacitors required on the DC bus are (m−1), the number of power electronic switches per phase are 2(m−1) and the number of diodes per phase are 2(m−2). This design formula is most common for all the neutral clamped multilevel inverters. The DC bus voltage is split into seven levels using six capacitors. The voltage across each capacitor is $V_{dc}/6$ and the voltage stress across each switch is limited to one capacitor voltage through clamping diodes. The midpoint of the six capacitors 'n' can be defined as the neutral point. As the number of levels increase the harmonic distortion decreases and efficiency of the inverter increases because of the reduced switching losses. The number of levels in multilevel inverters is limited because of the large number of clamping diodes required. The reverse recovery of these diodes is especially with multicarrier PWM techniques in a high voltage application is a major design challenge.

13.3 Multicarrier Overlapping PWM Technique

Multicarrier overlapping PWM technique is the widely adopted modulation technique for MLI. It is similar to that of the sinusoidal PWM technique except for the fact that several carriers are used. Multicarrier PWM is one in which several triangular carrier signals are compared with one sinusoidal modulating signal.

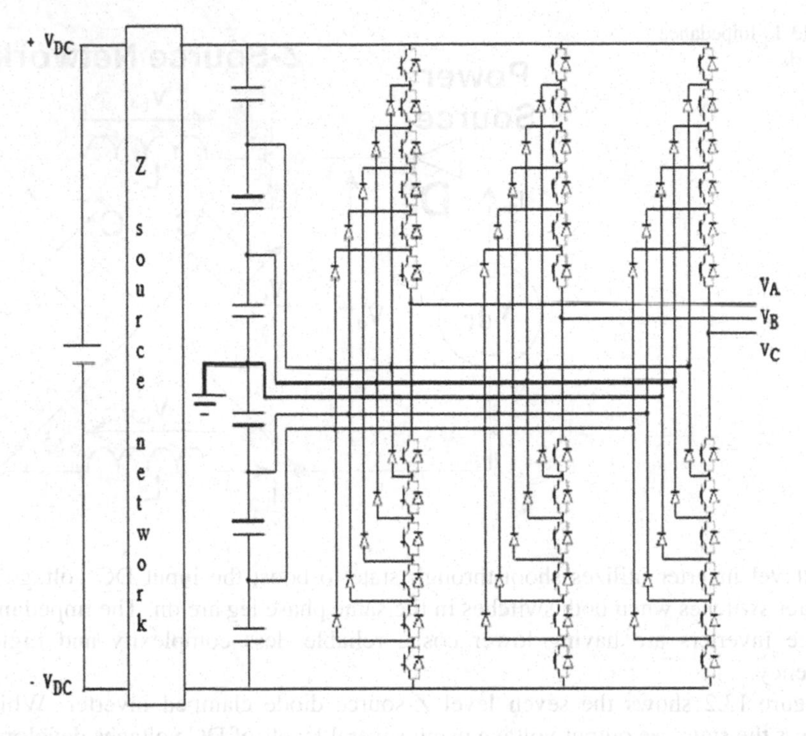

Fig. 13.2 Seven level Z-source diode clamped multilevel inverter

(m−1) carriers are required to produce m-level output. All carriers are having same frequency fc and same peak-to-peak amplitude AC are disposed such that the bands they occupy overlap each other; the overlapping vertical distance between each carrier is $A_c/2$. As far as the particular carrier signals are concerned, there are multiple CFD including Frequency, amplitude, phase of each carrier and offsets between carriers. The reference wave of multilevel carrier based PWM method can be sinusoidal. As far as the particular reference wave is concerned there is also multiple CFD including frequency, amplitude, phase angle of the reference wave and as in three phase circuits, the injected zero sequence signal to the reference wave. Therefore multilevel carrier based PWM methods can offer multiple CFD. These CFD combinations combined with the basic topology of multilevel inverters can produce many multilevel carrier based PWM methods. This paper focuses three new COPWM methods that utilize the CFD of vertical offsets among carriers. They are: COPWM-1, COPWM-2 and COPWM-3. The amplitude modulation index ma and the frequency ratio mf are defined in the carrier overlapping method as follows:

$$m_a = A_m / 2\ A_c. \tag{13.1}$$

$$m_f = f_c / f_m. \tag{13.2}$$

13.3.1 COPWM-1 Technique

The vertical offset of carriers for seven level inverter with COPWM-1 method is illustrated in Fig. 13.3. The carriers are divided into two groups according to the positive/negative average levels. The three carriers are overlapped in positive group and three are overlapped in the negative group and the reference sine wave is placed at the middle of the six carriers. In this technique all the carrier waveforms are in phase.

13.3.2 COPWM-2 Technique

Carriers for seven level inverter with COPWM-2 technique are shown in Fig. 13.4. It can be seen that they are divided equally into two groups according to the positive/negative average levels. In this technique the two groups are opposite in phase with each other while keeping in phase within the group [15].

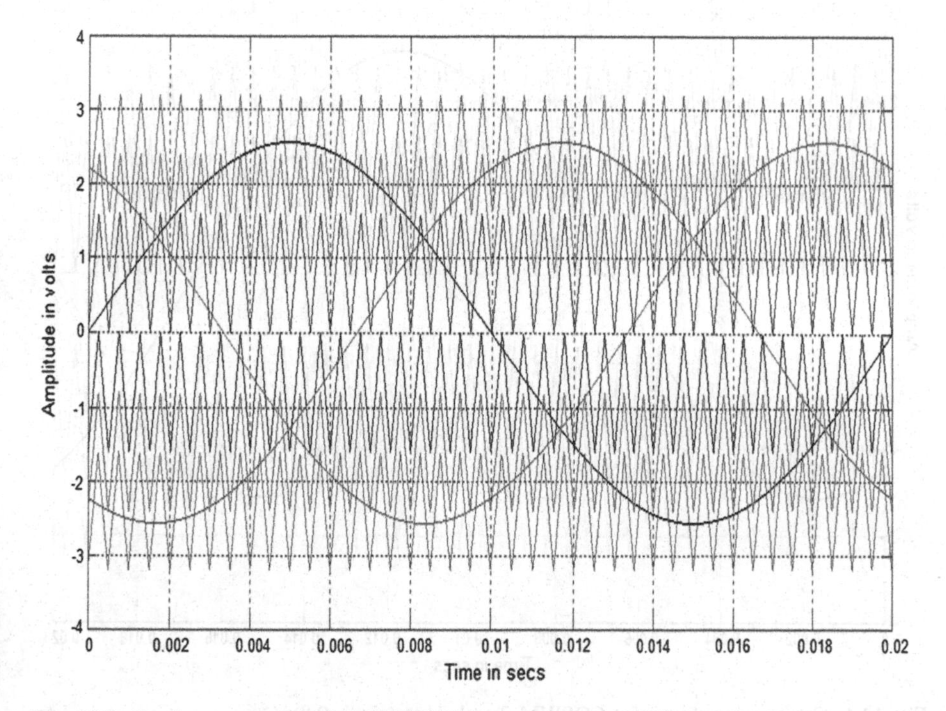

Fig. 13.3 Carrier arrangement for COPWM-1 technique (ma = 0.8)

13.3.3 COPWM-3 Technique

Carriers for seven level inverter with COPWM-3 technique are shown in Fig. 13.5. In this technique carriers invert their phase in turns from the previous one. It may be identified as PWM with amplitude-overlapped and neighbouring-phase-interleaved carriers. COPWM-2 and COPWM-3 have second control freedom change with the carriers horizontally phase shifted from COPWM-1 besides the offsets in vertical [15].

13.4 Simulation Results

The three phase Z-source diode clamped seven level inverter is modeled in SIMULINK using power system block set. Switching signals for diode clamped multilevel inverter using COPWM techniques are simulated. Simulations are performed for different values of ma ranging from 0.8 –1 and the corresponding

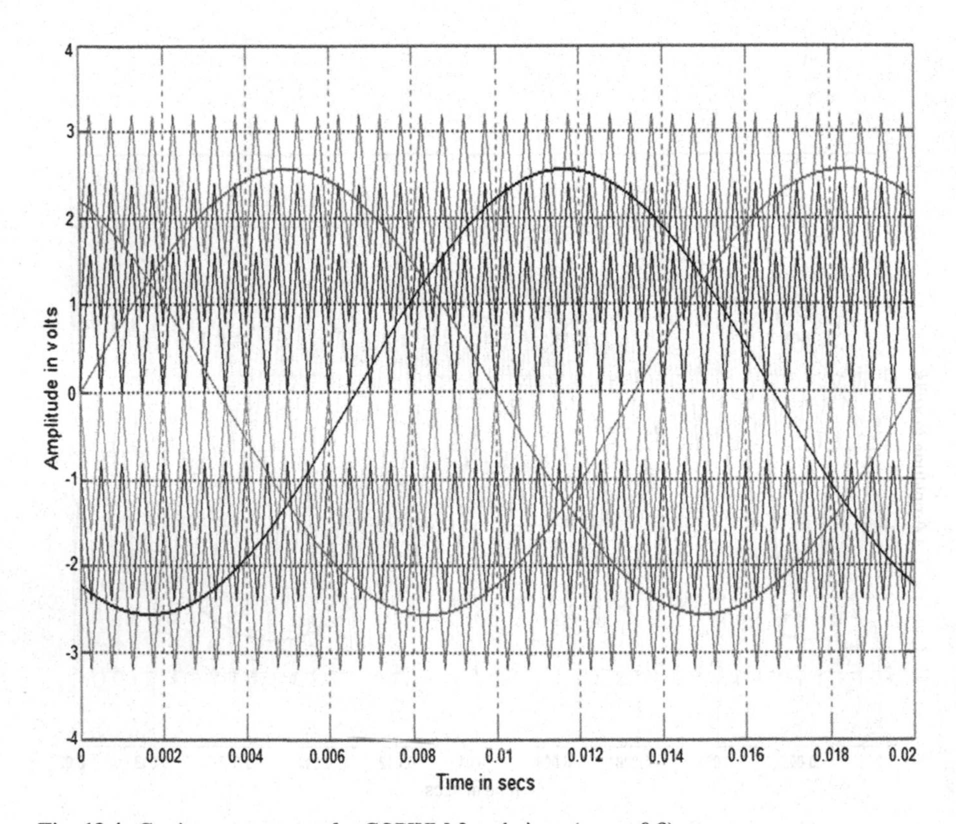

Fig. 13.4 Carrier arrangement for COPWM-2 technique (ma = 0.8)

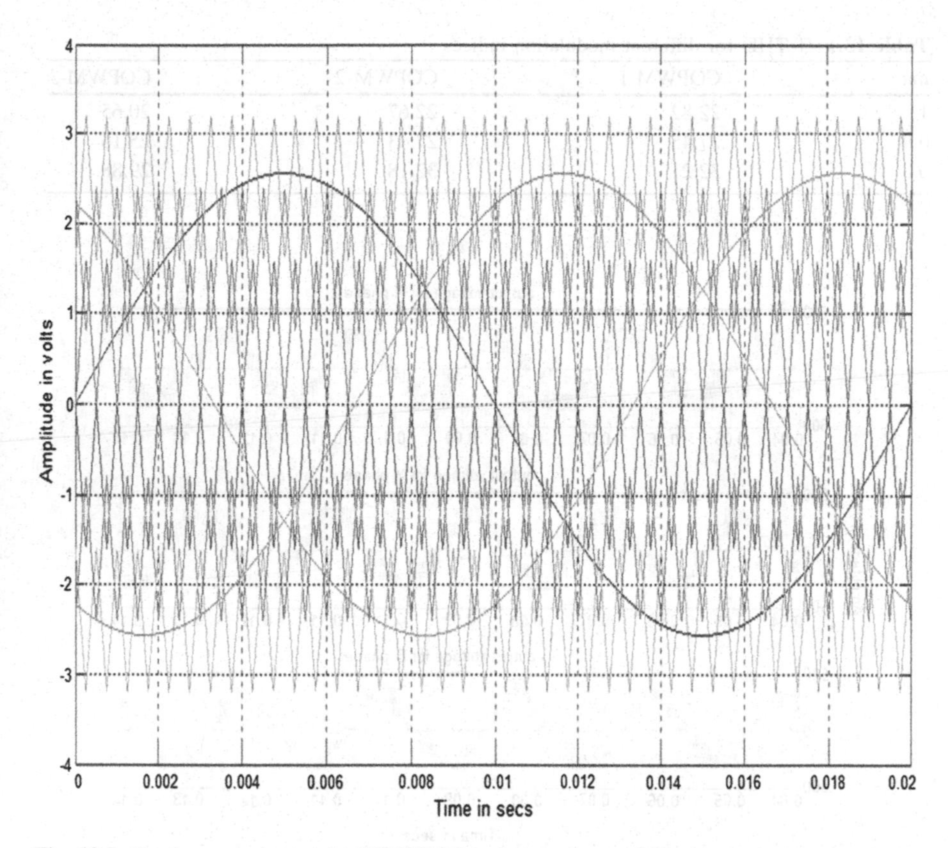

Fig. 13.5 Carrier arrangement for COPWM-3 technique (ma = 0.8)

% THD are measured using the FFT block and their values are shown in Table 13.1. Figures 13.6, 13.7, 13.8, 13.9, 13.10 and 13.11 show the simulated output voltage of Z-source DCMLI and their harmonic spectrum. Figure 13.6 displays the seven level output voltage generated by COPWM-1 switching technique and its FFT plot is shown in Fig. 13.7. Figure 13.8 shows the seven level output voltage generated by COPWM-2 switching technique and its FFT plot is shown in Fig. 13.9. Figure 13.10 shows the seven level output voltage generated by COPWM-3 switching technique and its FFT plot is shown in Fig. 13.11. Tables 13.2 and 13.3 show the Crest Factor (CF) of the output voltage and Vrms (fundamental) for various modulation indices of three phase Z-source MLI.

The following parameter values are used for simulation: VDC = 420 V, R (load) = 10 ohms, $A_c = 1.6$, $f_c = 2{,}000$ Hz and $f_m = 50$ Hz.

It is observed (from Table 13.1 and Fig. 13.11) the harmonic output voltage is least with COPWM-3 technique. The % CF is relatively equal for all the three strategies. From the Figs. (13.7, 13.9, 13.11) and Table 13.3 the following are

Table 13.1 % THD for different modulation indices

ma	COPWM-1	COPWM-2	COPWM-3
1	22.82	22.67	20.65
0.9	27.45	27.43	25.16
0.8	32.21	32.18	29.88

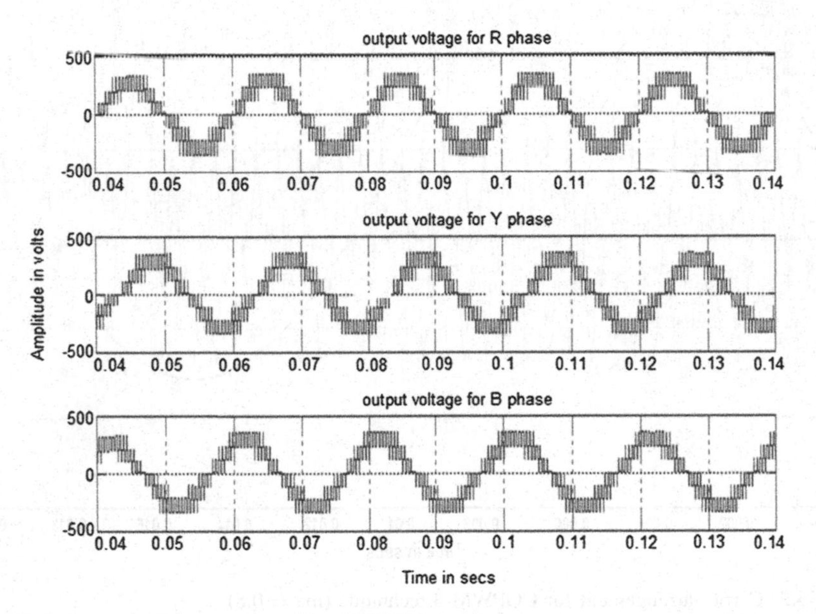

Fig. 13.6 Output voltage generated by COPWM-1 technique

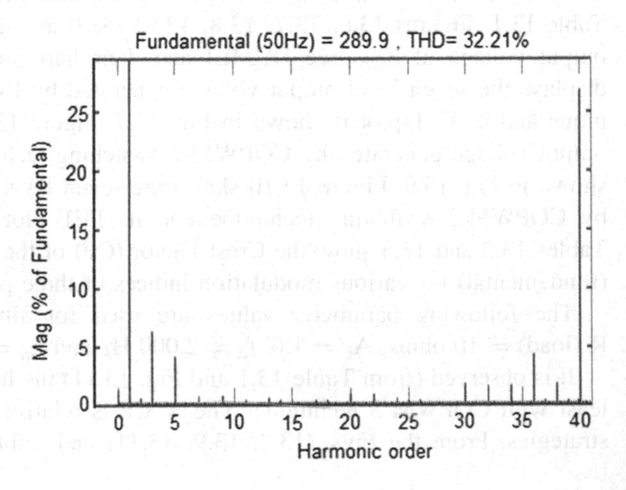

Fig. 13.7 FFT plot of COPWM-1 technique

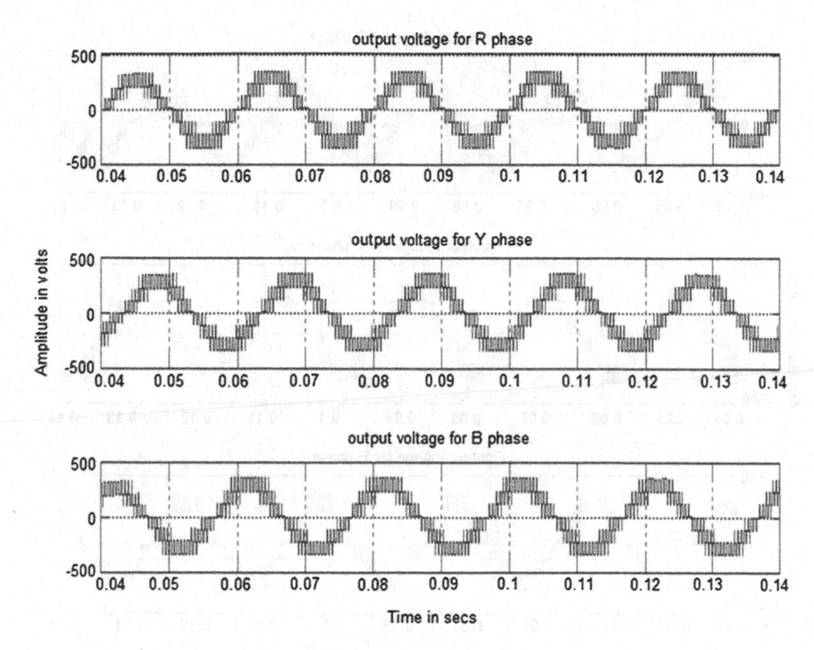

Fig. 13.8 *Output voltage* generated by COPWM-2 technique

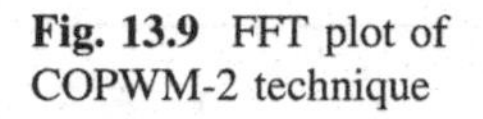
Fig. 13.9 FFT plot of COPWM-2 technique

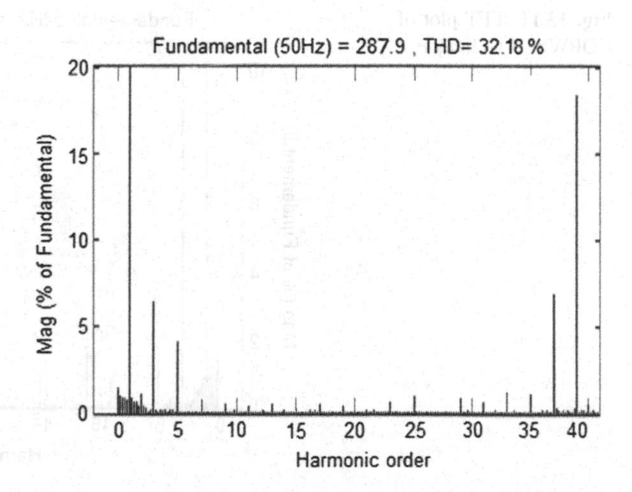

observed (1) lower order harmonics are relatively equal for all the three strategies, (2) 3rd order harmonic is dominant in all the strategies (3) Vrms value is high in COPWM-1 technique.

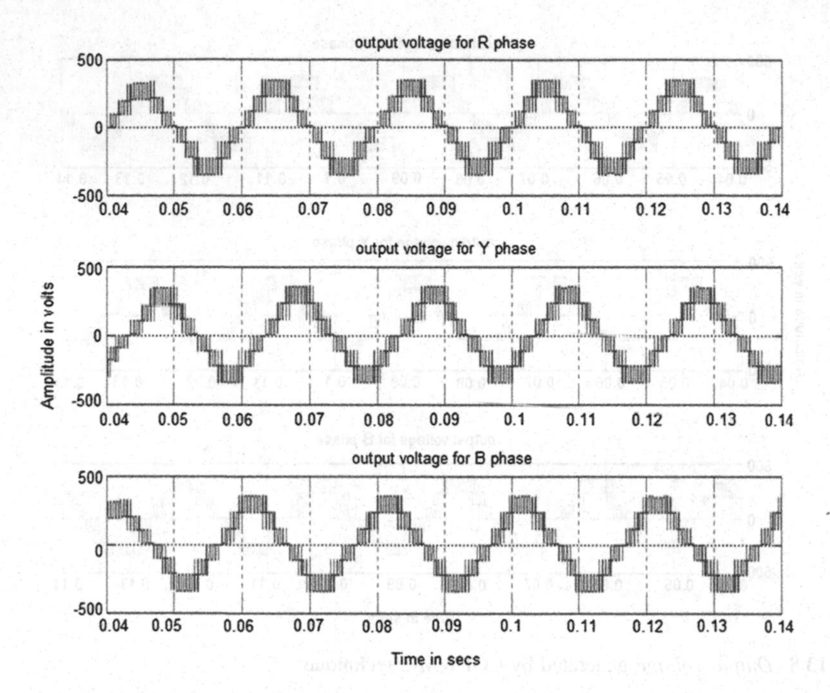

Fig. 13.10 *Output voltage* generated by COPWM-3 technique

Fig. 13.11 FFT plot of COPWM-3 technique

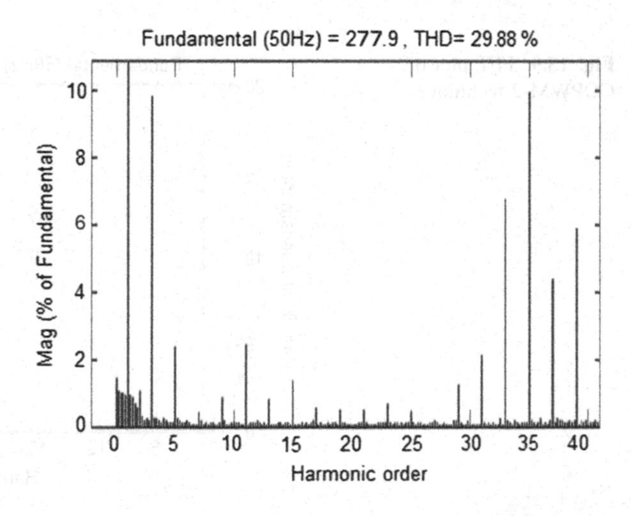

Table 13.2 % CF for different modulation indices

ma	COPWM-1	COPWM-2	COPWM-3
1	1.4143	1.4139	1.4143
0.9	1.4142	1.4140	1.4139
0.8	1.4141	1.4147	1.4142

Table 13.3 Vrms for different modulation indices

ma	COPWM-1	COPWM-2	COPWM-3
1	248.13	247.6	243.5
0.9	227.9	226.8	221.5
0.8	205	203.5	196.5

13.5 Conclusion

In this paper, COPWM techniques for three phase Z-source seven level diode clamped inverter have been presented. Z-source multilevel inverter gives higher output voltage through its Z network. Performance factors like % THD, Vrms and CF have been evaluated, presented and analyzed. It is found that the COPWM-3 technique provides lower % THD, COPWM-1 technique provide higher Vrms and less number of dominant harmonics than the other techniques. DC source can be replaced by PV source and this Z-source seven level diode clamped can be used for distributed generation systems. This work can be further extended to higher levels to minimize the THD.

References

1. Gajanayake CJ, Vilathgamuwa DM, Loh PC (2007) Development of a comprehensive model and a multiloop controller for Z-source inverter DG systems. IEEE Trans Ind Electron 54(4):1295–1308
2. Peng FZ, Joseph A, Wang J, Shen M, Chen L, Pan Z, Ortiz-Rivera E, Huang Y (2005) Z-source inverter for motor drives. IEEE Trans Power Electron 20(4):857–869
3. Kanimozhi S, Senthil Kumar V (2012) Z-source multilevel inverter for uninterruptible power supply application. Int J Power Syst Integr Circ 2(1):29–31
4. Kandasamy V, Manojchakkaravarthi RC (2012) A novel approach of impedance source cascaded multilevel inverter. Int J Mod Eng Res 2(2):394–397
5. Bakar MS, Rahim NA, GhaZali KH (2010) Analysis of various PWM controls on single-phase Z-source inverter. In: Proceedings of the IEEE conference record: 978-1-4244-8648-9/10 (SCORe2010)
6. Yousuf SM, Vijayadeepan P, Latha S (2012) The comparative THD analysis of neutral clamped multilevel Z-source inverter using novel PWM control techniques. Int J Mod Eng Res 2(3):1086–1091
7. Loh PC, Gao F, Blaabjerg F (2008) Topological and modulation design of three-level Z-source inverters. IEEE Trans Power Electron 23(5):2268–2277
8. Loh PC, Lim SW, Gao F, Blaabjerg F (2007) Three-level Z-source inverters using a single LC impedance network. IEEE Trans Power Electron 22(2):706–711
9. Loh PC, Blaabjerg F, Wong CP (2007) Comparative evaluation of pulse width modulation strategies for Z-source neutral-point-clamped inverter. IEEE Trans Power Electron 22(3):1005–1013
10. Chandrudu KR, Raju PS, Anjaneyulu GVP (2011) E Z-source fed induction motor drive: an experimental investigation. Int J Eng Sci Technol 3(8):6817–6823
11. Ali US, Kamaraj V (2011) Double carrier pulse width modulation control of Z-source inverter. Eur J Sci Res 49(2):168–176

12. Rajakaruna S, Jayawickrama L (2010) Steady-state analysis and designing impedance network of Z-source inverters. IEEE Trans Ind Electron 57(7):2483–2491
13. Sasikumar M, ChenthurPandian S (2010) Implementation and characteristics of induction generator fed three level ZSI for wind energy conversion scheme. Int J Advanced Eng Sci Technol 1(1):052–057
14. Dehghan SM, Mohamadian M, Yazdian A, Ashrafzadeh F (2010) A dual-input–dual-output Z-source inverter. IEEE Trans Power Electron 25(2):360–368
15. Shanthi B, Natarajan SP (2008) Carrier overlapping PWM methods for single phase cascaded five level inverter. Int J Sci Tech Autom Control Comput Eng 590–601
16. Johnson Uthayakumar R, Natarajan SP, Bensraj R (2012) A carrier overlapping PWM technique for seven level asymmetrical multilevel inverter with various references. IOSR J Eng 2(6):1301–1307

Chapter 14
Implementation of Sliding Mode Controller to Regulate the Speed for Series Connected Chopper Fed Separately Excited DC Motor Drive

N. Rathika, N. Sathya and A. Ezhilarasi

Abstract Multilevel static power conversion technology imbibes the ability to process high voltage and generate multi-tier voltage waveforms with high spectral quality. This technology is increasingly being used [1] in power converters and power conditioning circuits. Multilevel power converter is a general term applied to power converters with topologies capable of synthesizing multi-tier voltage waveforms and processing high voltages, by means of series connections of active devices to offer three or more discrete DC voltage [2] levels. Interconnection of power devices to split DC rail increases the voltage handling capability of these converters for the given power devices. Series-parallel DC–DC conversion systems in which multiple standardized converter modules are connected in series or parallel [3] at the output and input sides. Multiple connected DC–DC conversion systems attract more attention [4] in recent years, and are being to be widely used in various applications [5]. DC choppers are used to convert unregulated DC input voltage into a controlled DC output voltage at a desired level. They are widely preferred in the switched mode power supplies, and DC motor drives applications. Besides DC choppers find their role as interfaces between the DC systems [6] of different [7] voltage levels. The output voltage of PWM based [8] DC choppers are varied by varying the duty cycle.

N. Rathika (✉)
Department of Electrical and Electronics Engineering, M.A.M College of Engineering, Tiruchirappalli, Tamil Nadu, India
e-mail: rathika.111287@gmail.com

N. Sathya
Department of Electronics and Instrumentation Engineering, M.A.M College of Engineering, Tiruchirappalli, Tamil Nadu, India
e-mail: sathyaeie@gmail.com

A. Ezhilarasi
Department of Electrical Engineering, Annamalai University, Chidambaram, Tamil Nadu, India
e-mail: jee.ezhiljodhi@yahoo.co.in

R. Malathi and J. Krishnan (eds.), *Recent Advancements in System Modelling Applications*, Lecture Notes in Electrical Engineering 188,
DOI: 10.1007/978-81-322-1035-1_14, © Springer India 2013

Buck converter is a subset of DC–DC converters. It is desired to explore new methodologies, so as to enable DC choppers to elicit better performance of DC Drives.

Keywords Slide mode controller · DC drive · Chopper · Series motor

14.1 Introduction

DC Drive combines a DC motor with a Converter that regulates the DC voltage and current applied to the motor armature and shunt field. Direct current (DC) motors are widely used in many industrial applications such as electric vehicles, steel rolling mills, electric cranes, and robotic manipulators due to precise, wide, simple, and continuous control characteristics. Traditionally rheostatic armature control method is widely used for the speed control of low power dc motors. However, the controllability, cheapness, higher efficiency, and higher current carrying capabilities of static power converters effect a major change in the performance of electrical drives [9]. DC speed control is simpler, less costly, and offers more control over a greater range of speeds than AC speed control systems.

14.2 Existing Methods for Speed Control

The speed can be controlled by any of the following methods,

1. **Field flux control**: It is obtained by varying the field flux of a DC motor.
2. **Armature voltage control**: The voltage across the armature is varied by inserting an extra resistance (controller resistance) in series with the Armature. The base speed is defined as the speed at which the motor runs under rated armature voltage at the rated field current and rated armature current.
3. **Proportional-Integral Controller (PI Controller)**: PI Controller is a feedback controller which drives the plant to be controlled with a weighted sum of the error (difference between the output and desired set-point) and the integral of that value. It is a special case of the common PID controller in which the derivative (D) of the error is not used.
4. **Proportional–Integral–Derivative Controller (PID Controller)**: PID controller is a generic control loop feedback mechanism (controller) widely used in industrial systems. A PID is the most commonly used feedback controller. It calculates an "error" value as the difference between a measured process variable and a desired set-point. The controller attempts to minimize the error by adjusting the process control inputs.
5. **Intelligent control**: Intelligent control uses various AI computing approaches like neural networks, Bayesian probability, fuzzy logic, machine learning, evolutionary computation and genetic algorithms to control a dynamic system.

a. **Advantages of multiple connections**

It is preferred for large motor drives, especially if the inductor current requirements are large. However considering the additional complexity involved in increasing the number of DC choppers, there is not much reduction in the harmonics generated in the supply line [10].

It increases the effective chopper frequency there by reducing the component values and size of the input filter.

b. **Limitations of multiple connections**

Separate inductors are used for current sharing [11].

c. **Need for new schemes**

To make the DC motor more effective in several applications and to improve its performance characteristics, we go for several new schemes for regulating the speed.

14.3 Variable Structure Approach

Switched mode power supplies (SMPS) represent a particular class of variable structure systems (VSS) and they take advantage of nonlinear control techniques developed for this class of system. A VSS is based on the number of independent sub topologies, which are defined by the status of nonlinear elements (switches). SMPS, are nonlinear and time varying systems and thus the design of high performance control is usually a challenging issue. In fact, the control should ensure system stability in any operating condition and good static and dynamic performances in terms of rejection of input voltage disturbances and load changes. These characteristic of course are to be maintained in spite of large input voltage, output current and even parameter variations.

A classical control approach relies on the state space averaging method, which derives an equivalent mode by circuit-averaging all the system variables in a switching period. From the average model, a suitable small signal model is then derived by perturbation and linearization around a precise operating point. Finally, the small signal model is used to derive all the necessary converter transfer function to design a linear control system by using classical control techniques. The design procedure is generally not easy to account for the wide variation of system parameters, because of the strong dependence of small signal model parameters on the converter operating point.

Sliding mode control, which is derived for variable structure system theory, extends the properties of hysteresis control to multivariable environments, resulting in stability even for large supply and load variation and good dynamic response. Thus the sliding mode appropriately offers an alternative way to implement a control action that exploits the inherent variable structure nature of SMPS. In particular, the converter switches are driven as a function of the instantaneous values of the state variables to force the system trajectory to stay a suitable selected sliding surface on the phase space.

14.3.1 Problem Definition

The objective is to formulate a scheme through which three series connected choppers can share the input voltage equally to regulate the speed and feed to a DC drive of the desired power rating without exceeding [12] the current handling capability of the chopper and the drive [13]. The performance of the Algorithm is estimated using MATLAB based simulation to ensure its suitability for industrial applications.

14.3.2 Block Diagram

See Fig. 14.1.

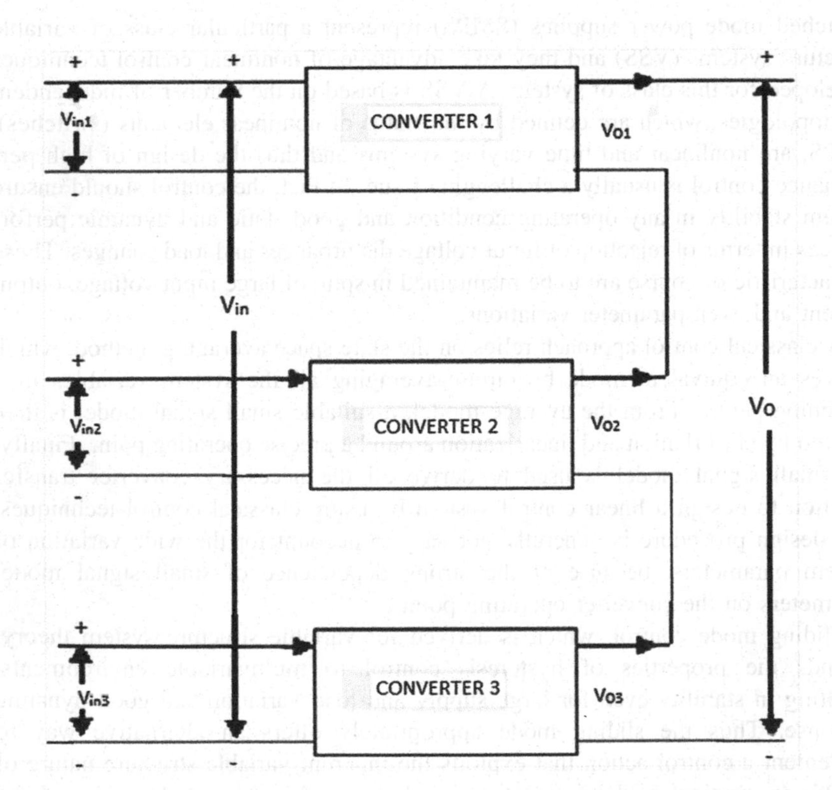

Fig. 14.1 Performance block diagram

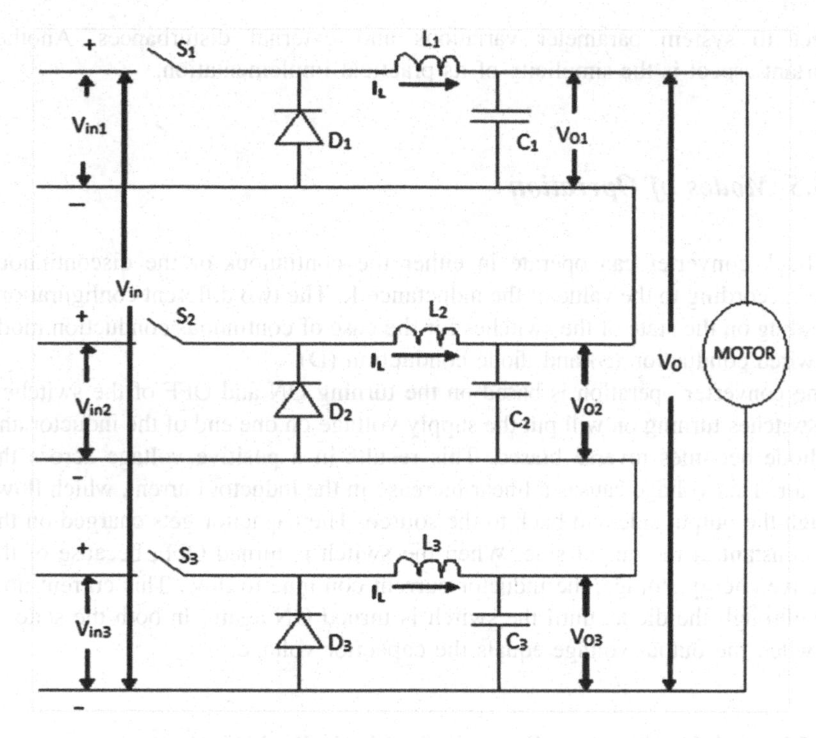

Fig. 14.2 Series connection of chopper fed DC drive

14.3.3 Chopper Module

See Fig. 14.2.

14.3.4 Control Algorithm

The variable structure systems (VSS) are characterized by a discontinuous control action which changes structure upon reaching a set of switching surface. The switch commutations of a static converter can be described by a discontinuous structure like VSS. The main problem in converters VSS control is to determine its switching instants. In VSS theory the sliding mode control (SMC) [14] is very attractive to solve this problem. The switching instants are determined by appropriate sliding surface or switching surface. If the switching surface is properly designed, the behavior of the sliding mode control can be guaranteed to be asymptotically stable.

It is recognized that robustness is the most important feature of SMC systems. Under certain conditions, they are invariant, more than just robust, [15] with respect to system parameter variations and external disturbances. Another important aspect is the simplicity of its practical implementation.

14.3.5 Modes of Operation

The buck converter can operate in either the continuous or the discontinuous modes according to the value of the inductance L. The two different configurations depending on the state of the switches for the case of continuous conduction mode are switch conduction (S) and diode conduction (D).

The converter operation is based on the turning ON and OFF of the switches. The switches turning on will put the supply voltage on one end of the inductor and the diode becomes reverse biased. This results in a positive voltage across the inductor. This voltage causes a linear increase in the inductor current, which flows through the output side and back to the source. The capacitor gets charged on the time constant at the output side. When the switch is turned OFF, because of the inductive energy storage, the inductor current continue to flow. This current now flows through the diode, until the switch is turned ON again. In both the state of the switch the output voltage equals the capacitor voltage.

14.3.5.1 Variable Structure Control of a Single Buck Converter

The state space model of the converter output stage equivalent circuit (Fig. 14.3) is

$$\begin{matrix} \dot{V}_o \\ \dot{i}_L \end{matrix} = \begin{matrix} -\frac{1}{R_o C} & \frac{1}{C} \\ -\frac{1}{L} & -\frac{rL}{L} \end{matrix} * \begin{matrix} V_o \\ i_L \end{matrix} + \begin{matrix} 0 \\ \frac{1}{L} \end{matrix} * V \tag{14.1}$$

where V_O and i_L are the state variables of the system.

U is the control input

$$U = \begin{cases} \in, \text{ if } S \text{ is } ON \\ 0, \text{ if } S \text{ is } OFF \end{cases} \tag{14.2}$$

S is a converter switch.

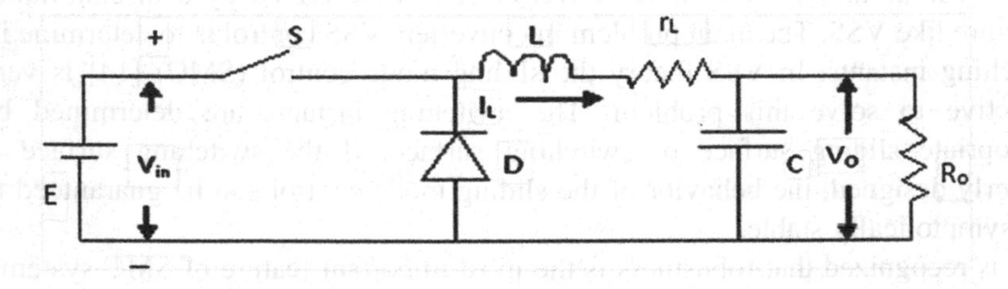

Fig. 14.3 Variable structure control of a single buck converter

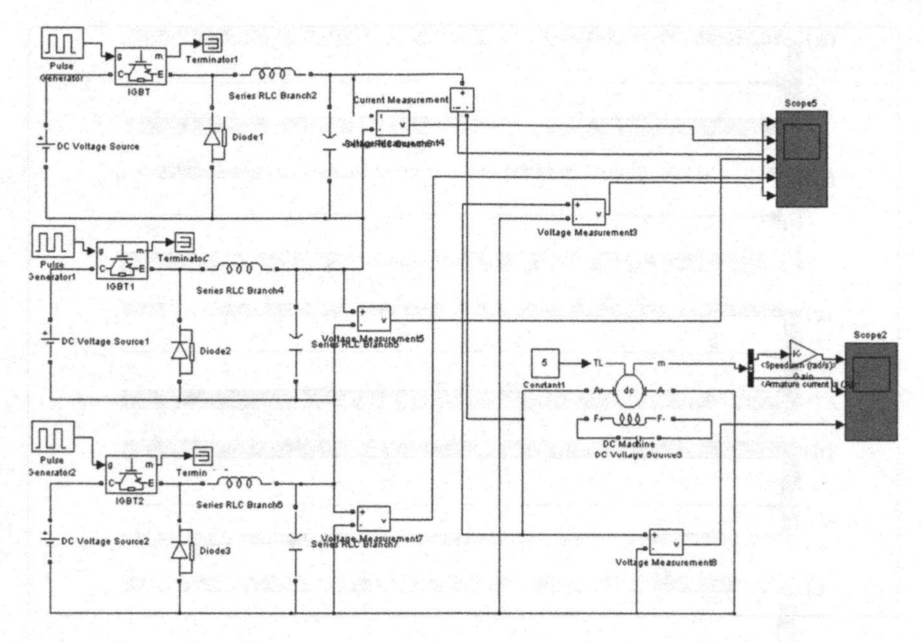

Fig. 14.4 Open loop schematic

In SMC the sliding mode surface gives the sequence and the duration of the ON and OFF switching states for the given dynamic specifications. The linear sliding surface is given by,

$$\sigma(e_v, \overset{\circ}{e}_v) = 0 \tag{14.3}$$

$$e_v = V_o - V_{referance} \tag{14.4}$$

$$\sigma = \overset{\circ}{e}_v + \alpha e_v = \frac{1}{C} i_c + \alpha e_v \tag{14.5}$$

In which, $\alpha > 0$ is a constant gain that guarantees stable converter behavior and the discontinuous control, (i.e.,) given by,

$$\mathrm{U} = \begin{cases} \in, \; if\ \sigma < 0 \\ 0, \; if\ \sigma > 0 \end{cases} \tag{14.6}$$

This sliding surface ensures that the output voltage (V_o) is controlled.

14.4 Results and Discussion

The open loop and closed loop simulation of series connected chopper fed DC drive was presented in this paper.

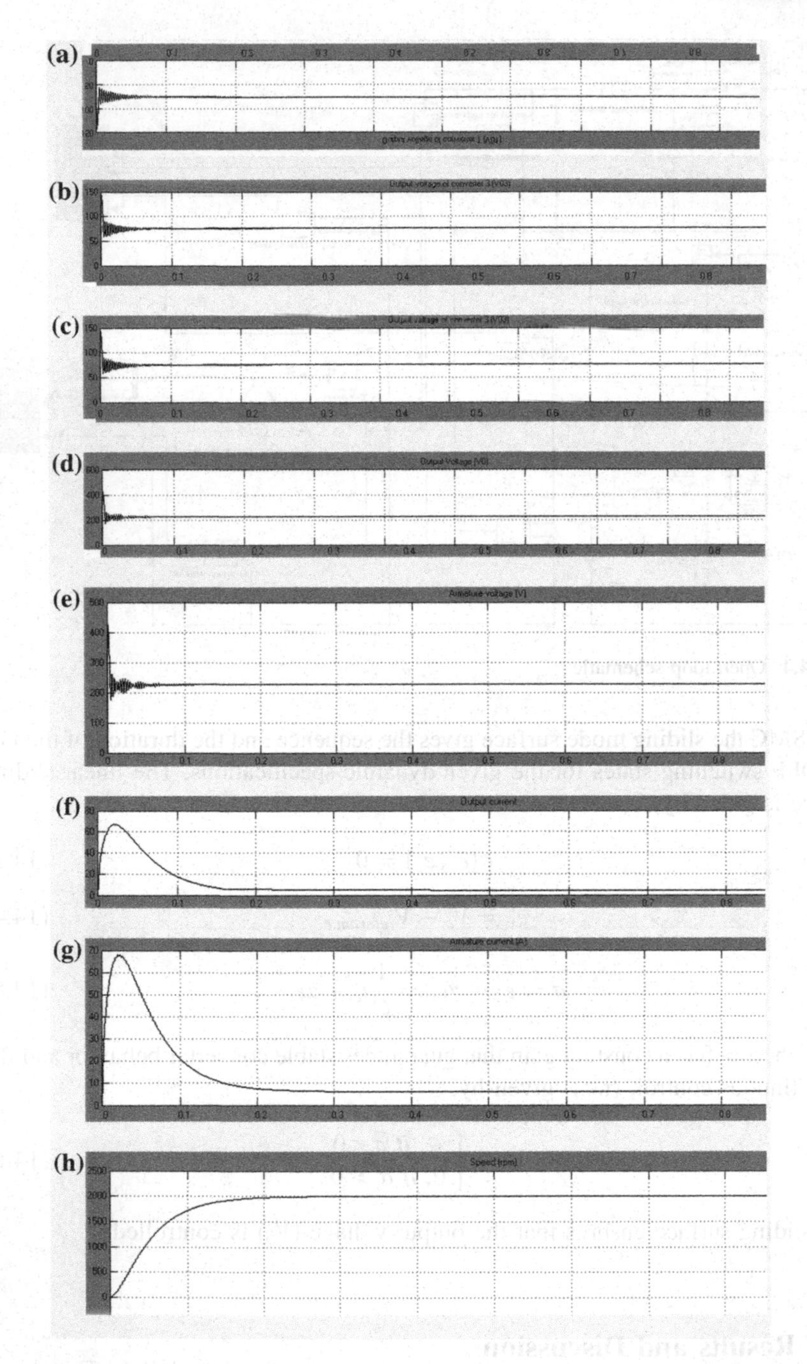

Fig. 14.5 **a** Open loop output voltage of converter 1. **b** Open loop output voltage of converter 2. **c** Open loop output voltage of converter 3. **d** Open loop output voltage. **e** Open loop armature voltage. **f** Open loop output current. **g** Open loop armature current. **h** Open loop speed

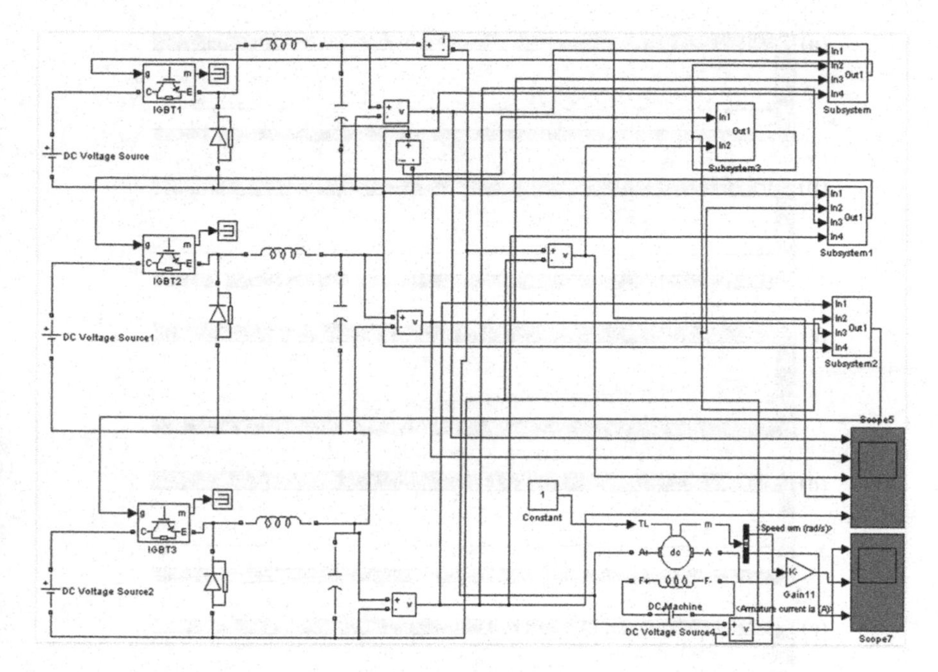

Fig. 14.6 Closed loop schematic without transient

14.4.1 Buck Converter Specification

Three buck converters are connected in series to a DC motor drive. The buck converter parameters are chosen as $L_1 = 300\ \mu H$, $C_1 = 940\ \mu F$, L_2, $L_3 = 250\ \mu H$, C_2, $C_3 = 840\ \mu F$, switching frequency = 20 kHz. The motor capacity is 5 HP, 240 V, 2,100 r.p.m. The input of 350 V is applied to each of the buck converter and the reference output voltage is fixed as 230 V, which can be shared equally across each converter module.

14.4.2 Open Loop Schematic

The open loop schematic of the MATLAB simulink is shown in the Fig. 14.4.

The output voltage across each converter, armature voltage, armature current and the speed of the open loop schematic is shown in the Fig. 14.5a–h. It is seen that the system is not regulated to the desired value with the pre designed duty cycle.

14.4.3 Closed Loop Schematic

The Fig. 14.6 shows the closed loop schematic with sliding mode controller.

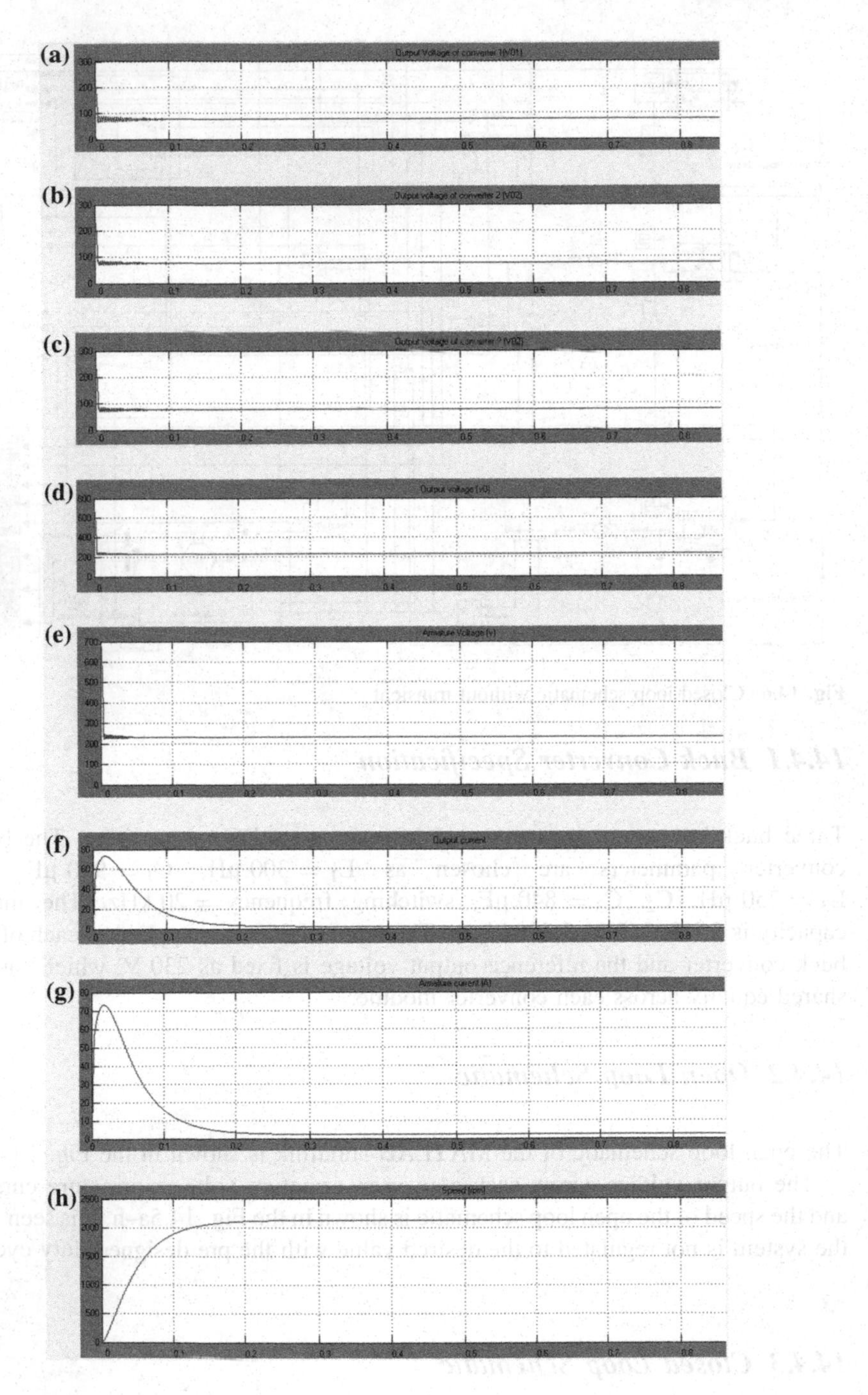
(a)
Output Voltage of converter 1 (V01)
(b)
Output voltage of converter 2 (V02)
(c)
(d)
Output voltage (V0)
(e)
Armature Voltage (v)
(f)
Output current
(g)
Armature current (A)
(h)
Speed (rpm)

◀ **Fig. 14.7** **a** Closed loop output voltage of converter 1. **b** Closed loop output voltage of converter 2. **c** Closed loop output voltage of converter 3. **d** Closed loop output voltage. **e** Closed loop Armature voltage. **f** Closed loop output current. **g** Closed loop armature current. **h** Closed loop speed

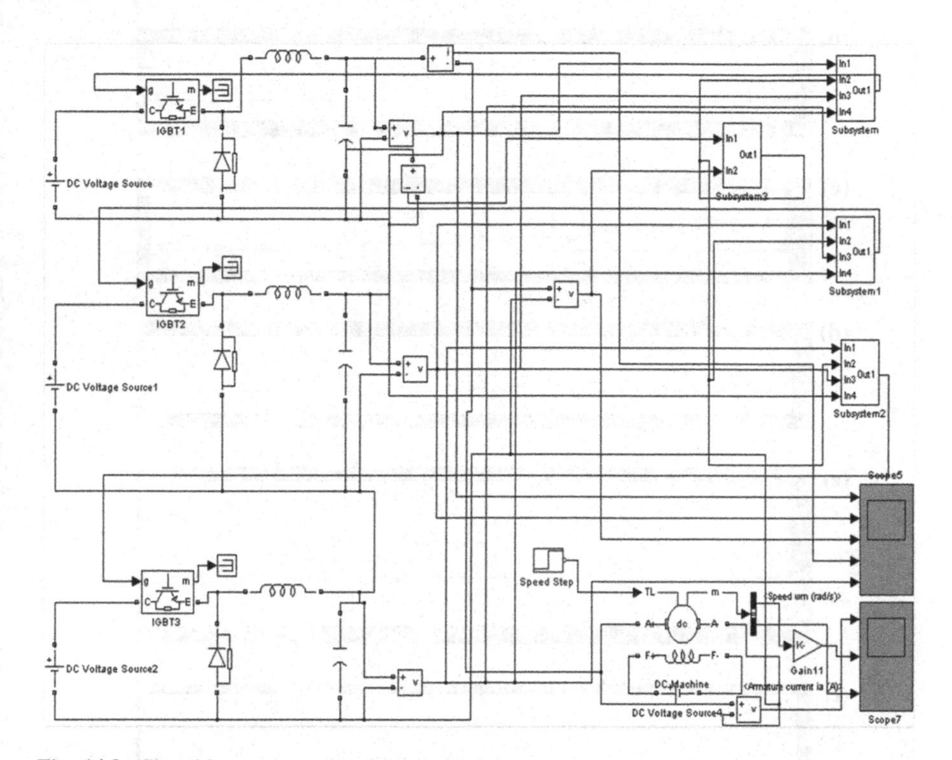

Fig. 14.8 Closed loop schematic with transient

The output voltage of converters, armature voltage, armature current and speed of the closed loop series connected buck converter is shown in Fig. 14.7a–h. It is seen that the desired output voltage of 230 V (speed 2,100 r.p.m) is obtained by the use of the variable structure controller (Figs. 14.8 and 14.9).

14.4.4 Closed Loop Transient Load Change

A sudden change in load is set at t = 0.5 s.

It can be seen that the controller modified the duty cycle to retain the stability thus regulates the output voltage. The controller ensures system stability even for large load variations.

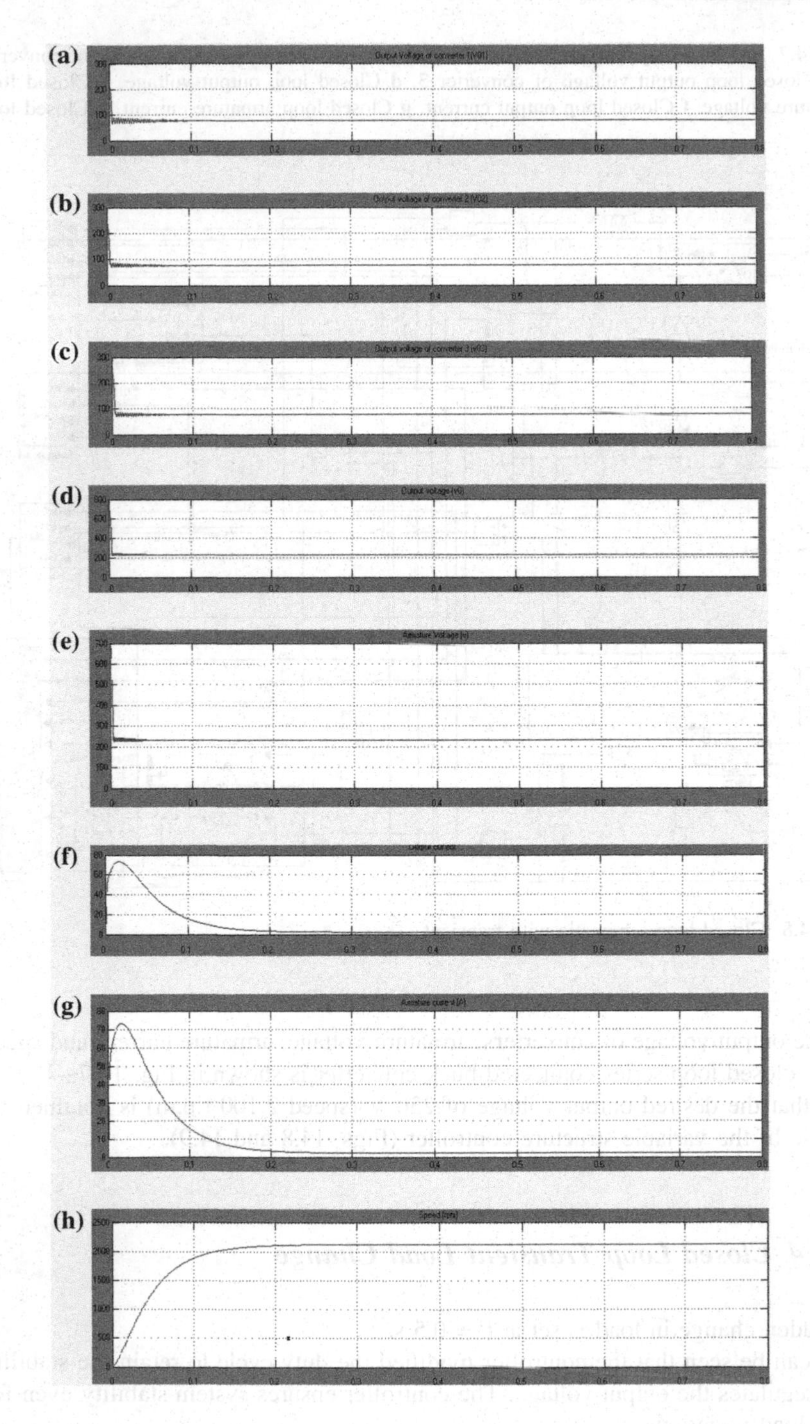
(a)
(b)
(c)
(d)
(e)
(f)
(g)
(h)

◀ **Fig. 14.9** **a** Closed loop output voltage of converter 1. **b** Closed loop output voltage of converter 2. **c** Closed loop output voltage of converter 3. **d** Closed loop output voltage. **e** Closed loop armature voltage. **f** Closed loop output current. **g** Closed loop armature current. **h** Closed loop speed

14.5 Conclusion

A new control algorithm has been modeled and inserted in the feedback path of the proposed series connected DC–DC converter fed separately excited DC motor drive. This system has been developed for stable speed regulation of a drive motor and uniform voltage distribution across each converter. The simulated open and closed loop responses have been compared. From the Fig. 14.5a–h it has been renowned that the system is not regulated to the desired value with the pre designed duty cycle. The variable structure controller is programmed to generate a value for the duty cycle that serves to regulate the output voltage. Perfect voltage tracking is achieved over a wide range of loading and input voltage by the usage of the controller. It can be illustrious from the Fig. 14.7a–h that the desired output voltage of 230 V (speed 2,100 r.p.m) is obtained by the use of the variable structure controller. The result shows that the proposed strategy will serve to explore innovative applications in this domain.

References

1. Manias SN, Kostakis G (1993) Modular DC–DC converter for high output voltage applications. IEEE Proc
2. Monteiro TC, Galassi M, Terrazas TM, Marafão FP, Matakas L Jr, Komatsu W (2009) Development of a DC–DC converter for DC bus voltage control of series connected device. IEEE Trans
3. Ruan X, Cheng L, Zhang T (2006) Control strategy for input-series output-paralleled converter. IEEE Power Electron pp 1–8
4. Cho IH, Yi KH, Cho KM, Moon GW (2010) High efficient multi-level half bridge converter. IEEE Trans Power Electr 25:943
5. Ruan X, Huang Y (2009) General control considerations for input-series connected DC/DC converters. IEEE Trans Circ Syst 56:1286–1296
6. Huang Y, Tse CK (2007) Circuit theoretic classification of parallel connected DC/DC converters. IEEE Trans Circ Syst I, Reg 54(5):1099–1108
7. Bai S, Huang N, Ioinovici A (2006) Small-signal modeling and dynamic analysis of a novel zvzcs three-level converter. IEEE Trans Circ Syst I, Reg Pap 53:1958–1965
8. Siri K, Conner KA (2006) Parallel connected converters with maximum power tracking. IEEE Trans
9. Yuan H, Chen W (2009) DC/DC conversion systems consisting of multiple converter modules: stability, control, and experimental verifications. IEEE Trans Power Electr 24:1463–1474
10. Middlebrook RD (1976) Input filter considerations in design and application of switching regulators. IEEE Proc

11. Kim JW, You JS, Cho BH (2001) Modeling, control and design of input-series, output parallel connected converter. IEEE Trans Ind Electr 48:536–544
12. Siri K (2007) Uniform voltage distribution control for series connected DC–DC converters. IEEE Trans Power Electr 22:1269–1279
13. Siri K, Truong C, Conner KA (2005) Uniform voltage distribution control for parallel-input, series-output connected converters. IEEE Trans
14. Giri R, Choudhary V, Ayyanar R, Mohan N (2006) Common-duty ratio control of input-series connected modular DC–DC converters with active input voltage and load-current sharing. IEEE Trans Ind Appl 42:1101–1111
15. Osada Y, Sakai R, Maruyama G, Matsuse K (2009) 5 level double converters with different DC divided link voltage. IEEE Trans

Chapter 15
New Modulation Strategies for Symmetrical Three Phase Multilevel Inverter with Reduced Number of Switches

C. R. Balamurugan, S. P. Natarajan and V. Vidhya

Abstract This work proposes new modulation strategies for three phase cascaded multilevel inverter topology with reduced number of switches and is able to create five level output. The main advantage of the proposed work is to reduce the number of switches when compared to the conventional MLIs. The reduced number of switches reduces the switching losses and improves the efficiency of the inverter. Variable amplitude variable frequency strategy provides output with relatively low distortion and better DC bus utilization is obtained with variable amplitude carrier overlapping phase disposition technique.

Keywords THD · VA · CMLI · PWM · CF · FF

15.1 Introduction

Multilevel converters offer a number of advantages when compared to the conventional two level converter counterpart. Corzine et al. [1] have presented control of cascaded multi-level inverters. Radan et al. [2] have discussed evaluation of carrier based pulse width modulation methods for multilevel inverters.

C. R. Balamurugan (✉) · V. Vidhya
Department of EEE, Arunai Engineering College, Tiruvannamalai, Tamilnadu, India
e-mail: crbalain2010@gmail.com

V. Vidhya
e-mail: vvidhya2010@gmail.com

S. P. Natarajan
Department of EIE, Annamalai University, Chidambaram, Tamilnadu, India
e-mail: spn_annamalai@rediffmail.com

R. Malathi and J. Krishnan (eds.), *Recent Advancements in System Modelling Applications*, Lecture Notes in Electrical Engineering 188,
DOI: 10.1007/978-81-322-1035-1_15, © Springer India 2013

Fundamentals of a new hybrid symmetrical multilevel inverter are proposed by Tehrani et al. [3]. Pablo lezana et al. [4] developed cascaded multilevel inverter with regeneration capability and reduced number of switches. A number of other multilevel converter topologies have been proposed including neutral-point-clamped, flying capacitor and cascaded H-bridge inverter Pan and Peng [5]. Caballero et al. [6] suggested a new asymmetrical hybrid multilevel voltage inverter. Konstantinou and Agelidis [7] presented performance evaluation of half-bridge cascaded multilevel converters operated with multicarrier sinusoidal PWM techniques. Bicrk and Nowicki [8] presented a modified cascaded multilevel inverter with reduced switch count employing bypass diodes. The steppcd approximation of the sinusoidal waveform using higher levels reduces the harmonic distortion of the output waveform, and the stresses across the semiconductor devices and allows higher voltage/current and power ratings. The reduced switching frequency of each individual switch of the converter also reduces the switching losses and improves the efficiency of the converter as suggested by Konstantinou et al. [9]. Malinowski et al. [10] made a survey on cascaded multilevel inverters. New multilevel inverter topology with reduced number of switches are suggested by Ahmed et al. [11]. Caballero et al. [12] discussed another symmetrical hybrid multilevel inverter concept based on multi stage switching cells. A five level inverter topology with single dc supply by cascading a flying capacitor inverter and an H-bridge have presented by Roshankumar et al. [13]. Luiz Batschauer et al. [14] have discussed Three-phase hybrid multilevel inverter based on half-bridge modules. Najafi and Yatim [15] introduced design and implementation of a new multilevel inverter. In this paper a new PWM strategies of symmetrical multilevel inverters have been developed to increase number of output voltage levels with less number of switches. Simulation results are presented for the validity of the proposed techniques.

15.2 Topology and Operation

The chosen five level hybrid cascaded inverter with reduced number of switches is shown in Fig. 15.1. The DC source is connected to all phase legs of the conventional three-phase, two level inverter and the H-bridge cell utilizes a capacitor as a voltage source. Assuming that the DC voltage is equal to V_{dc}, then the voltage of the capacitor of the H-bridge cell has to be maintained to V_{dc} so that a five-level waveform is synthesized in the output. Considering a split DC source, the output of the two level leg can be equal to either $+V_{dc}$ or V_{dc}. Table 15.1 shows the switching states and possible output voltages of the converter. The voltage of the capacitor is affected during the converter states when the capacitor is connected to the load. These converter states occur during when the output voltage levels are $+2\ V_{dc}$ and $2\ V_{dc}$ and during the zero voltage level. The first two cases can only be acquired by a single switching state combination as shown in Table 15.1.

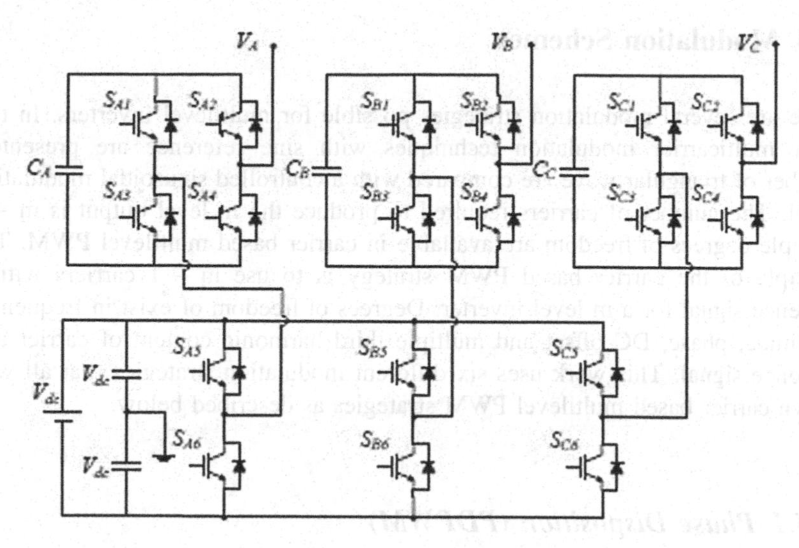

Fig. 15.1 Schematic of chosen three phase five level hybrid cascaded inverter with reduced number of switches

Table 15.1 Voltage output and switching states

$V_{phase\ a}$	S_{a1}	S_{a2}	S_{a3}	S_{a4}	S_{a5}	S_{a6}
2 V_{dc}	0	1	1	0	1	0
V_{dc}	1	1	0	0	1	0
0	0	1	1	0	0	1
$-V_{dc}$	1	1	0	0	0	1
$-2\ V_{dc}$	1	0	0	1	0	1

15.3 The Proposed Multilevel Inverter

The most important part in multilevel inverters are switches which define the reliability, circuit size, cost, installation area and control complexity. The number of required switches against required voltage levels is a very important element in the design. To provide a large number of output levels without increasing the number of bridges, a new power circuit topology and new modulation method for symmetrical multilevel converter is proposed in this paper Table 15.2 shows the comparison between existing systems and proposed system.

Table 15.2 Comparison between existing system and proposed system

Type	Conventional CMLI	Chosen hybrid cascaded inverter
No. of switches	24	18
No. of clamping diodes	24	18
No. of DC sources	6	4

15.4 Modulation Schemes

There are several modulation strategies possible for multilevel inverters. In this paper multicarrier modulation techniques with sine reference are presented. Number of triangular wave are compared with a controlled sinusoidal modulating signal. The number of carriers required to produce the m level output is m −1. Multiple degrees of freedom are available in carrier based multilevel PWM. The principle of the carrier based PWM strategy is to use m − 1 carriers with a reference signal for a m level inverter. Degrees of freedom of exist in frequency, amplitude, phase, DC offset and multiple third harmonic content of carrier and reference signal. This work uses six different modulation strategies that all well known carrier based multilevel PWM strategies as described below.

15.4.1 Phase Disposition (PDPWM)

The PD PWM signal generation for modulation index $m_a = 0.8$ is shown in Fig. 15.2. In this method all carriers have the same frequency, amplitude and phase but they are just different in DC offset to occupy contiguous bands. Since all carriers are selected with the same phase, the method is known as PD strategy.

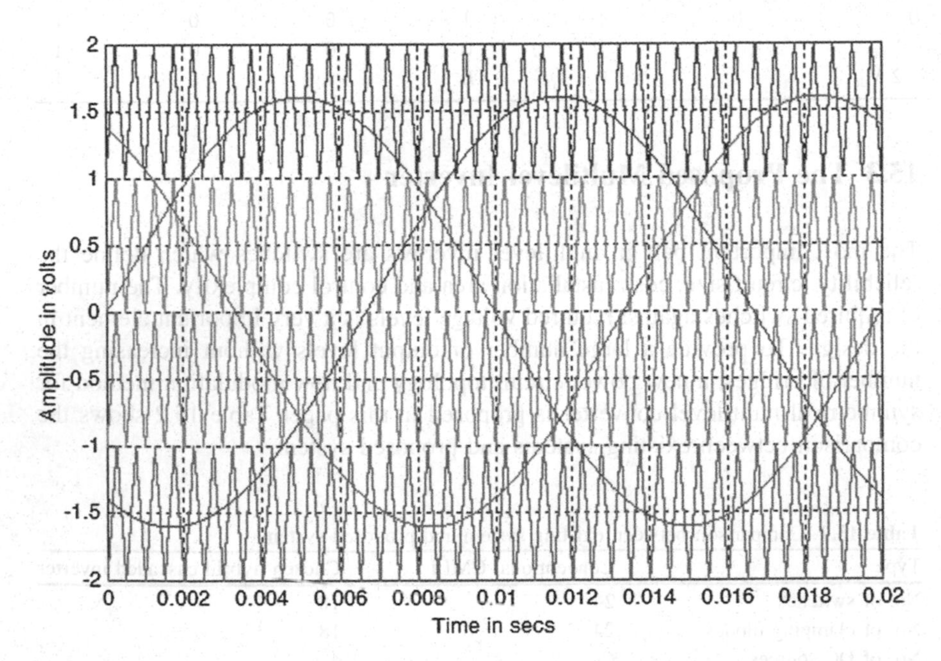

Fig. 15.2 Modulating and carrier waveform for PDPWM strategy ($m_a = 0.8$ and $m_f = 40$)

15.4.2 Variable Amplitude Phase Disposition (VAPDPWM)

The VAPDPWM signal generation for $m_a = 0.8$ is shown in Fig. 15.3. In this method all carriers have the same frequency, phase and different amplitude and but they are just different in DC offset to occupy contiguous bands. Since all carriers are selected with the same phase, the method is known as VAPD strategy.

15.4.3 Variable Frequency (VFPWM)

The VFPWM signal generation for $m_a = 0.8$ is shown in Fig. 15.4. The number of switchings for upper and lower devices of chosen MLI is much more than that of intermediate switches in PDPWM using constant amplitude constant frequency carriers. In order to equalize the number of switching for all the switches, variable frequency PWM strategy is used as in Fig. 15.4.

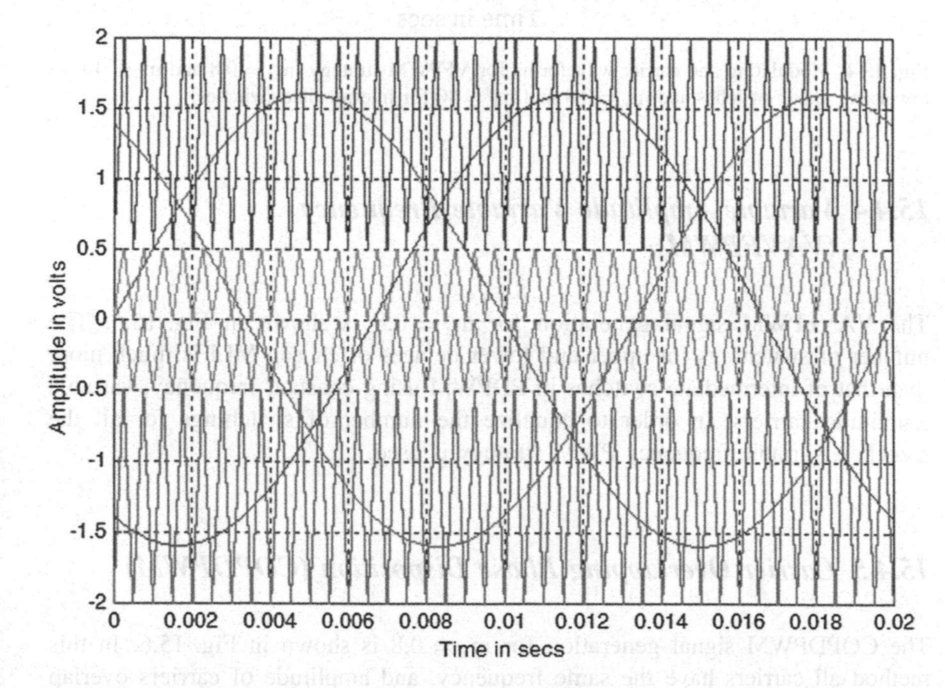

Fig. 15.3 Modulating and carrier waveforms for VAPDPWM strategy ($m_a = 0.8$ and $m_f = 40$)

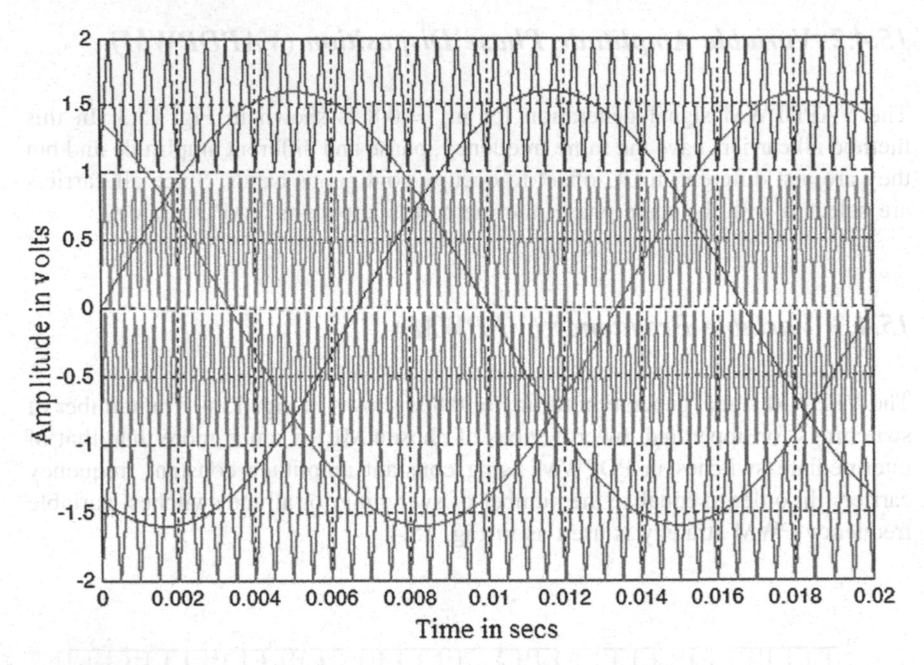

Fig. 15.4 Modulating and carrier waveforms for VFPWM strategy ($m_a = 0.8$ and $m_f = 40$ for lower and upper switches and $m_a = 0.8$ and $m_f = 80$ for intermediate switches)

15.4.4 Variable Amplitude Variable Frequency (VAVFPWM)

The VAVFPWM signal generation for $m_a = 0.8$ is shown in Fig. 15.5. The number of switchings for upper and lower devices of chosen MLI is much more than that of intermediate switches in PDPWM using constant frequency, variable amplitude carriers. In order to equalize the number of switchings for all the switches, variable frequency PWM strategy is used.

15.4.5 Carrier Overlapping Phase Disposition (COPDPWM)

The COPDPWM signal generation for $m_a = 0.8$ is shown in Fig. 15.6. In this method all carriers have the same frequency, and amplitude of carriers overlap with each other but they are just different in DC offset to occupy contiguous bands. Since all carriers are selected with the same phase, the method is known as COPD strategy.

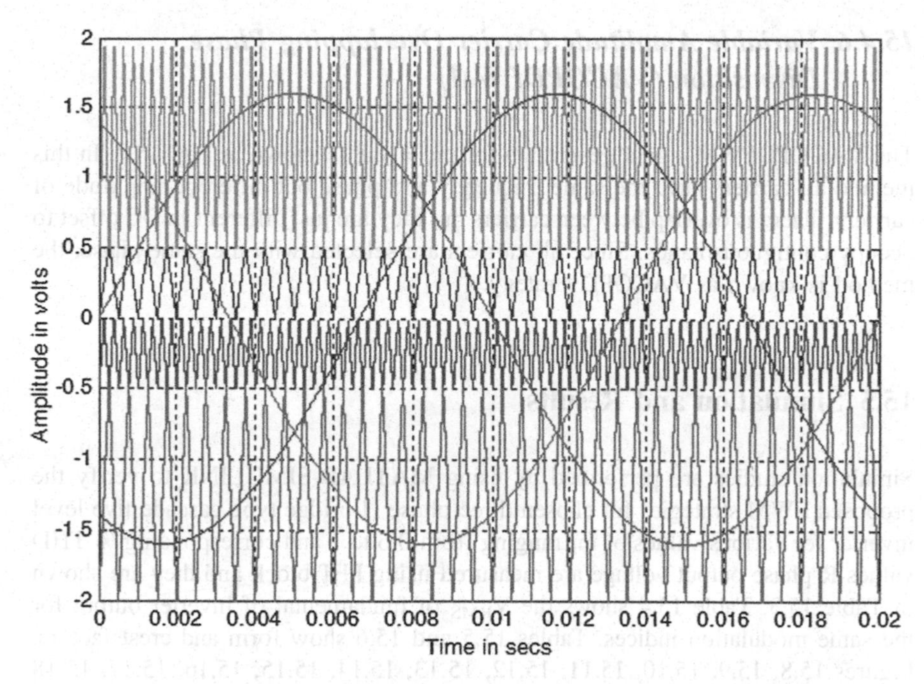

Fig. 15.5 Modulating and carrier waveforms for VAVFPWM strategy ($m_a = 0.8$ and $m_f = 40$)

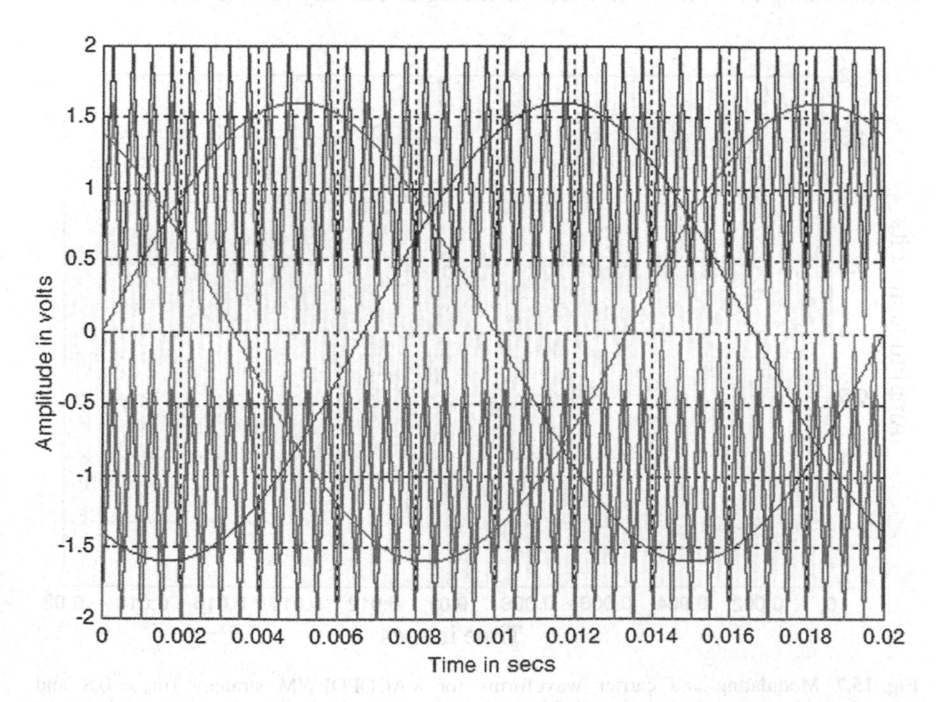

Fig. 15.6 Modulating and carrier waveforms for COPDPWM strategy ($m_a = 0.8$ and $m_f = 40$)

15.4.6 Variable Amplitude Carrier Overlapping Phase Disposition (VACOPDPWM)

The VACOPDPWM signal generation for $m_a = 0.8$ is shown in Fig. 15.7. In this method all carriers have the same frequency and phase but different amplitude of carriers. There is overlap between carriers but they are just different in DC offset to occupy contiguous bands. Since all carriers are selected with the same phase, the method is known as VACOPD strategy.

15.5 Simulation and Results

Simulation studies are performed by using MATLAB-SIMULINK to verify the proposed PWM strategies for chosen three phase H-bridge type cascade five level inverter for various values of m_a ranging from 0.6 to 1 and corresponding % THD values R-phase output voltage are measured using FFT block and they are shown in Table 15.3. Table 15.4 shows the V_{RMS} of fundamental of inverter output for the same modulation indices. Tables 15.5 and 15.6 show form and crest factors. Figures 15.8, 15.9, 15.10, 15.11, 15.12, 15.13, 15.14, 15.15, 15.16, 15.17, 15.18 show the simulated output voltage of chosen hybrid cascaded inverter and the corresponding FFT plots with different strategies but only for one sample value of

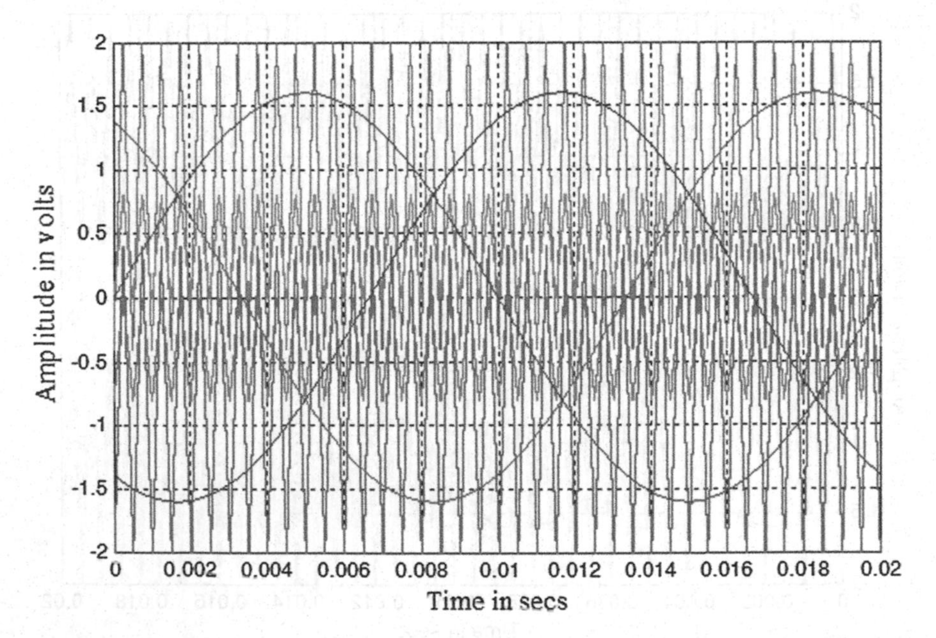

Fig. 15.7 Modulating and carrier waveforms for VACOPDPWM strategy ($m_a = 0.8$ and $m_f = 40$)

Table 15.3 % THD of output voltage of chosen hybrid CMLI for various values of modulating indices

m_a	PD	VAPD	VF	VAVF	COPD	VACOPD
1	34.54	32.16	34.55	32.12	41.61	39.91
0.9	44.83	39.65	44.77	39.57	50.86	45.33
0.8	52.82	45.55	52.67	45.60	62.09	50.32
0.7	57.98	51.08	58.00	51.07	78.19	55.56
0.6	56.64	54.83	56.61	55.38	95.08	61.35

Table 15.4 V_{RMS} (fundamental) of output of chosen hybrid CMLI for various values of modulating indices

m_a	PD	VAPD	VF	VAVF	COPD	VACOPD
1	215.3	239.7	215.3	239.2	231.4	257.6
0.9	215.3	218	185.5	218	209	243.7
0.8	155.9	197.2	156	197.7	183.9	230.64
0.7	127	177	126.7	176.7	153.5	216.1
0.6	99.62	155.8	99.82	155.2	126.1	200.8

Table 15.5 Form factor of output of chosen hybrid CMLI for various values of modulating indices

m_a	PD	VAPD	VF	VAVF	COPD	VACOPD
1	10765	4794	1656	7973	1652	1226
0.9	10765	3633	∞	778	2985	1624
0.8	742	2465	2228	637	∞	2560
0.7	1411	1770	3167	1963	852	1662
0.6	4981	1293	713	776	360	3067

Table 15.6 Crest factor of output of chosen hybrid CMLI for various values of modulating indices

m_a	PD	VAPD	VF	VAVF	COPD	VACOPD
1	1.4143	1.4142	1.4143	1.4142	1.4140	1.4145
0.9	1.4143	1.4142	1.4140	1.4142	1.4143	1.4144
0.8	1.4143	1.4143	1.4141	1.4142	1.4143	1.4140
0.7	1.4141	1.4145	1.4143	1.4146	1.4143	1.4146
0.6	1.1443	1.4149	1.4145	1.4143	1.4147	1.4143

$m_a = 0.8$ and $m_f = 40$. Figure 15.8 shows the five level output voltage generated by PDPWM strategy and its FFT plot is shown in Fig. 15.9. From Fig. 15.9, it is observed that the PDPWM strategy produces significant 3rd, 30th, 32nd, 34th, 36th, 38th and 40th harmonic energy. Figure 15.10 shows the five level output voltage generated by VAPDPWM strategy and its FFT plot is shown in Fig. 15.11. From Fig. 15.11, it is observed that the VAPDPWM strategy produces significant

3rd, 35th, 37th and 39th harmonic energy. Figure 15.12 shows the five level output voltage generated by VFPWM strategy and its FFT plot is shown in Fig. 15.13 From Fig. 15.13, it is observed that the VFPWM strategy produces significant 3rd, 34th, 38th and 40th harmonic energy. Figure 15.14 shows the five level output voltage generated by VAVFPWM strategy and its FFT plot is shown in Fig. 15.15. From Fig. 15.15, it is observed that the VAVFPWM strategy produces significant 3rd, 33rd, 35th, 36th, 38th, 39th and 40th harmonic energy. Figure 15.16 shows the five level output voltage generated by COPDPWM strategy and its FFT plot is shown in Fig. 15.17. From Fig. 15.17, it is observed that the COPDPWM strategy produces significant 3rd, 5th, 38th and 40th harmonic energy. Figure 15.18 shows the five level output voltage generated by VACOPDPWM strategy and its FFT plot is shown in Fig. 15.19. From Fig. 15.19, it is observed that the VACOPDPWM strategy produces significant 3rd, 5th, 38th and 40th harmonic energy.

15.5.1 Simulation of PDPWM Technique

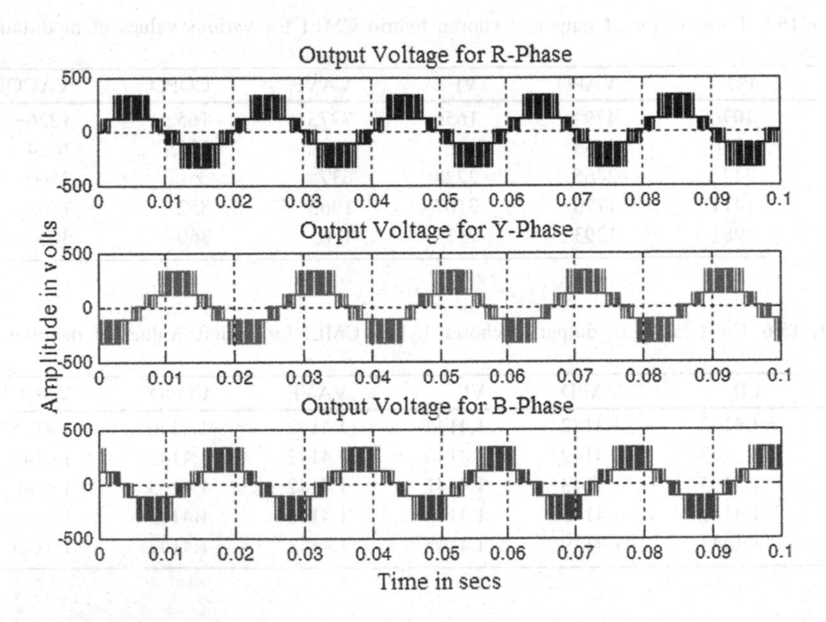

Fig. 15.8 Simulated output voltage generated by PDPWM technique for R-load

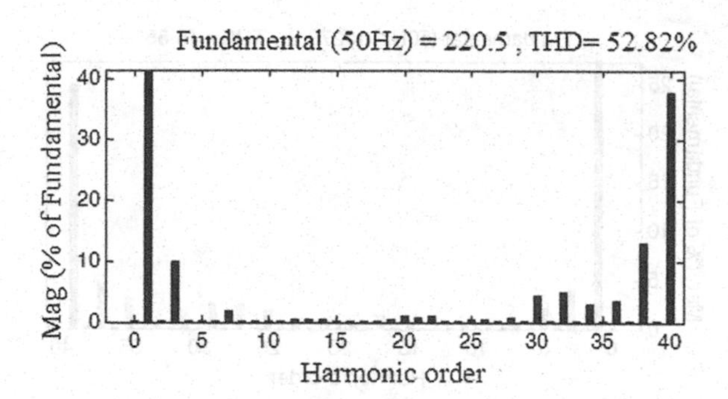

Fig. 15.9 FFT spectrum for PDPWM technique

15.5.2 Simulation of VAPDPWM Technique

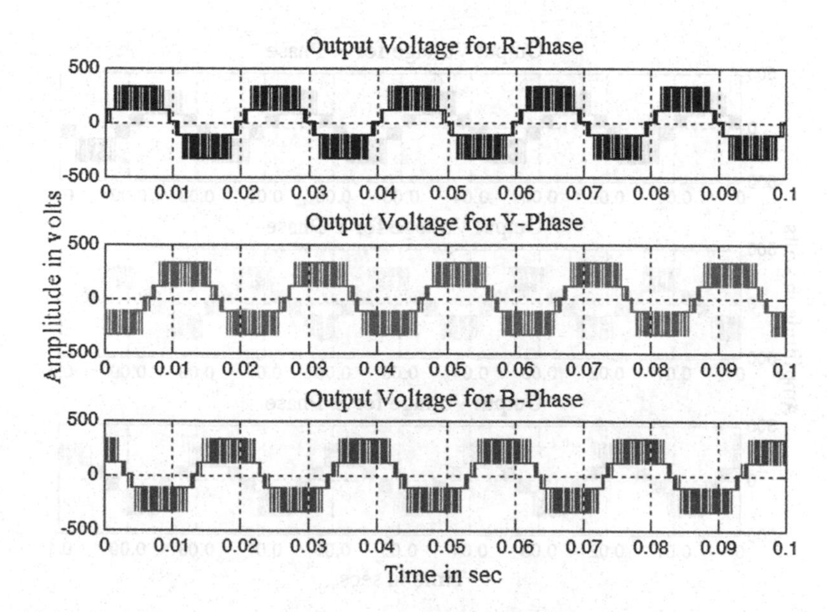

Fig. 15.10 Simulated output voltage generated by VAPDPWM technique for R-load

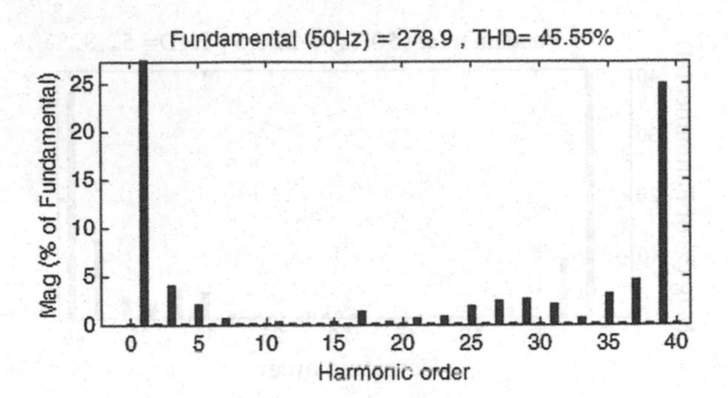

Fig. 15.11 FFT spectrum for VAPDPWM technique

15.5.3 Simulation of VFPWM Technique

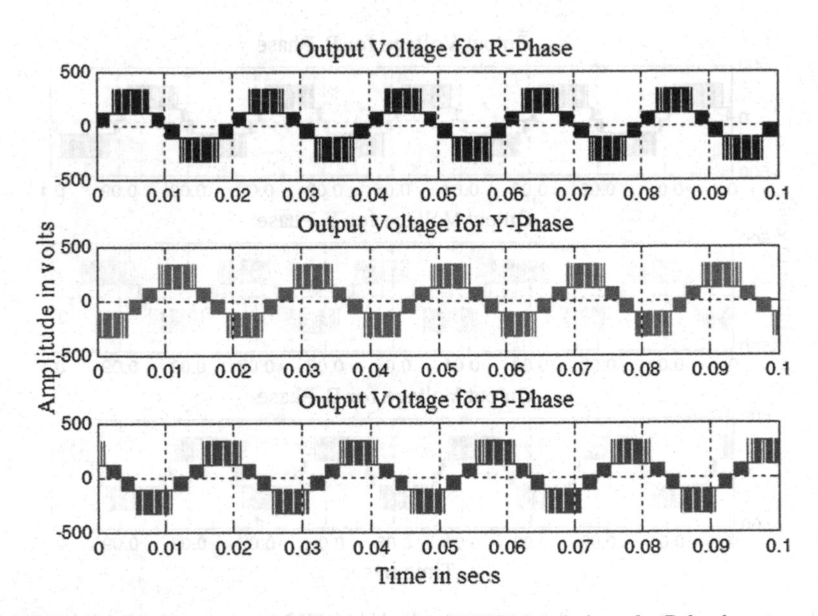

Fig. 15.12 Simulated output voltage generated by VFPWM technique for R-load

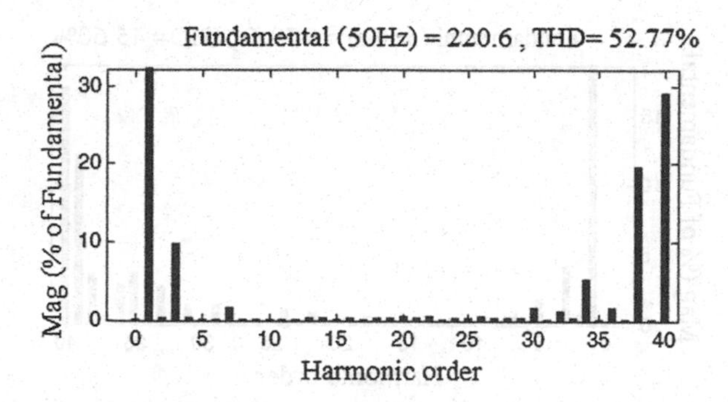

Fig. 15.13 FFT spectrum for VFPWM technique

15.5.4 Simulation of VAVFPWM Technique

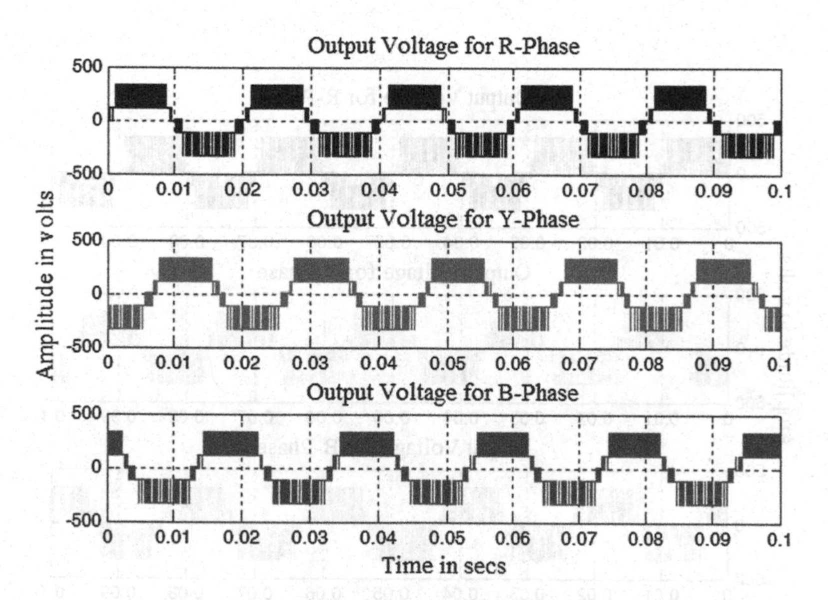

Fig. 15.14 Simulated output voltage generated by VAVFPWM technique for R-load

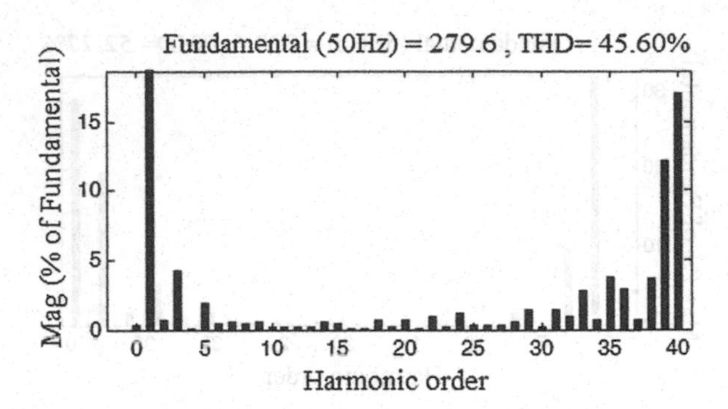

Fig. 15.15 FFT spectrum for VAVFPWM technique

15.5.5 Simulation of COPDPWM Technique

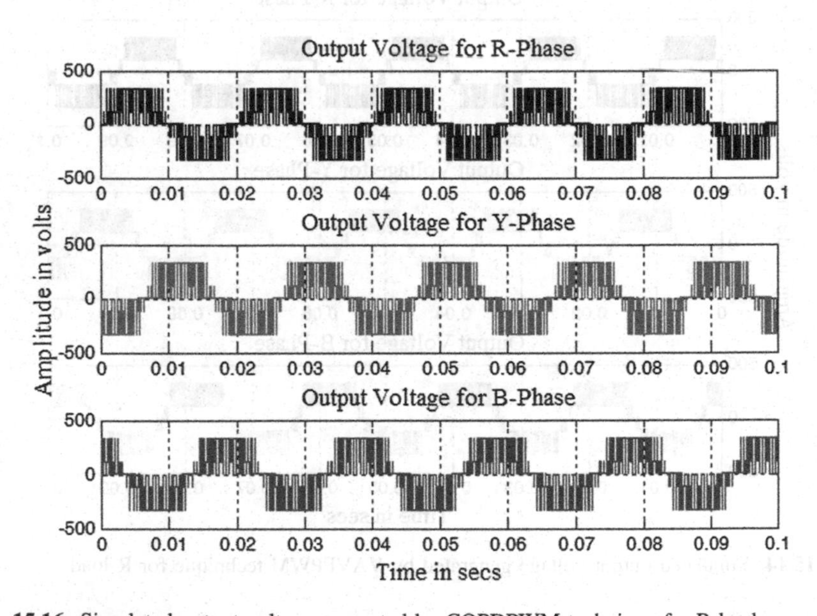

Fig. 15.16 Simulated output voltage generated by COPDPWM technique for R-load

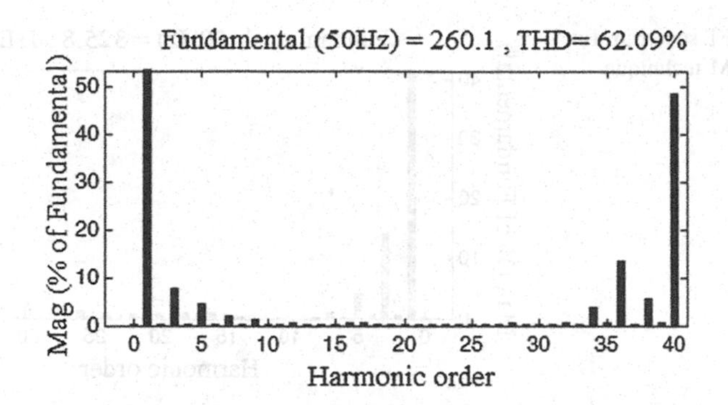

Fig. 15.17 FFT spectrum for COPDPWM technique

15.5.6 Simulation of VACOPDPWM Technique

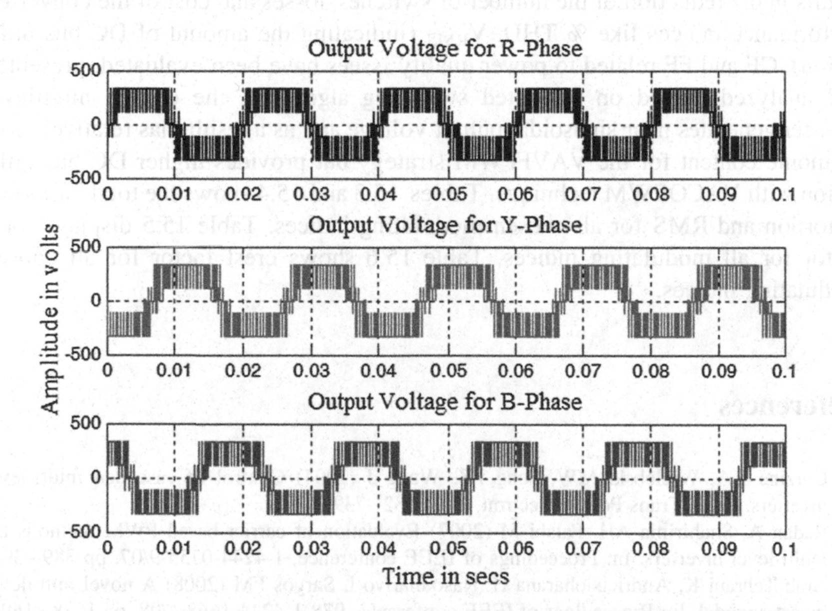

Fig. 15.18 Simulated output voltage generated by VACOPDPWM technique for R-load

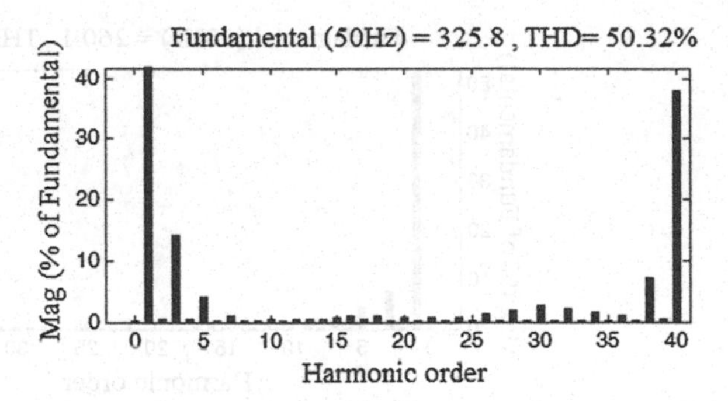

Fig. 15.19 FFT spectrum for VACOPDPWM technique

15.6 Conclusion

A hybrid cascaded symmetrical inverter has been proposed in this paper. The most important feature of the system is being convenient for expanding and increasing the number of output levels simply with less number of switches. This method results in the reduction of the number of switches, losses and cost of the converter. Performance indices like % THD, V_{RMS} (indicating the amount of DC bus utilization), CF and FF related to power quality issues have been evaluated, presented and analyzed. Based on presented switching algorithm, the chosen multilevel inverter generates near sinusoidal output voltage and as a result, has relatively low harmonic content for the VAVFPWM strategy but provides higher DC bus utilization with VACOPWM technique. Tables 15.3 and 15.4 shows the total harmonic distortion and RMS for all chosen modulating indices. Table 15.5 displays form factor for all modulating indices. Table 15.6 shows crest factor for all chosen modulating indices.

References

1. Corzine KA, Wielebski MW, Peng FZ, Wang J (2004) Control of cascaded multi-level inverters. IEEE Trans Power Electron 19(3):732–739
2. Radan A, Shahirinia AH, Falahi M (2007) Evaluation of carrier-based PWM methods for multi-level inverters. In: Proceedings of IEEE conference, 1-4244-0755-9/07, pp 389–394
3. Arab Tehrani K, Andriatsioharana H, Rasoanarivo I, Sargos FM (2008) A novel multilevel inverter model. In: Proceedings of IEEE conference, 978-1-4244-1668-4/08, pp 1688–1693
4. Lezana P, Rodríguez J, Oyarzún DA (2008) Cascaded multilevel inverter with regeneration capability and reduced number of switches. IEEE Trans Ind Electron 55(3):1059–1066
5. Pan Z, Peng FZ (2009) A sinusoidal PWM method with voltage balancing capability for diode clamped five level converters. IEEE Trans Ind Appl 45(3):1028–1034
6. Ruiz Caballero D, Martinez L, Reynaldo RA, Mussa SA (2009) New asymmetrical hybrid multilevel voltage inverter. In: Proceedings of IEEE conference, 978-1-4244-3370-4/09, pp 1–10
7. Konstantinou GS, Agelidis VG (2009) Performance evaluation of half-bridge cascaded multilevel converters operated with multicarrier sinusoidal PWM techniques. In: Proceedings of IEEE conference, 978-1-4244-2800-7/09, pp 3399–3404

8. Al-Judi A, Bierk H, Nowicki E (2009) A modified cascaded multilevel inverter with reduced switch count employing bypass diodes. In: Proceedings of IEEE conference, 978-1-4244-2601-0/09, pp 742–747
9. Ye M, Song P, Zhang C (2010) Study of harmonic elimination technology for multi-level inverters. In: Proceedings of IEEE conference, 978-1-4244-1718-6/10, pp 242–245
10. Malinowski M, Gopakumar K, Rodriguez J, Pérez MA (2010) A survey on cascaded multilevel inverters. IEEE Trans Ind Electron 57(7):2197–2206
11. Ahmed RA, Mekhilef S, Ping HW (2010) New multilevel inverter topology with minimum number of switches. In: Proceedings of international middle East power systems conference (TENCON'10), pp 1862–1867
12. Ruiz Caballero D, Sanhueza R, Arncibia S, Lopez M, Mussa SA, Heldwein ML (2011) Symmetrical hybrid multilevel inverter concept based on multi-stage switching cells. In: Proceedings of IEEE conference, 978-4577-1646-1/11, pp 776–781
13. Roshankumar P, Rajeevan PP, Mathew K, Gopakumar K, Leon JI, Franquelo LG (2012) A five level inverter topology with single dc supply by cascading a flying capacitor inverter and an H-bridge. IEEE Trans Power Electron 27(8):3505–3512
14. Batschauer AL, Mussa SA, Heldwein ML (2012) Three phase hybrid multilevel inverter based on half-bridge modules. IEEE Trans Ind Electron 59(2):668–678
15. Najafi E, Yatim AHM (2012) Design and implementation of a new multilevel inverter topology. IEEE Trans Ind Electron 59(11):4148–4154

Chapter 16
Comparative Study of Unipolar Multicarrier PWM Strategies for Five Level Diode Clamped Inverter

T. Sengolrajan and B. Shanthi

Abstract This paper presents the comparison of unipolar multicarrier Pulse Width Modulation (PWM) techniques for the Diode Clamped Multi Level Inverter (DCMLI). Due to switch combination redundancies, there are certain degrees of freedom to generate the five levels AC output voltage. The different types of unipolar PWM strategies for the chosen inverter are considered and the effectiveness of the developed strategies is demonstrated by the simulation. The results indicate that the multilevel inverter triggered by the developed sub-harmonic PWM strategy exhibits reduced harmonics. The results are presented and analysed.

Keywords DCMLI · PWM · Unipolar PWM · THD · COPWM

16.1 Introduction

Balamurugan et al. [1] analysed the performance of a three phase five level Bridge module type Diode Clamped Multilevel Inverter (DCMLI) using various modulating techniques for induction motor load. Behera et al. [2] described the harmonic analysis of the stepped output of the three level inverter based on number of triangular carriers and a sinusoidal modulating signal. Singh et al. [3] proposed the most relevant modulation methods such as multilevel sinusoidal pulse width

T. Sengolrajan (✉)
Department of EEE, Arunai Engineering College, Thiruvannamalai, India
e-mail: sengolmaha@gmail.com

B. Shanthi
Centralised Instrumentation and Service Laboratory,
Annamalai University, Chidambaram, India
e-mail: au_shan@yahoo.com

R. Malathi and J. Krishnan (eds.), *Recent Advancements in System Modelling Applications*, Lecture Notes in Electrical Engineering 188,
DOI: 10.1007/978-81-322-1035-1_16, © Springer India 2013

modulation, multilevel selective harmonic elimination, and space-vector modulation. The use of multilevel approach is believed to be a promising alternative in very high power conversion processing [4]. Chaturvedi et al. [5] described about the reduced switching loss pulse width modulation technique for three level diode clamped inverter. The simulation studies on different control techniques for three and five level DCMLI were presented in [6]. The performance of each technique has been investigated based upon reduction in THD. Aquila et al. [7] presented a digital implementation of DCMLI employing a fast microprocessor and logic circuits.

The idea of control degrees of freedom combination and the validity of the PWM strategies are demonstrated by simulation and experimentation of three and five level inverters by Wang et al. [8]. This is based on a direct extension of three level inverter to higher level. NagaHaskar Reddy et al. [9] described the advanced modulating techniques for diode clamped multilevel inverter fed induction motor. Kouro et al. [10] introduced multicarrier PWM with DC-link ripple feed-forward compensation for MLI. Sambath et al. [11] discussed a new modulation strategy for a single phase five level H-bridge type DCMLI with reduced components as compared to conventional DCMLI. Shanthi and Natarajan [12] presented the comparison of Carrier Overlapping Pulse Width Modulation (COPWM) techniques for the Diode clamped Multi Level Inverter (DCMLI) and also discussed the use of Control Freedom Degree (CFD) combination. A comparative study on various unipolar PWM strategies for single phase five-level cascaded inverter is described in [13]. Sule Ozdemir et al. [14] discussed the elimination of harmonics in a five level diode clamped multilevel inverter using fundamental modulation. Urmila and Subbarayudu [15] presented a comparative study of pulse width modulation techniques. This paper discusses a comparative study carried out on unipolar PWM strategies for chosen five level DCMLI by simulation.

16.2 Diode Clamped Multilevel Inverter

Multilevel inverters are being considered for an increasing number of applications due to their high power capability associated with lower output harmonics and lower commutation losses. Multilevel inverters have become an effective and practical solution for increasing power and reducing harmonics of AC load. Diode clamped multilevel inverter is a very general and widely used topology. DCMLI works on the concept of using diodes to limit voltage stress on power devices. A DCMLI typically consists of $(m - 1)$ capacitors on the DC bus where m is the total number of positive, negative and zero levels in the output voltage. Figure 16.1 shows a single phase half-bridge five level diode clamped inverter. The order of numbering of the switches is S1, S2, S3, S4, S1′, S2′, S3′ and S4′. The DC bus consists of four capacitors C1, C2, C3 and C4 acting as voltage divider. For a DC bus voltage V_{dc}, the voltage across each capacitor is $V_{dc}/4$ and voltage stress on each device is limited to $V_{dc}/4$ through clamping diode. The middle point of the four capacitors 'b' can be defined as the neutral point.

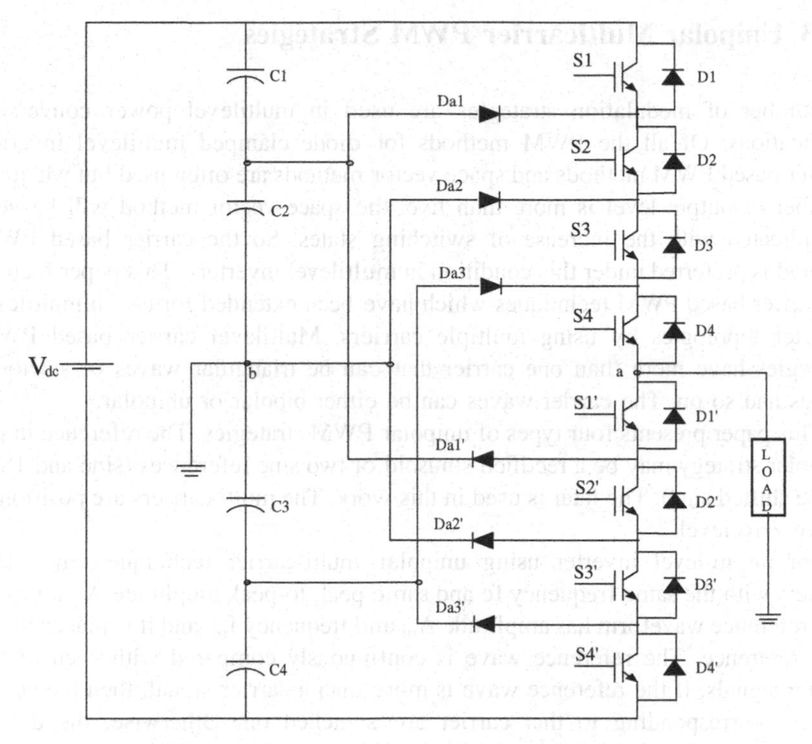

Fig. 16.1 Five level diode clamped inverter

The principle of diode clamping to DC-link voltages can be extended to any number of voltage levels. Since the voltages across the semiconductor switches are limited by conduction of the diodes connected to the various DC levels, this class of multilevel inverter is termed diode clamped MLI. The switches are arranged into 4 pairs (S1, S1′), (S2, S2′), (S3, S3′), (S4, S4′). If one switch of the pair is turned on, the complementary switch of the same pair must be off. Four switches are triggered at any point of time to select the desired level in the five level DCMLI.

Table 16.1 shows the output voltage levels and the corresponding switch states for the chosen five level DCMLI. The output voltage V_{ab} has five states: $V_{dc}/2$, $V_{dc}/4$, 0, $-V_{dc}/4$ and $-V_{dc}/2$. The gate signals for the chosen five level DCMLI are developed using MATLAB-SIMULINK. The gate signal generator model developed is tested for various values of modulation index.

Table 16.1 Switching scheme for single phase five level inverter

S1	S2	S3	S4	S1′	S2′	S3′	S4′	Vab
1	1	1	1	0	0	0	0	$+V_{dc}/2$
0	1	1	1	1	0	0	0	$+V_{dc}/4$
0	0	1	1	1	1	0	0	0
0	0	0	1	1	1	1	0	$-V_{dc}/4$
0	0	0	0	1	1	1	1	$-V_{dc}/2$

16.3 Unipolar Multicarrier PWM Strategies

A number of modulation strategies are used in multilevel power conversion applications. Of all the PWM methods for diode clamped multilevel inverter, carrier based PWM methods and space vector methods are often used but when the number of output level is more than five, the space vector method will be very complicated with the increase of switching states. So the carrier based PWM method is preferred under this condition in multilevel inverters. This paper focuses on carrier based PWM techniques which have been extended for use in multilevel inverter topologies by using multiple carriers. Multilevel carrier based PWM strategies have more than one carrier that can be triangular waves or sawtooth waves and so on. The carrier waves can be either bipolar or unipolar.

This paper presents four types of unipolar PWM strategies. The reference in the unipolar strategy may be a rectified sinusoid or two sine references (sine and 1800 phase shifted sine). The later is used in this work. The multi carriers are positioned above zero level.

For an m-level inverter using unipolar multi-carrier technique, (m − 1)/2 carriers with the same frequency fc and same peak-to-peak amplitude A_c are used. The reference waveform has amplitude A_m and frequency f_m. and it is placed at the zero reference. The reference wave is continuously compared with each of the carrier signals. If the reference wave is more than a carrier signal, then the active devices corresponding to that carrier are switched on. Otherwise, the device switches off. The frequency ratio m_f is defined as: $m_f = f_c/f_m$. In this paper, $m_f = 21$ and m_a is varied from 0.5 to 1.

16.3.1 *Unipolar Sub Harmonic PWM Strategy*

The principle of the USHPWM strategy is to use several triangular carriers with two modulation waves. For an m-level inverter, (m − 1)/2 triangular carriers of the same frequency fc and the same peak-to-peak amplitude A_c are disposed so that the bands they occupy are contiguous. The carrier set is placed above the zero reference.m_a for USHPWM strategy is $= A_m/(n \times A_c)$, where n is the number of carriers. Carrier arrangement for 5-level USHPWM are shown in Fig. 16.2 for $m_a = 0.8$.

16.3.2 *Unipolar Carrier Overlapping PWM Strategy*

The UCOPWM uses two carrier signals of peak-to-peak amplitude A_c and they overlap with each other. The gate signals for the MLI are derived by comparing the two overlapping carriers with two sinusoidal references. Figure 16.3 shows the

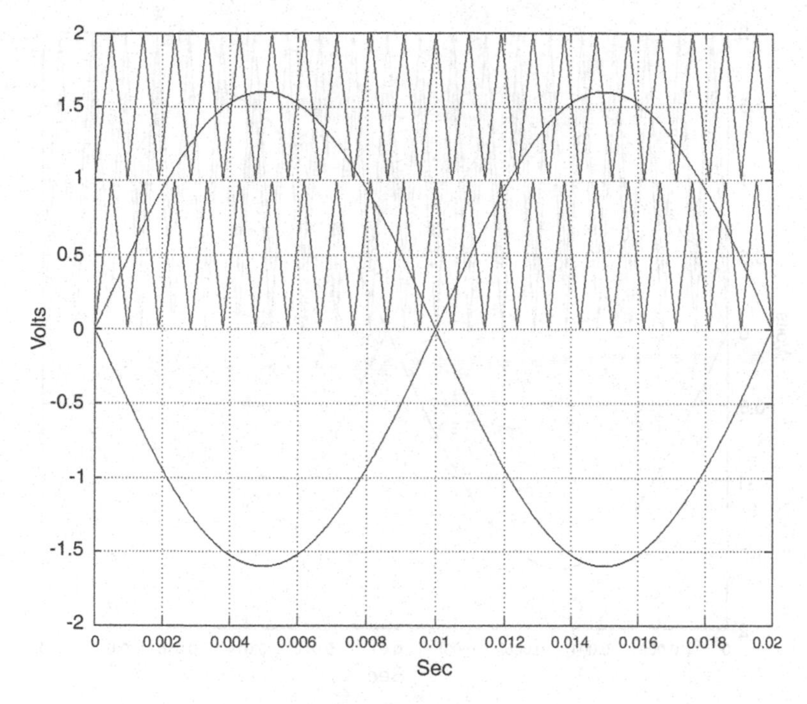

Fig. 16.2 Carrier arrangement for USHPWM strategy

carrier arrangement for the chosen MLI with UCOPWM strategy. The amplitude modulation index is defined for this strategy as follows:

$$m_a = A_m / \left((m/4) \times A_c\right)$$

16.3.3 Unipolar Inverted Sine Carrier Sub Harmonic PWM Strategy

This control scheme uses inverted (high frequency) sine carriers which are placed above zero reference. The fundamental frequency sinusoids of 0 and 180° phase are selected as the modulating waves. m_a for this strategy is same as that of USHPWM. Carriers for 5-level inverter with UISCSHPWM strategy are shown in Fig. 16.4 for $m_a = 0.8$.

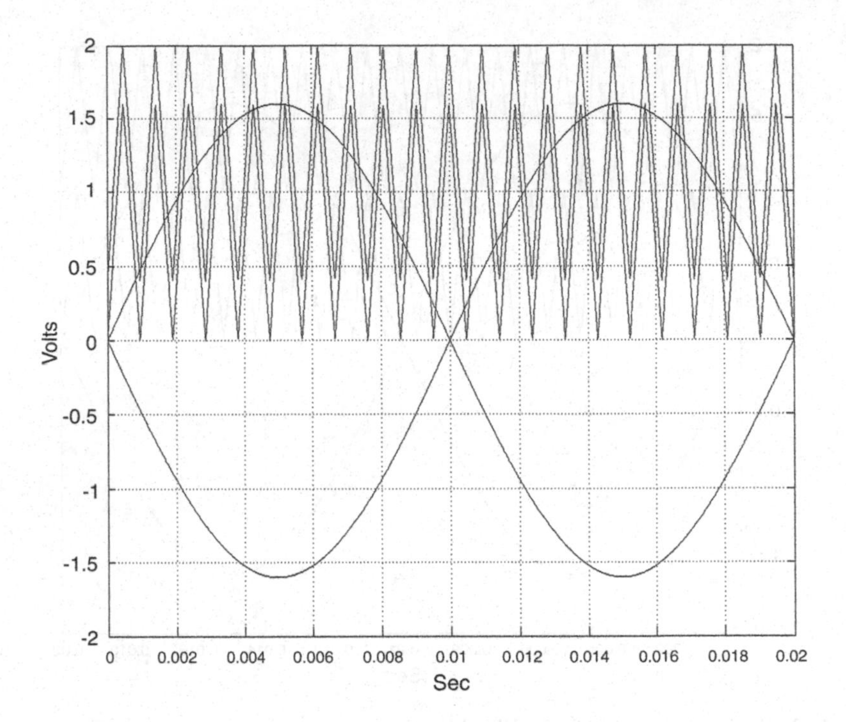

Fig. 16.3 Carrier arrangement of UCOPWM strategy

16.3.4 Unipolar Inverted Sine Carrier Overlapping PWM Strategy

In this pattern, the two inverted sine carriers are overlapped with other. Carriers for 5-level inverter with UISCOPWM strategy are shown in Fig. 16.5 for $m_a = 0.8$. Inverted sine carriers generated by using (i) sinusoidal voltage source and rectifier blocks in Simpower system module of Simulink or (ii) sine wave generator block and s-function based rectifier block.

16.4 Simulation Results

The diode clamped five level inverter is modeled in SIMULINK using power system block set. Switching signals for diode clamped multilevel inverter using unipolar PWM techniques arc simulated. Simulations are performed for different values of m_a ranging from 0.5 to 1 and the corresponding % THD is measured using the FFT block and their values are shown in Table 16.2. Figures 16.6, 16.7, 16.8, 16.9, 16.10, 16.11, 16.12, 16.13 show the simulated output voltage of

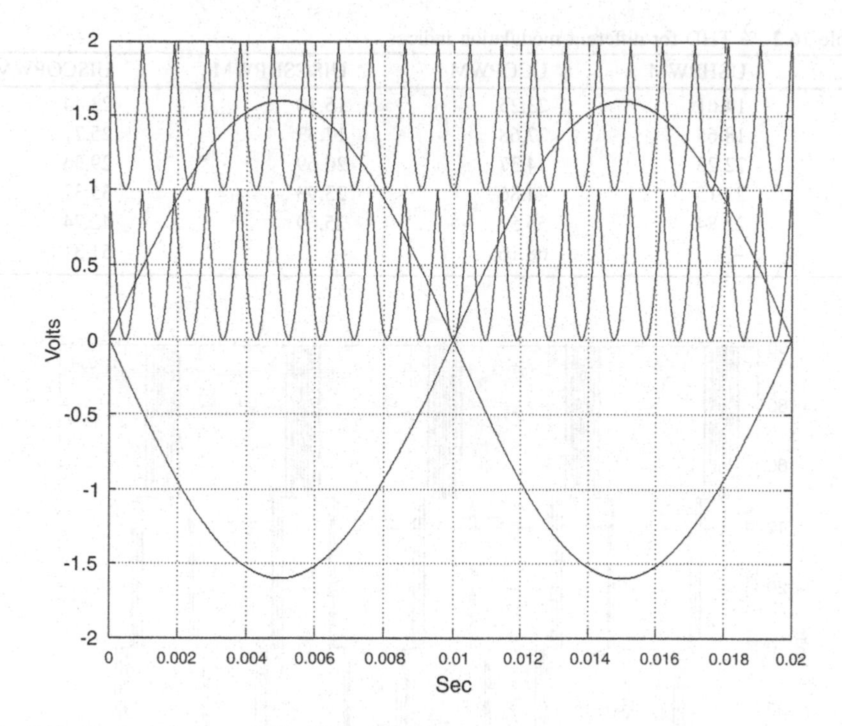

Fig. 16.4 Carrier arrangement for UISCSHPWM strategy

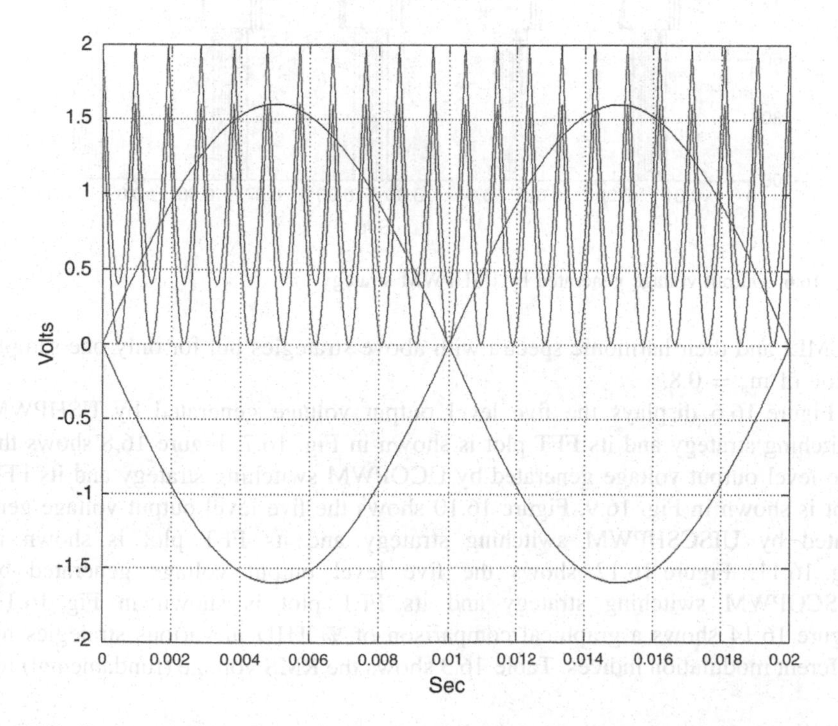

Fig. 16.5 Carrier arrangement for UISCOPWM strategy

Table 16.2 % THD for different modulation indices

m_a	USHPWM	UCOPWM	UISCSHPWM	UISCOPWM
1.0	14.61	23.25	15.10	23.44
0.9	18.65	27.68	17.33	25.71
0.8	22.23	34.37	20.69	29.36
0.7	23.15	44.86	23.93	35.32
0.6	22.84	56.52	25.69	42.74
0.5	–	68.16	–	51.31

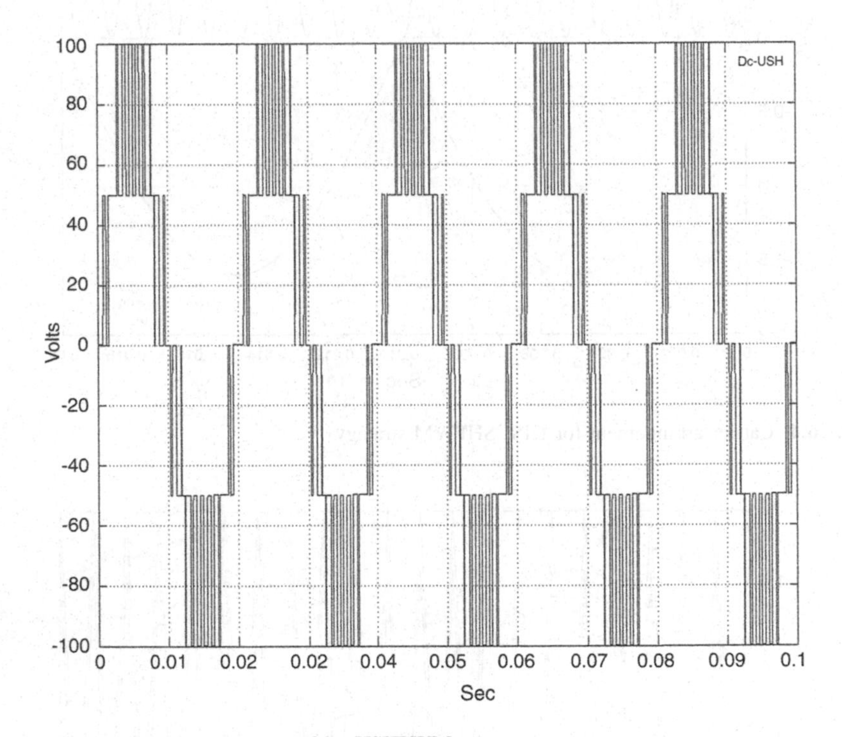

Fig. 16.6 Output voltage generated by USHPWM strategy

DCMLI and their harmonic spectra with above strategies but for only one sample value of $m_a = 0.8$.

Figure 16.6 displays the five level output voltage generated by USHPWM switching strategy and its FFT plot is shown in Fig. 16.7. Figure 16.8 shows the five level output voltage generated by UCOPWM switching strategy and its FFT plot is shown in Fig. 16.9. Figure 16.10 shows the five level output voltage generated by UISCSHPWM switching strategy and its FFT plot is shown in Fig. 16.11. Figure 16.12 shows the five level output voltage generated by UISCOPWM switching strategy and its FFT plot is shown in Fig. 16.13. Figure 16.14 shows a graphical comparison of % THD in various strategies for different modulation indices. Table 16.3 shows the RMS voltage (fundamental) for

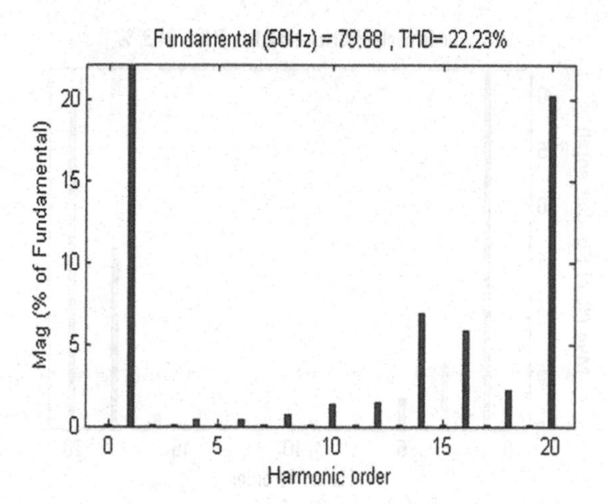

Fig. 16.7 FFT plot of output voltage for USHPWM strategy

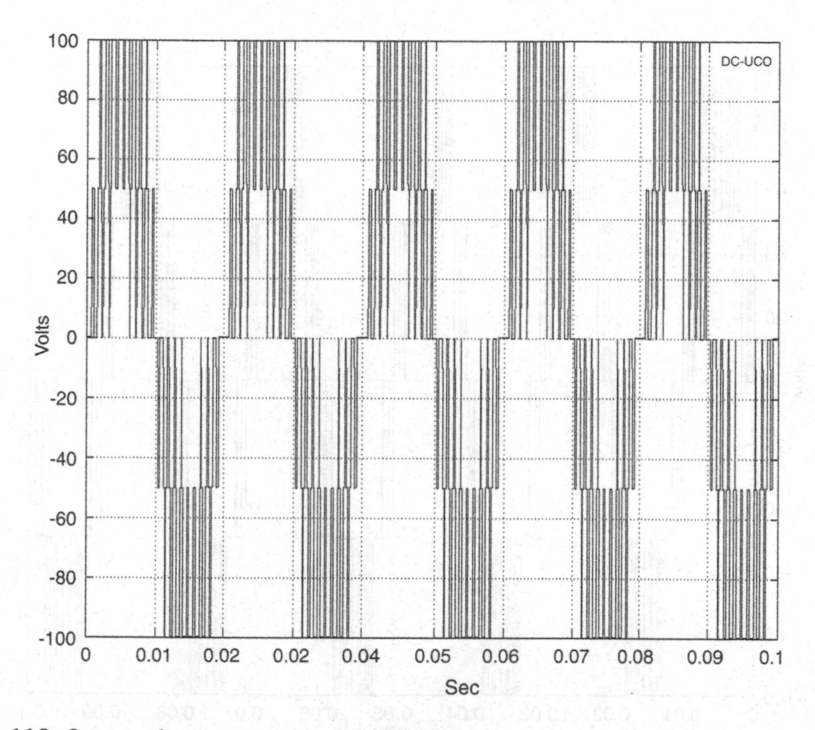

Fig. 16.8 Output voltage generated by UCOPWM strategy

various values of modulation index. Tables 16.4 and 16.5 show the Distortion Factor (DF) and Crest Factor (CF) of the output voltage of chosen MLI. The following parameter values are used for simulation: $V_{dc} = 200$ V, R (load) = 100 Ω, $f_c = 1050$ Hz, $f_m = 50$ Hz.

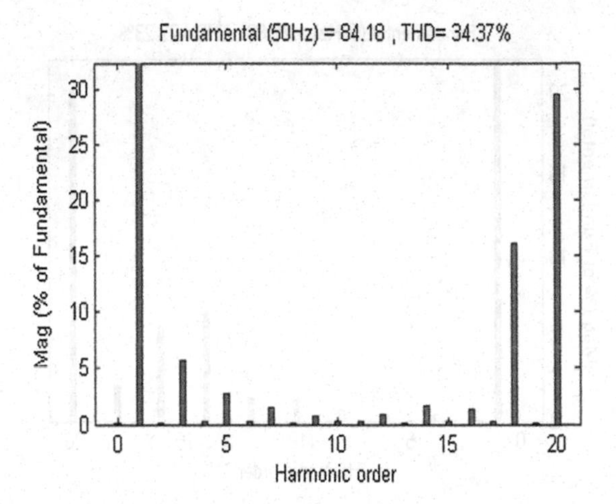

Fig. 16.9 FFT plot of output voltage for UCOPWM strategy

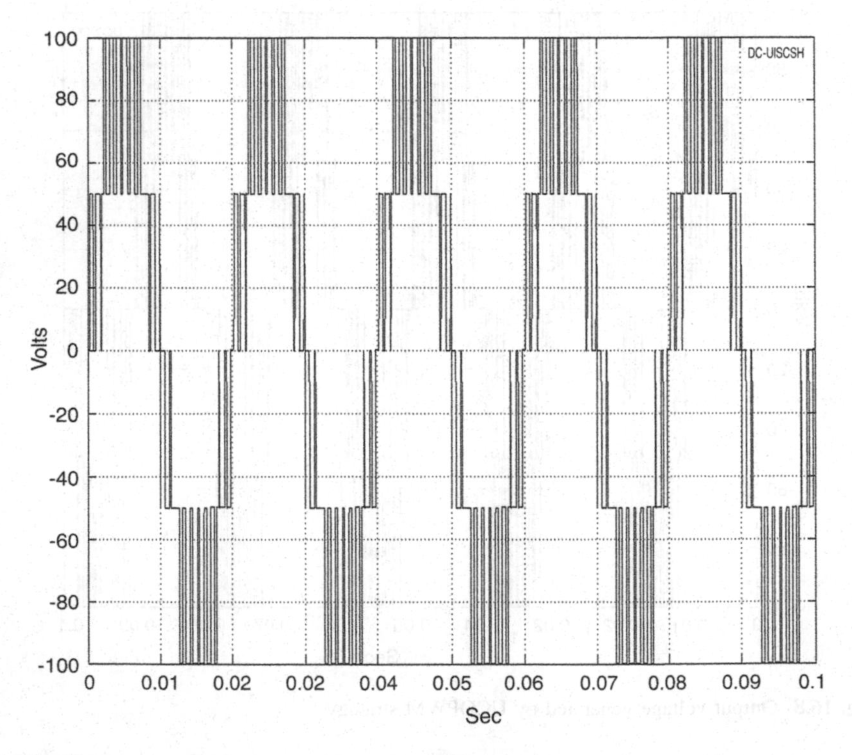

Fig. 16.10 Output voltage generated by UISCSHPWM strategy

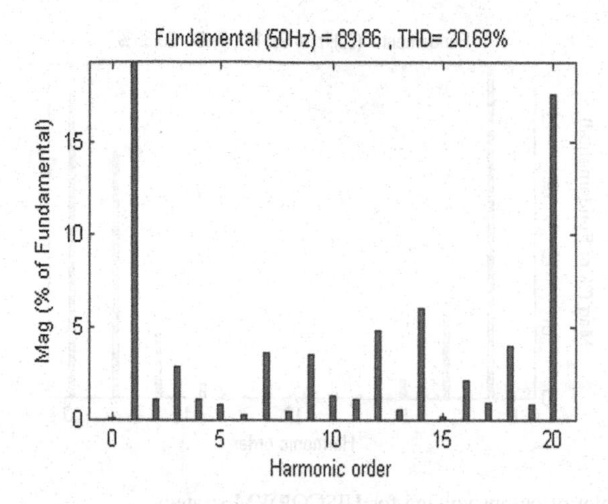

Fig. 16.11 FFT plot of output voltage for UISCSHPWM strategy

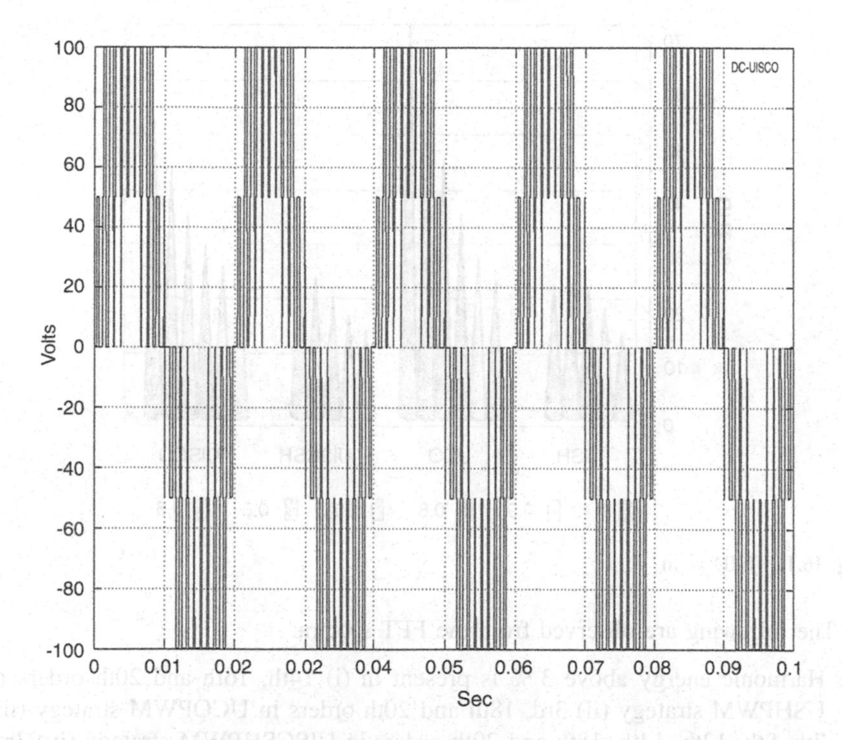

Fig. 16.12 Output voltage generated by UISCOPWM strategy

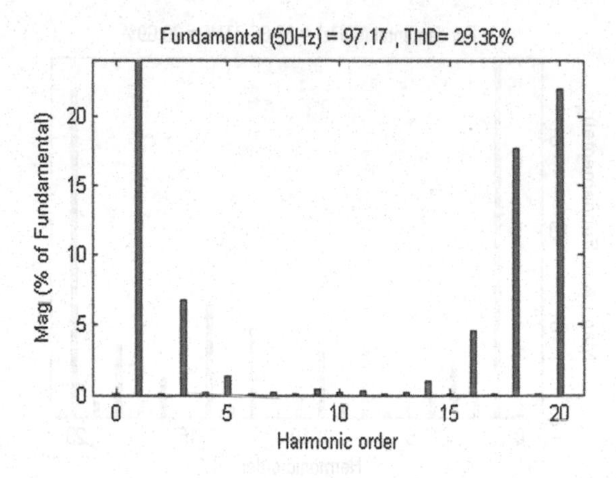

Fig. 16.13 FFT plot of output voltage for UISCOPWM strategy

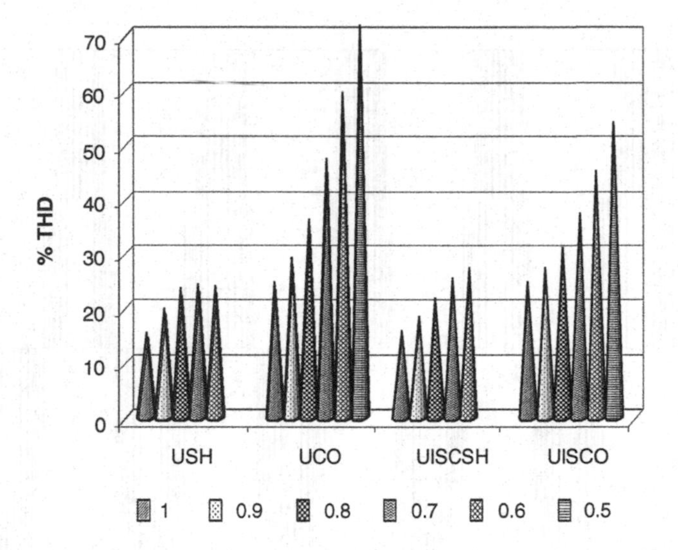

Fig. 16.14 THD vs m_a

The following are observed from the FFT spectra:

(a) Harmonic energy above 3 % is present in (i) 14th, 16th and 20th orders in USHPWM strategy (ii) 3rd, 18th and 20th orders in UCOPWM strategy (iii) 7th, 9th, 12th, 14th, 18th and 20th orders in UISCSHPWM strategy (iv) 3rd, 16th, 18th and 20th orders in UISCOPWM strategy.
(b) 3rd order harmonic is dominant in carrier overlapping PWM strategies.
(c) Dominant lower side band harmonic (20th order) is present in all unipolar PWM strategies.
(d) UISCSHPWM contains more number of dominant harmonics.

Table 16.3 V_{RMS} (fundamental) for different modulation indices

m_a	USHPWM	UCOPWM	UISCSHPWM	UISCOPWM
1.0	70.65	72.55	74.64	77.69
0.9	63.56	66.56	69.33	73.59
0.8	56.49	59.52	63.54	68.7
0.7	49.43	50.68	56.8	62.39
0.6	42.41	41.94	49.08	55.74
0.5	–	33.19	–	48.64

Table 16.4 % DF for different modulation indices

m_a	USHPWM	UCOPWM	UISCSHPWM	UISCOPWM
1.0	0.06	0.2	0.64	1.1
0.9	0.3	0.38	0.55	0.93
0.8	0.07	0.66	0.43	0.76
0.7	0.12	0.77	0.36	0.68
0.6	0.07	0.85	0.43	0.58
0.5	–	0.99	–	0.41

Table 16.5 % CF for different modulation indices

m_a	USHPWM	UCOPWM	UISCSHPWM	UISCOPWM
1.0	1.4	1.34	1.32	1.25
0.9	1.55	1.45	1.42	1.32
0.8	1.73	1.59	1.54	1.39
0.7	1.97	1.79	1.71	1.51
0.6	2.29	2.07	1.97	1.65
0.5	–	2.49	–	1.83

From Table 16.3 it is observed that DC bus utilization is higher with carrier overlapping strategies having inverted Sine carriers. It is found from Table 16.4 that % DF is least with USHPWM strategy. It is inferred from Table 16.5 that CF is lower for carrier overlapping strategies.

16.5 Conclusions

In this paper, various unipolar PWM strategies for chosen for DCMLI been developed using MATLAB SIMUINK model and analysed for different modulation indices ranging from 0.5 to 1. Various performance factors like (i) THD, DF, and harmonic spectra indicating purity of the output voltage (ii) CF which is a measure of the stress on the device and (iii) V_{RMS} indicating the amount of DC bus utilization have been evaluated, presented and analysed. It is observed that the sub harmonic strategies provide lower THD than the overlapping strategies

(Table 16.2 and Fig. 16.14). The maximum DC bus utilization is achieved in UISCOPWM strategy (Tables 16.3). The simulation results validated for various strategies are presented.

References

1. Balamurugan CR, Natarajan SP, Bensraj R (2012) Performance and evaluation of three phase bridge module type diode clamped multilevel inverter. Int J Eng Trends Technol 3(3):380–389
2. Behera RK, Dixit TV, Das SP (2006) Analysis of experimental investigation of various carrier-based modulation schemes for three level neutral point clamped inverter-fed induction motor drive. In: Proceedings of IEEE conference record, 0-7803-9771-1/06
3. Singh B, Mittal N, Verma KS, Singh D, Singh SP (2012) Multi-level inverter: a literature survey on topologies and control strategies. Int J Rev Comput 10:1–16
4. Bhagwat PM, Stefanovic VR (1983) Generalized structure of a multilevel PWM inverter. IEEE Trans Ind Appl 19(6):1057–1069
5. Chaturvedi PK, Jain S, Agarwal P (2011) Reduced switching loss pulse width modulation technique for three-level diode clamped inverter. IET Power Electron 4(4):393–399
6. Chaturvedi PK, Jain SK, Agrawal P, Modi PK (2006) Investigations on different multilevel inverter control techniques by simulation. In: Proceedings of IEEE conference on record, 0-7803-9771-1/06
7. Aquila D, Formosa R, Montaruli E, Zanchetta P (2000) Novel multilevel PWM inverter implementation. In: Proceedings of IEEE conference record, 0-7803-3932-0, pp 710–715
8. Wang H, Zhao R, Deng Y, He X (2003) Novel carrier-based PWM methods for multilevel Inverter. In: Proceedings of the IEEE conference record, 0-7803-7906-3/03, pp 2777–2782
9. Naga Bhaskar Reddy V, Babu CS, Suresh K (2011) Advanced modulating techniques for diode clamped multilevel inverter fed induction motor. ARPN J Eng Appl Sci 6(1):90–99
10. Kouro S, Lezana P, Angulo M, Rodriguez J (2008) Multicarrier PWM with DC-link ripple feed-forward compensation for multilevel inverters. IEEE Trans Power Electron 23(1):52–59
11. Sambath E, Natarajan SP, Balamurugan CR (2012) Performance evaluation of single phase H-bridge type diode clamped five level inverter. Int J Mod Eng Res (IJMER) 2(4):1908–1913
12. Shanthi B, Natarajan SP (2011) Comparative study on carrier overlapping PWM strategies for five level diode clamped inverter. Int J Electr Eng Inf 3:12–25
13. Shanthi B, Natarajan SP (2010) Comparative study on various unipolar PWM strategies for single phase Five-level cascaded inverter. Int J Power Electron 2(1):36–50
14. Ozdemir S, Ozdemir E, Tolfort LM, Khomfoi S (2007) Elimination of harmonics in a five level diode clamped multilevel inverter using fundamental modulation. In: Proceedings of the IEEE conference record, 1-4244-0645-5/07, pp 850–854
15. Urmila B, Subbarayudu D (2010) A comparative study of pulse width modulation techniques. J Scientific Eng Res 1(13):1–15

Chapter 17
A New Three-Level Zero Voltage Switching Converter

C. Karthikeyan and K. Duraiswamy

Abstract This paper deals with the design issues relevant to achieve ZVS for three level converters. It shows the method of designing a three level converter and to achieve ZVS in the wide range of load current and input voltage by employing coupled inductor. This converter overcomes the drawbacks presented by the conventional zero-voltage switching (ZVS) three-level converter, such as high circulating energy, severe parasitic ringing on the rectifier diodes, loss of duty cycle, high conduction loss and limited ZVS load range for the primary switches. This converter employs a coupled inductor to achieve zero-voltage switching of the primary switches in the entire line and load range is described. Because the coupled inductor does not appear as a series inductance in the load current path, it does not cause a loss of duty cycle or severe voltage ringing across the output rectifiers.

Keywords Zero voltage switching · Three level converter · Switching converter

17.1 Introduction

In recent years, multilevel power converters have received a lot of attention due to their suitability for applications with high input voltages [1]. Specifically, multilevel inverters and DC–DC converters can be implemented with semiconductor switches rated at a fraction of the input voltage, which are typically less expensive

C. Karthikeyan (✉)
Department of EEE, K.S.R. College of Engineering, Tiruchengode, Tamil Nadu, India
e-mail: karthykeya@yahoo.com

K. Duraiswamy
K.S.R College of Technology, Tiruchengode, Tamil Nadu, India

R. Malathi and J. Krishnan (eds.), *Recent Advancements in System Modelling Applications*, Lecture Notes in Electrical Engineering 188,
DOI: 10.1007/978-81-322-1035-1_17, © Springer India 2013

and more efficient than their high-voltage-rated counterparts. Because the implementation complexity of multilevel converters is increased dramatically by the number of levels, which diminishes the benefits of multilevel conversion, the majority of development efforts in DC–DC multilevel conversion have been focused on three-level converters.

Generally, three-level DC–DC converters feature power conversion with semiconductor switches rated at one-half of the input voltage. The major deficiencies of the ZVS implementations described in [2, 3] are brought about by an increased inductance in the primary circuit that is required to achieve a complete ZVS of all primary switches down to light loads. This inductance, which is obtained by intentionally increasing leakage inductance of the transformer and/or by adding an external inductance in the transformer and/or by adding an external inductance in series with the primary of the transformer, has a detrimental effect on the performance. It introduces a circulating current on the primary side, causes a secondary—side loss of duty cycle, and produces severe parasitic ringing on the secondary side of the transformer as it resonates with the rectifier's junction capacitance.

The circulating current caused by excessive energy stored in the inductance employed to extend the ZVS range down to light loads increases the current stress of the primary switches and the primary-side conduction losses at heavy load. The primary side conduction losses are further increased due to the secondary side duty cycle loss which must be compensated by reducing the turn's ratio of the transformer. Furthermore, a smaller turn's ratio of the transformer also increases the voltage stress on the secondary-side rectifiers so that rectifiers with a higher voltage rating that typically exhibit a higher Conduction loss may be required [4, 5]. Finally, to control the ringing voltage across the output rectifiers, a lossy snubber circuit is required on the secondary side which also reduces the conversion efficiency.

The proposed three-level ZVS converter employs a coupled inductor on the primary side to achieve ZVS in the entire line and load range. Since this coupled inductor does not appear as a series inductance in the load current path, it does not cause loss of duty cycle or severe voltage ringing across the output diode. As a result, the proposed circuit exhibits increased conversion efficiency.

17.2 Three-Level ZVS Converter with Coupled Inductor

It shows a circuit diagram of the proposed three-level soft-switched DC–DC converter that employs a coupled inductor on the primary side to extend the ZVS range of the primary switches with a minimum circulation energy and conduction loss. The three-level converter consists of a series connection of four primary switches Q_1 through Q_4, rail-splitting capacitors C_{IN1} and C_{IN2}, "flying capacitors" C_{S1} and C_{S2}, isolation transformer TR, and coupled inductor L_C. In this circuit, the load is coupled to the converter through a full-wave rectifier connected to the center-tapped secondary of the transformer. In addition, clamping diodes D_{C1} and D_{C2} are used to clamp the voltage of outer switches Q_1 and Q_4, respectively, to $V_{IN}/2$ after the switches are turned off. Finally, blocking capacitor C_B is employed to prevent transformer saturation.

17.3 Principle of Operation

In the simplified circuit, it is assumed that inductance of output filter L_F is large enough so that during a switching cycle the output filter can be modeled as a constant current source with the magnitude equal to output current i_o. Also, it is assumed that the capacitances of capacitors C_{IN1} and C_{IN2}, which form a capacitive divider that splits the input voltage in half, are large so that capacitors C_{IN1} and C_{IN2} can be modeled by voltage sources $V_1 = V_{IN}/2$ and $V_2 = V_{IN}/2$, respectively. Similarly, it is assumed that the capacitances of capacitors C_{S1} and C_{S2} are large enough so that the capacitors can be modeled as constant voltage sources V_{CS1} and V_{CS2}, respectively.

The average voltages of the coupled inductor windings and the transformer windings during a switching cycle are zero and for the phase-shift control the outer pair of switches and the inner pair of switches operate with 50 % duty cycle, the magnitude of voltage across flying capacitor V_{CS1} and V_{CS2} is equal to half of the voltage across the rail-splitting capacitors, i.e., $V_{CS1} = V_{CS2} = V_{IN}/4$. To further simplify the analysis of operation of the circuit, it is assumed that the resistance of the conducting semiconductor switches is zero, whereas the resistance of the non conducting switches is infinite. In addition, the leakage inductances of both transformer TR and coupled inductor L_C, as well as the magnetizing inductance of transformer TR [6] are neglected since their effect on the operation of the circuit is not significant. However, the magnetizing inductance of coupled inductor L_C and output capacitances C_1–C_4 of primary switches are not neglected in this analysis since they play a major role in the operation of the circuit. Consequently, coupled inductor L_C is modeled as the ideal transformer with turns ratio $n_{LC} = 1$ and with parallel magnetizing inductance L_{MC} across the series connection of windings AC and CB [7], where as transformer TR is modeled only by the ideal transformer with turns ratio n_{TR}. It should be noted that magnetizing inductance of inductor represents the inductance measured between terminals A and B with terminal C open. The following relationships between currents can be established [6]:

$$i_P = i_{P1} + i_{P2} \tag{17.1}$$

$$i_1 = i_{P1} + i_{MC}. \tag{17.2}$$

$$i_2 = i_{P2} - i_{MC} \tag{17.3}$$

Since the number of turns of winding AC and winding CB of coupled inductor L_C are the same, it must be that

$$i_{P1} = i_{P2}. \tag{17.4}$$

Substituting (17.4) into (17.1)–(17.3) gives

$$i_{P1} = i_{P2} = \frac{i_P}{2} \tag{17.5}$$

$$i_1 = \frac{i_P}{2} + i_{MC} \tag{17.6}$$

$$i_2 = \frac{i_P}{2} + i_{MC}. \tag{17.7}$$

As can be seen from (17.6) and (17.7), currents and are composed of two components: (1) primary-current component $i_P/2$ and (2) magnetizing-current component i_{MC}. The primary-current component directly depends on the load current, whereas the magnetizing current does not directly depend on the load, but rather on the volt-second product across the magnetizing inductance. Namely, a change of the magnetizing current with a change in the load current occurs only if the phase shift between the turn on instants of outer switches S_1 and S_4 and respective inner switches S_2 and S_3 is changed to maintain the output regulation. Usually, the change of phase shift with a load change is greater at light loads, i.e., as the load decreases toward no load than at heavier loads. Since in the circuit in Fig. 17.1 the phase shift increases as the load approaches zero, the volt-second product of L_{MC} also increases so that the circuit in Fig. 17.1 exhibits the maximum magnetizing current at no load [5], which makes it possible to achieve ZVS at no load. Because magnetizing current i_{MC} does not contribute to the load current, it represents a circulating current. Generally, this circulating current and its associated energy should be minimized to reduce losses and maximize the conversion efficiency. Due to an inverse dependence of the volt-second product of L_{MC} on the load current, circuit in Fig. 17.1 circulates less energy at full load than at light load, and, therefore, features ZVS in a wide load range with a minimum circulating current. It can be seen that

$$V_{AB} = V_{AC} + V_{CB} \tag{17.8}$$

Since both windings of coupled inductor L_C have the same number of turns, i.e., since the turns ratio of L_C is $n_{LC} = 1$, it must be that

$$V_{AC} = V_{CB} \tag{17.9}$$

Or

$$V_{AC} = V_{CB} = \frac{V_{AB}}{2} \tag{17.10}$$

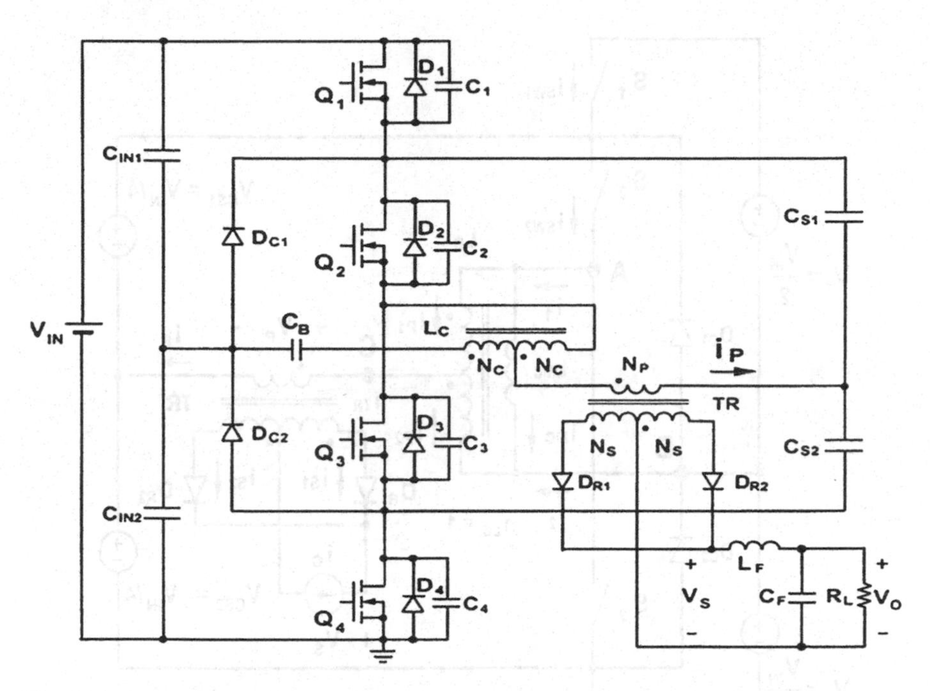

Fig. 17.1 Proposed three-level ZVS converter with coupled inductor

Generally for constant-frequency phase-shift control, voltage V_{AB} is a square wave voltage consisting of alternating positive and negative pulses of magnitude $V_{IN}/2$ that are separated by time intervals with $V_{AB} = 0$. According to (17.10) and with reference to Fig. 17.2 during the time intervals when either of inner switches S_2 and S_3 is closed and when $V_{AB} = 0$, the primary voltage magnitude is $|V_P| = V_{IN}/4$, whereas during time intervals when $|V_{AB}| = V_{IN}/2$, the primary voltage magnitude is $|V_P| = 0$. As shown in Fig. 17.3 Since during time interval T_0–T_1 switches S_1 and S_2 are closed while switches S_3 and S_4 are open, voltage $V_{AB} = V_1 = V_{IN}/2$ so that primary voltage $V_P = 0$ In addition, during this topological stage, output current i_O flows through output rectifier D_{R2} and the corresponding secondary of the transformer so that primary current $i_P = -i_O/n_{TR}$, where $n_{TR} = N_P/N_S$ is the turns ratio of the transformer, N_P is the number of primary winding turns, and N_S is the number of secondary winding turns.
The primary current is negative, both currents i_1 and i_2 are also negative as shown in Fig. 17.2. At the same time, magnetizing current i_{MC} is linearly increasing with slope $V_{IN}/(2L_{MC})$, since voltage V_{AB} is positive and equal to half of the input voltage, i.e., $V_{AB} = V_{IN}/2$. As a result, current i_1 increases while current i_2 decreases. During this interval, voltage V_S which is equal to the secondary winding voltage is zero because primary winding voltage V_P is zero. This stage ends at $T = T_1$ when switch is turned off. After switch is turned off at, the current which was flowing through the transistor of switch S_1 is diverted to switch's output capacitance C_1. In this topological stage, current i_2 charges capacitor C_1 and C_4

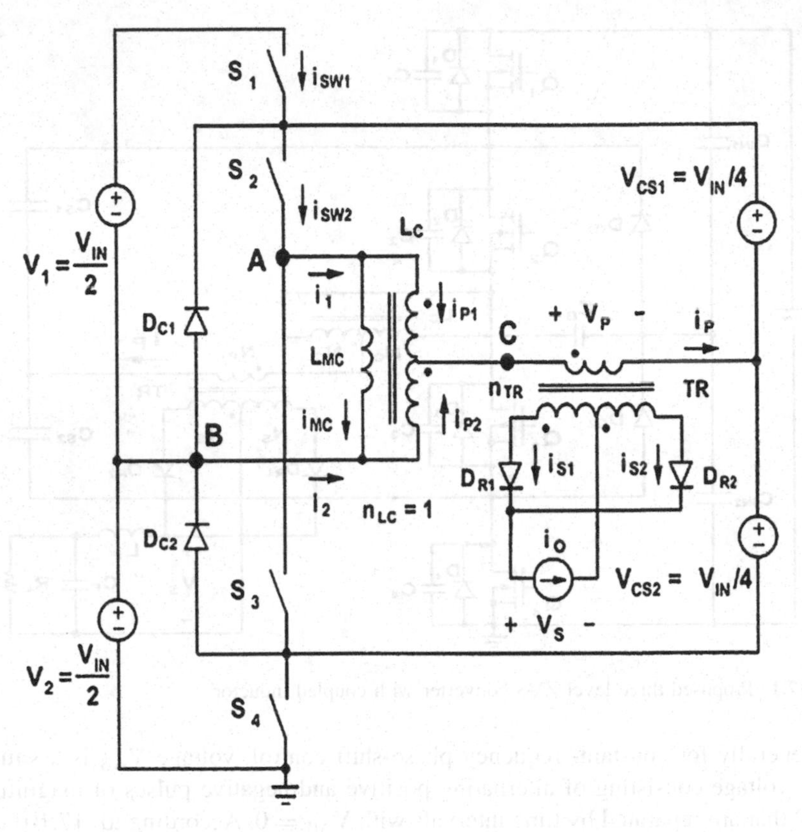

Fig. 17.2 Simplified circuit diagram of proposed three-level ZVS converter

discharges capacitor at the same rate [4] since the sum of the voltages across capacitors and is equal to constant voltage $V_{IN}/2$. As a result, voltage across switch S_1 increases while voltage across switch S_4 decreases, as illustrated in Fig. 17.2. In addition, during this stage the potential of point A decreases causing a decrease of voltage V_{AB} from $V_{IN}/2$ toward zero and the simultaneous increase of primary voltage from zero toward $V_{IN}/4$, as illustrated in Fig. 17.2. The positive primary voltage initiates the commutation of output current i_O from rectifier D_{R2} to rectifier D_{R1}. Since the [8] leakage inductance of transformer TR neglected, this commutation is instantaneous. However, in the presence of leakage inductance, the commutation of current from one rectifier to the other takes time. Because during this commutation time both rectifiers are conducting, i.e., the secondary windings of the transformer are shorted, voltage V_S is zero, as shown in Fig. 17.3.

After capacitor C_4 is fully discharged at $T = T_2$, i.e., after voltage V_{S4} reaches zero, current i_2 continues to flow through antiparallel diode D_4 of switch and clamp diode D_{C1} instead of through capacitors C_1 and C_4. Since it is desirable to minimize the leakage inductance of transformer TR to minimize the secondary-side parasitic ringing, the energy stored in its leakage inductances is relatively small, i.e., much

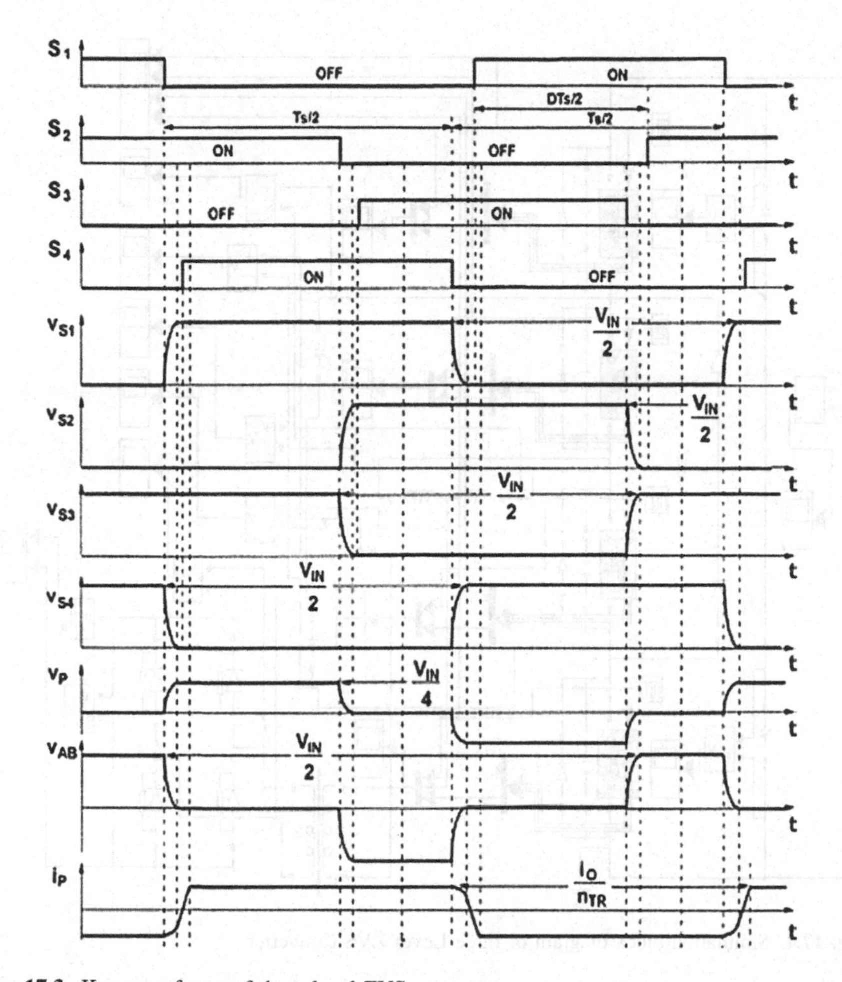

Fig. 17.3 Key waveforms of three-level ZVS converter

smaller than the energy stored in output-filter inductance. As a result, in the circuit in Fig. 17.1, it is easy to achieve ZVS of inner switches S_2 and S_3 in the entire load range, where as ZVS of the outer switches S_1 and S_3 requires a proper sizing of the magnetizing inductance [9] since at light loads almost the entire energy required to create ZVS condition of outer switches and is stored in the magnetizing inductance.

17.4 Design Consideration

In the proposed three-level ZVS circuit with a coupled inductor, it is more difficult to achieve ZVS of the outer pair of switches than the inner pair of switches because the available energies for creating the ZVS conditions in the two pairs of switches are different. Generally, to achieve ZVS this energy must be at least equal

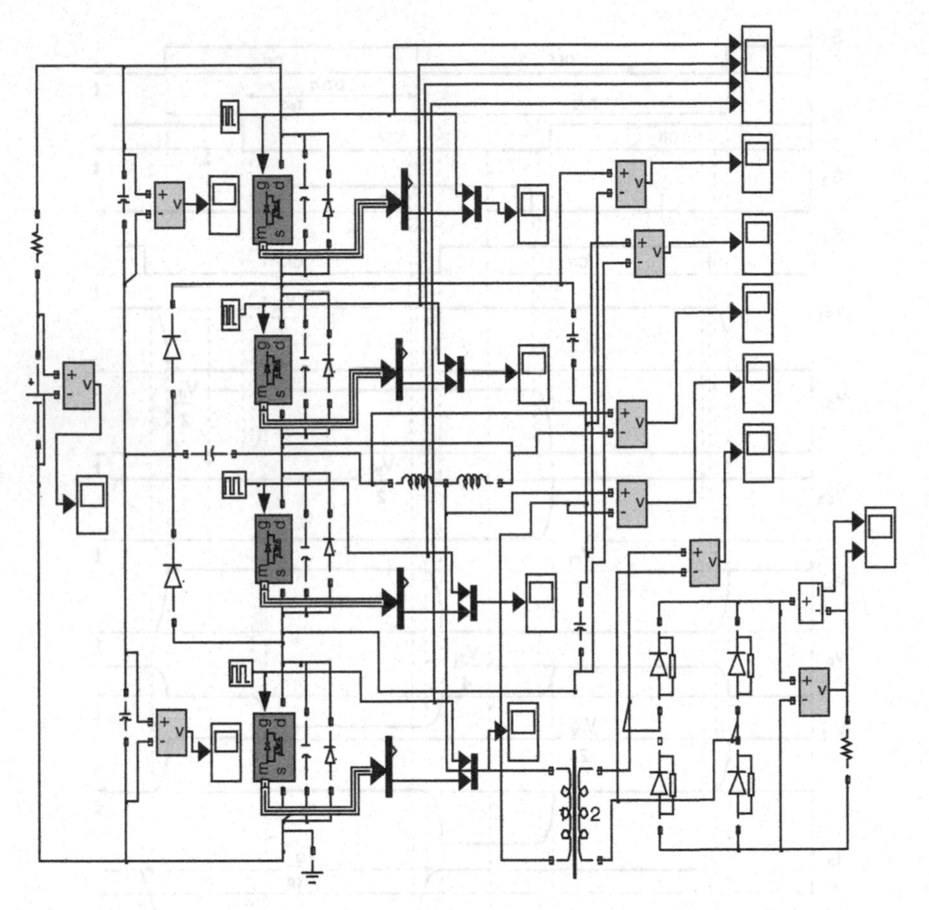

Fig. 17.4 Simulation block diagram of three Level ZVS Converter

to the energy required to discharge the capacitance of the switch which is about to be turned on and at the same time charge the capacitance of the switch that just has been turned off. At heavier load currents, ZVS is primarily achieved by the energy stored in the leakage inductances of transformer TR. As the load current decreases, the energy stored in the leakage inductances also decreases, whereas the energy stored in inductance L_C increases so that at light loads inductance L_C provides an increasing share of the energy required for ZVS [10]. In fact, at no load, this inductance L_C provides the entire energy required to create the ZVS condition. Therefore, if the value of inductance L_C is selected so that ZVS is achieved at no load and maximum input voltage $V_{IN(max)}$, ZVS is achieved in the entire load and input-voltage range.

Neglecting the capacitances of the transformer's windings, magnetizing inductance L_{MC} necessary to achieve ZVS of the outer switches in the implementations in Fig. 17.1 is,

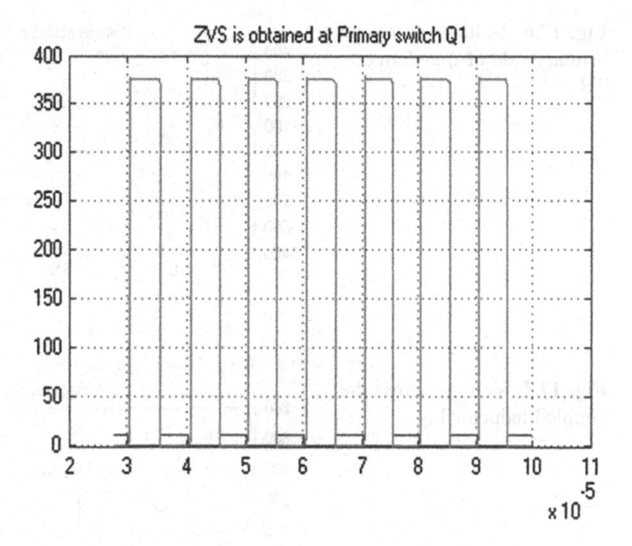

Fig. 17.5 Input gate pulse and voltage across the switch Q_1 at ON and OFF states

$$L_{MC} \leq \frac{1}{32Cf_S^2}$$

where, C is the total capacitance across the primary switches (parasitic and external capacitance, if any) in the corresponding switch pairs.

Finally, it should be noted that the magnitude of primary current i_P of the proposed converter is approximately two times larger than, for example, that of the conventional three-level converter described in [3] if these converters are designed to meet the same specifications. Namely, because during the energy delivery period voltage V_P across the primary winding of transformer TR in the proposed converter is one half of that of the conventional converter, turns ratio n_{TR} of transformer TR of the proposed converter is one half of that in [3]. However, the switch currents of the proposed converter are similar to those of the conventional converter in [3] because each switch in the converter in Fig. 17.1 carries approximately one half of primary current i_P therefore, if the transformer is designed to have the primary winding resistance much smaller than the on-resistance of the primary switches, the conduction losses on the primary side of the converter in Fig. 17.1 are approximately the same as those in the converter described in [3].

17.5 Simulation Results

The simulation circuit shown in Fig 17.4 The simulation results shown in Figs. 17.5, 17.6, 17.7. Figure 17.5 demonstrates that ZVS is achieved with input gate pulse of the switch's Q_1, Fig. 17.6 shows voltage across the primary side of transformer TR. Figure 17.7 shows voltage across the coupled inductor L_C.

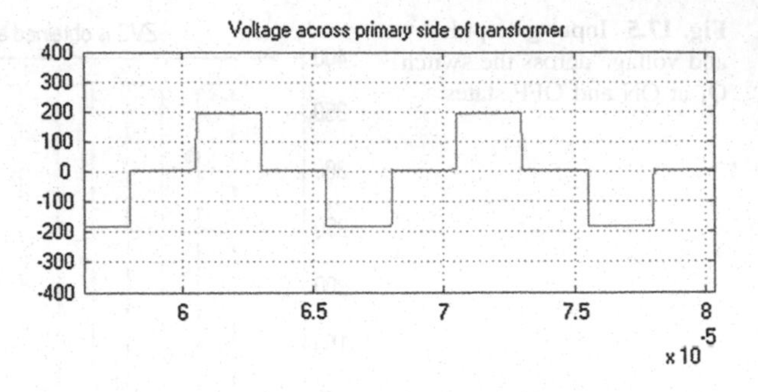

Fig. 17.6 Voltage across the primary side of transformer TR

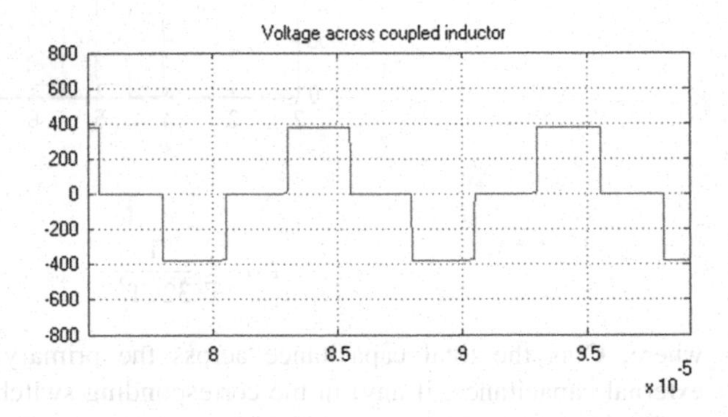

Fig. 17.7 Voltage across the coupled inductor L_C

17.6 Conclusion

A new isolated, constant-frequency, three-level ZVS converter which employs a coupled inductor on the primary side to achieve ZVS in a wide range of load current and input voltage with reduced circulating energy and conduction losses has been described. Since this coupled inductor does not appear as a series inductance in the load current path, it does not cause a loss of duty cycle or severe voltage ringing across the output rectifiers.

Acknowledgments I would like to express my heartiest thanks to Prof K.Duraiswamy., Professor of the Department of computer science and Engineering, who has been the source of inspiration and behind my every move in achieving the goal of my work.

References

1. Lai JS, Peng FZ (1996) Multilevel converters-a new breed of power converters. IEEE Trans Ind Appl 32(3):509–517
2. Pinheiro JR, Barbi I (1993) The three-level ZVS-PWM DC-to-DC converter. IEEE Trans Power Electron 8(4):486–492
3. Canales F, Barbosa PM, Burdio JM, Lee FC (2000) A zero-voltage switching three-level DC/DC converter. In: Proceedings of IEEE international telecommunications energy conference (INTELEC), pp 512– 517

4. Jin M, Weiming M (2006) Power converter EMI analysis including IGBT nonlinear switching transient model. IEEE Trans Ind Electron 53(5):1577–1583
5. Gonzalez D, Gago J, Balcells J (2003) Analysis and simulation of conducted EMI generated by switched power converters: application to a voltage source inverter. IEEE Trans Ind Electron 50(6):801–806
6. Krug F, Russer P (2005) Quasi-peak detector model for a time-domain measurement system. IEEE Trans Electromagn Compat 47(2):320–326
7. Ruan X, Zhou L, Yan Y (2001) Soft-switching PWM three-level converters. IEEE Trans Power Electron 16(5):612–622
8. Liu Q, Wang F, Boroyevich D (2004) Model conducted EMI emission of switching modules for converter system EMI characterization and prediction. In: Proceedings of IEEE IAS annual meeting, pp 1817–1823
9. Barbi I, Gules R. Redl R, Sokal NO (1998) DC/DC converter for high input voltage: four switches with peak voltage of V = 2, capacitive turn-off snubbing, and zero-voltage turn-on. In: Proceedings of IEEE power electronics specialists conference (PESC), pp 1–7
10. Ran L, Gokani S et al (1998) Conducted electromagnetic emissions in induction motor drive systems—part I: time domain analysis and identification of dominant modes. IEEE Trans Power Electron 13(4):757–767
11. Jeon SJ, Canales F, Barbosa PM, Lee FC (2002) A primary-side-assisted zero-voltage and zero-current switching three-level DC–DC converter with phase-shift control. In: Proceedinds of IEEE applied power electronics conference (APEC), pp 641–647

Chapter 18
Common Mode Injection PWM Scheme with Equal Zero Vector Placement for Three Level NPC Inverter

S. Nageswari and V. Suresh Kumar

Abstract This paper proposes a variable common mode injection pulse width modulation (VCMIPWM) scheme for three level neutral point clamped (NPC) inverter. Using suitable PWM control technique the waveform quality of the inverter output can be improved. Optimal harmonic profile may be obtained by having equal zero vectors in a switching cycle. If the expression of common mode injection as calculated for two level inverters is directly applied for three level inverters the dwell times for zero vectors will be unequal and the harmonic performance will be poor. The amount of common mode injection required to have equal zero vectors depends on the magnitude of sinusoidal reference vectors. The amount of common mode injection is calculated for various cases of reference sinusoidal vectors and is injected with the sinusoidal reference vectors to produce modified reference vectors. Simulation has been carried out in MATLAB/SIMULINK for three level inverter with the VCMIPWM and the results are compared with the fixed common mode injection pulse width modulation (FCMIPWM).

S. Nageswari (✉)
Department of Electrical and Electronics Engineering, A.C College of Engineering and Technology, Karaikudi, Tamil nadu, India
e-mail: mahabashyam@gmail.com

V. Suresh Kumar
Department of Electrical and Electronics Engineering, Thiagarajar College of Engineering, Madurai, Tamil nadu, India
e-mail: vskeee@tce.edu

R. Malathi and J. Krishnan (eds.), *Recent Advancements in System Modelling Applications*, Lecture Notes in Electrical Engineering 188,
DOI: 10.1007/978-81-322-1035-1_18, © Springer India 2013

18.1 Introduction

Multilevel inverters can meet the increasing demand of power ratings and power quality associated with reduced harmonic distortion and lower electromagnetic interference. Among various multilevel inverters only three level NPC inverter has found wide applications in high power medium voltage drives. The medium voltage drives cover power ratings from 0.4 to 40 MW at the medium voltage level of 2.3–13.8 kV [1]. The main features of the NPC inverter include reduced dv/dt and THD in its ac output voltages in comparison to the two level inverter. The switches in the NPC inverter do not have dynamic voltage sharing problem.

PWM techniques have been the subject of intensive research during the last few decades. A large variety of PWM methods, different in concept and performance have been compared based on different merit factors such as total harmonic distortion (THD) of the output voltage or current at the inverters, switching losses of the inverters, maximum inverter output voltage as a function of a given DC bus voltage [2–7]. The PWM methods are classified into carrier based PWM, space vector PWM (SVPWM) and selective harmonic elimination (SHEPWM). In carrier based PWM the popular method is sinusoidal PWM. Sinusoidal pulse width modulation (SPWM) is used to control the inverter output voltage and maintains good performance of the drive in the entire range of operation between zero and 78 % of the value that would be reached by square operation. If the modulation index exceed this value, linear relationship between modulation index and output voltage is not maintained and the over modulation methods are required. And in recent years, an inherently digital modulation technique known as SVPWM which is based on space vector theory, has been developed [8]. Space vector modulation techniques have been increased by using in last decade, because they allow reducing commutation losses and the harmonic content of output voltage, and to obtain higher amplitude modulation indexes if compared with conventional SPWM techniques [5].

The selective harmonic elimination (SHE) is an off-line PWM scheme and may not be able to provide a fast adjustment of its modulation index. The selective harmonic elimination is considered as an optimum modulation scheme, which provides a superior harmonic profile with a minimum switching frequency. The harmonic performance for SHEPWM is better when compared to other PWM but dynamic performance is poor [1].

This paper addresses the calculation of variable common mode for VCMIPWM scheme to have equal zero vector placement in a switching cycle. And the simulation investigations are carried out if the same common mode expression as in the two level inverter is applied to three level inverter (FCMIPWM). The harmonic performance comparison is carried out and presented.

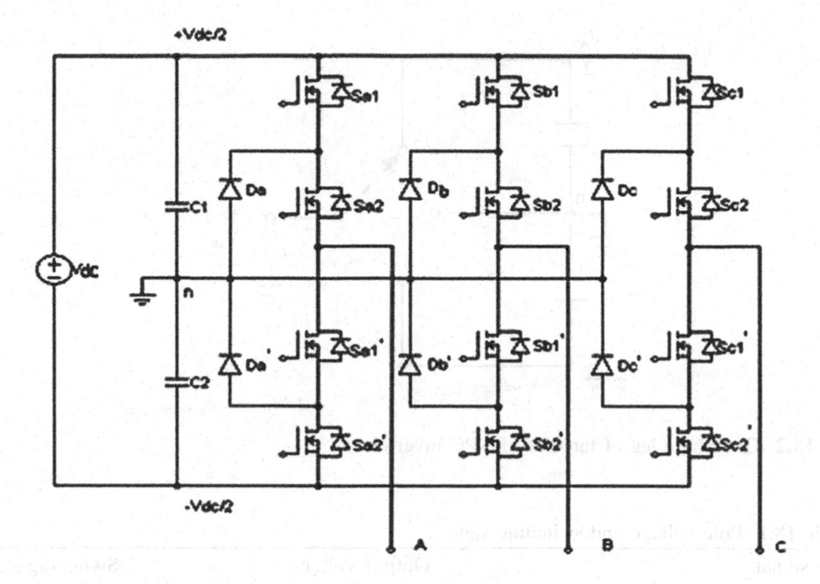

Fig. 18.1 Schematic diagram of three level NPC inverter

18.2 Operation of Three Level NPC Inverter

Figure 18.1 shows the circuit diagram of three level NPC inverter. The circuit acts as a single pole triple throw switch as in Fig. 18.2. The circuit has four switches per phase. The upper leg switches are named as S_{x1}, S_{x2} and lower leg switches are their complementary S_{x1}', S_{x2}'. Where x is any of the phases a, b or c. Phase pole voltage measured with respect to the DC neutral n depends on the state of the individual switches in the leg as shown in Table 18.1.

In sinusoidal PWM, a sinusoidal reference v_x is compared with two carriers v_{cr1} and v_{cr2}. It is shown in Fig. 18.3.

If $v_x > v_{cr1}$ gate pulse to S_{x1}, S_{x2} and $v_{xn} = \frac{V_{dc}}{2}$

If $v_{cr2} < v_x < v_{cr1}$ gate pulse to S_{x2}, S_{x1} and $v_{xn} = 0$

If $v_x < v_{cr2}$ gate pulse to S_{x1}, S_{x2} and $v_{xn} = -\frac{V_{dc}}{2}$

The above switching logic is explained in Table 18.1.

18.3 Review of Carrier Based PWM

The carrier based PWM is broadly classified into phase shifted multi carrier modulation and Level shifted multi carrier modulation. When frequency modulation index m_f which is defined as the ratio of the frequency of carrier to the frequency of modulating signal, is an integer the modulation scheme is known as synchronous PWM which is more suitable for implementation with a digital

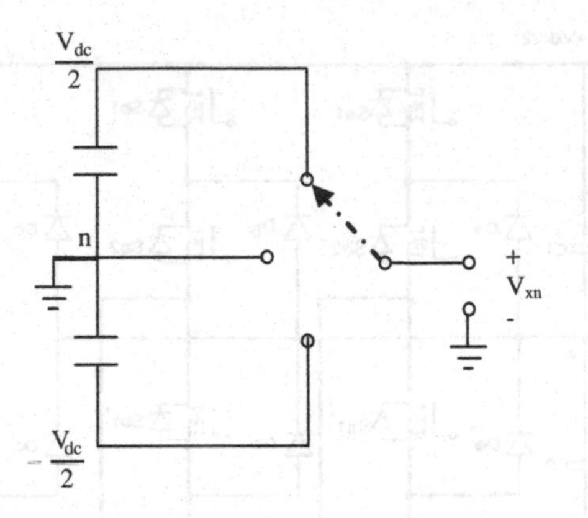

Fig. 18.2 One phase leg of three level NPC inverter

Table 18.1 Pole voltage and switching state

Gate signal		Output voltage	Switching state
S_{x1}	S_{x2}		
1	1	$\frac{V_{dc}}{2}$	(+)
0	1	0	(0)
0	0	$-\frac{V_{dc}}{2}$	(−)

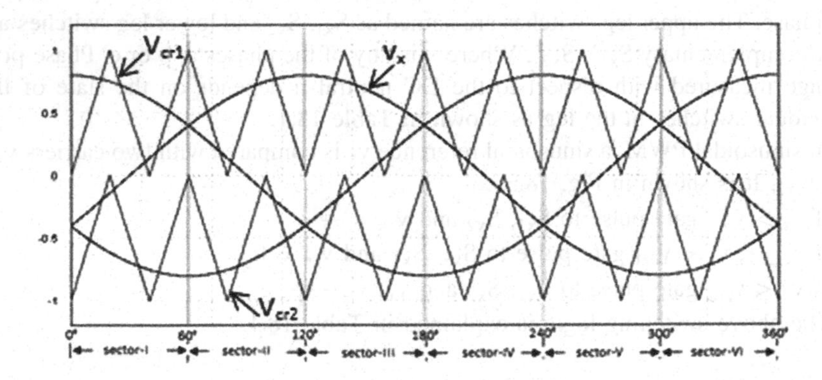

Fig. 18.3 Modulating signal and carrier waves for three level NPC inverter in sinusoidal PWM

processor. If m_f is ≥9 and it is multiple of three then all the harmonics in line voltage with the order lower than (m_f − 2) are eliminated. The harmonics are centred on m_f and its multiples. In-phase disposition (IPD) modulation where all carriers are in phase provides best harmonic profile [2]. Phase shifted multi carrier modulation and level shifted multicarrier modulation schemes can be employed for multilevel inverters. It should be noted that phase shifted modulation schemes cannot be utilized for diode clamped multilevel inverters.

The great advantage of the PWM algorithm is its ability to control the content in fundamental voltage across the load. The pulse width variations are based not only on the shape of the control reference but also on the modulation index. Addition of 1/4th of third harmonic minimizes THD. It is only slightly better than SVPWM and has narrower voltage linearity range. Addition of 1/3rd of third harmonic maximizes the fundamental voltage [9]. Addition of 1/6th of third harmonic which is equivalent to SVPWM increases the amplitude of fundamental phase voltage by 15.5 % and therefore in the line voltage waveforms [10].

A properly selected zero sequence signal can extend the volt second linearity range of SVPWM. It can improve waveform quality and reduce the switching losses significantly [11]. Parameters of PWM converter depends on zero vector placement in SVPWM and on the shape of the zero sequence signal in carrier based PWM [12]. If the pulse width modulation centres the duty cycles of active vectors in each half carrier period the harmonic performance will be better [13]. To have optimum harmonic performance the start and redundant vector periods should be made equal. Many authors revealed that the relation between carrier based PWM and SVPWM for two level inverters [5, 6, 9–11, 13] and three level inverters [7, 8, 14, 15]. Average voltage concept is used in SVPWM techniques. For two level inverters with carrier based SVPWM techniques to have equal placement of zero vectors, the common mode voltage is added to the sinusoidal reference signal [7]. Due to that the reference signals are modified. The modified reference signals are then compared with the carriers to produce pulses.

In this paper level shifted PWM is preferred to have better harmonic performance. The modulation task is simplified by considering the relation between the time duration and the output voltage. Using this effective time concept [5], inverter output voltage can be directly synthesized. In this paper the above said time concept is used for three level inverters. To improve the harmonic performance, zero voltage applying time is distributed symmetrically, at the start and end of the half carrier period without affecting the dwell times of active pulse width. Doing this the active pulse width is relocated at the centre of the half carrier period. For this a proper common mode injection is to be calculated. The calculation of variable common mode injection is explained in Sect. 18.4.

18.4 Common Mode Voltage Calculation

The reference signal and the carrier waves are shown in Fig. 18.3a. The three phase reference signals are defined using Eqs. (18.1)–(18.3) respectively. The sampled value of the modulating signal and triangle carrier comparison in one rising carrier period is shown in Fig. 18.4. The highest of the three phase references in one sampling period is V_{max}, the lowest as V_{min}, and the middle valued one as V_{mid}. The reference phase voltages cross the triangular carrier at different instants in a sampling period T_s. If reference voltage crosses the triangular carrier it causes a change in inverter state. The sampling time interval is divided into four

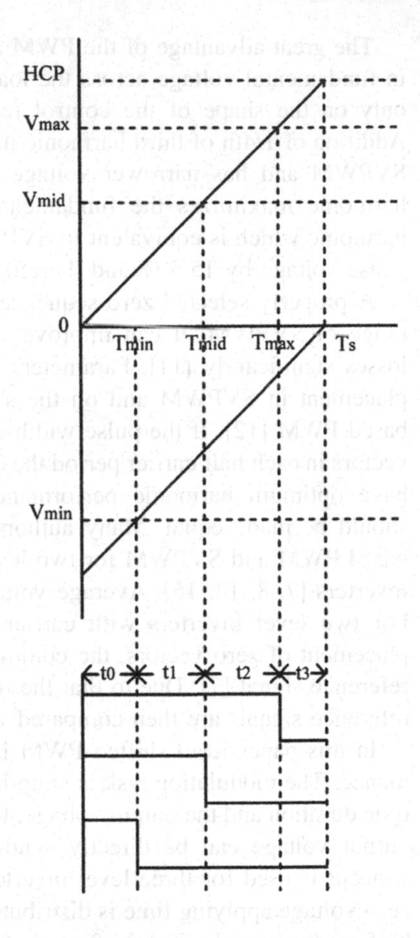

Fig. 18.4 Switching instants and inverter states in a rising carrier period

intervals t_0, t_1, t_2 and t_3. t_1 and t_2 are the time durations of active vectors. t_0 and t_3 are the time durations of zero vectors. The effective time is $t_1 + t_2$. The three phase difference which has maximum value with in a sampling period is V_{max} defined in Eq. (18.4) and the instant at which V_{max} crosses the carrier is T_{max}. Similarly V_{min}, V_{mid}, T_{min} and T_{mid} can be defined. It is necessary to identify which phase reference produce max, mid and min values.

Referring Fig. 18.3,

$$v_a = v_m \cos \omega t \tag{18.1}$$

$$v_b = v_m \cos(\omega t - 2\pi/3) \tag{18.2}$$

$$v_c = v_m \cos(\omega t + 2\pi/3) \tag{18.3}$$

$$V_{max} = \begin{cases} v_a & \text{if } |v_a| \geq |v_b|, |v_c| \\ v_b & \text{if } |v_b| \geq |v_a|, |v_c| \\ v_c & \text{if } |v_c| \geq |v_a|, |v_b| \end{cases} \tag{18.4}$$

Table 18.2 Switching sequence of phases

ωt	Sector number	V_{max}	V_{mid}	V_{min}
$0° < \omega t < 60°$	I	v_a	v_b	v_c
$60° < \omega t < 120$	II	v_b	v_a	v_c
$120° < \omega t < 180°$	III	v_b	v_c	v_a
$180° < \omega t < 240°$	IV	v_c	v_b	v_a
$240° < \omega t < 300°$	V	v_c	v_a	v_b
$300° < \omega t < 360°$	VI	v_a	v_c	v_b

Table 18.3 Zero states for different cases

Case	Condition	T_{mid}	Effective time	Zero states	
				t_0	t_3
A	$V_{mid} < 0$, $(V_{max} - V_{min}) > T_s$, $(T_s - V_{max}) < (-V_{mid})$	$T_s + V_{mid}$	$T_{max} - T_{min}$	T_{min}	$T_s - T_{max}$
B	$V_{mid} < 0$, $(V_{max} - V_{min}) > T_s$, $(T_s - V_{max}) > (-V_{mid})$	$T_s + V_{mid}$	$T_{mid} - T_{min}$	T_{min}	$T_s - T_{mid}$
C	$V_{mid} < 0$ and $(V_{max} - V_{min}) < T_s$	$T_s + V_{mid}$	$T_{mid} - T_{max}$	T_{max}	$T_s - T_{mid}$
D	$V_{mid} > 0$ and $(V_{max} - V_{min}) < T_s$	V_{mid}	$T_{min} - T_{mid}$	T_{mid}	$T_s - T_{min}$
E	$V_{mid} > 0$, $(V_{max} - V_{min}) > T_s$ and $(T_s + V_{min}) > V_{mid}$	V_{mid}	$T_{max} - T_{mid}$	T_{mid}	$T_s - T_{max}$

Consider sector-I. ($0° < \omega t < 60°$). The instantaneous value of v_a is greater than the instantaneous values of v_b and v_c and therefore the $V_{max} = v_a$. Similarly for all the sectors V_{max}, V_{mid} and V_{min} values are find out and tabulated in Table 18.2.

The common mode voltage V_{CM} may be added to the three references depends on modulation index.

For two level inverters the addition of common mode voltage as in Eq. (18.5) to the three phase sequences centres the active vectors in a half carrier period to have SVPWM like performance.

$$V_{CM} = -\left[\frac{V_{max} + V_{min}}{2}\right] \tag{18.5}$$

This is not the case for three level inverters. A properly selected common mode voltage V_{CM} centres the active vectors and produces equal zero state vectors in a switching cycle. The value of common mode voltage V_{CM} depends on modulation index. From the examination of the waveforms, for various modulation indices according to the occurrence of T_{max}, T_{mid} and T_{min} values in a rising carrier period, five different cases are identified and tabulated in Table 18.3. For all the cases $T_{max} = V_{max}$, $T_{min} = T_s + V_{min}$ and T_{mid} differs for different cases and they are tabulated in Table 18.3.
From Fig. 18.4, $V_{max} + V_{mid} + V_{min} = 0$.

Table 18.4 New zero states for different cases

Case	V_{CM}	New zero states ($t_0 = t_3$)
A	$-\left[\frac{V_{max}+V_{min}}{2}\right]$	$T_s - \left[\frac{V_{max}-V_{min}}{2}\right]$
B	$-\left[\frac{T_s-V_{max}}{2}\right]$	$\left[\frac{T_s+V_{min}-V_{mid}}{2}\right]$
C	$\frac{V_{min}}{2}$	$\frac{V_{max}-V_{mid}}{2}$
D	$\frac{V_{max}}{2}$	$\frac{V_{mid}-V_{min}}{2}$
E	$\left[\frac{T_s+V_{min}}{2}\right]$	$\left[\frac{[T_s+V_{mid}-V_{max}]}{2}\right]$

For example the common mode calculation for case A is derived using Eqs. (18.6)–(18.11).

$$t_0 = T_{min} = T_s + V_{min} \tag{18.6}$$

$$t_3 = T_s - T_{max} \tag{18.7}$$

The zero states are not equal. In order to have equalled zero states common mode voltage V_{CM} is added.

$$t_0 = T_s + (V_{min} + V_{CM}) \tag{18.8}$$

$$t_3 = T_s - (V_{max} + V_{CM}) \tag{18.9}$$

From Eqs. (18.8) and (18.9)

$$T_s + (V_{min} + V_{CM}) = T_s - V_{max} - V_{CM} \tag{18.10}$$

$$V_{CM} = -\frac{(v_{max} + v_{min})}{2} \tag{18.11}$$

Similarly for different cases the common mode voltages to be added to the reference sinusoidal wave are calculated. The common mode voltage V_{CM} and new zero states are tabulated in Table 18.4.

For different cases the zero vector durations t_0, t_3 and zero vector durations t_1, t_2 are determined as shown in Fig. 18.5a. The zero vector durations are not same. After the addition of common mode voltage V_{CM}, the new zero vector durations are equalized without affecting the dwell times of the active vectors and they are shown in Fig. 18.5b.

18.5 Simulation Analysis

The three level inverter is simulated in MATLAB/SIMULINK platform. The various conditions given in Table 18.3 are simulated and the common mode

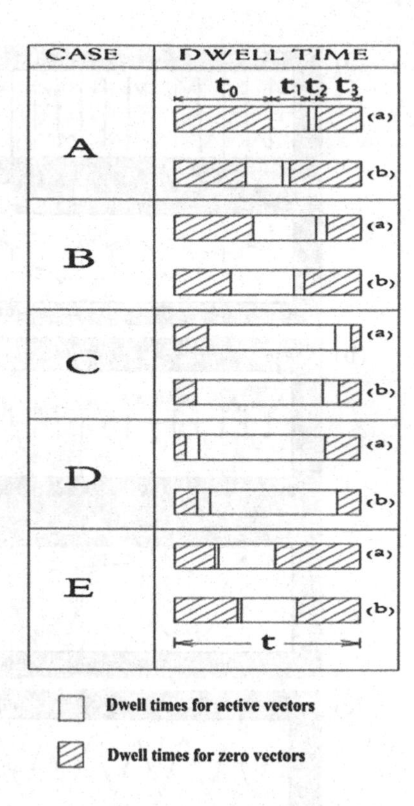

Fig. 18.5 Zero vector and active vector durations. **a** Before common mode addition. **b** After common mode addition

voltages are calculated and the modified reference waveforms are obtained as shown in Fig. 18.6. For various modulation indices 0.3, 0.65, 0.8 and 1, the common mode voltages and modulating signals have different wave shapes to have equalled zero vectors in a switching cycle. To have better harmonic performance the frequency modulation index (m_f) is taken as an odd triplen integer and being kept at 21. FCMIPWM and VCMIPWM are tested on a three level NPC inverter model. Figure 18.7 shows the line voltage waveforms at modulation index of 0.6 and 100 V dc bus voltage. For FCMIPWM, the inverter output line voltage has not half wave symmetry and due to that it has even harmonics as shown in Fig. 18.8a. But for VCMIPWM the waveform has half wave symmetry and has better harmonic performance as shown in Fig. 18.8b. For various modulation indices the three level inverter is applied with both the PWMs and harmonic performances are studied and THD values after filtering are obtained and plotted as shown in Fig. 18.9. From the comparison results it is observed that for VCMIPWM the three level inverter has better harmonic performance than FCMIPWM.

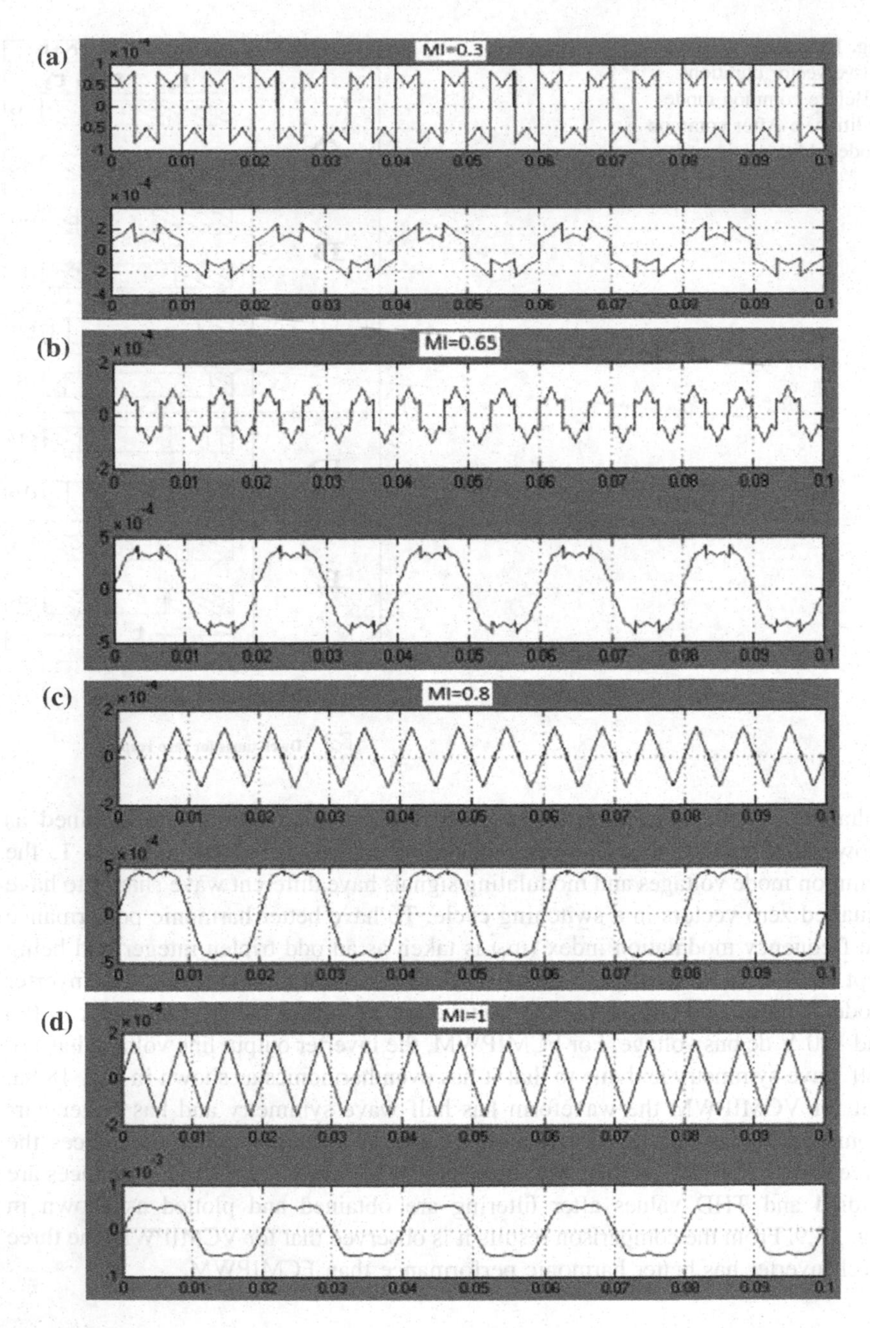

Fig. 18.6 Common mode voltages and modified reference waveforms for various modulation indices MI. **a** MI = 0.3. **b** MI = 0.65. **c** MI = 0.8. **d** MI = 1

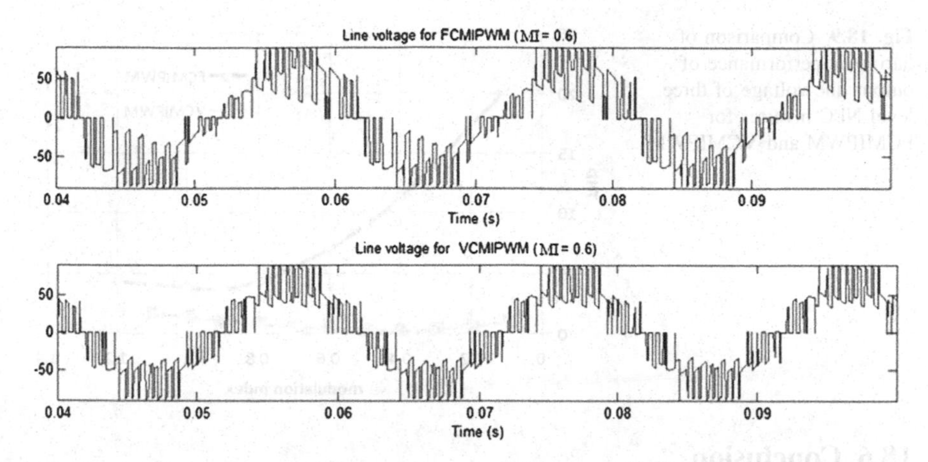

Fig. 18.7 Line voltage waveforms of three level NPC inverter at modulation index 0.6 **a** for FCMIPWM and **b** for VCMIPWM

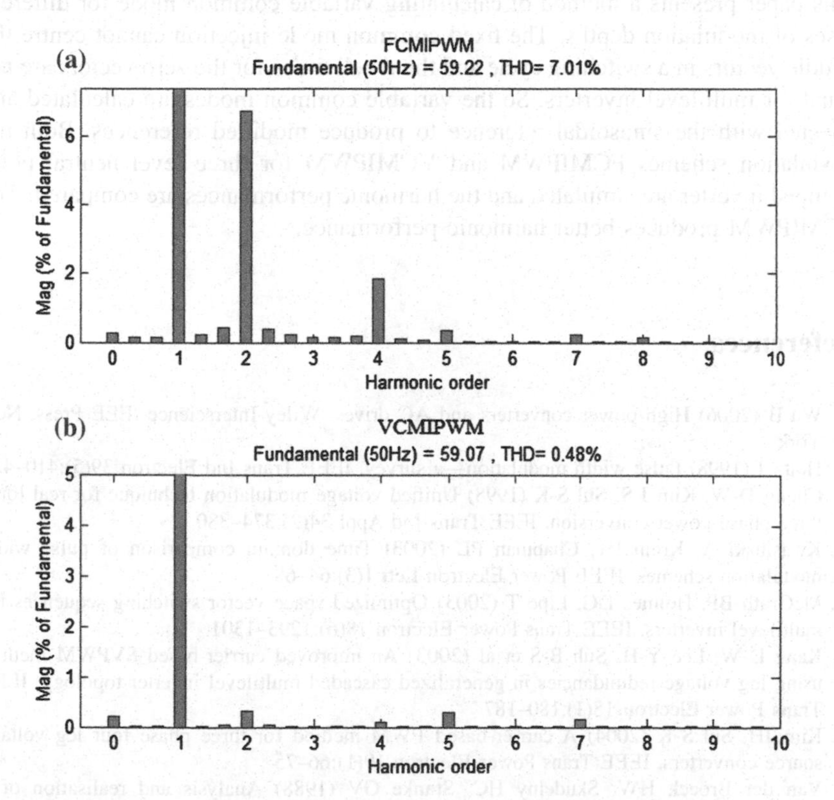

Fig. 18.8 Frequency spectra of line voltages at modulation index 0.6 **a** for FCMIPWM and **b** for VCMIPWM

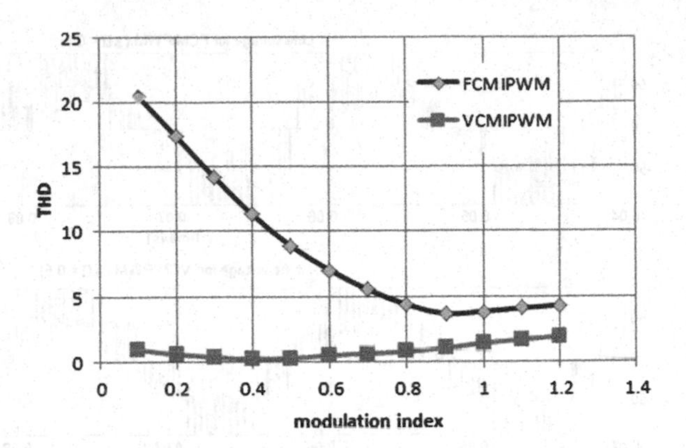

Fig. 18.9 Comparison of harmonic performance of output line voltage of three level NPC inverter for FCMIPWM and VCMIPWM

18.6 Conclusion

This paper presents a method of calculating variable common mode for different cases of modulation depths. The fixed common mode injection cannot centre the middle vectors in a switching cycle and the dwell times for the zero vectors are not equal for multilevel inverters. So the variable common modes are calculated and injected with the sinusoidal reference to produce modified references. Both the modulation schemes FCMIPWM and VCMIPWM for three level neutral point clamped inverter are simulated and the harmonic performances are compared. The VCMIPWM produces better harmonic performance.

References

1. Wu B (2006) High power converters and AC drives. Wiley-Interscience-IEEE Press, New York
2. Holtz J (1998) Pulse width modulation—a survey. IEEE Trans Ind Electron 39(5):410–420
3. Chung D-W, Kim J-S, Sul S-K (1998) Unified voltage modulation technique for real tome three phase power conversion. IEEE Trans Ind Appl 34(2):374–380
4. Kwasinski A, Krein PT, Chapman PL (2003) Time domain comparison of pulse width modulation schemes. IEEE Power Electron Lett 1(3):64–68
5. McGrath BP, Holmes DG, Lipo T (2003) Optimized space vector switching sequences for multilevel inverters. IEEE Trans Power Electron 18(6):1293–1301
6. Kang D-W, Lee Y-H, Suh B-S et al (2003) An improved carrier-based SVPWM method using leg voltage redundancies in generalized cascaded multilevel inverter topology. IEEE Trans Power Electron 18(1):180–187
7. Kim JH, Sul S-K (2004) A carrier-based PWM method for three phase four leg voltage source converters. IEEE Trans Power Electron 19(1):66–75
8. Van der Broeck HW, Skudelny HC, Stanke GV (1988) Analysis and realisation of a pulsewidth modulator based on voltage space vectors. IEEE Trans Ind Appl 24(1):142–150
9. Neacsu DO (2006) Power switching converters. Taylor & Francis, CRC Press, Boca Raton
10. Houldsworth JA, Grant DA (1984) The use of harmonic distortion to increase the output voltage of a three-phase PWM inverter. IEEE Trans Ind Appl IA-20(5):1224–1228

11. Hava AM, Sul S-K, Kerkman RJ, Lipo TA (1999) Dynamic overmodulation characteristics of triangle intersection PWM methods. IEEE Trans Ind Appl 35(4):896–907
12. Kazmierkowski MP, Krishnan R, Blaabjerg F, Irwin JD (2002) Control in power electronics: selected problems. Academic Press, New York
13. Holmes DG (1996) The significance of zero space vector placement for carrier-based PWM schemes. IEEE Trans Ind Appl 32(5):1122–1129
14. Wang F (2002) Sine triangle versus space vector modulation for three level PWM voltage source inverters. IEEE Trans Ind Appl 38(2):500–506
15. Yao W, Haibing H, Zhengyu L (2008) Comparisons of space vector modulation and carrier based modulation of multilevel inverters. IEEE Trans Power Electron 23(1):45–51

Chapter 19
Fuzzy Based Harmonic Reduction Strategy for DC Link Inverters

N. Radhakrishnan and M. Ramaswamy

Abstract The paper develops a fuzzy based methodology to improve the frequency spectrum of the output voltage and almost eliminates the need of a filter in a DC link inverter. The presence of harsh loads attempt to deteriorate the quality of the current drawn from the mains. It therefore orients to simultaneously reshape the input current phasor to a nearly sinusoidal waveform in the sense it drastically reduces the amplitude of the multiple frequency harmonic components of the input AC current. The scheme derives the reference for the PWM pulses through the use of fuzzy principles and arrives at the appropriate width for the pulses to the power switches. The frequency of the carrier is allowed to vary randomly with a view to distinctly decrease the current frequency magnitudes over the operating range. It includes the simulation results obtained using MATLAB to illustrate its potential to extract a variable magnitude output voltage in tune with the change in the modulation index. The regulated output voltage over a range of operating loads and the sinusoidal shape of the supply current substantiate the merits of the strategy and claim its applicability in the utility world.

Keywords DC link inverter · Fuzzy control · RPWM strategy · Harmonic reduction · Input power factor

N. Radhakrishnan (✉) · M. Ramaswamy
Annamalai University, Chidambaram, India
e-mail: nradhaa75@yahoo.co.in

M. Ramaswamy
e-mail: aupowerstaff@gmail.com

R. Malathi and J. Krishnan (eds.), *Recent Advancements in System Modelling Applications*, Lecture Notes in Electrical Engineering 188,
DOI: 10.1007/978-81-322-1035-1_19, © Springer India 2013

19.1 Introduction

Switch-mode DC–AC converters espouse a renewed interest in AC power supplies and motor drives to meet the emerging industrial challenges. The utilities exhibit a need to generate a sinusoidal AC output whose magnitude and frequency can be controlled. However, the inverter interfaces powered through a front end rectifier is subjected to sub-harmonics owing to the presence of ripple in the DC link. The modulation between the inverter switching function and the DC link ripple is the primary cause of deteriorating the quality of output voltage.

The ripples in the inverter input voltage are a cause of harmonics that limit the amplitude of the fundamental frequency component of the supply current. Besides the elaborate use of motor loads along with the pulsations on account of the distorted nature of the supply further add to interleave undesired harmonic constituents in the source current.

The effect of switching the desired power to the load is another reason for the generation of harmonic components both in the input current and output voltage. Load harmonics can cause overheating of the magnetic cores of distribution transformer and drive motors. On the other hand voltage and current source harmonics increase the power losses, electromagnetic interference (EMI) and torque pulsations in AC motor drives. The frequency of each harmonic component is an integral multiple of its fundamental. The most widely used measure to estimate the quantity of harmonic content is the total harmonic distortion (THD), which is defined in terms of the amplitudes of the harmonics, *Hn*, at frequency nw_O, where w_O is frequency of the fundamental component whose amplitude H_1 and *n* is an integer.

A random pulse width modulation based on programmed PWM has been articulated to eliminate lower order harmonics [1]. A RPWM technique has been developed to spread the noise spectrum over a wide range and thus reduce the amplitude of the harmonics generated in the operation of active filters [2]. A selective harmonic elimination strategy has been proposed to mitigate the harmonics in seven levels MLI by generating appropriate negative harmonics [3]. The effect of spectral power density of a randomized pulse width modulated waveforms has been investigated through the use of different probability distribution loss [4]. A carrier based PWM strategy has been developed to ensure a balance of the DC link capacitor voltage in diode clamped converters [5, 6]. A vector based PWM strategy has been presented to synthesize the desired output voltage in the four switch three phase inverter [7].

A mathematical model of space vector modulated three phase inverter has been generated and a method formulated to eliminate the most significant line voltage harmonic components [8]. A harmonic separation process has been suggested to reduce the power loss and achieve a lower THD in inverters [9]. Three single phase pulse width modulated inverters have been connected in series with a low frequency square wave inverter to shape its output voltage to a sinusoidal form [10]. A single phase two wire inverter system has been proposed for photo voltaic power injection and filtering of active power in a non-linear inductor frame work [11].

Though a host of harmonic reduction techniques are in existence, most of them result in the reappearance of the lower order harmonics in the output waveform. It is in this perspective that a new effort through the use of a fuzzy based approach that can offer a higher quality of output voltage and improve the quality of power is explored.

19.2 Proposed Strategy

The objective echoes to evolve a scheme to lower the magnitude of the multiple frequency components, significantly increase the fundamental and end up with a reduced value for the THD of the output voltage in a three phase DC link inverter. The methodology ascribes the use of fuzzy principles to derive the reference along with a randomly varied carrier for generating PWM pulses to the power switches.

The front end rectifier seen in Fig. 19.1 acquires the DC input from a three phase AC source to power the three phase 400 V inverter and in turn support a 3 kW star connected resistive load. The inverter is constituted of insulated gate bipolar transistors (IGBTs) that can switch on and off several thousand times per second and precisely control the power delivered to the load. The capacitor in the DC bus serves to sustain the required level of DC voltage.

19.3 Fuzzy Control Algorithm

The philosophy of fuzzy theory corners to translate human-like thinking into actions in a control plane. A fuzzy controller extracts control signals through deductive reasoning and emulates human intelligence in a precise sense. Fuzzy control continues to find applications over almost all fields to suit emerging sophisticated nature of the utilities. It essays the action of the theory using linguistic descriptions and relies on the role of an inference engine to meet the task.

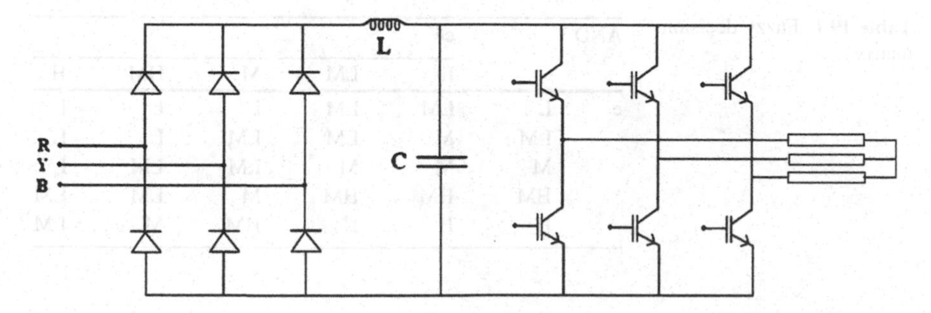

Fig. 19.1 Power circuit

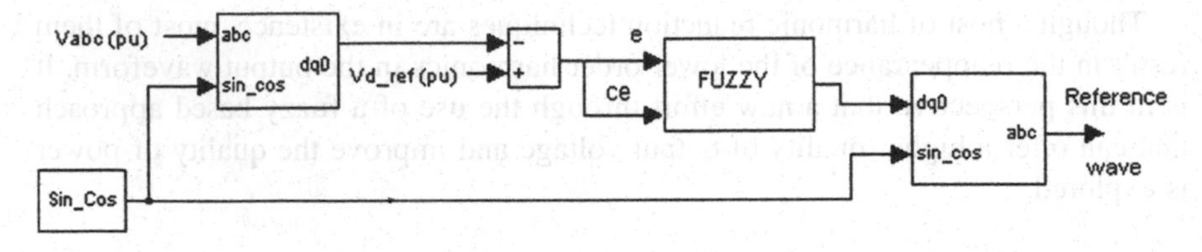

Fig. 19.2 Fuzzy controller

The procedure involves the choice of input linguistic variables as 'e', the voltage error and 'ce', the change in error and the output linguistic variable to be 'u', the reference and rigs out a combinatorial triangular and trapezoidal structure for both. It identifies the normalized variables to lie across a range defined with labels low (L), low-medium (LM), medium (M), high-medium (HM) and high (H). The sets defining the e, ce and u are expressed as:

$$e = \{L, LM, M, HM, H\}$$

$$ce = \{L, LM, M, HM, H\}$$

$$u = \{L, LM, M, HM, H\}$$

The FLC seen in Fig. 19.2 is designed to calculate the current reference for the controller. The two inputs to the fuzzy algorithm are (1) the error (e) between a reference voltage and the actual load voltage, which droops because of the increase in current, and (2) the change in error (ce) which is precisely the difference between the error at a certain operating point and the preceding error.

The rules are summarised in the fuzzy decision matrix in Table 19.1, where the consequents of the rules are shown in the shaded part of the matrix. The fuzzy results must be defuzzified through what is called a defuzzification process, to achieve a crisp numerical value. The most commonly used centroid or centre of gravity defuzzification strategy is adopted.

Table 19.1 Fuzzy decision matrix

AND		ce				
		L	LM	M	HM	H
e	L	LM	LM	L	L	L
	LM	M	LM	LM	L	L
	M	M	M	LM	LM	L
	HM	HM	HM	M	LM	LM
	H	H	H	HM	M	LM

Table 19.2 Closed loop performance comparison

S No	Load kW	Load Ct. (A)		Load voltage(V)		Input power factor		%THD	
		PI	Fuzzy	PI	Fuzzy	PI	Fuzzy	PI	Fuzzy
1	1	2.43	2.51	400	400	0.931	0.949	43.2	40.5
2	1.5	3.62	3.77	400	400	0.946	0.955	35.9	31.6
3	2	4.95	5.05	400	400	0.968	0.981	25.6	20.0
4	2.5	6.19	6.27	400	400	0.985	0.995	11.9	8.3
5	3	7.41	7.53	400	400	0.991	0.999	4.7	1.5

19.4 Simulation

The scheme is simulated using a three phase 400 V, 3 kW DC link inverter in a MATLAB–SIMULINK platform. The performance of the RPWM strategy is evaluated over a range of load powers and the results validated with the existing schemes to establish the viability of the proposed algorithm.

The closed loop results seen in Table 19.2 relate to the improvement in input power factor owing allegiance to the reshaping action, the reduction in THD and the regulation of the load voltage through the range of operating loads. It brings out the merits of the fuzzy philosophy over the traditional PI controller in its attempt to enhance the power quality of the system.

The PWM pulses, input current, load voltage and its frequency spectrum corresponding to a load of 1 kW are depicted in Figs. 19.3, 19.4, 19.5, 19.6, and 19.7. The sinusoidal shape of the input current and PWM nature of the load voltage waveforms subscribe to meet the realistic needs of the present day applications.

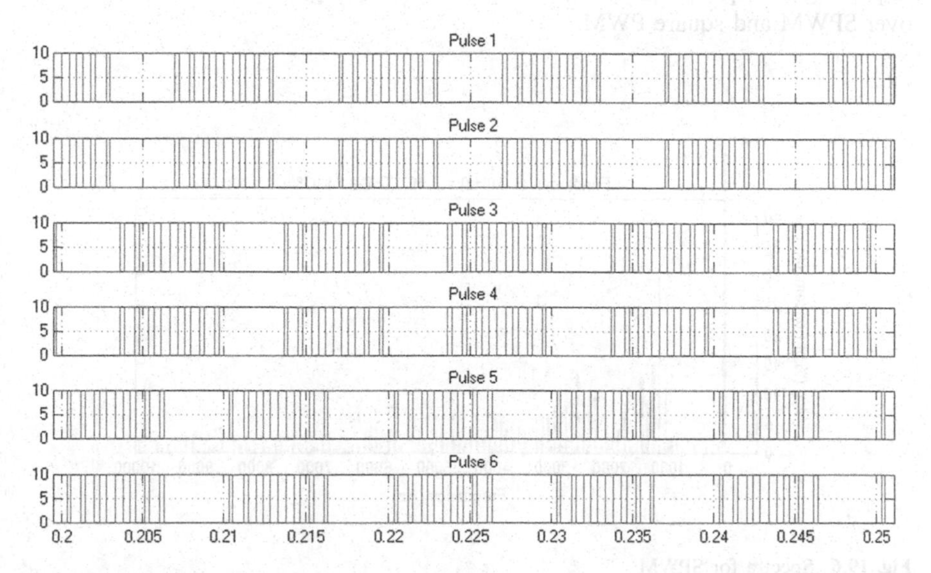

Fig. 19.3 PWM pulses

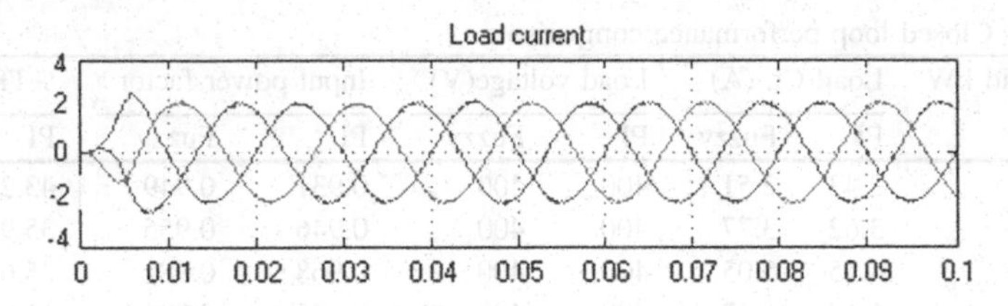

Fig. 19.4 Input current

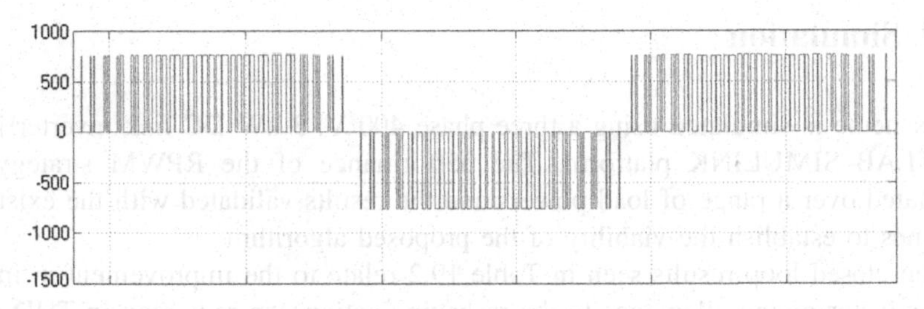

Fig. 19.5 Output voltage waveform

The numerical values of percentage THD and input power factor obtained using a fuzzy controller are compared for three different PWM strategies across the range of load powers. The bar diagram shown in Fig. 19.8 establishes the supremacy of RPWM over other approaches in terms of much lower THD. The graph portrayed in Fig. 19.9 explains the role of RPWM in significantly improving the input power factor through the entire range and elaborates its merits over SPWM and square PWM.

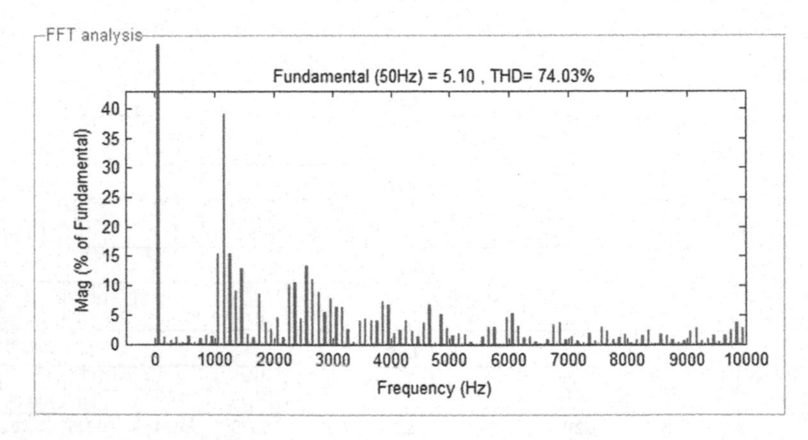

Fig. 19.6 Spectra for SPWM

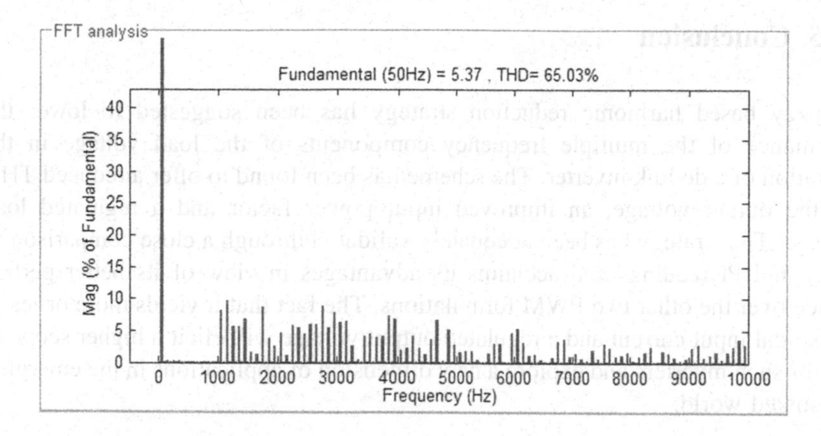

Fig. 19.7 Spectra for RPWM

Fig. 19.8 %THD comparison

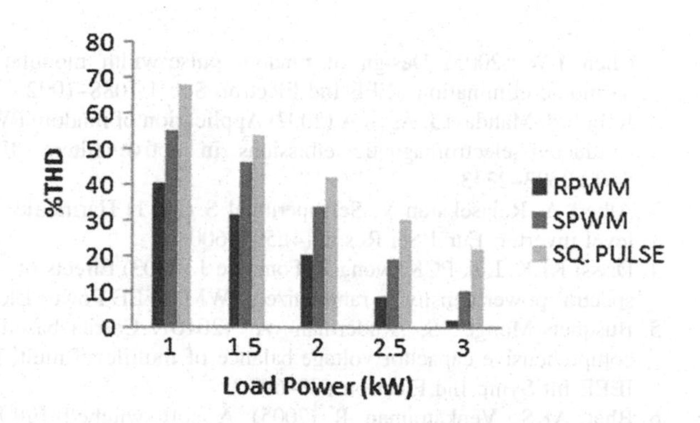

Fig. 19.9 Power factor comparison

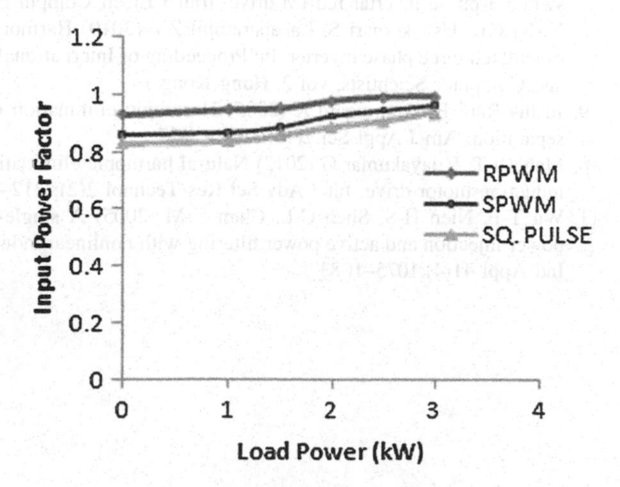

19.5 Conclusion

A fuzzy based harmonic reduction strategy has been suggested to lower the dominance of the multiple frequency components of the load voltage in the operation of a dc link inverter. The scheme has been found to offer a reduced THD for the output voltage, an improved input power factor and a regulated load voltage. The strategy has been adequately validated through a close comparison of fuzzy and PI readings and acclaims its advantages in view of its better performance over the other two PWM formulations. The fact that it yields more or less a sinusoidal input current and a regulated output voltage will elicit a higher scope of use for such inverters and explore a new dimension of applications in the emerging automated world.

References

1. Chen J-W (2005) Design of random pulse-width modulated inverter with lower-order harmonic elimination. IEEE Ind Electron Soc 1:1088–1092
2. Kaboli S, Mahdavi J, Agah A (2007) Application of random PWM technique for reducing the conducted electromagnetic emissions in active filters. IEEE Trans Industr Electron 54(4):2333–2343
3. Albert A, Rajasekaran V, Selvaperumal S (2011) Harmonic elimination of H-bridge seven level inverter. Eur J Sci Res 65(4):594–600
4. Drissi KEK, Luk PCK, wong B, Fontaine J (2003) Effects of symmetric distribution laws on spectral power density in randomized PWM. IEEE Power Electron Lett 1(2):41–44
5. Busquets-Monge S, Ruderman A (2010) Carrier-based PWM strategies for the comprehensive capacitor voltage balance of multilevel multi leg diode-clamped converters. IEEE Int Symp Ind Electron 688–693
6. Bhat AKS, Venkatraman R (2005) A soft-switched full-bridge single- stage AC-DC converter with low-line-current harmonic distortion. IEEE Trans Ind Electron 52(40)
7. Uddin MN, Radwan TS, Rahman MA (2006) Performance analysis of a cost effective 4-switch 3-phase inverter fed IM drive. Iran J Electr Comput Eng 5(2):97–102
8. Nisha GK, Ushakumari S, Lakaparampil ZV (2010) Harmonic elimination of space vector modulated three phase inverter. In: Proceeding of International multi conference of Engineers and Computer Scientists, vol 2, Hong Kong
9. Justus Rabi B, Arumugam R (2005) Harmonic elimination of inverters using blind signal separation. Am J Appl Sci 2(10):1434–1437
10. Mahesh T, Vijayakumar G (2012) Natural harmonic elimination of square-wave inverter for induction motor drive. Int J Adv Sci Res Technol 2(2): 212–218
11. Wu T-F, Nien H-S, Shen C-L, Chen T-M (2005) A single-phase inverter system for PV power injection and active power filtering with nonlinear inductor consideration. IEEE Trans Ind Appl 41(4):1075–1083

Chapter 20
Analysis on Electrical Distance in Deregulated Electricity Market

S. Prabhakar Karthikeyan, C. Abirami, P. Devi, I. Jacob Raglend and D. P. Kothari

Abstract In the deregulated environment, the economic related problems like generation dispatch, unit commitment, network cost allocation, congestion management, market clearing prices etc. are all associated with the geographical location of the generation companies and the consumers. The number of transmission lines, the distance and the voltage levels are some of the predominant factors in deciding the solution for the above problems for which electrical distance between the utilities is widely used as a tool by the power system researchers. In this paper, a complete analysis has been made on the concept of electrical distance and on the various factors affecting it. A sample five bus system is taken for analysis. The results are quite encouraging and help the researchers to extend the application of electrical distance in various domains.

Keywords Electrical distance · Z_{bus}-transmission lines · Line resistance · Power flows

S. Prabhakar Karthikeyan (✉) · C. Abirami · P. Devi
School of Electrical Engineering, VIT University, Vellore, Tamil Nadu, India

I. Jacob Raglend
Department of Electrical Engineering, NI University, Thakkalai, Kanyakumari, Tamil Nadu, India

D. P. Kothari
Fellow IEEE, FNAE, FNASc, FIE (India), Raisoni Group of Institutions, Nagpur, Maharashtra, India

R. Malathi and J. Krishnan (eds.), *Recent Advancements in System Modelling Applications*, Lecture Notes in Electrical Engineering 188,
DOI: 10.1007/978-81-322-1035-1_20,

20.1 Introduction

Owing to various reasons, viz., high demand growth coupled with inefficient system management, irrational tariff policies and to allow the power suppliers function competitively and consumers to choose suppliers of electric energy, the structure of the market slowly paved way to a competitive market called *Deregulated electricity market.*

Liu et al. proposes a new method to determine electrical distance using the time and size of energy transfer method. The authors have also discussed about the existing method namely, sensitivity method and impedance method and brings out its limitations [1].

In [2], the authors describe a new method to divide power network into zones such that within a zone, all buses are closely located geographically. To evaluate the quality of their solution, four indices were introduced. They are Clustering Tightness Index (CTI), Cluster Count Index (CCI), Cluster Size Index (CSI) and Cluster Connectedness (CC). The advantages of this approach is that it does not depend on any time varying variables like current, voltages, stability limits, locational prices. The authors have also derived a methodology to measure the "electrical centre" of any power systems. This is very much essential to study the topology of a system electrically rather physically in the area of system security and reliability [3].

Wang et al. presents a new tracing method for power flows by means of network partitioning by considering electrical distance between the busses. The author claims that the conventional tracing methods either produce unreasonable loss allocation or negative [4].

Thukaram et al. has proposed a noval approach in solving most of the issues in deregulated structured market using relative electrical distance (RED). In this approach, the relative locations of load nodes with respect to the generator nodes are computed in the form of a matrix. The authors have extensively used this matrix as a tool for various problems like optimal placement of Distributed Generation (DG) [5], to compute the reactive power contribution by the sources [6], generation planning [7]. They have proposed various indices like T-index to quantify the amount of generation expansion in the planning studies [7], L-index which gives the information about the voltage stability margin [6].

Conejo et al. proposed new methodology for transmission network cost allocation using Z_{bus} which applies kirchoffs' law and basic network theory. In this paper, authors have introduced a new definition called "electrical distance" i.e. the electrical distance between a line and the nodes. This helps in determining the usage of transmission line by both the generators and the loads [8].

20.2 Problem Definition

In this paper, a complete analysis has been made on the concept of electrical distance as proposed in [8]. Various factors like line parameters, bus and line location which influence the distance are thoroughly examined using standard test system.

20.3 Methodology

Consider the complex power flow S_{jk} computed at bus j and flowing through the line connecting bus j to bus k as shown in Fig. 20.1. As the power flow solution is known, we select the direction of the complex power flow so that $P_{jk} > 0$.

The complex power flow S_{jk} is

$$S_{jk} = V_j I_{jk}^* \tag{20.1}$$

Using the Z_{bus} matrix, the voltage at node j is given by

$$V_j = \sum_{i=1}^{n} Z_{ji} I_i \tag{20.2}$$

The current through the line jk, I_{jk} is obtained as

$$I_{jk} = (V_j - V_k) Y_{jk} + V_j Y_{jk}^{sh} \tag{20.3}$$

Substituting (20.2) in (20.3)

$$I_{jk} = \left(\sum_{i=1}^{n} z_{ji} I_i - \sum_{i=1}^{n} z_{ki} I_i\right) Y_{jk} + \sum_{i=1}^{n} z_{ji} I_i Y_{jk}^{sh} \tag{20.4}$$

Rearranging (20.4)

$$I_{jk} = \sum_{i=1}^{n} [(z_{ji} - z_{ki}) Y_{jk} + z_{ji} Y_{jk}^{sh}] I_i \tag{20.5}$$

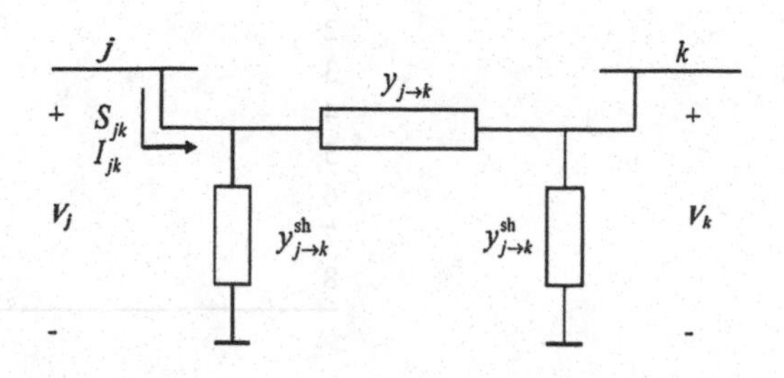

Fig. 20.1 Π equivalent circuit of line section jk

Note that the first term of the product in (20.5) is constant, as it depends only on network parameters. Thus, (20.5) can be written as

$$I_{jk} = \sum_{i=1}^{n} a^{i}_{jk} I_i \tag{20.6}$$

where

$$a^{i}_{jk} = (z_{ji} - z_{ki})Y_{jk} + z_{ji}Y^{sh}_{jk} \tag{20.7}$$

Observe that the magnitude of parameter a^{i}_{jk} provides a measure of the electrical distance between the bus *i* and line *jk*. The above derivation is applicable when the power flow, $P_{jk} \geq 0$, i.e. in the direction of power flow. It is to note that the distance parameters are not symmetrical with respect to line indexes, i.e. $a^{i}_{jk} \neq a^{i}_{kj}$.

20.4 Simulation and Results

A standard five bus system (shown in appendix) is taken as test system and simulations are carried out for various cases. MATLAB (R2009a) version is used for all simulations.

20.4.1 Case 1: Electrical Distance Between Bus i and Line jk

Figure 20.1 shows the distance between each transmission lines and the busses. Branch numbers are shown in Table 20.1. The magnitude of a_{jk} gives the information about the distance between the busses and the lines.

From Fig. 20.2, the distance between branch 1 and bus 3 is smaller than the distance with respect to all other busses. Similarly, bus 2 is located farther

Table 20.1 Branch details

Branch number	From bus	To bus
1	1	2
2	1	3
3	1	4
4	2	3
5	2	4
6	2	5
7	3	4
8	4	5

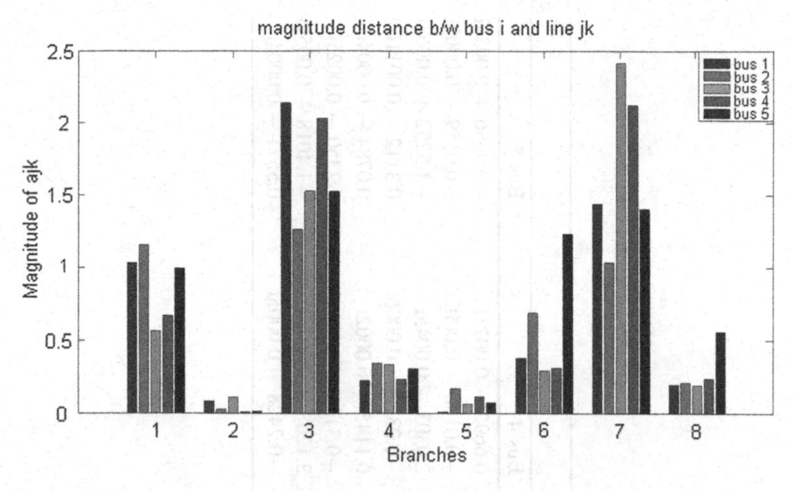

Fig. 20.2 Electrical distance (magnitude of a_{jk}) between the branch and a given bus

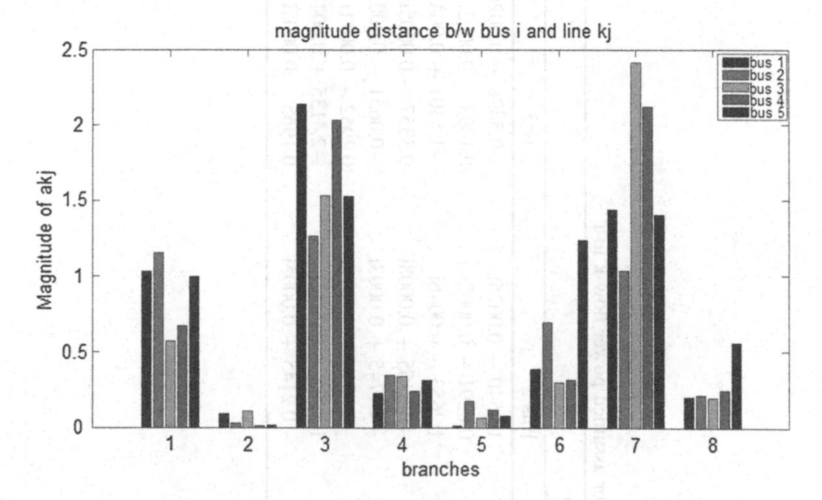

Fig. 20.3 Electrical distance (magnitude of a_{kj}) between the branch and a given bus

electrically for the given direction of power flow. The difference between Fig. 20.2 and Fig. 20.3 is the direction in which the power flow is assumed in the line i.e. from j to k or k to j i.e. a^i_{jk} *or* a^i_{jk} respectively. It is to note that $a^i_{jk} \neq a^i_{jk}$ and the actual power flow varies line to line so that the electrical line distance has to be considered as per the power flow (Table 20.2).

Table 20.2 Electrical distance details for assumed power flow k to j

Branch	Electrical distance (a^i_{kj})				
	Bus 1	Bus 2	Bus 3	Bus 4	Bus 5
1–2	−1.0322 + 0.0024i	1.1540 − 0.0025i	−0.5678 + 0.0020i	0.6692 − 0.0021i	−0.9956 + 0.0034i
1–3	−0.0879 + 0.0002i	0.0291 − 0.0002i	0.1093 + 0.0003i	−0.0106 − 0.0002i	−0.0159 + 0.0002i
1–4	−2.1392 + 0.0048i	1.2653 − 0.0046i	−1.5301 + 0.0043i	2.0303 − 0.0045i	−1.5252 + 0.0063i
2–3	0.2274–0.0007i	−0.3465 + 0.0008i	0.3357 − 0.0005i	−0.2377 + 0.0006i	0.3112 − 0.0011i
2–4	−0.0118 − 0.0002i	−0.1745 + 0.0003i	−0.0651 − 0.0001i	0.1145 + 0.0002i	0.0784 − 0.0004i
2–5	0.3835 − 0.0013i	−0.6916 + 0.0017i	0.2952 − 0.0011i	−0.3135 + 0.0012i	1.2350 − 0.0025i
3–4	−1.4399 + 0.0044i	1.0355 − 0.0042i	−2.4135 + 0.0047i	2.1208 − 0.0046i	−1.4018 + 0.0059i
4–5	0.2006 − 0.0005i	−0.2145 + 0.0005i	0.1965 − 0.0005i	−0.2428 + 0.0005i	0.5571 − 0.0004i

20.4.2 Case 2: Electrical Distance a_{12}^1 with respect to all Buses

Figure 20.4 shows the electrical distance between the line (branch 1) and the busses. In other words, the plot represents the locus of electrical distance. Blue, red, green, black and pink colour circles are the locus of the electrical distance from bus 1, 2, 3, 4 and 5 respectively.

20.4.3 Case 3: Electrical Distance a_{12} with respect to the 'R' of a Line (i.e. Line 1)

From Fig. 20.5, it is inferred that as the resistance of a transmission line (for example, line 1 as shown in Fig. 20.4) increases, the electrical distance between bus 1 and the line 1 decrease. Similar inference can be made for other line with respect to the busses.

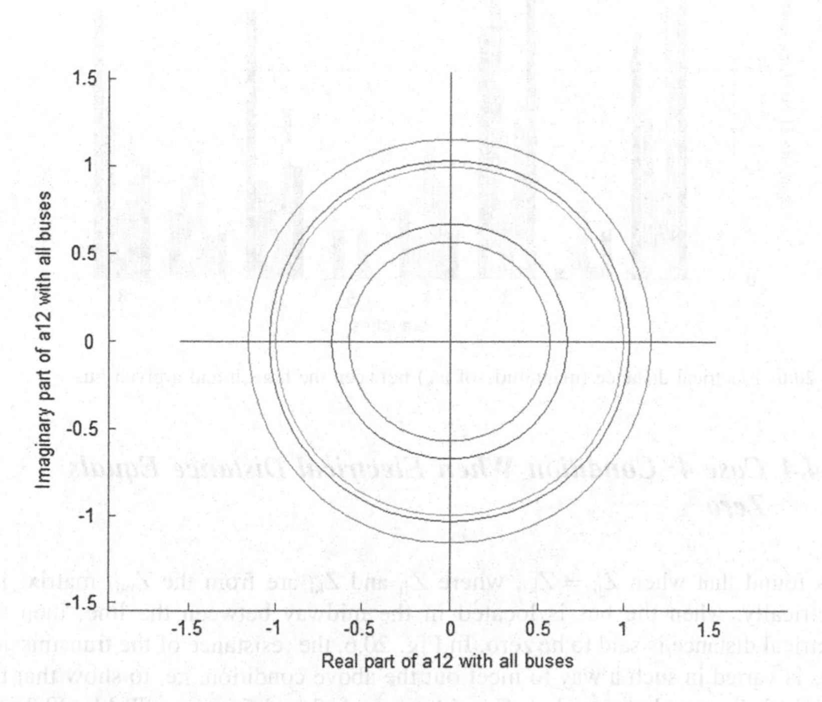

Fig. 20.4 Locus of a_{12} with respect to various buses

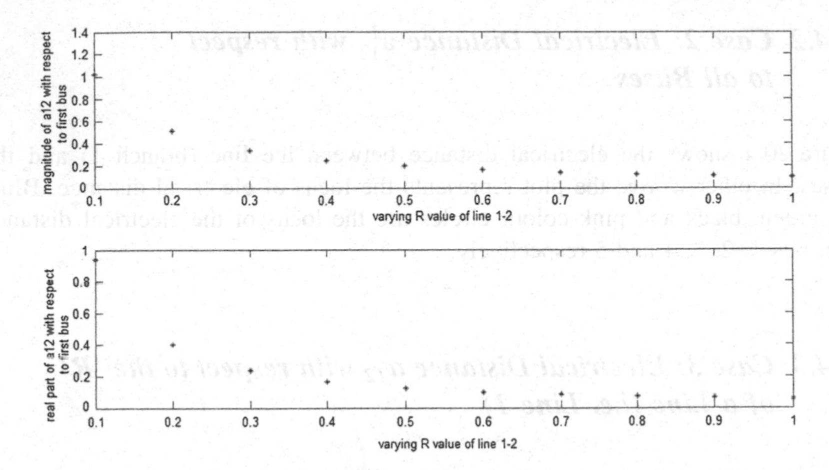

Fig. 20.5 Plot showing real part of a_{12} and magnitude of a_{12} for different values of resistance R of line 1

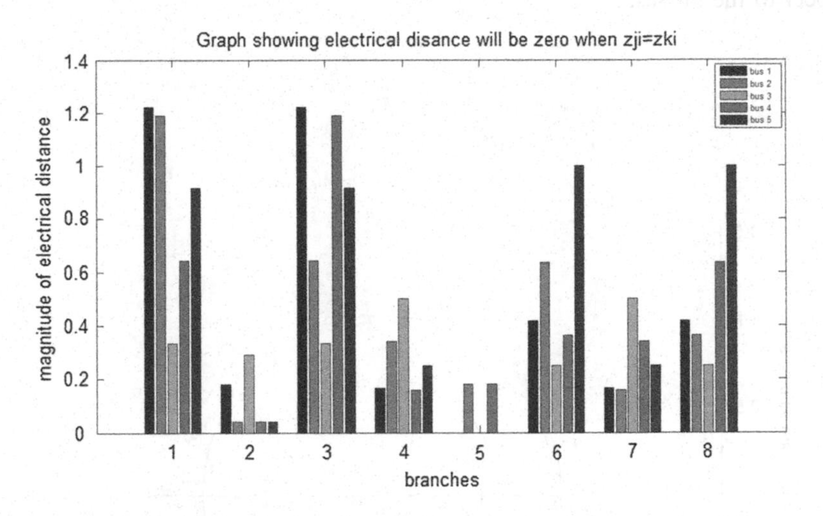

Fig. 20.6 Electrical distance (magnitude of a^{i}_{jk}) between the branch and a given bus

20.4.4 Case 4: Condition When Electrical Distance Equals Zero

It is found that when $Z_{ji} = Z_{ki}$, where Z_{ji} and Z_{ki} are from the Z_{bus} matrix. i.e. electrically, when the bus is located in the midway between the line, then the electrical distance is said to be zero. In Fig. 20.6, the resistance of the transmission lines is varied in such a way to meet out the above condition. i.e. to show that the electrical distance between line 5 and bus no. 1, 3 and 5 is zero (Table 20.3).

Table 20.3 Electrical distance between various buses and the lines when $Z_{21} = Z_{41}$; $Z_{25} = Z_{45}$; $Z_{23} = Z_{43}$

Branches	Electrical distance (a^{i}_{jk})				
	Bus 1	Bus 2	Bus 3	Bus 4	Bus 5
1–2	1.2222 + 0.0000i	−1.1894 − 0.0000i	0.3333 + 0.0000i	−0.6439 − 0.0000i	0.9167 + 0.0000i
1–3	0.1806	−0.0417 − 0.0000i	−0.2917 − 0.0000i	−0.0417 − 0.0000i	0.0417 − 0.0000i
1–4	1.2222 + 0.0000i	−0.6439 − 0.0000i	0.3333 + 0.0000i	−1.1894 − 0.0000i	0.9167 + 0.0000i
2–3	−0.1667–0.0000i	0.3409 + 0.0000i	−0.5000 − 0.0000i	0.1591 + 0.0000i	−0.2500 − 0.0000i
2–4	0	0.1818 + 0.0000i	0	−0.1818 − 0.0000i	0
2–5	−0.4167–0.0000i	0.6364 + 0.0000i	−0.2500 − 0.0000i	0.3636 + 0.0000i	−1.0000 − 0.0000i
3–4	0.1667 + 0.0000i	−0.1591 − 0.0000i	0.5000 + 0.0000i	−0.3409 − 0.0000i	0.2500 + 0.0000i
4–5	−0.4167–0.0000i	0.3636 + 0.0000i	−0.2500 − 0.0000i	0.6364 + 0.0000i	−1.0000 − 0.0000i

20.5 Conclusion

In this paper, a complete analysis has been made on the electrical distance which is measured between a bus and a transmission line. Firstly, the distance is graphically shown by the magnitude of the term a^i_{jk}. It is proved that $a^i_{jk} \neq a^i_{kj}$ as the concept stands on the basis of power flow. Secondly, the relation between the line parameter and the electrical distance is visualized and its inverse relation is proved. Finally, the condition for electrical distance is also found with an assumption that the shunt component of the line is neglected. It is sure that the results obtained will give a better understanding about the electrical distance so that the application of this terminology is left opened widely for various applications.

Acknowledgments The authors sincerely thank the Management of VIT University, Tamil Nadu, INDIA for their encouragement and support during the course of research.

A.1 Appendix

Line data

From bus number	To bus number	Resistance p.u	Reactance p.u	Grounding admittance B/2	Tap setting value
1	2	0.02	0.06	0.03	1
1	3	0.08	0.24	0.025	1
1	4	0.01	0.03	0.015	1
2	3	0.06	0.18	0.02	1
2	4	0.06	0.18	0.02	1
2	5	0.04	0.12	0.015	1
3	4	0.01	0.03	0.010	1
4	5	0.08	0.24	0.025	1

References

1. Liu H, Bao H, Liu L (2011) A new method about calculating electrical distance. In: Proceeding of the IEEE, School of Electrical and Electronics Engineering, North China Electric Power University, Beijing, China. Power engineering and automation conference (PEAM), 2011, IEEE, vol 1. pp 382–385. doi:10.1109/PEAM.2011.6134879
2. Blumsack S, Hines P, Patel M, Barrows C, Cotilla Sanchez E (2009) Defining power network zones from measures of electrical distance Power and Energy Society general meeting, 2009, PES'09. IEEE, pp 1–8. doi:10.11.09/PES.2009.5275353

3. Hines P, Blumsack S (2008) A centrality measures for electricity networks. In: Proceedings of the 41st Hawali international conference on system sciences
4. Wang H, Liu R, Li W (2009) Power flow tracing with consideration of the electrical distance. Department of Electrical and Electronics Engineering, Dailon University of Technology, Dailon, China. Power and Energy Engineering conference, 2009. APPEEC 2009, Asia-Pacfic, pp 1–4. doi:10.1109/APPEEC.2009.4918606
5. Thukaram D, Vyjayanthi C, Surendra S (2009) Optimal placement of distributed generation for a projected load increases using relative electrical distance approach, Department of Electrical Engineering, Indian Institute of Sciences, Bangalore, India, 2009. In: 3rd international conference on power systems, kharagpur, India, paper #109, December 27–29
6. Thukaram D, Vyjayanthi C (2009) Relative electrical distance concepts for evaluation of network reactive power and loss contribution in a deregulated systems. Indian Institute of Sciences, Bangalore, India. IET Generation, Transmission and distribution, vol. 3, Issue 11, pp. 1000–1019
7. Thukaram D, Vyjayanthi C (2008) Ranking of prospective new generation location for a power network in a deregulated system. Department of Electrical Engineering, Indian Institute of Science, Bangalore, India. Power System Technology and IEEE Power India conference, 2008. POWERCON 2008. Joint International conference, pp 1–8. doi:10.1109./ICPST.2008.4745257
8. Conejo AJ, Contreras J, Lima DA, Padilha-Feltrin A (2007) Z_{bus} transmission network cost allocation. IEEE Trans Power Syst 22(1):342–349

Chapter 21
A Novel Fuzzy Control Approach for Load Frequency Control

Rahul Umrao and D. K. Chaturvedi

Abstract Power system frequency regulation entitled load frequency control (LFC), as a major function of automatic generation control (AGC), has been one of the important control problems in electric power system design and operation. Normally, for the control of frequency conventional control methodology is used. In the recent years, the soft computing techniques such as artificial neural network (ANN), Fuzzy systems and evolutionary algorithms are used to develop an intelligent load frequency controller. The Performance of a Fuzzy logic controller is limited by its large number of rules and if the rules are large then computation time and requirement of memory is large. This problem is compensated by using Polar Fuzzy logic controller. So a Polar Fuzzy logic controller is proposed for the load frequency control problem. The aim of the Polar Fuzzy Controller is to restore the frequency and tie-line power in a smooth way to its nominal value in the shortest possible time if any load disturbance is applied. System performance is examined and compared with a standard Fuzzy logic controller, and conventional PI controller.

Keywords Load frequency control · Polar fuzzy logic · Power systems · AGC

R. Umrao (✉) · D. K. Chaturvedi
Department of Electrical Engineering, Dayalbagh Educational Institute,
Dayalbagh, Agra 282005, India
e-mail: rahulumrao@gmail.com

D. K. Chaturvedi
e-mail: dkc.foe@gmail.com

R. Malathi and J. Krishnan (eds.), *Recent Advancements in System Modelling Applications*, Lecture Notes in Electrical Engineering 188,
DOI: 10.1007/978-81-322-1035-1_21,

21.1 Introduction

Frequency stability problems, related control solutions and long-term dynamic simulation programs have been emphasized in the 1970s and 1980s following some major system events [1–7]. Frequency instability is the inability of a power system to maintain system frequency within the specified operating limits. Generally, frequency instability is a result of a significant imbalance between load and generation, and it is associated with poor coordination of control and protection equipment, insufficient generation reserves and inadequacies in equipment responses [8, 9]. Load-frequency control is a very important issue in power system operation and control for supplying sufficient and reliable electric power with good quality. Whenever there is a sudden load perturbation from customer side results a changes in tie-line power and frequency deviates. Large frequency deviation can damage equipment, degrade load performance, cause the transmission lines to be overloaded and can interfere with system protection schemes, ultimately leading to an unstable condition for the power system. Maintaining frequency and power interchanges with neighbouring control areas at the scheduled values are the two main primary objectives of a power system LFC [10, 11]. These objectives are met by measuring a control error signal, called the area control error (ACE), which represents the real power imbalance between generation and load, and is a linear combination of net interchange and frequency deviations.

As the operating point of a power system changes continuously, a fixed controller may no longer be suitable for all operating conditions. In recently years, modem control techniques, especially adaptive control configurations [12], are applied to load-frequency control like ANN, Fuzzy logic controllers etc. Fuzzy logic has always been a promising field for process control and process modelling over the past 2 decades [13, 14].

In this paper a novel Polar Fuzzy logic rule based load frequency controller, is used for two-area thermal power system, after testing on single area system.

21.2 Conventional Controller

Normally in industry PI controller is used as conventional controller. The proportional gain provides stability and high frequency response. The integral term insures that the average error is driven to zero. Advantages of PI are; only two gains need to be tuned, there is no long-term error, and the method normally provides highly responsive system. The predominant weakness is that PI controllers often produce excessive overshoot to a step command [15].

21.3 Fuzzy Controller

Fuzzy controllers are preferred over conventional controllers because:

- Higher robustness.
- To develop a fuzzy controller is cheaper.
- It is easy to understand because it is expressed in natural linguistic terms.
- It is easy to learn how fuzzy controller operates and how to design.

In last 2 decades the fuzzy set theory is a new methodological tool for handling ambiguity and uncertainty. Load frequency control has the main goal to maintain the balance between production and consumption. The power system is very complex in nature because of there are several variable conditions. So the fuzzy logic helps to develop robust and reliable controller for load frequency control problem [15, 16].

Functional module of a simple fuzzy logic controller (FLC) is shown in Fig. 21.1. These controllers are useful for solving a wide range of control problems [17]. For load frequency control problem two inputs, area control error (e) and change in area control error (ce), are chosen. Five triangular, for two inputs and output (u) both, membership functions are taken for design of FLC. The functional block model of fuzzy logic controller is shown in Fig. 21.1.

Set of linguistic variables e and ce can be represented as:

L (e, ce) = {NB, NM, Z, PM, PB}

Fuzzy controller output (u) can also be represents as:

L (u) = {NB, NM, Z, PM, PB}

where, NB = Negative Big, NM = Negative medium, Z = Zero, PM = Positive Medium, PB = Positive Big.

So here the number of rules is 25. Fuzzy inference rule for fuzzy controller is shown as Table 21.1

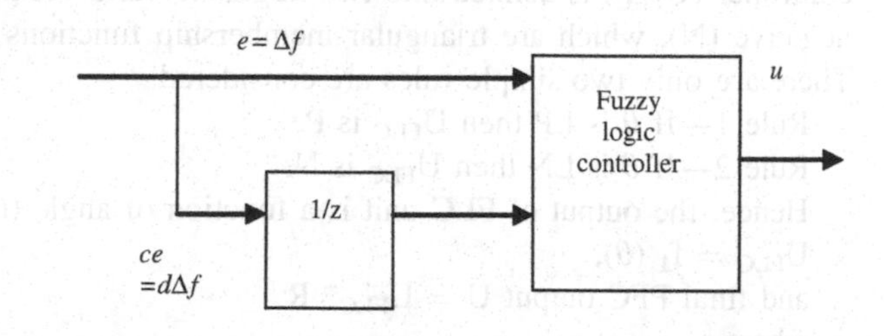

Fig. 21.1 Functional block model of fuzzy logic controller

Table 21.1 Fuzzy Inference Rules

Input		Change in frequency (e)				
Change in frequency deviation (ce)		NB	NM	ZE	PM	PB
	NB	NB	NB	NM	NM	ZE
	NM	NB	NB	NM	ZE	ZE
	ZE	NM	NM	ZE	PM	PM
	PM	ZE	PM	PM	PB	PB
	PB	ZE	ZE	PM	PB	PB

21.4 Polar Fuzzy Controller

Polar fuzzy is a new fuzzy control methodology. As the name suggest, in this new control methodology fuzzy sets are defined in polar coordinate. The (x, y) co-ordinates of a point in the plane are called its Cartesian co-ordinates. But there is another way to specify the position of a point, and that is to use polar co-ordinates (r, θ) as shown in Fig. 21.2. For polar co-ordinates we take an origin (or pole) O, and a fixed initial line OA. A point P is then described by specifying a distance r, the distance O to P along the radius direction, and the angle θ we have had to turn from the initial line to be looking along the radius direction. The point then has polar co-ordinates (r, θ)

The linguistic values are formed to vary with θ, the angle defined on the unit circle and their membership values are on μ (θ). Polar fuzzy is useful in situations that have a natural basis in polar coordinates or in situations where the value of a variable is cyclic. Polar fuzzy sets are applied in quantitative description of linguistic variables known truth-values. Polar fuzzy sets differ from standard fuzzy sets only in their Polar fuzzy sets are defined on a universe of angle and hence repeat shapes every 2π radian [18–20].

In this section, the working of PFC is described. The block diagram of polar fuzzy logic controller is shown in Fig. 21.3. Primarily frequency deviation and cumulative error define in complex plane and this complex quantity (consisting of real and imaginary part) is then converted into equivalent polar co-ordinates (i.e. angle and magnitude). The input to polar fuzzy controller is angle and its output is intermediate control action. Two fuzzy sigmoid membership functions are used which are large positive (LP) and large negative (LN) as shown in Fig. 21.4 for angle as input. These two membership functions are complimentary to each other.

Control action should be such that system attains desired frequency as early as possible with minimum deviation and oscillations. Output of the fuzzy logic controller (U_{FLC}) is defined into two linguistic variables namely, positive (P) and negative (N), which are triangular membership functions as shown in Fig. 21.5. There are only two simple rules are considered.

Rule 1—If θ is LP then U_{FLC} is P.

Rule 2—If θ is LN then U_{FLC} is N.

Hence, the output of FLC unit is a function of angle (θ) i.e.,

$U_{FLC} = f_1(\theta)$,

and final PFC output $U = U_{FLC} * R$

where

θ—angle in degree $= \tan^{-1}(ce/e)$;

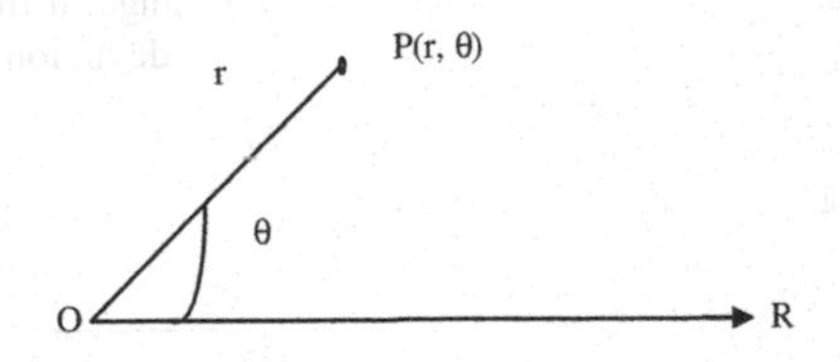

Fig. 21.2 Polar co-ordinate representation

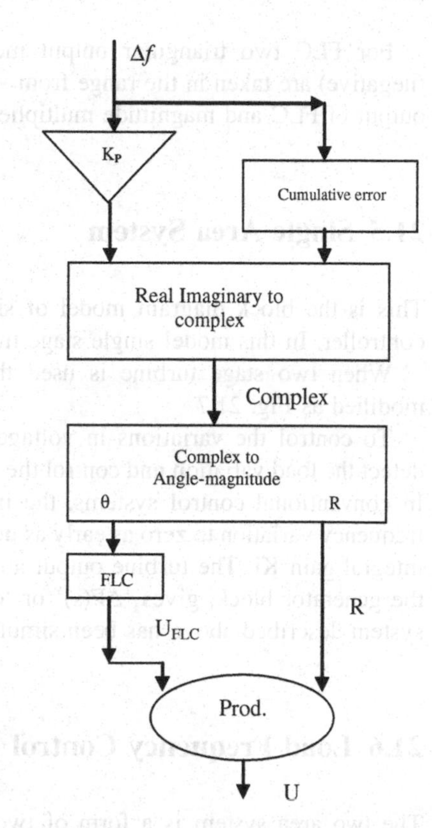

Fig. 21.3 Working block diagram model of polar fuzzy controller

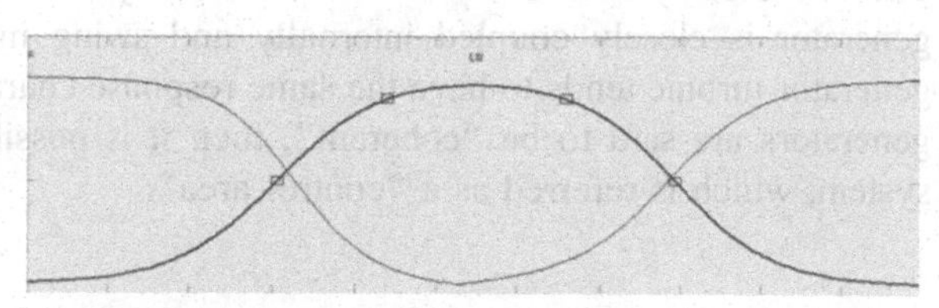

Fig. 21.4 Input membership functions

Fig. 21.5 Output membership functions

R—Magnitude $= \sqrt{(e^2 + ce^2)}$;
$e = K_o * \Delta f$ and ce—cumulative frequency error.

The working block diagram model of Polar fuzzy controller is shown as Fig. 21.3.

For FLC two triangular output membership functions P (positive) and N (negative) are taken in the range from −0.15 to +0.15 as shown in Fig. 21.5. The output of FLC and magnitude multiplied together to get the final output 'U'.

21.5 Single Area System

This is the block diagram model of single area thermal power system without controller. In this model single stage turbine model is used Fig. 21.6.

When two stage turbine is used then the turbine block diagram model is modified as Fig. 21.7

To control the variations in voltage and frequency, sophisticated controllers detect the load variation and control the input to counter balance the load variation. In conventional control systems, the integral part of the controller controls the frequency variation to zero as early as possible by getting the optimum value of the integral gain Ki. The turbine output and the electrical load perturbation given in the generator block, gives 'ΔF(s)' or 'df' as the output. The single area thermal system described above has been simulated in Matlab 7.0/Simulink environment.

21.6 Load Frequency Control of Two Area System

The two area system is a form of two-area system of AGC, where a group of generator is closely coupled internally and swing in unison. Furthermore, the generator turbine tends to have the same response characteristics. Such a group of generators are said to be "coherent", then it is possible to represent the whole system, which is referred as a "control area".

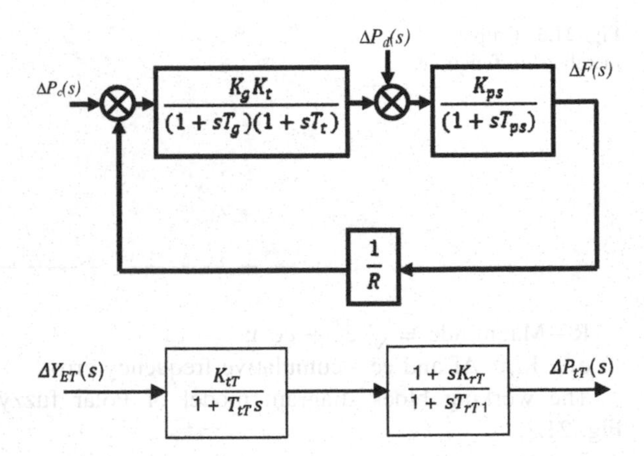

Fig. 21.6 Block diagram model of single area thermal power system without controller

Fig. 21.7 Two stage turbine block diagram model (with reheat)

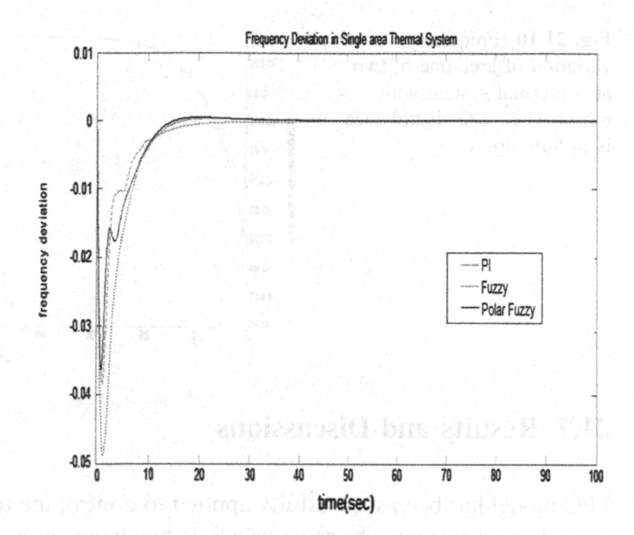

Fig. 21.8 Frequency variation of single area with reheat unit

The objective is to regulate the frequency of each area and simultaneously regulate the tie line power as per inter-area power contracts. Since a tie line transport power in or out of area, this fact must be accounted for in the incremental power balance equation of each area. The Simulink model of the two area system has been developed to study the system under different perturbations. With the help of this model system performance can be checked for different controllers for different perturbations. Performance of the PFC has been compared with conventional controller for with reheat unit in the two area thermal system. The results are shown in Figs. 21.9 and 21.10.

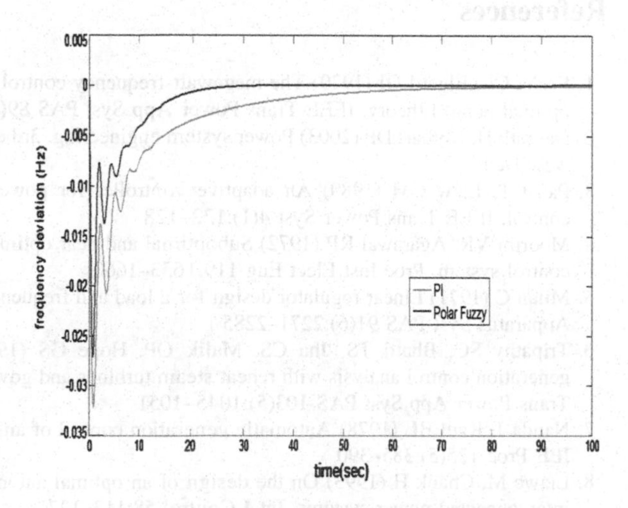

Fig. 21.9 Frequency variation of area one of two area thermal system with reheat when 1 % disturbance is in area 1

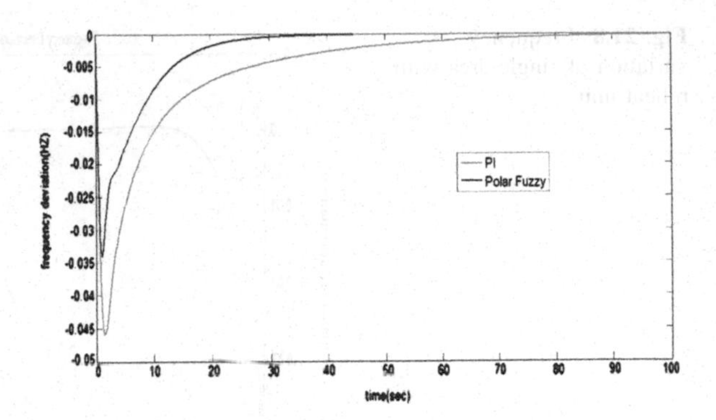

Fig. 21.10 Frequency variation of area one of two area thermal system with reheat when 1 % disturbance is in both areas

21.7 Results and Discussions

PFC model has been successfully applied to control the turbine reference power of a single area system. The same principle has been applied to a simulated to a two-area system. Each PFC controller receives only local information about the system (frequency in that specific area). Figure 21.8 shows the comparison of frequency deviation in the single area thermal system when 1 % disturbance is applied. The performance of PFC controller is better than conventional PI and fuzzy controllers. Similarly, the performance of PFC has been compared with conventional PI controller to control the frequency deviation in two area system as shown in Figs. 21.9 and 21.10. A comparison of the system dynamic response with the various controllers shows that the Polar Fuzzy controller yields improved control performance when compared to the PI and fuzzy logic controllers.

References

1. Fosha CE, Elgerd OI (1970) The megawatt-frequency control problem: a new approach via optimal control theory. IEEE Trans Power App Syst PAS 89(4):556–563
2. Nagrath IJ, Kothari DP (2003) Power system engineering, 3rd edn. McGraw-Hill Publication, New Delhi
3. Pan CT, Liaw CM (1989) An adaptive controller for power system and load frequency control. IEEE Trans Power Syst 4(1):122–128
4. Moorthi VR, Agarawal RP (1972) Suboptimal and near optimal control of a load frequency control system. Proc Inst Elect Eng 119:1653–1660
5. Milan C (1971) Linear regulator design for a load and frequency control. IEEE Trans Power Apparatus Syst PAS 91(6):2271–2285
6. Tripathy SC, Bhatti TS, Jha CS, Malik OP, Hope GS (1984) Sampled data automatic generation control analysis with reheat steam turbines and governor dead band effects. IEEE Trans Power App Syst PAS 103(5):1045–1051
7. Nanda J, Kaul BL (1978) Automatic generation control of an interconnected power system. IEE Proc 125(5):385–390
8. Liawc M, Chaok H (1993) On the design of an optimal automatic generation controller for interconnected power systems. Int J Control 58:113–127

9. Kumar P, Ibraheem (1996) AGC strategies: a comprehensive review. Int J Power Energy Syst 16(1):371–376
10. Kundur P (2008) Power system stability and control, 3rd edn. TMH Publication, New Delhi (5th reprint)
11. Ibraheem, Kumar P, Kothari DP (2005) Recent philosophies of automatic generation control strategies in power systems. IEEE Trans Power Syst 20(1):346–357
12. Ibraheem, Kumar P (2006) Overview of power system operation and control philosophies. Int J Power Energy Syst 26(1):203–214
13. Anand B, Ebenezer Jeyakumar A (2009) Load frequency control with fuzzy logic controller considering non-linearities and boiler dynamics. ICGST-ACSE J 8:15–20
14. Aravindan P, Sanavullah MY (2009) Fuzzy logic based automatic load frequency control of two area power system with GRC. Int J Comput Intell Res 5:37–44
15. Moon YH, Heon-Su R, Baik K, Park SC (2001) Fuzzy logic based extended integral control for load frequency control. In: Proceeding of IEEE winter meeting, vol 1, pp 1289–1293
16. Mathur HD, Manjunath HV (2006) A fuzzy based improved intelligent controller for automatic generation control. Int J Eng Simul 7(3):29–35
17. Chaturvedi DK, Satsangi PS, Kalra PK (1999) Load frequency control: a generalized neural network approach. Elect Power Energy Syst 21(6):405–415
18. Chaturvedi DK (2007) Soft computing: applications to electrical engineering problem, 1st edn. Springer, Berlin
19. Chaturvedi DK, Malik OP, Choudhury UK (2009) Polar fuzzy adaptive power system stabilizer. Inst Eng (India) 90:35–45
20. Umrao R, Chaturvedi DK (2010) Load frequency control using polar fuzzy controller. In: IEEE conference, Tencon, Fukuoka, Japan, 21–24 Nov 2010, pp 557–562

Chapter 22
Order Reduction of Interval Systems Using Alpha and Factor Division Method

D. Kranthi Kumar, S. K. Nagar and J. P. Tiwari

Abstract The paper proposes a new mixed method for reducing the order of interval systems i.e., systems having uncertain but bounded parameters. The denominator of the reduced order model is obtained by α table and numerator is derived by applying factor division and Cauer second form. A numerical example has been discussed to illustrate the procedures. The errors between the original higher order and reduced order models have also been highlighted to support the effectiveness of the proposed methods.

Keywords Cauer second form · α table · Factor division · Interval systems · Mixed method

22.1 Introduction

The approximation of the high order systems by low order models is an active area of research in many engineering applications especially in control system analysis and design. The analysis and design of a high order control system is both tedious and costly. Therefore, it is highly desirable to replace such a high order system by a system of lower order. A wide variety of model reduction techniques [1–9] have proposed by several authors in frequency domain, no one gives the best results.

The analysis and design problems in engineering and science involve uncertainty to varying degree. The uncertainty may be specified in a number of ways,

D. Kranthi Kumar (✉) · S. K. Nagar · J. P. Tiwari
Department of Electrical Engineering, Indian Institute of Technology-Banaras Hindu University, Varanasi, Uttar Pradesh, India
e-mail: kranthi.kumar.eee08@itbhu.ac.in

S. K. Nagar
e-mail: sknagar.eee@itbhu.ac.in

R. Malathi and J. Krishnan (eds.), *Recent Advancements in System Modelling Applications*, Lecture Notes in Electrical Engineering 188,
DOI: 10.1007/978-81-322-1035-1_22, © Springer India 2013

such as probabilistic, convex or fuzzy descriptions. However, in many systems the coefficients are constant but uncertain within a finite range. Such systems are classified as interval systems.

To combine the advantages of the above subgroup, mixed methods were proposed practically for fixed parameter systems. However, in view of parameter uncertainties for entire range of operation in many practical system models, like flexible manipulator system or nuclear reactor systems, lead to another area of research with the pioneering circle of Kharitonov [10], where in the stability of reduced order model was of great concern.

Many researchers have attracted the attention on interval systems and to study the stability and the transient analysis of interval systems [10, 11]. For reduction of continuous interval systems, Routh approximation [12], γ-δ Routh approximation [13], and also to reduce the complexity of calculations a simple direct method using γ table [14] have also been proposed. Dolgin and Zeheb [15, 16] modified the Bandypadhyay [13] method and suggested new mixed method to overcome the instability of the system. Unfortunately, Hwang et al. [17] commented on Dolgin's method and proved the failure of stability. Apart from these several other mixed methods [18–23] were proposed recently to reduce the complexity and to increase the accuracy of the system.

In this paper, some mixed methods for model order reduction of interval systems is proposed. The denominator of the interval reduced model is obtained by differentiation method. The numerator is obtained by different mixed methods, such as direct truncation of differentiation, factor division, Cauer second form and Routh approximation. These methods ensure that the reduced order model is stable if the higher order interval system is asymptotically stable. The outline of this paper is as follows: Section 22.2 contains problem statement. Section 22.3 contains proposed method and integral square error. Numerical example is presented in Sect. 22.4 and conclusions in Sect. 22.5.

22.2 Problem Statement

Let the transfer function of a higher order interval systems be:

$$G_n(s) = \frac{[C_{21}^-, C_{21}^+] + [C_{22}^-, C_{22}^+]s + \ldots + [C_{2n-1}^-, C_{2n-1}^+]s^{n-1}}{[C_{11}^-, C_{11}^+] + [C_{12}^-, C_{12}^+]s + \ldots + \left[C_{1,n}^-, C_{1,n}^+\right]s^n} = \frac{N(s)}{D(s)} \tag{22.1}$$

where $\left[C_{2j}^-, C_{2j}^+\right]$; $1 \le j \le n-1$ and $\left[C_{1j}^-, C_{1j}^+\right]$; $1 \le j \le n$ are known as scalar constants.

The reduced order model of a transfer function be considered as

$$R_k(s) = \frac{[d_{21}^-, d_{21}^+] + [d_{22}^-, d_{22}^+]s + \ldots + [d_{2k-1}^-, d_{2k-1}^+]s^{k-1}}{[d_{11}^-, d_{11}^+] + [d_{12}^-, d_{12}^+]s + \ldots + \left[d_{1,k}^-, d_{1,k}^+\right]s^k} = \frac{N_k(s)}{D_k(s)} \tag{22.2}$$

where $\left[d_{2j}^-, d_{2j}^+\right]; 1 \leq j \leq k-1$ and $\left[d_{1j}^-, d_{1j}^+\right]; 1 \leq j \leq k$ are known as scalar constants.

The rules of the interval arithmetic have been defined in [24], and are presented below.

Let [e, f] and [g, h] be two intervals.

Addition:

$$[e, f] + [g, h] = [e + g, f + h] \tag{22.3}$$

Subtraction:

$$[e, f] - [g, h] = [e - h, f - g] \tag{22.4}$$

Multiplication:

$$[e, f] \times [g, h] = [Min(eg, eh, fg, fh), Max(eg, eh, fg, fh)] \tag{22.5}$$

Division:

$$\frac{[e,f]}{[g,h]} = [e,f] \times \left[\frac{1}{h}, \frac{1}{g}\right] \tag{22.6}$$

22.3 Proposed Methods

The proposed method consists of the following steps for obtaining reduced order model.

Determination of the denominator polynomial of the kth order reduced model as given in Eq. (22.2) by using α table:

Reciprocal of higher order denominator gives

$$D(s) = \frac{1}{s}\hat{D}\left(\frac{1}{s}\right) \tag{22.7}$$

Differentiate the above Eq. (22.2) into (n−k) times

$$\hat{D}(s) = \left[c_{11}^-, c_{11}^+\right]s^n + [c_{12}^-, c_{12}^+]s^{n-1} + \ldots + \left[c_{1,n+1}^-, c_{1,n+1}^+\right] \tag{22.8}$$

The α table is shown in Table 22.1

Let $A_k(s)$ denote the denominator and numerator of Routh convergent respectively.

$$A_1(s) = \alpha_1 s + 1$$

Table 22.1 Alpha table

	$a_0^0 = [c_{11}^-, c_{11}^+]$ $a_2^0 = [c_{13}^-, c_{13}^+]\ldots$ $a_0^1 = [c_{12}^-, c_{12}^+]$ $a_2^1 = [c_{14}^-, c_{14}^+]\ldots$
$\alpha_1 = \frac{a_0^0}{a_0^1}$ $\alpha_2 = \frac{a_0^1}{a_0^2}$ $\alpha_3 = \frac{a_0^2}{a_0^3}$	$a_0^2 = a_2^0 - \alpha_1 a_2^1$ $a_2^2 = a_4^0 - \alpha_1 a_4^1\ldots$ $a_0^3 = a_2^1 - \alpha_2 a_2^2\ldots$ $a_0^4 = a_2^2 - \alpha_3 a_2^3\ldots$

$$A_2(s) = \alpha_1\alpha_2 s^2 + \alpha_2 s + 1$$

$$\ldots\ldots\ldots\ldots\ldots\ldots\ldots\ldots\ldots\ldots$$

$$A_k(s) = \alpha_1 A_{k-1} s + A_{k-2} s$$

The reduced order depends upon the order of the system reciprocal of $A_k(s)$.

Algorithm

Determine the reciprocal $\hat{D}(s)$ of the full model $D(s)$.

Construct α table corresponding to $\hat{D}(s)$.

Determine kth denominator by Routh convergent $\hat{D}(s) = A_k(s)$.

Reciprocate $\hat{D}(s)$ to determine denominator by Routh convergent $D_k(s)$.

Determine the numerator coefficients of the kth order reduced model by using factor division method and cauer second form method.

Case 1 Determination of the denominator polynomial of the kth order reduced model by using factor division method:

Any method of reduction which relies upon calculating the reduced denominator first and then the numerator, where $D_k(s)$ has already been calculated.

$$G_n(s) = \frac{N(s)D_k(s)/D(s)}{D(s)} \tag{22.9}$$

$$N(s)D_k(s) = [u_{11}^-, u_{11}^+] + [u_{12}^-, u_{12}^+]s + \ldots + \left[u_{1,k-1}^-, u_{1,k-1}^+\right]s^{k-1} \tag{22.10}$$

$$\frac{N(s)D_k(s)}{D(s)} = \frac{[u_{11}^-, u_{11}^+] + [u_{12}^-, u_{12}^+]s + \ldots + \left[u_{1,k-1}^-, u_{1,k-1}^+\right]s^{k-1}}{[d_{11}^-, d_{11}^+]] + [d_{12}^-, d_{12}^+]s + \ldots + \left[d_{1,k+1}^-, d_{1,k+1}^+\right]s^k} \tag{22.11}$$

Therefore,

$$[\alpha_{11}^-, \alpha_{11}^+] = \frac{[u_{11}^-, u_{11}^+]}{[d_{11}^-, d_{11}^+]}\left\{\begin{matrix}[u_{11}^-, u_{11}^+] & [u_{12}^-, u_{12}^+] & \ldots \\ [d_{11}^-, d_{11}^+] & [d_{12}^-, d_{12}^+] & \ldots\end{matrix}\right\}$$

$$[\alpha_{12}^-, \alpha_{12}^+] = \frac{[r_{11}^-, r_{11}^+]}{[d_{11}^-, d_{11}^+]}\left\{\begin{matrix}[r_{11}^-, r_{11}^+][r_{12}^-, r_{12}^+] \ldots \\ [d_{11}^-, d_{11}^+][d_{12}^-, d_{12}^+] \ldots\end{matrix}\right\}$$

$$[\alpha_{13}^-, \alpha_{13}^+] = \frac{[s_{11}^-, s_{11}^+]}{[d_{11}^-, d_{11}^+]}\left\{\begin{matrix}[s_{11}^-, s_{11}^+][s_{12}^-, s_{12}^+] \ldots \\ [d_{11}^-, d_{11}^+][d_{12}^-, d_{12}^+] \ldots\end{matrix}\right\}$$

$$\ldots\ldots\ldots\ldots\ldots\ldots\ldots\ldots\ldots\ldots\ldots\ldots$$

$$\left[\alpha_{1,k-2}^-, \alpha_{1,k-2}^+\right] = \frac{[x_{11}^-, x_{11}^+]}{[d_{11}^-, d_{11}^+]}\left\{\begin{matrix}[x_{11}^-, x_{11}^+][x_{12}^-, x_{12}^+] \\ [d_{11}^-, d_{11}^+][d_{12}^-, d_{12}^+]\end{matrix}\right\}$$

$$\left[\alpha_{1,k-1}^-, \alpha_{1,k-1}^+\right] = \frac{[y_{11}^-, y_{11}^+]}{[d_{11}^-, d_{11}^+]}\left\{\begin{matrix}[y_{11}^-, y_{11}^+] \\ [d_{11}^-, d_{11}^+]\end{matrix}\right\}$$

where

$$[r_{1,i}^-, r_{1,i}^+] = [u_{1,i+1}^-, u_{1,i+1}^+] - [\alpha_{11}^-, \alpha_{11}^+][d_{1,i+1}^-, d_{1,i+1}^+]; \quad i = 0, 1, 2, \ldots, k-2.$$

$$\left[s_{1,i}^-, s_{1,i}^+\right] = \left[r_{1,i+1}^-, r_{1,i+1}^+\right] - [\alpha_{12}^-, \alpha_{12}^+]\left[d_{1,i+1}^-, d_{1,i+1}^+\right]; \quad i = 0, 1, 2, \ldots, k-3$$

$$\ldots\ldots\ldots\ldots\ldots\ldots\ldots\ldots\ldots\ldots\ldots\ldots$$

$$[y_{1,i}^-, y_{1,i}^+] = [x_{11}^-, x_{11}^+] - [\alpha_{1,k-2}^-, \alpha_{1,k-2}^+][d_{11}^-, d_{11}^+]$$

The reduced transfer function given by

$$R_k(s) = \frac{[\alpha_{11}^-, \alpha_{11}^+] + [\alpha_{12}^-, \alpha_{12}^+]s + \ldots + [\alpha_{1,k-1}^-, \alpha_{1,k-1}^+]s^{k-1}}{D_k(s)} \tag{22.12}$$

where

$$[\alpha_{11}^-, \alpha_{11}^+] = [d_{21}^-, d_{21}^+]$$

$$[\alpha_{12}^-, \alpha_{12}^+] = [d_{22}^-, d_{22}^+]$$

$$\ldots\ldots\ldots\ldots\ldots$$

$$[\alpha_{1,k-1}^-, \alpha_{1,k-1}^+] = [d_{2k}^-, d_{2k}^+]$$

Case 2 Determination of the denominator polynomial of the *Kth* order reduced model by using Cauer second form:

Coefficient values from Cauer second form $[h_i^-, h_i^+]$ (i = 1, 2, 3,…., k) are evaluated by forming Routh array as

$$[h_1^-, h_1^+] = \frac{[c_{11}^-, c_{11}^+]}{[c_{21}^-, c_{21}^+]}\left\{\begin{matrix}[c_{11}^-, c_{11}^+][c_{12}^-, c_{12}^+] \ldots \\ [c_{21}^-, c_{21}^+][c_{22}^-, c_{22}^+] \ldots\end{matrix}\right\}$$

$$[h_2^-, h_2^+] = \frac{[c_{21}^-, c_{21}^+]}{[c_{31}^-, c_{31}^+]} \begin{Bmatrix} [c_{21}^-, c_{21}^+][c_{22}^-, c_{22}^+] & \dots \\ [c_{31}^-, c_{31}^+][c_{32}^-, c_{32}^+] & \dots \end{Bmatrix}$$

$$[h_3^-, h_3^+] = \frac{[c_{31}^-, c_{31}^+]}{[c_{41}^-, c_{41}^+]} \begin{Bmatrix} [c_{31}^-, c_{31}^+][c_{32}^-, c_{32}^+] & \dots \\ [c_{41}^-, c_{41}^+][c_{42}^-, c_{42}^+] & \dots \end{Bmatrix}$$

$$\dots\dots\dots\dots\dots\dots\dots\dots\dots\dots\dots\dots\dots$$

The first two rows are copied from the original system numerator and denominator coefficients and rest of the elements are calculated by using well known Routh algorithm.

$$[c_{1,j}^-, c_{1,j}^+] = [c_{i-2,j+1}^-, c_{i-2,j+1}^+] - [h_{i-2}^-, h_{i-2}^+][c_{i-1,j+1}^-, c_{i-1,j+1}^+] \tag{22.13}$$

where i = 3, 4,….., and j = 1, 2,…

$$[h_i^-, h_i^+] = \frac{[c_{i,1}^-, c_{i,1}^+]}{[c_{i+1,1}^-, c_{i+1,1}^+]} \tag{14}$$

The coefficient values of $[d_{ij}^-, d_{ij}^+]$ (j = 1, 2,….., (k + 1)) and Cauer quotients $[h_i^-, h_i^+]$(i = 1, 2,…, k) are matched for finding the coefficients of numerator of the reduced model $R_k(s)$.

The inverse Routh array is constructed as

$$[d_{i+1,1}^-, d_{i+1,1}^+] = \frac{[d_{i,1}^-, d_{i,1}^+]}{[h_i^-, h_i^+]} \tag{22.15}$$

where i = 1, 2,…, k and k ≤ n.

Also,

$$[d_{i+1,j+1}^-, d_{i+1,j+1}^+] = \frac{[d_{i,j+1}^-, d_{i,j+1}^+] - [d_{i+2,j}^-, d_{i+2,j}^+]}{[h_i^-, h_i^+]} \tag{16}$$

where i = 1, 2,…., (k − j) and j = 1, 2,…., (k − 1)

Using the above equations, the numerator coefficients of the reduced model will be obtained.

To minimize the steady state error the zeros are adjusted by multiplying the numerator polynomial with the gain correction factor η. It can be calculated using the relation

$$\eta = \left.\frac{G_k(s)}{R_k(s)}\right|_{s=0} \tag{22.17}$$

For interval systems, η is calculated after converting the interval coefficients G(s) and $R_k(s)$ into the fixed coefficients by taking their means. Thus the gain correction factor is

$$\eta = \left(\frac{c_{21}}{c_{11}}\right)\left(\frac{d_{11}}{d_{21}}\right)$$

where $c_{21} = \frac{c_{21}^- + c_{21}^+}{2}; \frac{c_{11}^- + c_{11}^+}{2}\ d_{21} = \frac{d_{21}^- + d_{21}^+}{2}; d_{11} = \frac{d_{11}^- + d_{11}^+}{2}$

22.3.1 Integral Square Error

The integral square error (ISE) between the transient responses of higher order system (HOS) and Lower order system (LOS) is determined to compare different approaches of model reduction. This is given by

$$ISE = \int_0^{\infty} [y(t) - y_r(t)]^2 \tag{22.18}$$

where, $y(t)$ and $y_r(t)$ are the unit step responses of original system $G_n(s)$ and reduced order system $R_k(s)$.

22.4 Numerical Examples

This section includes example to illustrate the method.
Example: Consider a third order system described by the transfer function [13]

$$G_3(s) = \frac{[2,3]s^2 + [17.5,18.5]s + [15,16]}{[2,3]s^3 + [17,18]s^2 + [35,36]s + [20.5,21.5]}$$

Reduction by α method
Step 1: Apply reciprocal $\hat{D}(s)$ to get

$$\hat{D}(s) = [20.5,21.5]s^3 + [35,36]s^2 + [17,18]s + [2,3]$$

Table 22.2 Alpha table

	[20.5, 21.5] [17, 18]
	[35, 36] [2, 3]
$\alpha_1 = [0.5694, 0.6143]$	[15.1571, 16.8612]
$\alpha_2 = [2.07582.3751]$	[0, 0] [2, 3]
$\alpha_3 = [5.2524, 8.4306]$	

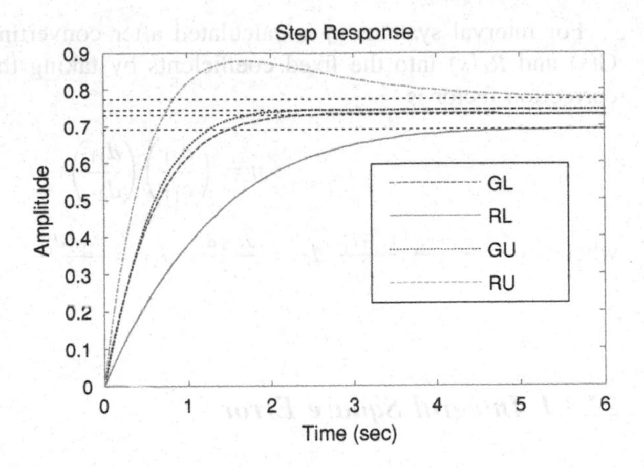

Fig. 22 1 Step response of original model and ROM

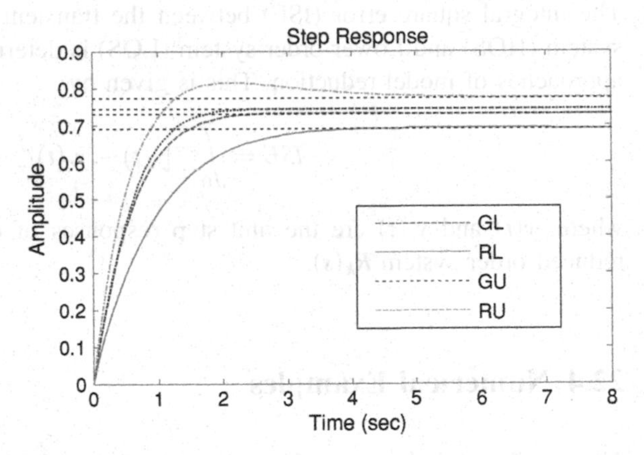

Fig. 22 2 Step response of original model and ROM

Step 2: Construct α Table 22.2.
Step 3: Denominator for second order

$$D_2(s) = s^2 + [2.0758, 2.3751]s + [1.1896, 1.459]$$

Method 1: Numerator is reduced by factor division method
Step 1: Using the factor division method

$$\frac{N(s)D_2(s)}{D(s)} = \frac{[17.844, 23.344] + [531.955, 64.9931]s + \ \dots}{[20.5, 21.5] + [35, 36]s + \ \dots}$$

Step 2: Finding the values of $[\alpha_{11}^{-}, \alpha_{11}^{+}]$ and $[\alpha_{12}^{-}, \alpha_{12}^{+}]$

$$[\alpha_{11}^{-}, \alpha_{11}^{+}] = [0.8299, 1.1387]$$

$$[\alpha_{12}^{-}, \alpha_{12}^{+}] = [0.4861, 1.8389]$$

$$N_2(s) = [0.8299, 1.1387] + [0.4861, 1.8389]s$$

Table 22.3 Comparision of reduced order models

Method of order reduction	ISE for lower limit	ISE for upper limit
α and factor division	0.0125	0.0082
α and Cauer second form	0.0094	0.0073
G. V. K. Sastry [14]	0.2256	0.0095

Step 3: Using gain correction factor ($\eta = 0.993$)

$$N_2(s) = [0.8241, 1.1307] + [0.4838, 1.826]s$$

Step 4: The reduced transfer function and Fig. 22.1 is Shown below

$$R_2(s) = \frac{[0.4838, 1.826]s + [0.8241, 1.1307]}{s^2 + [2.0758, 2.3751]s + [1.1896, 1.459]}$$

Method 2: Reduction by Cauer second form method
Step 1: Using the Cauer second form is obtained as

$$[h_1^-, h_1^+] = [1.2812, 1.434]$$

$$[h_2^-, h_2^+] = [1.1046, 1.8859]$$

$$[d_{21}^-, d_{21}^+] = [0.8295, 1.1387]$$

$$[d_{22}^-, d_{22}^+] = [0.7286, 1.5105]$$

Step 2: Numerator of second order system is written as

$$N_2(s) = [0.7286, 1.5105]s + [0.8295, 1.1387]$$

Step 3: Using gain correction factor ($\eta = 0.9932$)

$$N_2(s) = [0.7236, 1.5]s + [0.8238, 1.1309]$$

Step 4: The reduced order model and Fig. 22.2 is shown below

$$R_2(s) = \frac{[0.7236, 1.5]s + [0.8238, 1.1309]}{s^2 + [2.0758, 2.3751]s + [1.1896, 1.459]}$$

The original and reduced order model are compared by the step response as shown in Figs. 22.1, 22.2. The results obtained by the proposed methods have been compared with other existing order reduction methods for a second order reduced interval system model, as shown in Table 22.3.

22.5 Conclusions

In this paper new mixed methods are employed for order reduction. The denominator polynomial of reduced model is obtained by using α table. The numerator is obtained by factor division method and Cauer second form. The proposed method guarantees the stability of reduced model if the original system is stable. The method is conceptually simple and it yields comparatively better results than those obtained by existing methods proposed in the literature.

References

1. Aoki M (1968) Control of large-scale dynamic systems by aggregation. IEEE Trans Autom Control 13:246–253
2. Shamash Y (1974) Stable reduced order models using pade type approximation. IEEE Trans Autom Control 19:615–616
3. Hutton MF, Friedland B (1975) Routh approximation for reducing order of linear time invariant system. IEEE Trans Autom Control 20:329–337
4. Krishnamurthy V, Seshadri V (1978) Model reduction using routh stability criterion. IEEE Trans Autom Control 23:729–730
5. Sinha NK, Kuszta B (1983) Modelling and identification of dynamic systems. New York Van Nostrand Reinhold, Chapter 8, pp 133–163
6. Glover K (1984) All optimal hankel-norm approximations of linear multivariable systems and their error bounds. Int J Control 39(6):1115–1193
7. Shamash Y (1975) Model reduction using routh stability criterion and the pade approximation. Int J Control 21:475–484
8. Parmar G (2007) A mixed method for large-scale systems modelling using eigen spectrum analysis and cauer second form. IETE J Res 53(2):93
9. Bai-Wu W (1981) Linear model reduction is using mihailov criterion and pade approximation technique. Int J Control 33(6):1073
10. Kharitonov VL (1978) Asymptotic stability of an equilibrium position of a family of systems of linear differential equations. Differentsialnye Uravneniya 14:2086–2088
11. Bhattacharyya SP (1987) Robust stabilization against structured perturbations (lecture notes in control and information sciences). Springer, New York
12. Bandyopadhyay B, Ismail O, Gorez R (1994) Routh pade approximation for interval systems. IEEE Trans Autom Control 39:2454–2456
13. Bandyopadhyay B (1997) γ-δ Routh approximations for interval systems. IEEE Trans Autom Control 42:1127–1130
14. Sastry GVK, Raja Rao GR, Rao PM (2000) Large scale interval system modelling using routh approximants. Electron Lett 36(8):768
15. Dolgin Y, Zeheb E (2003) On routh pade model reduction of interval systems. IEEE Trans Autom Control 48(9):1610–1612
16. Dolgin Y (2005) Author's reply. IEEE Trans Autom Control 50(2):274–275
17. Hwang C, Yang S-F (1999) Comments on the computation of interval routh approximants. IEEE Trans Autom Control 44(9):1782–1787
18. Choo Younseok (2007) A note on discrete interval system reduction via retention of dominant poles. Int J Control Autom Syst 5(2):208–211
19. Saraswathi G (2007) A mixed method for order reduction of interval systems. Int Conf Intell Adv Syst pp 1042–1046

20. Ismail O, Bandyopadhyay B (1995) Model order reduction of linear interval systems using pade approximation. IEEE Int Symp Circ Syst
21. Singh VP, Chandra D (2010) Routh approximation based model reduction using series expansion of interval systems. IEEE Int Conf Power Control Embed Syst (ICPCES) 1:1–4
22. Singh VP, Chandra D (2011) Model reduction of discrete interval system using dominant poles retention and direct series expansion method. In: Proceedings of the IEEE 5th International power engineering and optimization conference (PEOCO), vol 1. pp 27–30
23. Kranthi Kumar D, Nagar SK, Tiwari JP (2011) Model order reduction of interval systems using modified routh approximation and factor division method. In: Proceedings of 35th national system conference (NSC), IIT Bhubaneswar, India
24. Hansen E (1965) Interval arithmetic in matrix computations. Part I Siam J Numer Anal pp 308–320

Chapter 23
Impulse Fault Detection and Classification in Power Transformers with Wavelet and Fuzzy Based Technique

N. Vanamadevi and S. Santhi

Abstract Impulse testing of transformers after assembly is a routine procedure carried out for the assessment of their winding insulation. During impulse test insulation failure may result in two classes of winding faults in a transformer namely series faults and shunt faults. Series faults are due to the short between turns in the section and the shunt faults are due to the short between turns in the section and the ground. This paper aims at deriving a technique for the detection and classification of impulse faults in a transformer winding using wavelet transform and a fuzzy Inference system. A specially designed 6.6 kV model layer winding is considered for the study. The entire winding comprising ten sections are divided into three regions namely sections near line end, sections near the neutral end and the sections in the middle of the winding. The neutral currents are recorded with series faults and shunt faults introduced in the sections belonging to the three regions. Continuous wavelet transform is applied on these neutral current records to extract the discriminating features. The features extracted from the wavelet transformed signal are the second most predominant frequency, the time range at which it occurs and the corresponding wavelet coefficient. A fuzzy Inference system is designed and implemented using Matlab software with these three features extracted from the wavelet transformed signal as inputs and generates an output that classifies the fault and no fault conditions. It is observed that the results are satisfactory.

Keywords Transformer · Impulse faults · Wavelet transform · Fuzzy inference system

N. Vanamadevi (✉) · S. Santhi
Department of Instrumentation Engineering, Annamalai University,
Annamalai Nagar, Chidambaram 608002, India
e-mail: santhi.sathyamurthy@gmail.com

R. Malathi and J. Krishnan (eds.), *Recent Advancements in System Modelling Applications*, Lecture Notes in Electrical Engineering 188,
DOI: 10.1007/978-81-322-1035-1_23, © Springer India 2013

23.1 Introduction

Early detection and diagnosis of transformer faults are essential for the assessment of its condition, to prepare a maintenance schedule, and to provide improved operational efficiency. Transformer windings are frequently exposed to short-duration impulse voltages of high amplitude produced by lightning or switching transients, and by insulator flashover. These impulse voltages may produce permanent damage to the insulation of windings. Transformers are subjected to standardized impulse test to check its dielectric strength. It is a routine test as explained in standards such as IEC-60076, Part IV, 2000 [1, 2]. The failure during the impulse test is diagnosed conventionally by observing the oscilloscopic recording of the neutral currents at reduced and full voltage levels. Any deviation between the current waveforms indicates that there is a failure in the transformer winding. But comparison of current waveforms in time domain under normal and fault condition is not sufficient to discriminate minor faults such as a turn to turn short. The transfer function method exists as an improved assessment technique that involves frequency domain approach [3, 4]. Two popular time–frequency tools, the short-time Fourier transform and wavelet transforms were investigated for analyzing the neutral currents, for fault diagnosis during impulse tests on power transformers [5, 6, 7].

The wavelet transform (WT) is an efficient and powerful technique for the analysis of transient phenomena in power transformers [4]. It is capable of discriminating the time and frequency features of a non-stationary signal with good resolution in both time and frequency than the conventional Fourier transform that provides only frequency domain information. It is an extension to short time Fourier transforms (STFT) that has constant window length. WT allows high frequency components to be analyzed with short time interval and low frequency components to be analyzed with long time intervals. Recently wavelets have been widely used in various fault detection schemes [8].

The continuous wavelet transform of a signal x(t), as given in [7] is

$$W_x(a,\ b) = 1/\sqrt{a} \int x(t)\Psi((t-b)/a)dt \tag{23.1}$$

where Ψ is the basis function or the mother wavelet, ‘a’ and ‘b’ are the dilation and translation parameters respectively. The choice of mother wavelet is crucial and it depends on a particular application. There are many types of mother wavelets such as Harr, Coiflet, Symmlet, and Daubichies. But Daubichies, one of the most popular and widely used wavelet in fault detection applications [9, 14] is chosen as the mother wavelet in this work to extract the discriminating features from the measured neutral current.

Artificial neural networks, fuzzy inference systems and adaptive network based fuzzy inference systems have been used to achieve automatic fault classification [10–14]. In this work a fuzzy inference system is considered for fault classification since in fuzzy logic, knowledge representation is explicit, expressed using simple

IF–THEN rules. This paper demonstrates a method to extract discriminating features under various fault conditions using continuous wavelet transform from the neutral current records and classification of the faults using a fuzzy inference system.

23.2 Objective and Scope

The objective of the work is to detect the type and location of faults during an impulse test by developing the detection scheme using wavelet transform and fuzzy inference system for the classification of the various faults. The detection and classification strategy is developed by considering the series faults and shunt faults at the line-end, middle of the winding and the neutral-end. These faults are described in the Table 23.1. The scope of the work is limited to detection and classification of series and shunt faults in the winding structure alone.

23.3 Device Under Test

A specially designed 6.6 kV voltage transformer model winding is considered as the device under test (DUT) due to the following reasons.

- It comprises of a multilayer winding as used mostly in large power transformers.
- Its dimensions are compact.
- The topology of the winding is same from 6.6 to 400 kV.

Figure 23.1 shows the photograph of the DUT. The high voltage (HV) winding of the DUT comprises of 20 layers with 250 turns per layer wound on a former of diameter 100 mm. The axial length of the HV winding is 100 mm. A single turn metallic tube of height 340 mm is treated as the secondary (low voltage LV) winding which is shorted to ground. The HV winding is constructed to have ten sections with each section comprising of 500 turns and two layers per section. The thickness of the conductor is 0.4 mm and the spacer thickness is 1 mm.

Table 23.1 Description of the faults

Fault name	Description of the fault
NF	No fault
SFL	Short within sections near the line end (sections 1, 2 and 3) of the winding
SFM	Short within sections near the middle end (sections 4, 5, 6 and 7) of the winding
SFN	Short within sections near the neutral end (sections 8, 9 and 10) of the winding
SHFL	Short between sections near the line end and ground of the winding
SHFM	Short between sections near the middle end and ground of the winding
SHFN	Short between sections near the neutral end and ground of the winding

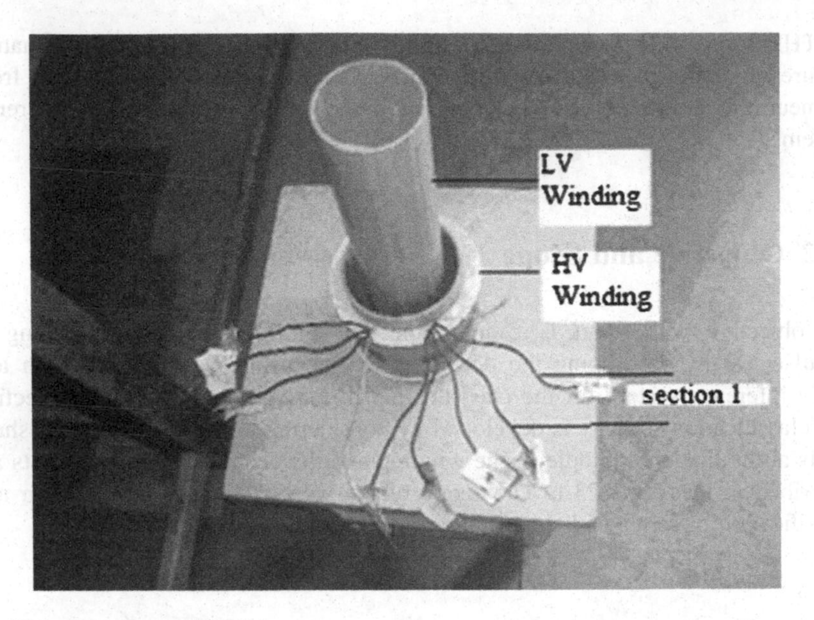

Fig. 23.1 Photograph of DUT

23.4 Experimentation for Feature Extraction

The experiment is conducted with the DUT in a systematic manner and the experimental arrangement is as shown in Fig. 23.2. The device under test is excited with a low voltage impulse similar to a standard lightning impulse (LI) of 1.2/50 μs and 1 V amplitude as shown in Fig. 23.3 from an Agilent arbitrary function generator. The HV winding current is measured through a current viewing resistor using Agilent DSO and the necessary post processing and analysis is carried out with Pentium IV PC connected through RS232 interface. The winding current is measured under no fault condition and with various faults introduced as described below.

In order to introduce series fault across two sections, the beginning of a section and the end of next section is shorted. Figures 23.4, 23.5 and 23.6 show the neutral currents recorded with series faults introduced in sections 1–10. Series fault current differ much from the no fault current but quite difficult to distinguish series faults in the various sections.

Similarly a shunt fault between a section and ground is introduced by shorting the end of the section to ground. Figures 23.7, 23.8 and 23.9 show the neutral currents recorded with shunt faults introduced in sections 1–10. These fault current plots are shown along with the current recorded under no fault and again there is clear distinction between no fault and the shunt fault currents but it is difficult to distinguish the shunt fault current records in the various sections.

Thus in order to distinguish the fault and no fault condition resonant frequency which is characteristic of a winding is considered as the discriminating feature.

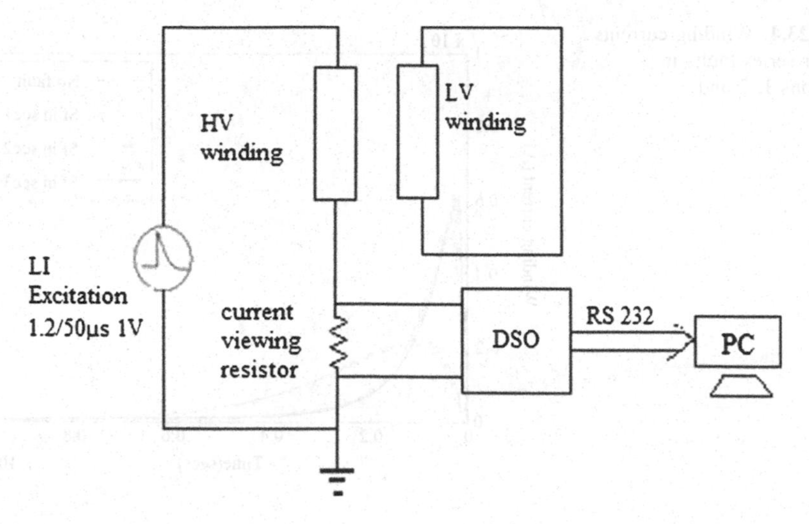

Fig. 23.2 Experimental arrangement

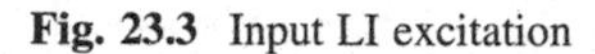

Fig. 23.3 Input LI excitation

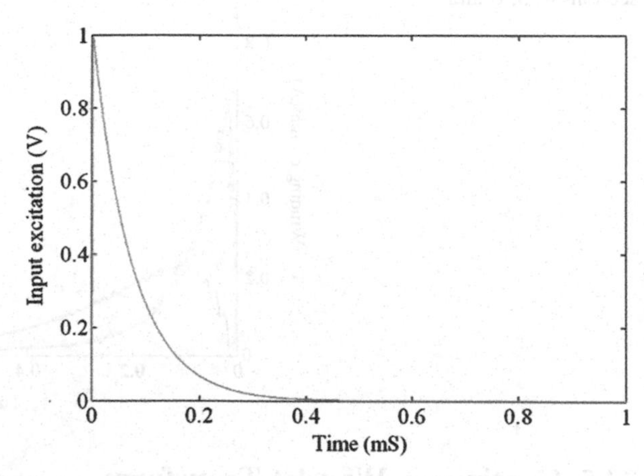

Winding currents under various series and shunt fault conditions are acquired through a personal computer for fault detection and classification. The measured winding currents are transformed to frequency domain using FFT to initially identify the dominant resonant frequencies and to study the effect of series and shunt fault conditions on them. Figures 23.10 and 23.11 show the respective frequency domain plots of winding current under no fault and various fault conditions indicating shifts in resonant frequencies. It is observed that the shift in the first anti resonant frequency shows clear distinction among the series faults and shunt faults in the three different regions. The distinction can be further enhanced by carrying out wavelet analysis.

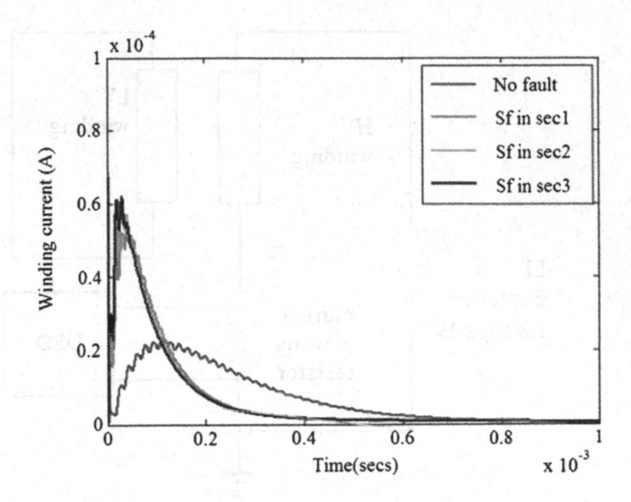

Fig. 23.4 Winding currents under series faults in sections 1, 2 and 3

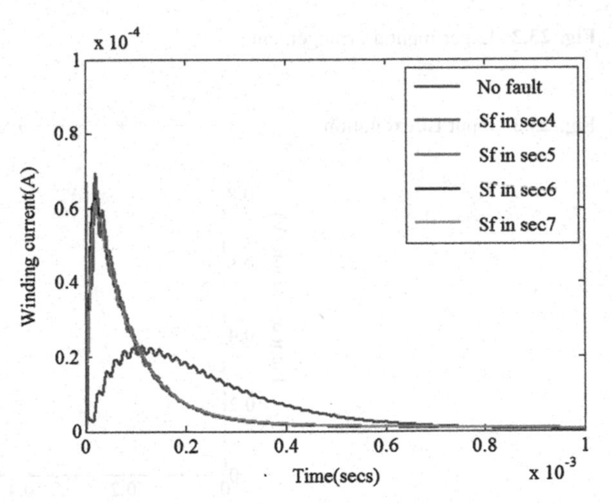

Fig. 23.5 Winding currents under series faults in sections 4, 5, 6 and 7

23.5 Continuous Wavelet Transform

Wavelet analysis is very much suitable for detecting local variations in signals. The continuous wavelet transform of the neutral current records obtained using DB5 with the scale ranging from 1 to 150 showed greater amount of distinction among the series and shunt faults in the three different regions. The 3D surface plots shown in Figs. 23.12, 23.13 and 23.14 show the CWT of the winding currents under no fault condition, under series fault introduced in section 6 and under shunt fault in section 1 respectively.

The most predominant frequency component identified as that frequency component which has the highest wavelet transform coefficient had been identified to distinguish various faults in this work. The 3D CWT plots obtained for the neutral currents recorded for the six different fault cases and no fault are found to

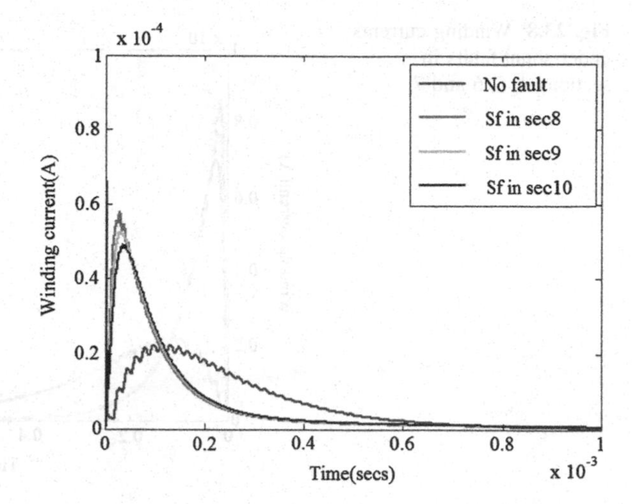

Fig. 23.6 Winding currents under series faults in sections 8, 9 and 10

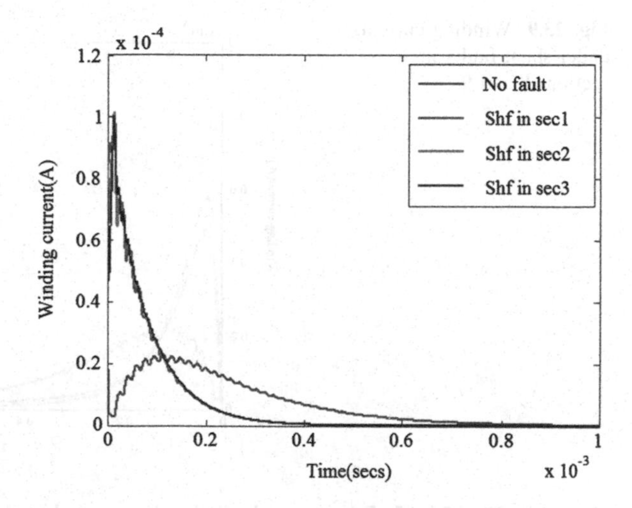

Fig. 23.7 Winding currents under shunt faults in sections 1, 2 and 3

have greater amount of distinction when the second most predominant frequency component is considered and hence the features are extracted by considering this frequency component. The translation along the X-axis is converted into time in μs and the scale along the Y-axis is converted into frequency (pseudo). Table 23.2 shows the time frequency characteristics of the winding currents under different fault conditions along with the normalized coefficients. From the nineteen neutral current records obtained, the ranges for the three discriminating features are decided and are mentioned in the above table.

The features such as time range, the second most predominant frequency and the normalized CWT coefficient which are extracted for series faults and shunt faults within sections near the line-end (section 1–3), middle of the winding (section 4–7) and neutral end (section 8 and 9) are plotted in the 3D scatter plot as

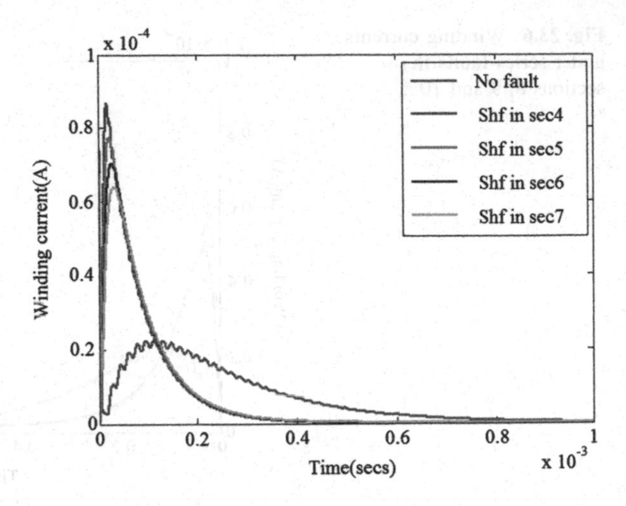

Fig. 23.8 Winding currents under shunt faults in sections 4, 5, 6 and 7

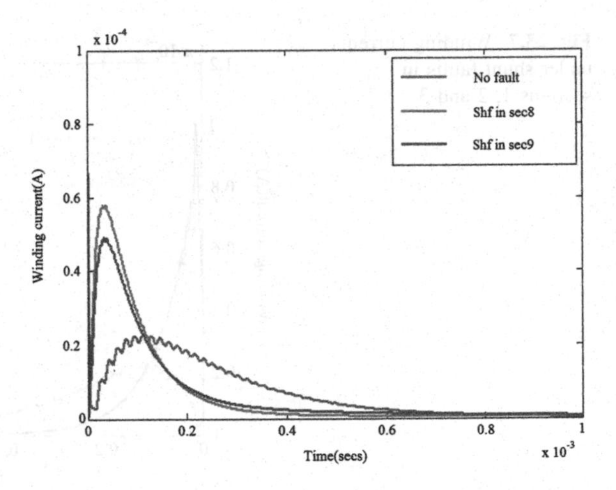

Fig. 23.9 Winding currents under shunt faults in sections 8 and 9

shown in Fig. 23.15. From the plot it is observed that the series faults SFL, SFM and SFN and the shunt faults SHFL, SHFM and SHFN fall in distinctive clusters.

23.6 Fuzzy Inference Systems for Impulse Fault Classification

The fuzzy inference system is a popular computing framework based on the concepts of fuzzy set theory, fuzzy if–then rules, and fuzzy reasoning. It has found successful applications in a wide variety of fields such as automatic control, data classification [12–14], decision analysis, expert systems, time series prediction, robotics, and pattern recognition. A fuzzy inference system (FIS) required for fault

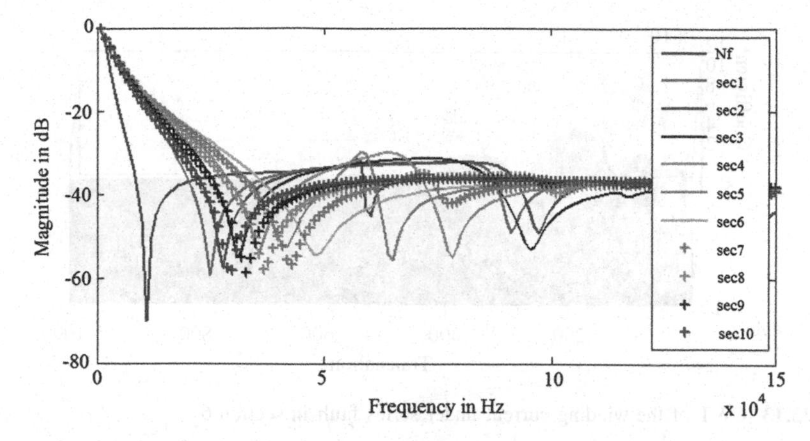

Fig. 23.10 FFT of winding currents under series faults in the 10 sections

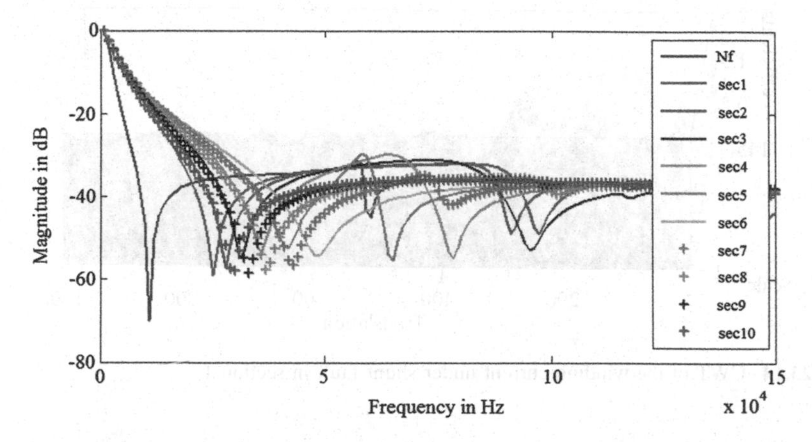

Fig. 23.11 FFT of winding currents under shunts faults in the 10 sections

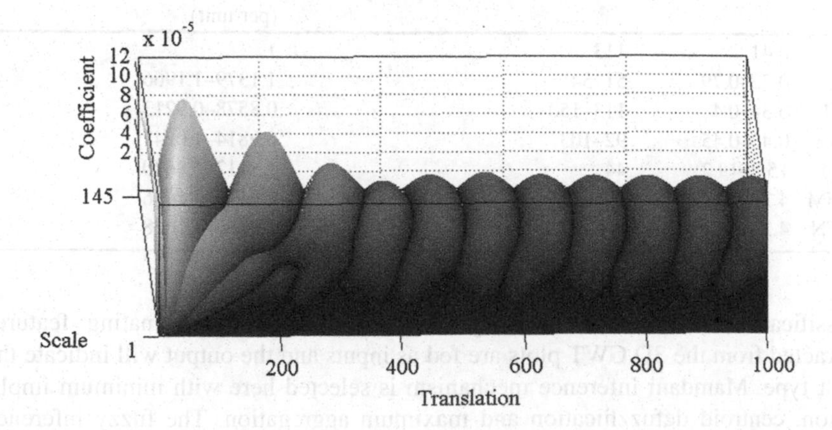

Fig. 23.12 CWT of the winding current under no fault

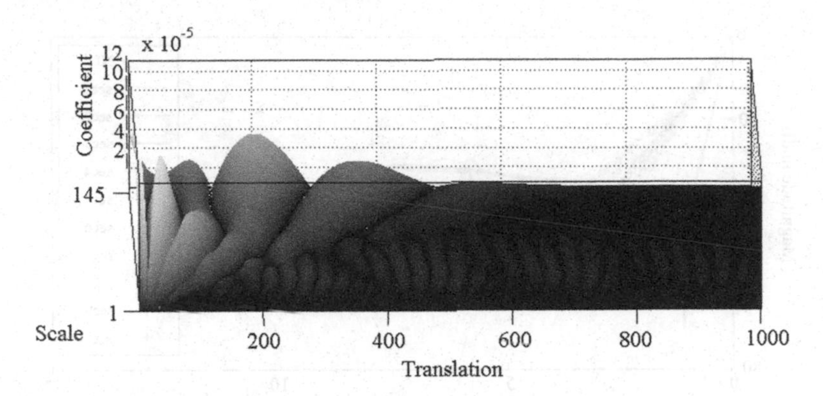

Fig. 23.13 CWT of the winding current under series fault in section 6

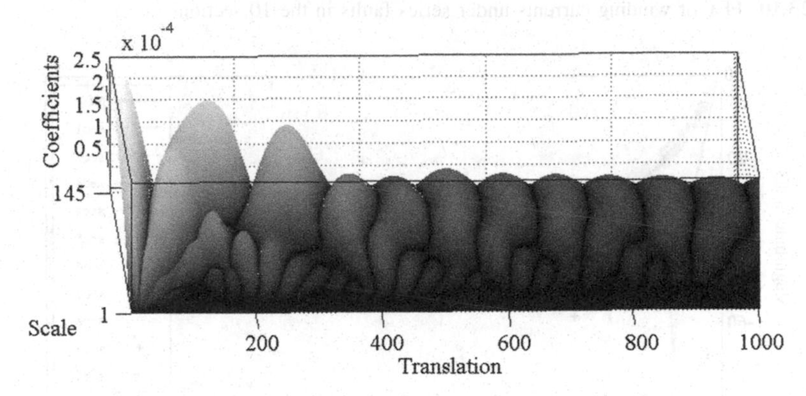

Fig. 23.14 CWT of the winding current under shunt fault in section 1

Table 23.2 Time frequency characteristics of winding current under different fault conditions

Fault	Time range (μs)	Predominant frequency range (kHz)	Normalized Wavelet coefficient (per unit)
NF	0.41	113	1
SFL	0.72–0.79	51–54	1.1379–1.1966
SFM	0.34–0.4	113–150	0.8578–0.9213
SFN	0.42–0.45	92–103	0.9814–1.0247
SHFL	15.8–19.2	44	1.3117–1.8640
SHFM	4.3–4.7	101–111	1.0176–1.1476
SHFN	4.3–4.4	95–101	0.9766–1.0028

classification is shown in the Fig. 23.16. The three discriminating features extracted from the 3D CWT plots are fed as inputs and the output will indicate the fault type. Mamdani inference mechanism is selected here with minimum implication, centroid defuzzfiication and maximum aggregation. The fuzzy inference

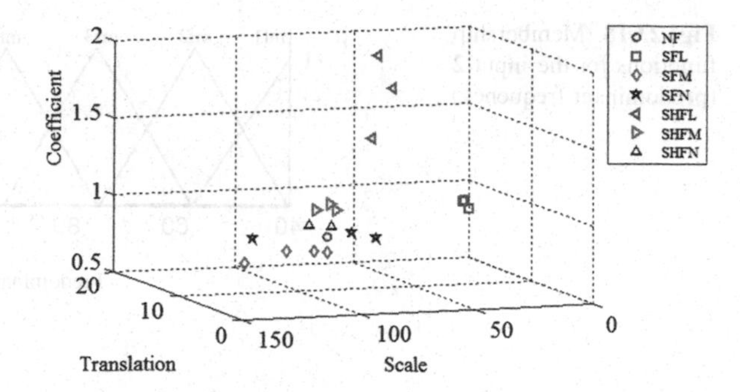

Fig. 23.15 Scatter plot of parameters used for impulse fault classification

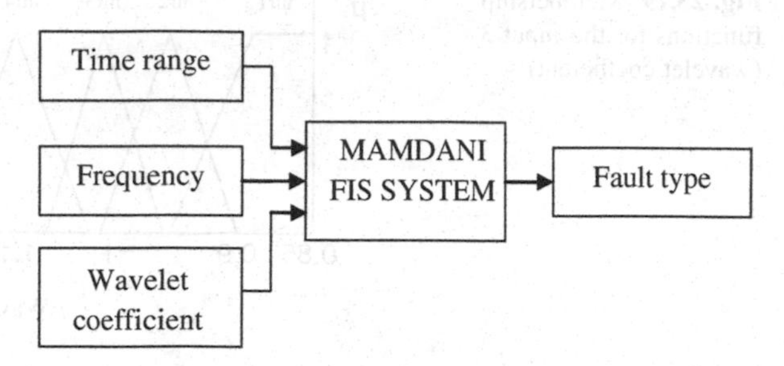

Fig. 23.16 Fuzzy inference system for impulse fault classification

process consists of the following steps: Fuzzy set definition, Fuzzification, Rule base, Fuzzy operators, Implication, Aggregation and Defuzzification.

These steps are implemented as follows:

1. Fuzzy sets are defined for the three input variables namely the second most predominant frequency, its time range of occurrence and the corresponding normalized wavelet coefficient and for the only output variable representing the no fault and six other fault types. They are shown in Figs. 23.17, 23.18, 23.19, and 23.20.
2. The crisp input values are converted into fuzzy sets.
3. Rule base is created as shown in Table 23.3 comprising of seven rules representing the no fault and the six fault types [13].
4. The centre of gravity method of defuzzification is adopted to map a fuzzy output to crisp output.

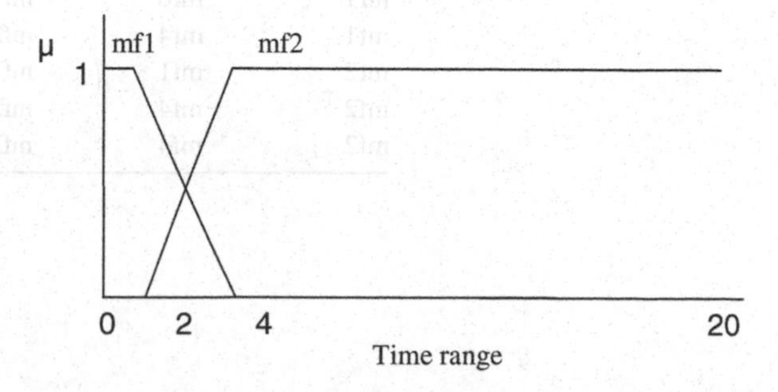

Fig. 23.17 Membership functions for the input 1 (time range)

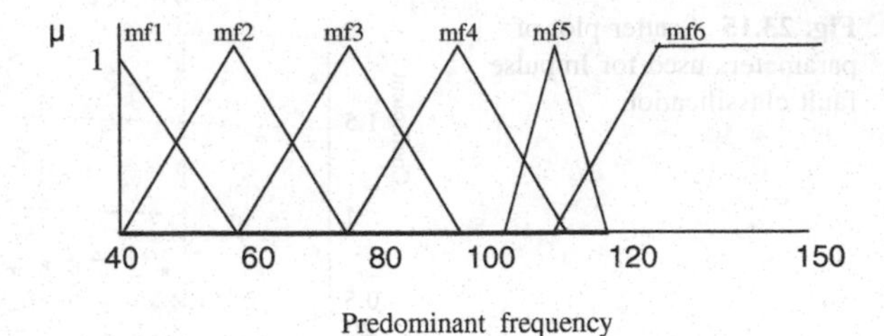

Fig. 23.18 Membership functions for the input 2 (predominant frequency)

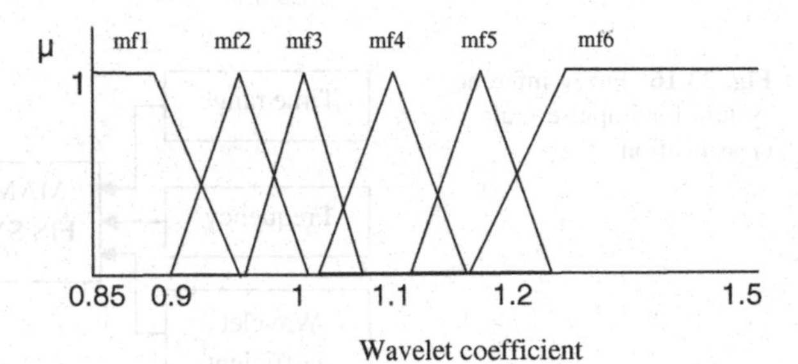

Fig. 23.19 Membership functions for the input 3 (wavelet coefficient)

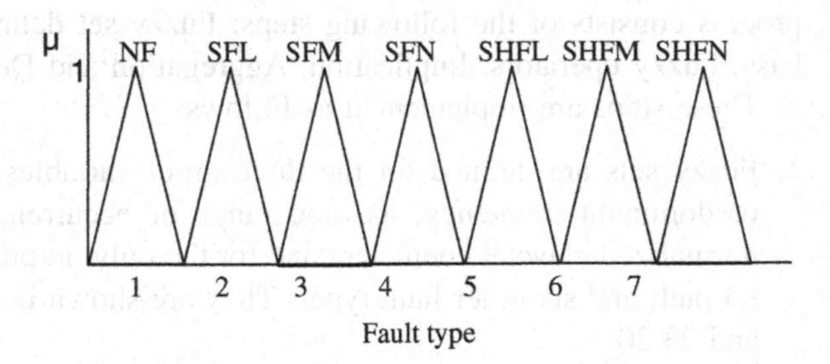

Fig. 23.20 Membership functions for the output (Fault type)

Table 23.3 Fuzzy rule matrix

Time range input 1	Frequency input 2	Wavelet coefficient input 3	Fault output	Weight
mf1	mf5	mf6	NF	1
mf1	mf2	mf4	SFL	1
mf1	mf6	mf1	SFM	1
mf1	mf4	mf2	SFN	1
mf2	mf1	mf5	SHFL	1
mf2	mf4	mf3	SHFM	1
mf2	mf4	mf2	SHFN	1

23.7 Conclusion

This paper demonstrated a systematic approach for the detection and classification of impulse faults in power transformer by considering a specially designed 6.6 kV VT model winding. This fault detection and classification system implemented has been shown to work satisfactorily for the chosen fault conditions. The localized nature of wavelets is highly useful in discriminating the faults. The trained fuzzy Inference system classified the no fault and six different faults with 100 % efficiency.

Acknowledgments The authors are very much thankful to the authorities of Annamalai University for their constant encouragement in conducting the research work.

References

1. IEEE C57.98–1993, Guide for transformers impulse tests
2. Electrical Insulation and Dielectric Phenomena. (2004) Annual report conference
3. Malewski R, Poulin B (1988) Impulse testing of power transformer using the transfer function method. IEEE Trans Power Deliv 3(2):476–489
4. Leibfried T, Feser K (1999) Monitoring of power transformers using the transfer function method. IEEE Trans Power Deliv 14(4):1333–1341
5. Satish L (1998) Short-time fourier and wavelet transforms for fault detection in power transformers during impulse tests. IEEE Proc Sci Meas Tecnol 145(2):77–84
6. Duncan ARL, Mills W (2004) Signal analysis -Time, Frequency, Scale and Structure. Wiley-Interscience publication, IEEE Press
7. Degeneff RC (1977) A general method for determining resonances in transformer windings. IEEE Trans Power App Syst PAS-96:423–430
8. Santhi S, Jayashankar V, Jagadeesh Kumar V (2008) Time frequency analysis of method for the detection of winding deformation in transformers during short circuit test. In: 2MTC-08, Victoria,Vancouver Island, Canada, 12–15 May 2008
9. Youssef OAS (2004) Combined fuzzy-Logic wavelet-based fault classification technique for power system relaying. IEEE Trans Power Deliv 19(2):582–589
10. Li-Xin, Mendel JM (1992) Generating fuzzy rules by learning from examples. IEEE T Syst Man Cyb 22(6):1414–1427
11. Jayalalitha S, Jayashankar V (2005) Fuzzy logic based Impulse test Analysis, SMCia/05. In: IEEE Mid-summer workshop on Soft Computing in Industrial Applications, Helsinki University of Technology, Espoo, Finland, 28–30 June 2005
12. Purkait P, Chatterjie A, Chakravorti S, Bhattacharya K (2003) Translationally adaptive fuzzy classifier for transformer Impulse fault identification. IEE Proc Gener Transmn Distrib 150(1):33–39
13. Hung CP, Wang MH (2004) Diagnosis of incipient faults in power transformers using CMAC neural network approach. Sci Direct, Electr Power syst Res 71:235–244
14. Youssef OAS (2004) Applications of fuzzy-logic wavelet-based techniques for transformers Inrush currents identification and ique for power system faults classfication. In: IEEE conference on power systems and exposition, vol 1. pp 553–559

Chapter 24

Implementation of Neural Network Based V/F Control for Three Phase Induction Motor Drive with Power Factor Improvement

K. Prakasam, R. Poornima and S. Ramesh

Abstract In this study, we discuss about a neural system for three-phase induction-motor speed control and power factor correction. The speed control strategy consists in keeping constant voltage–frequency ratio of the induction-motor supply source. A neural-control system uses speed error and speed-error variation to change both the fundamental voltage amplitude and frequency of a sinusoidal pulse width modulation inverter. The controller performance in relation to reference and load-torque variations is considered. A high-performance single-phase AC to DC rectifier with active power factor correction technique is used for line power factor correction. Single phase supply is converted to three phase and it is fed to three phase induction motor. The proposed approach has many advantages over conventional fuzzy based induction motor speed control such as less settling time, accuracy and improved efficiency. In this approach we obtain power factor correction in single phase source. According to the simulation results, proposed method has achieved better results by suppressing speed overshoot and ripple as compared to the conventional method and able to correct speed error from load-torque variations.

Keywords Neural network · V/F control · Three phase induction motor · Power factor improvement

K. Prakasam (✉) · S. Ramesh
Department of Electrical and Electronics Engineering, K.S.R. College of Engineering, Thiruchengode, India
e-mail: prakasameee@yahoo.co.in

S. Ramesh
e-mail: rameshksrce@gamil.com

R. Poornima
Department of Electronics and Communication Engineering, Erode Sengunthar Engineering College, Thudupathi, India
e-mail: poorniesec@yahoo.com

R. Malathi and J. Krishnan (eds.), *Recent Advancements in System Modelling Applications*, Lecture Notes in Electrical Engineering 188,
DOI: 10.1007/978-81-322-1035-1_24, © Springer India 2013

24.1 Introduction

Over the past 30 years, concern for effective use of energy resources has grown in response to increased fuel cost, increased demand upon energy supply systems and concern that growing energy use may be causing irreparable damage to the environment. In response to these concerns, there has been a desire to increase the efficiency of electrical motors, since these devices constitute a significant fraction of the total use of electrical energy. A considerable amount of research has been done in the area of the design of squirrel cage induction motor itself as well as the control for the purpose of optimizing the efficiency. As a result, the market for energy efficient electrical motors has expanded.

Three-phase induction motors are widely applied in several industrial sectors. The extensive use of this motor is frequently associated with its simple and rugged structure, adaptation to several load situations, and low cost acquisition and maintenance [1–3]. Several studies have been carried out in the field of vector control system due to its better dynamic response [4]. However, scalar control [5–8] presents a simple structure characterized by low steady-state error. Therefore, the constant voltage–frequency (*V/f*) scalar control system is considered in this paper due to its wide application in industrial fields. Proportional–integral (PI) control methodology is commonly applied in constant voltage–frequency control strategy for induction motors. However, in addition to the fact that a mathematical model is desirable for a systematic controller design with conventional methods, the difficulty of identifying the precise parameters for a complex nonlinear and time-varying behavior of real plants may render, in some cases, its fine tuning procedure very time-consuming, even if the empirical methods are used [8]. Advanced pulse width modulation (PWM) techniques are employed to guarantee high quality output voltage with reduced harmonics and sinusoidal input current irrespective of the load. To obtain sinusoidal input current at the terminal of single phase source a high performance active input power factor correction technique for single phase boost switch mode rectifier operating with discontinuous current conduction is used. The operation is based on variable turn-on time. Thus, three phase AC drives using single phase supply with improved power factor is an approach to implement high frequency induction boosting along with the three phase pulse width modulated inverter for controlling the speed of three phase induction motor by maintaining voltage-frequency ratio at constant value. This scheme can be used in lathe machines, small cranes, lifts etc., which are frequently switched ON and OFF. From the practical and mathematic analysis, it is demonstrated that the three-phase induction motor, which is working in an over-dimensioned way, presents a reduction in its power factor and a decrease in its efficiency. On the other hand, three-phase induction motors working in an under-dimensioned way present overheat and drastic reduction in its useful life. A research made at CEMIG (Electric Energy Company of the Minas Gerais State—Brazil), with 3,425 three-phase induction motors, in several industry sectors, has shown that 28.7 % of them were over dimensioned and 5.9 % of them worked under dimensioned. Another research made at COPEL (Electric Energy Company of the Parana State—Brazil) with 6,108 three-phase induction motors has shown that 37.75 % were working over dimensioned. Loads connected to the TIM shaft can be pre-sorted according to the torque characteristics in four main categories: constant, linear, quadratic and inverse.

24.2 Proposed Method

In recent years artificial neural networks (ANNs) have gained a wide attention in control applications. It is the ability of the artificial neural networks to model nonlinear systems that can be the most readily exploited in the synthesis of nonlinear controllers. Artificial neural networks have been used to formulate a variety of control strategies.

The non-ideal character of the input current drawn by these rectifiers creates number of problems like increase in reactive power, high input current harmonics, low input power factor, lower rectifier efficiency, large input voltage distortion etc. To compensate for the higher reactive power demand by the converters at high power transfer levels, power factor correction becomes mandatory. To overcome these problems number of passive and active current wave shaping techniques have been suggested in the literature. But the passive power factor correction techniques have the disadvantages like large size of reactive elements, power factor improvement for a narrow operating region, large output dc voltage ripple [9]. Active current wave shaping techniques overcome these disadvantages and significantly improve the performance of rectifiers [10, 11]. Hysteresis current control is a simple active current wave shaping technique that gives close to unity power factor operation while delivering near sinusoidal currents. The rectifiers using discontinuous conduction of input current with a single boost switch gives close to unity power factor at constant turn-on time and frequency of the boost switch.

Current control technique may use continuous conduction mode or discontinuous conduction mode. The popular continuous mode of conduction with switch mode rectifiers are hysteresis current control with constant hysteresis window, Bang–bang hysteresis current control and constant switching frequency current control with error triangulation. Discontinuous mode of conduction operates with constant switching frequency and variable turn-on time using one or two switches.

24.3 Methodology

Scalar control means that variables are controlled only in magnitude and the feedback and command signals are proportional to dc quantities. A scalar control method can only drive the stator frequency using a voltage or a current as a command. Among the scalar method known to control an induction motor, one assumes that by varying the stator voltages in proportion with frequency the torque is kept constant.

The Voltage-Frequency method is based on steady-state characteristics of the motor and the assumption that the stator voltages and currents are sinusoidal. Applied to the majority of existing variable-speed AC drives by mean of an open-loop constant voltage-frequency voltage source converters, this standard control method has no inner current controller. The advantages of this control technique are its simplicity, easy and fast to program and require only few calculation capabilities.

This paper proposes an alternative method for simplifying a general-purpose embedded neural networks algorithm so that it can be built in hardware with reduced memory space and low computational power. This procedure significantly reduces the memory space required for a membership function. The proposed method was applied to embed a neural networks control algorithm in a digital signal processor (DSP) for real time voltage–frequency (*V/f*) induction-motor speed control. The system was responsible for measuring the three phase induction motor (TIM) shaft angular speed with an optical encoder, achieving the neural networks control algorithm, and finally generating the sinusoidal-modulated pulse width modulation (PWM) signal in order to turn on six insulated-gate bipolar transistors (IGBTs) of a three phase inverter.

24.4 Speed Control System

The speed control of the induction motor was carried out by maintaining constant the voltage–frequency ratio in order to avoid the air-gap flux variations. If the supply voltage is varied without frequency adjustment, the induction motor can operate in the flux saturation region or with a weakened field. The block diagram of the proposed neural network-control system is shown in Fig. 24.1.

The speed signal of the induction motor was compared with the speed reference (ωRef), providing the inputs of the neural networks controller with the speed error $\omega er(k)$. This last signal was acquired by computing the difference in speed value. The neural networks system algorithm processes the inputs and provides the sinusoidal PWM fundamental frequency variation [$df(k)$] in its output. Afterward, this signal is added to the last frequency value [$f(k-1)$], resulting in the actual fundamental-frequency reference. The generation logic of pulse width modulation (PWM) signals, which turns on the six switches, was also built on the processor. The three phase induction motor was supplied by a pulse width modulation (PWM) inverter with a fundamental frequency and equivalent voltage, such that the *V/f* ratio was kept constant. The circuit representation is shown in Fig. 24.2. The embedded neural system was developed taking into account the memory reduction and code optimization to be implemented in a microprocessor.

The main requirement we have specified is maintaining the desired speed of the induction motor. Considered for the neural controller output were the voltage

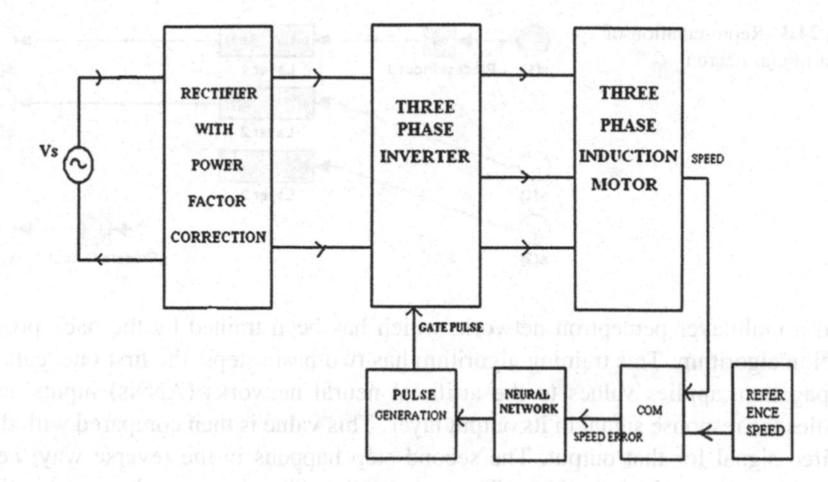

Fig. 24.1 Block diagram of proposed method

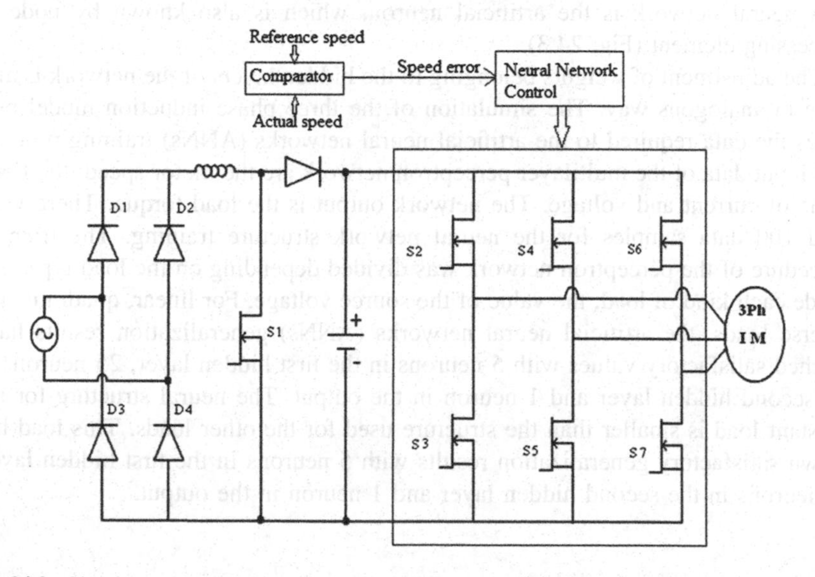

Fig. 24.2 Circuit configuration of AC motor drive

components that would present an action intervention for pulse width modulation (PWM), which would eventually produce the stator voltage desired values from the mains voltage (rectified via using an uncontrolled rectifier).

Identification using artificial neural networks has shown promising in solving a series of problems involving power systems. In this work, artificial neural networks (ANNs) were applied to estimate torque demanded by the load coupled on the induction motor shaft. The main objective here is in using artificial neural networks to estimate the load behavior on the motor shaft. In this work, it will be

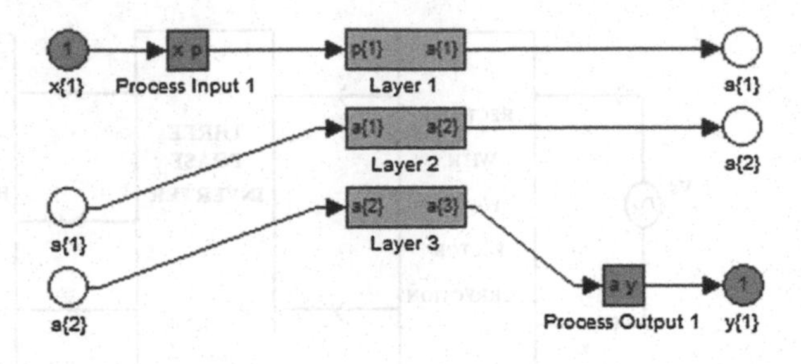

Fig. 24.3 Representation of the artificial neuron

used a multilayer perceptron network, which has been trained by the back propagation algorithm. This training algorithm has two basic steps: the first one, called propagation, applies values to the artificial neural networks (ANNs) inputs and verifies the response signal in its output layer. This value is then compared with the desired signal for that output. The second step happens in the reverse way, i.e., from the output to the input layer. The error produced by the network is used in the adjustment process of its internal parameters (weights and bias). The basic element of a neural network is the artificial neuron, which is also known by node or processing element (Fig. 24.3).

The adjustment of weights belonging to the hidden layers of the network is also done in analogous way. The simulation of the three phase induction model produces the data required to the artificial neural networks (ANNs) training process. The input data of the multilayer perceptron network are the motor speed, the RMS value of current and voltage. The network output is the load torque. There were used 100 data samples for the neural network structure training. The training procedure of the perceptron network was divided depending on the load type and, inside each kind of load, the value of the source voltage. For linear, quadratic and inverse loads, the artificial neural networks (ANNs) generalization results have reached satisfactory values with 5 neurons in the first hidden layer, 25 neurons in the second hidden layer and 1 neuron in the output. The neural structure for the constant load is smaller than the structure used for the other loads. This load has shown satisfactory generalization results with 5 neurons in the first hidden layer, 15 neurons in the second hidden layer and 1 neuron in the output.

24.5 Experimental Results

To verify the effectiveness of the speed control and identification of the speed using tacho-generator, comparator is used to compute speed error from the difference between the actual speed value and set speed value. The speed error is given to the input of neural networks and the output is sinusoidal pulse width modulation (PWM) fundamental frequency. The output of the neural network is provided to the input of the subsystem. The subsystem generate corresponding gate triggering pulse for the switches in three phase inverter. We control the gate pulse

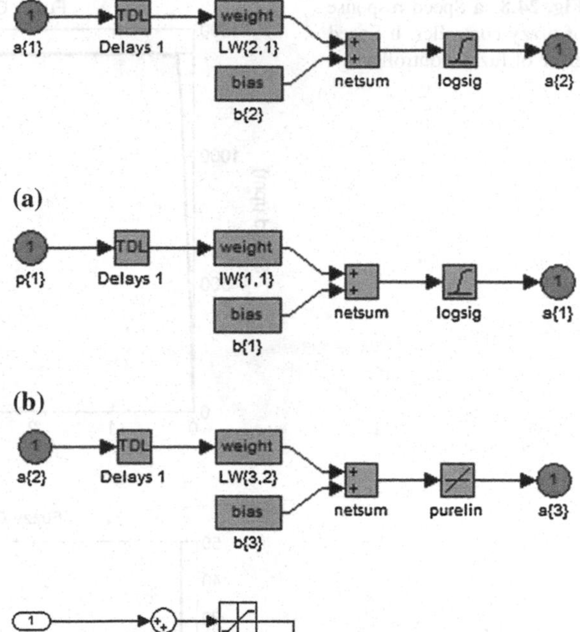

Fig. 24.4 Comparison between two different layers (2, 1) and levels with time delay input using logarithmic signals

Fig. 24.5 **a** Comparison between single layer (1, 1) different levels with time delay input using logarithmic signals. **b** Comparison between two different layers (3, 2) and levels with time delay input using pure line

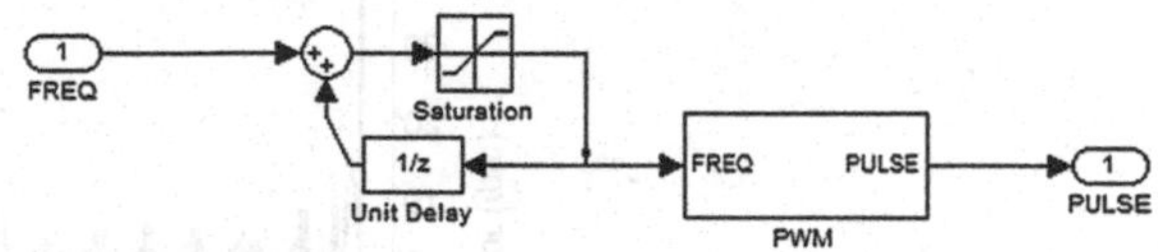

Fig. 24.6 PWM control technique

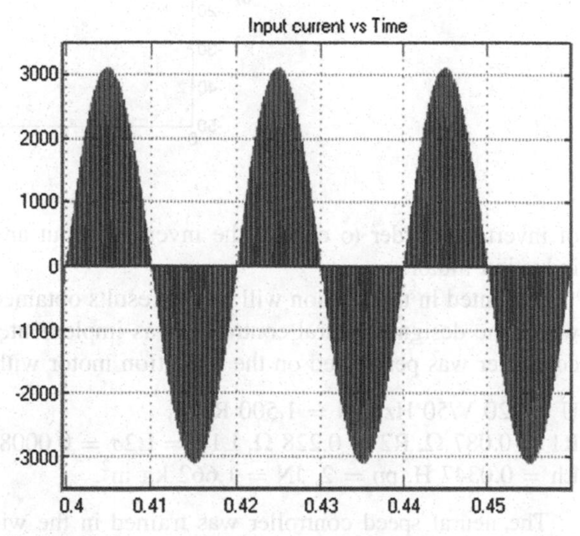

Fig. 24.7 Input current waveform

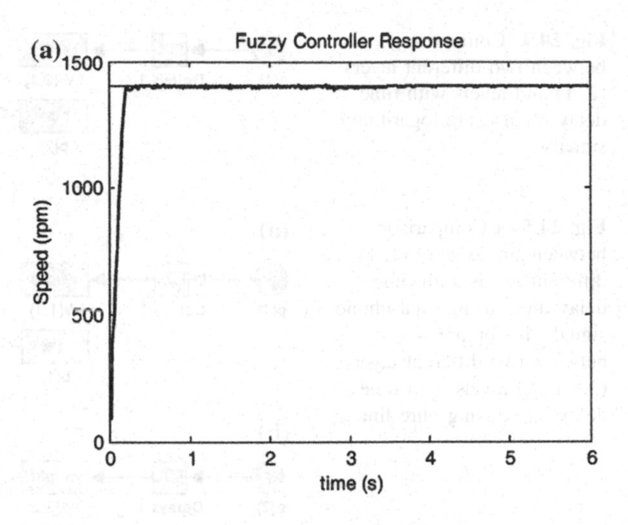

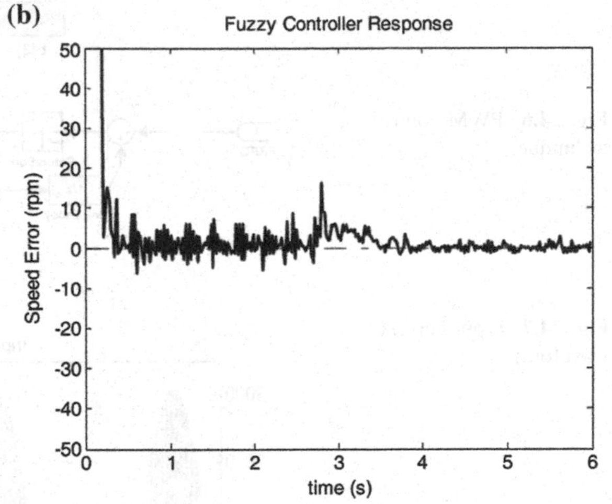

Fig. 24.8 **a** Speed response of fuzzy controller. **b** Speed error of fuzzy controller

of inverter in order to control the inverter output and hence to control speed of induction motor.

Presented in this section will be the results obtained in MATLAB environment where the designed neural controller was implemented. The testing of the neural controller was performed on the induction motor with the following parameters:

U = 220 V/50 Hz, nN = 1,500 RPM,
R1 = 0.087 Ω, R2 = 0.228 Ω, $L1\sigma = L2\sigma = 0.0008$ H,
Lh = 0.0347 H, pp = 2, JN = 1.662 kg m^2.

The neural speed controller was trained in the wide range of speed and load torque changes. Then the trained controller was tested for speed reference signal different than used in the training procedures. The layer of neural network with weight and bias adjustments is illustrated in Fig. 24.4.

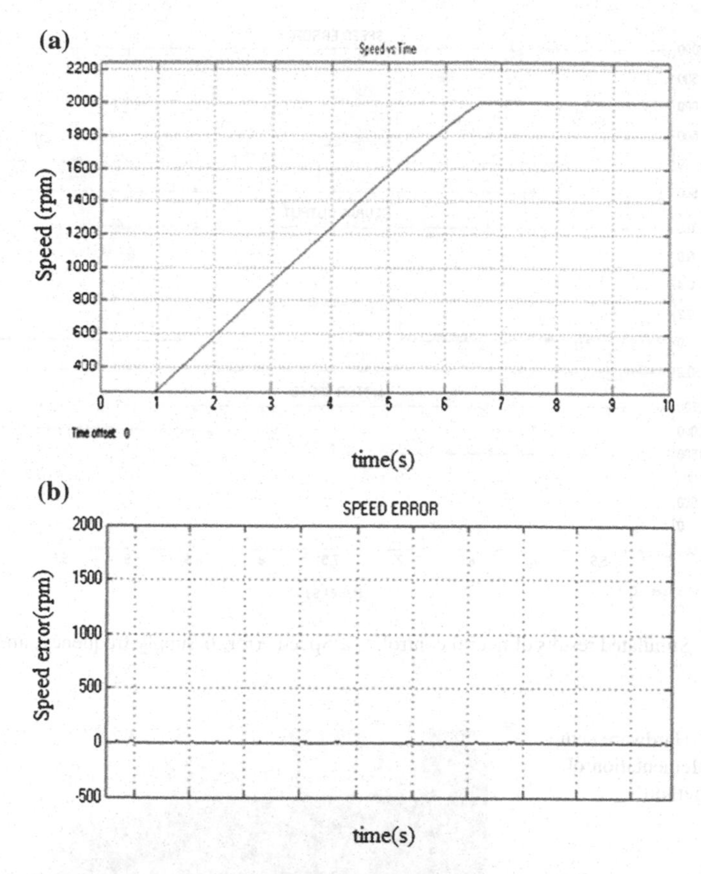

Fig. 24.9 **a** Speed response of neural controller. **b** Speed error of neural controller

Training samples for the speed controller were attained via simulation of an induction motor model in MATLAB-Simulink environment. Simulation results using MATLAB verify the effectiveness of proposed controller.

The methods of generating logarithmic signals includes the layers a{1} and a{2}, and comparing the weight of two different layers and levels is shown in Figs. 24.5a and b. The layer output and bias output combined get single output using net sum block. The net sum value will be provided at pure line generator to generate corresponding logarithmic signals.

The neural networks system algorithm processes the inputs and provides the sinusoidal PWM fundamental frequency variation in its output. Afterward, this signal is added to the last frequency value [$f(k-1)$], resulting in the actual fundamental-frequency reference as shown in Fig. 24.6. The generation logic of pulse width modulation (PWM) signals, which turns on the six Insulated gate bipolar transistors (IGBTs), was also built on the processor. The three phase induction motor was supplied by a pulse width modulated inverter with a fundamental frequency and equivalent voltage, such that the voltage-frequency ratio was kept

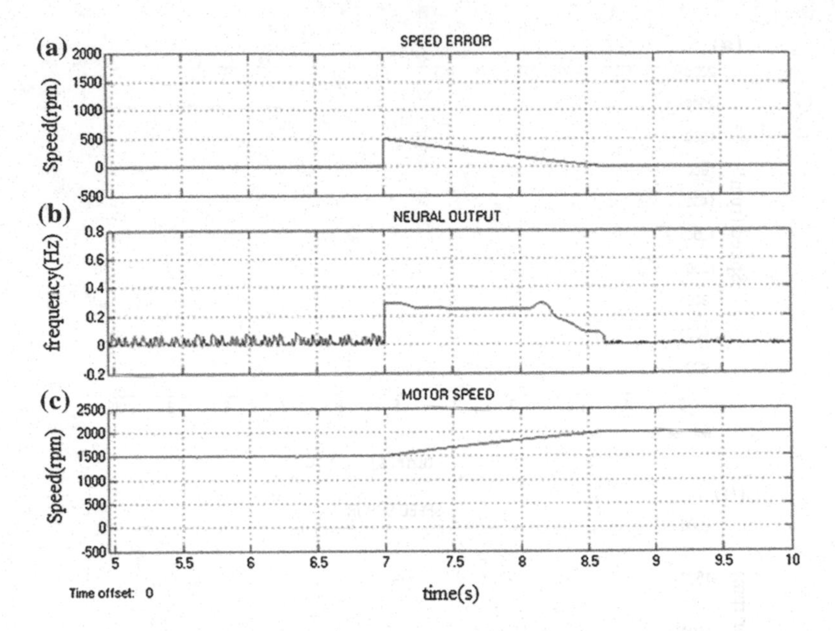

Fig. 24.10 Simulated results of neural controller **a** Speed error, **b** Supply frequency, and **c** Speed response

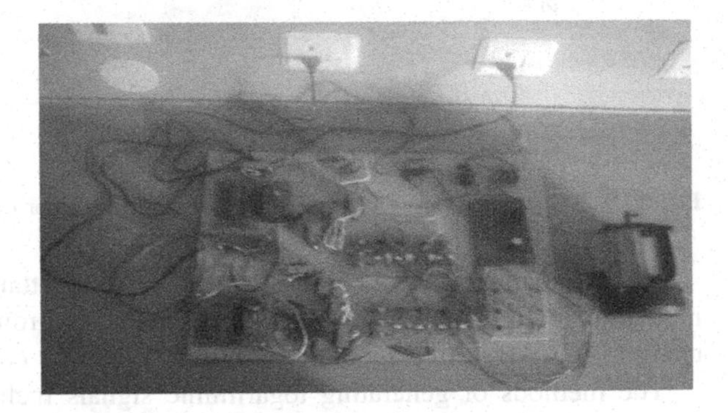

Fig. 24.11 Hardware setup for the implementation of proposed method

constant. The input current waveform with power factor correction is shown in Fig. 24.7 and simulation output waveforms for fuzzy controller are shown in Fig. 24.8. These results are taken to compare with the results of neural controller as depicted in Fig. 24.9.

The embedded neural system was able to correct the error by increasing the frequency. The result of the speed-error performance is shown in Fig. 24.9b. The speed error achieved by the neural controller was less than 10 r/min.

In this section, a comparative analysis with conventional fuzzy controller is carried out in order to validate the proposed neural controller. Therefore, step reference and load-torque variation experiments were carried out for the performance analyses. Figure 24.10 shows speed response waveforms in a case of speed

Fig. 24.12 Testing of hardware

variation. Hardware implementation for the proposed method of speed control is done with the neural network controller programmed in PIC microcontroller. The setup is shown in Fig. 24.11. It is tested for different reference speeds and the setup with testing equipments is shown in Fig. 24.12.

According to the results, the neural controller eliminated the overshoot and ripple in relation to the speed response of the fuzzy controller.

24.6 Conclusion

The simulation work for the proposed system has been completed. The proposed system configuration is capable of controlling the speed of three phase induction motor drives using voltage-frequency control with power factor improvement. The neural-based control methodology has the ability to cope with system nonlinearity and its control performance is less affected by system parameter variations. The advantages of proposed topology are reduced settling time, fast response and low overshoot in speed response of induction motor and improved power factor. The simulation results confirmed that the performance of neural network-control system is satisfactory, incrementing robustness in relation to load-torque variations while achieving the reference speed. Thus, the proposed neural network-control system is an acceptable alternative method for voltage-frequency common conventional control applications. The comparative analysis with conventional controllers demonstrated that the proposed method achieved better results by suppressing speed overshoot and ripple. Future work of this project is to implement direct torque control and vector control using neural networks.

References

1. Suetake M, da Silva IN, Goedtel A (2011) Embedded DSP-based compact fuzzy system and its application for induction-motor v/f speed control. IEEE Trans Ind Electron 58(3):750–760
2. Goedtel A, da Silva IN, Serni PJA (2007) Load torque identification in induction motor using neural networks technique. Elect Power Syst Res 77(1):35–45
3. Lu B, Habetler TG, Harley RG (2006) A survey of efficiency-estimation methods for in-service induction motors. IEEE Trans Ind Appl 42(4):924–933
4. Shi D, Unsworth PJ, Gao RX (2006) Sensorless speed measurement of induction motor using Hilbert transform and interpolated fast Fourier transform. IEEE Trans Instrum Meas 55(1):290–299
5. Maiti S, Chakraborty C, Hori Y, Ta MC (2008) Model reference adaptive controller-based rotor resistance and speed estimation techniques for vector controlled induction motor drive utilizing reactive power. IEEE Trans Ind Electron 55(2):594–601
6. Tae-Chon A, Yang-Won K, Hyung-Soo H, Pedricz W (2001) Design of neuro-fuzzy controller on DSP for real-time control of induction motors. In Proceedings of IFSA world congress and 20th NAFIPS international conference, vol 5. pp 3038–3043
7. Islam N, Haider M, Uddin MB (2005) Fuzzy logic enhanced speed control system of a VSI-fed three phase induction motor. In: Proceedings of 2nd international conference on electrical and electronics engineering, pp 296–301
8. El-Saady G, Sharaf AM, Makky A, Sherbiny MK, Mohamed G (1994) A high performance induction motor drive system using fuzzy logic controller. In: Proceedings of 7th mediterranean electrotechnical conference, vol 3. pp 1058–1061
9. Prasad AR, Ziogas PD, Manias S (1991) A Passive input current wave shaping method for three-phase diode rectifiers. In: IEEE APEC-91 conference record, pp 319–330
10. Patil PM (2002) An active power factor correction technique for single phase AC to DC boost converters. J Inst Electron Telecommun Eng 43(4):34–38
11. Patil PM, Kyatanavar DN, Zope RG, Jadhav DV (2001) Three-phase ac drive using single phase supply. J Inst Eng (India) 82:43–47

Chapter 25
Model Based Approach for Fault Detection in Power Transformers Using Particle Swarm Intelligence

M. Arivamudhan and S. Santhi

Abstract Transformer is an essential device in power systems. Winding deformation due to short circuit is one of the faults that require serious attention. Model based approaches for winding deformation detection have attracted researchers widely. This paper aims at determination of distributed parameters of the lumped element model of a transformer winding using particle swarm intelligence. A specially designed layer winding model is used to carry out the frequency response experiment. Difference between the simulated frequency response and experimental frequency response is defined as the fitness function that is minimized using particle swarm optimization technique.

Keywords Transformer · Lumped parameter model · FRA · PSO · Winding deformation detection

25.1 Introduction

Nowadays reliability is an inevitable part of power system studies and operation due to significant increase in the number of industrial electrical consumers. Power transformer is one of the major and critical elements in power industry, and their outage may result in costly and time-consuming repair and replacement in addition to human safety related problems. For each transformer, windings and insulations are the most critical elements and hence an effective condition monitoring is essential.

M. Arivamudhan · S. Santhi (✉)
Department of Instrumentation Engineering, Annamalai University, Annamalai Nagar, Chidambaram 608002, India
e-mail: santhi.sathyamurthy@gmail.com

R. Malathi and J. Krishnan (eds.), *Recent Advancements in System Modelling Applications*, Lecture Notes in Electrical Engineering 188,
DOI: 10.1007/978-81-322-1035-1_25, © Springer India 2013

Different faults modes in transformers are classified into two main groups; the mechanical type, and the electrical types. Different methods have been applied for the detection of each of these faults, such as oil analysis (dissolved gas analysis), partial discharge measurements, dielectric response analysis, transfer function analysis, and frequency response analysis (FRA). Among these methods, transfer function method is a well-known method for the detection of winding mechanical faults [1]. The mechanical faults such as axial displacement of winding, radial deformation, and disc space variation, due to external or internal short circuits in the winding affects the transfer functions and cause shift in resonant frequencies and moreover will decrease or increase the magnitude of resonant frequencies in the transfer functions.

Investigation shows that about 70–80 % of transformer failures such as turn to turn short circuit, turn to ground short circuit are caused by internal winding short circuit faults. One important reason for these faults is erosion of the winding and conductor insulation due to vibrations initiated by the electromechanical forces at service current and over currents. This problem leads to over-current in windings that result in terrible damages such as severe hot-spots, oil heating, winding deformation, damage to the clamping structure, core damage, and may even cause explosion of the transformer. Hence there is a necessity for continuous observation of the winding to protect the device from severe damages.

In order to detect and distinguish faults in the winding, many methods are suggested which can be classified into two major categories: experimental methods and model based methods [2]. It is obvious that the experimental detection of internal fault is essential and important, but applying long term and high cost experimental tests to find the best detection method is impossible because of the transformer cost. Therefore, due to economical limitations in experimental tests, model based approach has been considered as the best approach to perform innovative analysis, to compare with other methods and finally, to choose the best method for the fault detection in transformers. This paper demonstrates a method for model identification using particle swarm optimization (PSO) and for subsequent winding deformation detection.

25.2 Motivation and Objective

The winding structures of power transformers are considered to be the heart of the transformer. Transformers are required to be certified for their structural integrity through short circuit test, which is a special test as specified in standard IEC 60076 Part V [3]. The conventional method to demonstrate short circuits withstand capability of windings is based on reactance measurement before and after the short circuit test. A deviation of 2 % for transformers with rating >1 MVA is considered to be a failure [4]. Concurrent high frequency excitation for online analysis of structural faults has been proposed in [5, 6]. Transfer function method by applying low voltage impulse and swept frequency response method exist but do not provide sufficient information regarding the cause of failure. Model based diagnosis methods [7] have attracted researchers and this paper has emerged with the objective of attempting an intelligent learning technique namely PSO to determine the distributed parameters of lumped parameter model of transformer winding.

25.3 Description of Particle Swarm Optimization

PSO technique is an optimization technique used in data mining that attempts to find some optimal solution from a large space of possible solutions [8, 9]. It gets its name from the fact that a number of possible solutions are maintained and they are made to swarm towards an optimal solution by a computer program.

The technique is part of a general set of techniques such as ant colony optimization that are biologically inspired. Particle swarm optimization has the potential of efficient computation with very large number of concurrently operating processors [10, 11]. The population of PSO is called swarm and each individual is called a particle. For the ith particle at iteration *k*, it has the following two attributes namely current position and current velocity.

A current position in an N-dimensional search space

$$X_i = \left(X^k_{i,1\ldots\ldots}, X^k_{i,n\ldots\ldots}, X^k_{i,N}\right) \tag{25.1}$$

where $x^k_{i,n} \varepsilon [l_n, u_n]$, $1 << n << Nl_n$ and u_n are the lower and upper bounds for the nth dimension respectively. A current velocity $v_i{}^k$,

$$V^k_i = \left(V^k_{i,1\ldots\ldots}, V^k_{i,n\ldots\ldots}, V^k_{i,N}\right) \tag{25.2}$$

which is clamped to a maximum velocity

$$V^k_{max} = \left(V^k_{max,1\ldots\ldots}, V^k_{max,n\ldots\ldots}, V^k_{max,N}\right).$$

At each iteration, the swarm is updated by the following equations.

$$V^{k+1}_i = \omega V^k_i + c_1 r_1 \left(P^k_i - X^k_i\right) + c_2 r_2 \left(P^k_g - X^k_i\right) \tag{25.3}$$

$$X^{k+1}_i = X^k_i + V^{k+1}_i \tag{25.4}$$

where p_i is the best previous position of the ith particle i.e. p best and P_g is the global best position among all the particles in the swarm i.e. g best, r_1 and r_2 are elements from two uniform random sequence on the interval [0, 1] and ω is inertia weight which is typically chosen in the range of [0, 1] and is critical for convergence of the algorithm. c_1 and c_2 are acceleration constants that controls particle's movement in a single iteration. The maximum velocity V_{max} is set to be half the length of the search space.

The fitness function is defined as the difference between the experimental frequency response and the model frequency response as given below.

$$F_{obj} = \sum \|F_{oi} - \alpha F_{mi}\|$$

where F_{oi} represents i number of the original frequency responses and F_{mi} represents i number of model frequency responses and α is a vector [0,…,0 1].

The objective of PSO algorithm is to minimize the fitness function F_{obj} to determine appropriate model parameters of the transformer. Under this condition there is a close agreement between the responses of the model and the device. Now fault detection strategy could be applied to the model since it resembles the device under consideration.

25.4 Model Based Fault Detection Using PSO

The model based approach for fault detection using PSO can be best explained as shown in Fig. 25.1. This requires that the device under test (DUT) and the model need be excited through the same source. The measured frequency responses of the DUT and the frequency responses of the model are the inputs to PSO algorithm. The algorithm identifies the model parameters by minimizing the fitness function. Once the exact model is identified the fault detection system detects and distinguishes the respective fault occurred.

25.5 Validation of PSO Algorithm

Particle swarm optimization algorithm is implemented through matlab coding. The efficiency of the algorithm to identify the model parameters is verified initially by considering a simple RLC resonant circuit as shown in Fig. 25.2. Frequency response experiment is conducted and the response is shown in Fig. 25.3. Next the same circuit is simulated using circuit simulation package namely pspice orcad to obtain the simulated model frequency response as shown in Fig. 25.4. The model

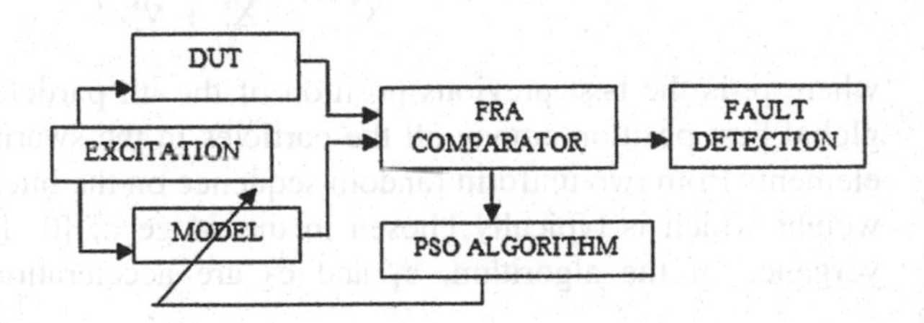

Fig. 25.1 Model based approach for fault detection using PSO

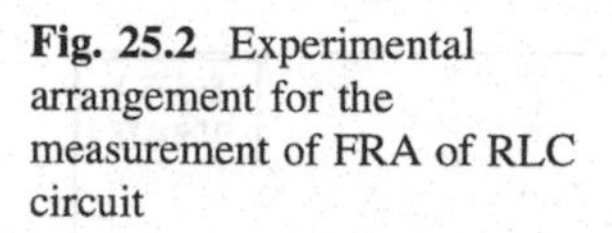

Fig. 25.2 Experimental arrangement for the measurement of FRA of RLC circuit

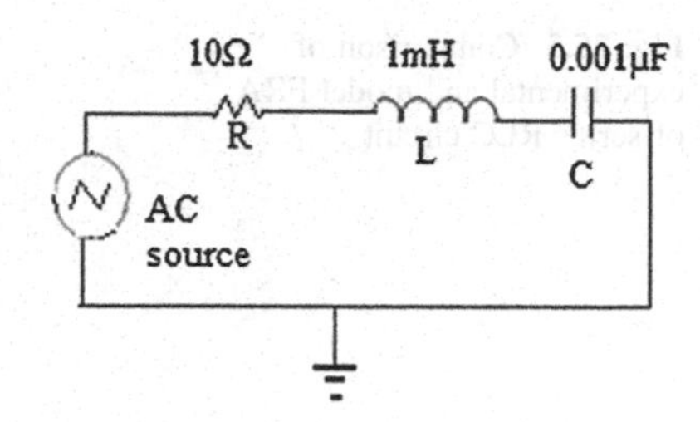

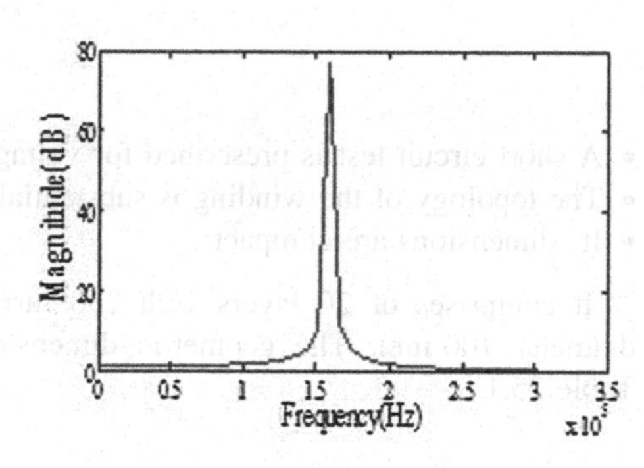

Fig. 25.3 Experimental FRA of series RLC circuit

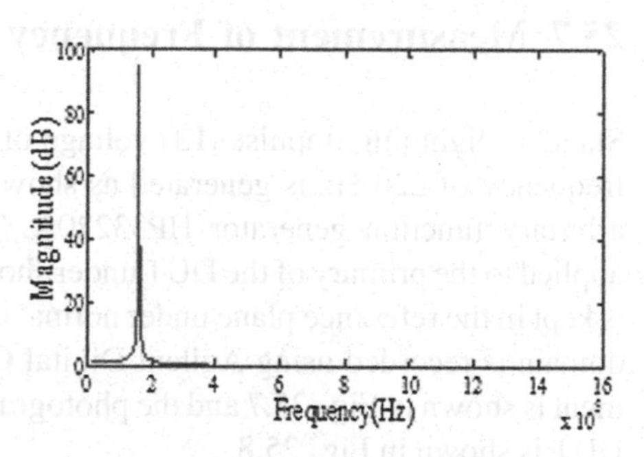

Fig. 25.4 Simulated FRA of model of RLC circuit

parameters are changed and the respective simulated frequency responses are determined and applied as the population to PSO. The algorithm optimizes difference between the experimental and model frequency response and determines the parameters the RLC circuit. Figure 25.5 shows the experimental and model frequency response of the RLC circuit. It is observed that the parameters of the RLC circuit model are identified when the algorithm converges.

25.6 Transformer Model Identification Using PSO

The Device under Test is a multilayer winding of a 6.6 kV voltage transformer. The reasons for the choice of voltage transformer winding as the device under test are as follows. It comprises of a layer winding which is also used in large power transformers.

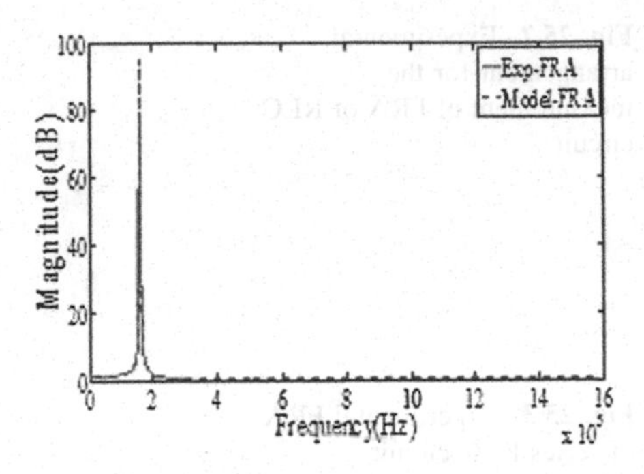

Fig. 25.5 Comparison of experimental and model FRA of series RLC circuit

- A short circuit test is prescribed for voltage transformers.
- The topology of the winding is substantially the same from 6.6–420 kV.
- Its dimensions are compact.

It comprises of 20 layers with 250 turns per layer wound on a former of diameter 100 mm. The geometric dimensions of the DUT are shown in the Table 25.1.

25.7 Measurement of Frequency Response of DUT

Standard lightning impulse (LI) voltage of 1.2/50 μs with a magnitude of 1 V and frequency of 220 Hz is generated as shown in Fig. 25.6 and stored in the Agilent arbitrary function generator HP33220A. This sharp fronted unidirectional LI is applied to the primary of the DUT under shorted secondary condition. The secondary is kept in the reference plane under normal condition and the winding current in time domain is recorded using Agilent Digital Oscilloscope. The experimental arrangement is shown in Fig. 25.7 and the photograph of the experimental arrangement with DUT is shown in Fig. 25.8.

The measured winding current in time domain as shown in Fig. 25.9 is transformed to frequency domain using fast Fourier transform implemented through

Table 25.1 Dimensions of winding of 6.6 kV voltage transformers

Number of sections	10
Number of layers	20
Number of turns/section	500
Total number of turns	5,000
Thickness of the conductor	0.4 mm
Space between two conductors	1 mm
Axial length	100 mm
LV winding diameter	85.3 mm
Outer diameter of HV winding	116 mm
Axial length of LV winding	340 mm

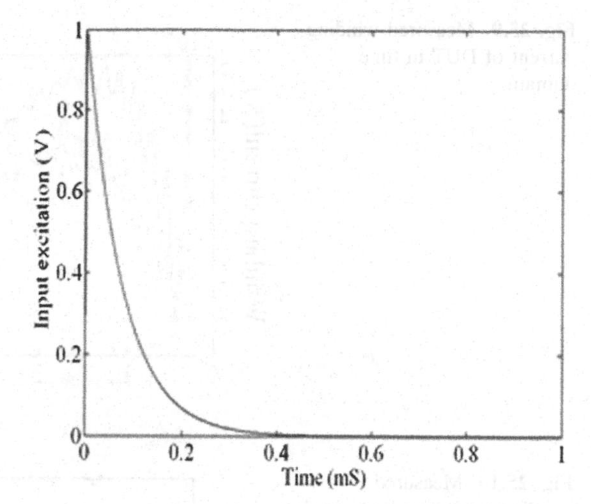

Fig. 25.6 LI excitation in time domain

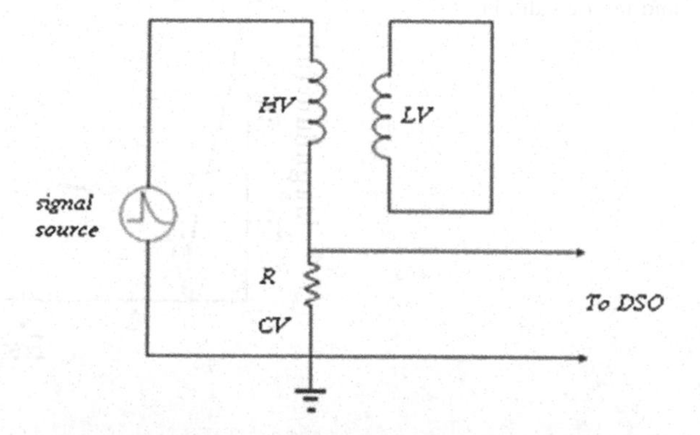

Fig. 25.7 Experimental arrangement for FRA measurement

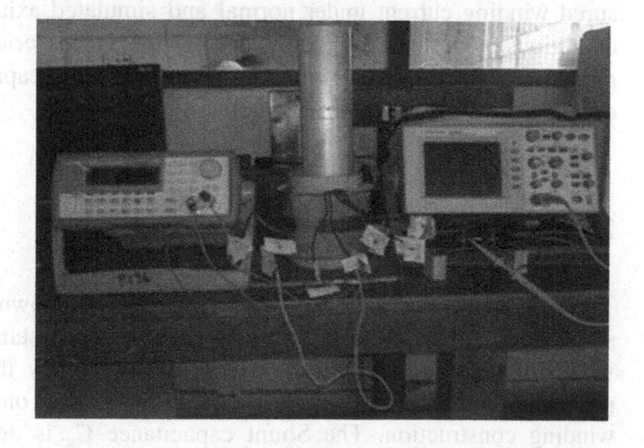

Fig. 25.8 Photograph of the experimental setup

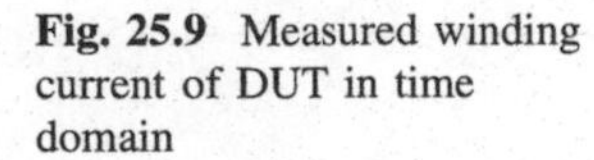

Fig. 25.9 Measured winding current of DUT in time domain

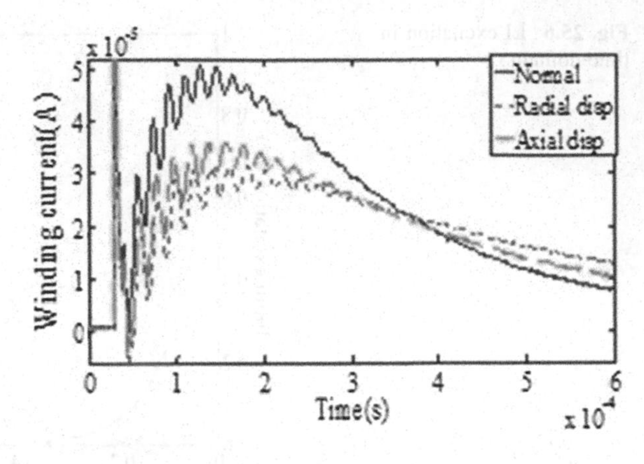

Fig. 25.10 Measured winding current under normal and fault condition

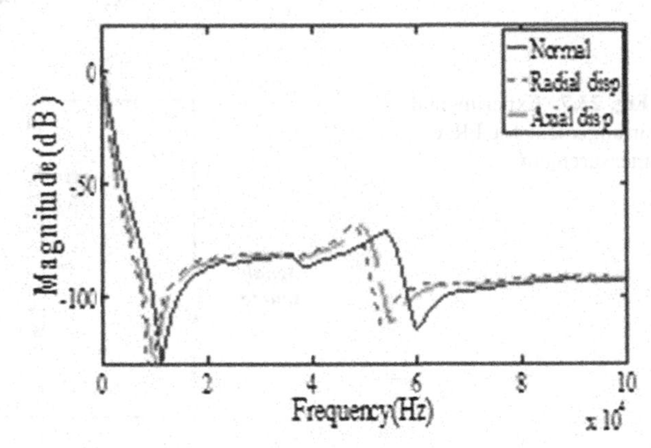

matlab coding. Figure 25.10 shows the frequency domain representation of measured winding current under normal and simulated axial and radial deformation conditions by introducing change in self inductance, series capacitance to simulate radial deformation and by effecting change in shunt capacitance to simulate axial deformation fault condition respectively.

25.8 Simulation of FRA

The lumped parameter model of the winding as shown in Fig. 25.11 is used to simulate the model frequency responses subjected to standard LI excitation of 1 V with 1.2/50 μs front time and fall time respectively through circuit simulation package. The model parameters are determined based on the geometry of the layer winding construction. The Shunt capacitance C_g is defined as the capacitance between the HV winding and the structures that are earthed. Since the LV winding

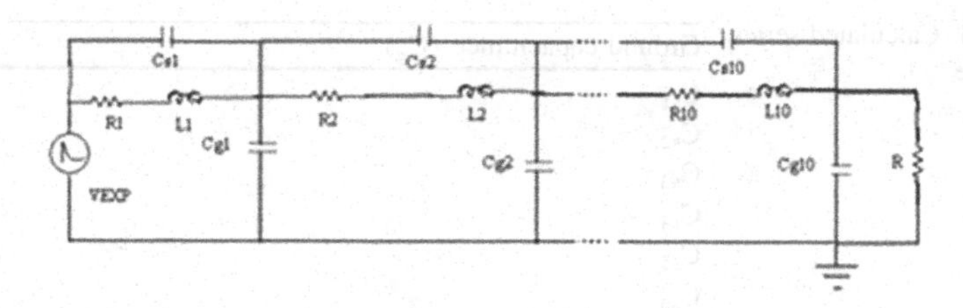

Fig. 25.11 Lumped parameter model of the DUT

Table 25.2 Calculated shunt capacitance

Shunt capacitance (C_s)	Pf
C_{g1}	614
C_{g2}	625
C_{g3}	636
C_{g4}	647
C_{g5}	658
C_{g6}	669
C_{g7}	681
C_{g8}	692
C_{g9}	703
C_{g10}	714

is earthed during short circuit test the shunt capacitance is considered to be capacitance between HV winding and LV.

Winding and is calculated using the formula for the capacitance between concentric cylinders as given below.

$$Cg = 2\pi\varepsilon_0\varepsilon_r L/\ln(D/d)$$

where, ε_r is the relative permittivity of insulation between HV and LV winding, ε_0 is permittivity of free space.

L is the axial length of the winding.

D is the outer diameter of LV winding.

d is the inner diameter of outer HV winding.

The calculated shunt capacitances are shown in Table 25.2.

The series capacitance represents the inter turn capacitance of winding of a section and is calculated as capacitance between sections using the same formula that has been used for the calculation of shunt capacitance since the sections in this design are also concentric as it is a layer winding construction. The calculated series capacitance of each section is shown in Table 25.3.

The self inductance of each section is calculated by treating it like a single layer coil. This calculation of Self Inductance of a single layer coil is based on the formula for cylindrical current sheet that is where the current flows around the cylinder. Except in the case of open helical windings the inductance of a single

Table 25.3 Calculated series capacitance

Ground capacitance (C_g)	Pf
C_{s1}	35.4
C_{s2}	31.8
C_{s3}	28.9
C_{s4}	26.5
C_{s5}	24.5
C_{s6}	22.9
C_{s7}	21.4
C_{s8}	20.4
C_{s9}	19.1
C_{s10}	18.1

layer coil is closely equal to that of the cylindrical current sheet having the same number of turns as the coil, the same mean radius and a length equal to the number of turns in the coil times the distance between the center of adjacent wires. The Nagaoka's formula [12] gives greater degree of accuracy and is given by

$$L = 0.002\pi^2 a(2a/b)n^2 k$$

where 'N' is the number of turns, 'b' is the axial length and 'a' is the radius of the coil. The values for k are given by Nagaoka based on the shape ratio b/2a and 2a/b depending on whether the coil is a short coil or a long coil respectively. Table 25.4 shows the calculated self inductances of each section of HV winding.

The mutual inductance between sections is calculated using the formula derived (as a case for concentric coils having equal axial length) from the general formula for the mutual inductance of coaxial cylindrical current sheet as mentioned in [12].

Let the two coils, coil 1 and 2 have radius, axial length, winding density as 'a', '$2m_1$', 'n_1' and 'a', '$2m_2$', 'n_2' respectively. Formula for mutual inductance M of concentric coils of equal axial length as described in [12] is given by

$$M = 0.004\pi^2 a^2 n_1 n_2 [r_1 B_1 - r_2 B_2]$$

Table 25.4 Calculated self inductance

Self inductance	MH
L_1	16.79
L_2	17.36
L_3	18.06
L_4	18.46
L_5	19.09
L_6	19.71
L_7	20.32
L_8	20.93
L_9	21.54
L_{10}	22.16

Table 25.5 Calculated section resistance

Resistance (R)	Ohms
R_1	38.5
R_2	36.4
R_3	37.0
R_4	37.7
R_5	38.4
R_6	39.2
R_7	39.5
R_8	40.2
R_9	40.9
R_{10}	41.5

$$\alpha = a/A$$

$$n_1 = 2m_1/N_1 \text{ and } n_2 = 2m_2/N_2$$

$$pn_2 = A_2/m_2 \text{ and } r_1 = \left(A^2 + X_1^2\right)^{1/2}$$

where N_1, N_2 are the number of turns of coil 1 and 2 respectively.

B_n values based on α and pn_2 are obtained from the table as in [12]. The mutual inductances between sections are calculated as per [12]. The coupling coefficients between Sects. 25.1 and 25.2 is

$$K_{12} = M_{12}/(L_1L_2)^{1/2}$$

Resistances are of each section using the formula mentioned below and tabulated in Table 25.5.

$$R = \rho l/A$$

and

$$A = \pi r^2 h$$

where

$h = l = 100$ mm
r = radius of the section
ρ = specific resistivity.

The calculated values of resistance of each section are tabulated as shown in table 25.5.

Circuit simulation of the lumped element model is carried out with the calculated parameters of the lumped element model using circuit simulation package to determine simulated FRA and is shown in Fig. 25.12. The difference between the simulated FRA and the experimental FRA is defined as the objective function and

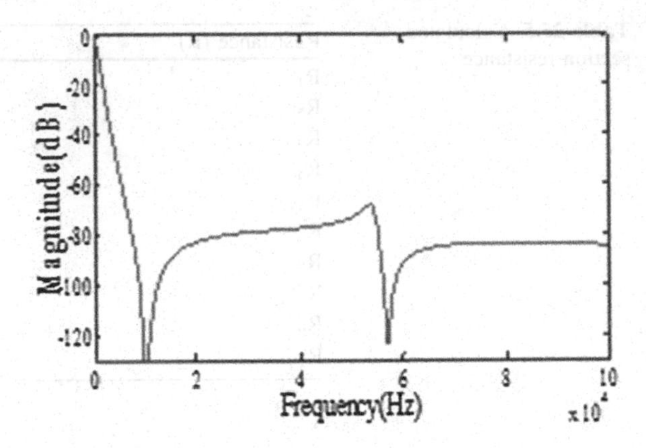

Fig. 25.12 Simulated FRA of lumped parameter model of DUT

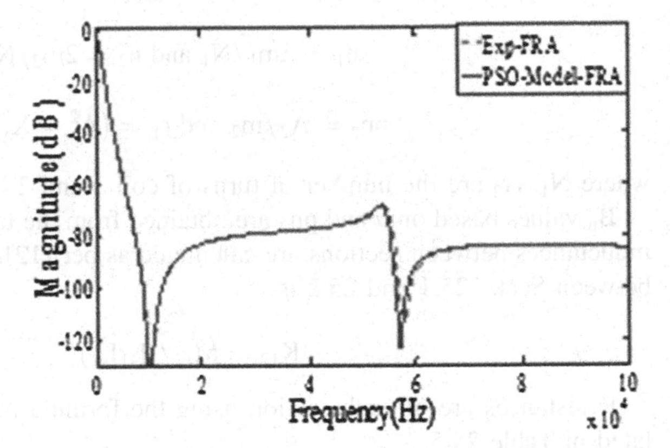

Fig. 25.13 Experimental and PSO identified model FRA of DUT

the PSO algorithm implemented through matlab coding minimizes the objective function and the corresponding distributed parameters of the lumped parameter model is considered as the PSO identified model parameters. Circuit simulation of the lumped element model is carried out once again with the PSO identified model parameters and the experimental FRA of DUT and the PSO identified model FRA are compared. It is observed from Fig. 25.13 that the algorithm identifies the model parameters satisfactorily.

25.9 Results and Discussion

The objective of transformer model identification using particle swarm optimization technique is attempted in this paper. The previous section has demonstrated the approach systematically by considering a simple series RLC circuit and it has been shown that the algorithm identifies the R, L, and C values that produce simulated FRA to coincide with experimental FRA.

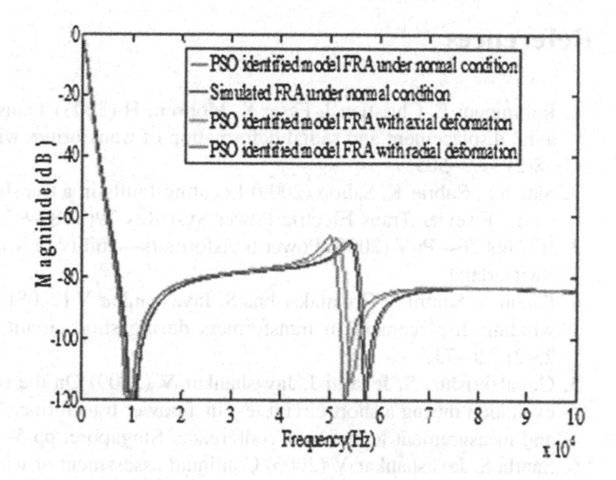

Fig. 25.14 PSO model validation under normal, fault conditions

In order to validate the PSO algorithm for transformer model identification and structural fault detection the work proceeded as described below. The lumped element model is simulated by effecting change in self inductance Ls (decrease of 0.5 mH) and series capacitance Cs (decrease of 3.5 pF) in order to simulate radial deformation and by varying cg (decrease of 1.5 pF)the shunt capacitance to simulate axial deformation as the DUT has a layer winding configuration. The FRA is simulated and the corresponding model parameters contribute to PSO swarm. The distributed parameters of the DUT are identified by the PSO algorithm by minimizing the objective function. Figure 25.14 shows the model identification of the DUT under normal and fault condition through PSO algorithm.

25.10 Conclusion

This paper has demonstrated in a systematic manner the application of particle swarm intelligence in respect of model identification and subsequent structural fault detection namely the winding deformation in axial and radial direction. The algorithm can be further improved by introducing passive congregation as in natural information sharing mechanism. Further, other intelligent learning techniques could be studied and compared with PSO in order to determine an efficient learning approach for model identification and fault diagnosis.

Acknowledgments The authors are very much thankful to the authorities of Annamalai University for their constant encouragement in conducting the research work.

References

1. Rahimpour E, Christian J, Feser K, Mohseni H (2003) Transfer function method to diagnose axial displacement and radial deformation of transformer winding. IEEE Trans Power Deliv 18(2): 493–505
2. Satish L, Subrat K, Sahoo (2009) Locating faults in a transformer winding: An experimental study. Elsevier Trans Electric Power Syst Res 79(1):89–97
3. IEC 60076—Pt V (2000) Power transformers—ability to withstand shortcircuit, IEC Geneva, Switzerland
4. Palani A, Santhi S, Gopalakrishna S, Jayashankar V (2008) Real-time techniques to measure winding displacement in transformers during short-circuit tests. IEEE Trans Power Deliv 23(2):726–732
5. Gopalakrishna S, Jeyaraj J, Jayashankar V (2009) On the use of concurrent high frequency excitation during a short circuit test in a power transformer. In: International instrumentation and measurement technology conference, Singapore, pp 5–7
6. Santhi S, Jayashankar V (2006) Continual assessment of winding deformation during a short circuit test. IEE Japan Trans PE 126(7):712–713
7. Christian J, Feser K (2004) Procedures for detecting winding displacements in power transformers by the transfer function method. IEEE Trans Power Deliv 19(1): 214–220
8. Kennedy J, Eberhart RC (1995) Particle swarm optimization, vol 4. In: IEEE international conference on neural networks. IEEE Press, pp 1942–1948
9. Kennedy J (1997) The particle swarm: social adaptation of knowledge. In: Proceedings IEEE international conference on evolutionary computation indianapolis, pp 303–308
10. Rashtchi V, Shayeghi H, Mahdavi M, Kimiyaghalam A, Rahimpour E (2008) Using an improved PSO algorithm for parameter identification of transformer detailed model. Int J Electr Electron Eng 666–672
11. Tang WH, He S, Wu QH, Richardson ZJ (2006) Winding deformation identification using particle swarm optimizer with passive congregation for power transformers. IEEE Trans Int J Innov Energy Syst Power 1(1):46–52
12. Grover FW (1946) Inductance calculations: working formulas and tables. Dover Publications Inc, New York

Chapter 26
Finite Element Method Magnetics Based Demonstration of Rotating Field in 4-Pole Induction Motor

Gururaj S. Punekar, D. Harimurugan and Gautham H. Tantry

Abstract The explanation related to the concept of Rotating Magnetic Field (RMF) in 3-phase induction motor (IM) and its visualization is a tricky issue in teaching–learning process. The complexity increases with the number of poles. Hence visualization of RMF for a 4-pole Induction motor is attempted via magnetic field distribution pattern(s). The aim of this paper is to explore and utilize the capability of Finite Element Method Magnetics (FEMM) as a tool for demonstrating rotating magnetic field effect produced in the stator of a 3-phase induction motor. In addition to the RMF demo, visual correlation between angular rotations of electrical wave with mechanical degree is reported. Also, the effect of phase sequence reversal is incorporated in the graphical exhibition.

Keywords Finite element analysis · Induction motors · Number of poles · Phase sequence · Rotating magnetic fields · Stator magnetic fields

26.1 Introduction

Three phase induction motor is the most commonly used motor in any industrial fields. The operation of induction motor is due to the Rotating Magnetic Field (RMF) produced in the stator winding. When a three phase winding is energized

G. S. Punekar (✉) · D. Harimurugan
Department of EED, NITK, Mangalore 575025, India
e-mail: gsp652000@yahoo.com

D. Harimurugan
e-mail: had.11ps08f@nitk.edu.in

G. H. Tantry
BMS College of Engineering, Bangalore, India

R. Malathi and J. Krishnan (eds.), *Recent Advancements in System Modelling Applications*, Lecture Notes in Electrical Engineering 188,
DOI: 10.1007/978-81-322-1035-1_26, © Springer India 2013

from a 3-phase supply, the three-phase sinusoidal (time varying) currents will produce fluxes in each phase, which varies sinusoidally. The magnitude of flux due to each phase surrounding it depends on the value of instantaneous current magnitude in the corresponding phase. The resultant flux at any instant will be the vector sum of all the three phases at that instant. The resultant flux retains its sinusoidal form and its peak shifts progressively around the air gap; the net result can be seen to be a flux of constant amplitude rotating at a uniform angular velocity [1, 2]. For a two pole machine this would give rise to two fictitious poles (north–south pole pair) which role as function of time. Complexity in explaining the concept of RMF increases, as the number of poles becomes greater than two.

For the sake of better understanding, the magnetic field produced due to the current in the stator winding of a 4-pole, 3-phase IM is simulated using FEMM software (open source) [3, 4]. The magnetic field at different time instances of current are simulated and snapshots of these FEMM simulations are used in a sequential fashion (combo pictures; a chain of images that are linked together to form a complete subject). Such a process and presentation to visualize the RMF in 4-pole IM is discussed in this paper. In addition to the RMF, the correlation between the electrical degree and mechanical degree is briefed. The phase reversal which results in the change in direction of rotation of magnetic field is also explained. Computer programming techniques have been used for teaching EM theory and RMF effects [5–7]. But RMF demonstration using FEMM described in this paper is a unique attempt which is expected to give the students clear understanding of the phenomena, with visual magnetic field distribution patterns inside the motor. As the process uses the FEM based package (FEMM), a brief description of its usage oriented towards the present objective of 3-phase IM is given below.

26.2 Model Implementation in FEMM

FEMM is a finite element package for solving two dimensional planar and axis symmetric problems in low frequency magnetics, heat flow, current flow and electrostatics. Finite element method is a numerical process of solving ordinary and partial differential equations [3]. Finite element method involves following steps: Discretization (dividing the solution region into number of small regions called elements), deriving equations for each element, solving the system of equations and the interpretation of the results [8]. The FEMM program consists of three sections namely Preprocessing, Processing and Post processing [3].

In the preprocessing module, the data specific to a problem like current, material properties and boundary conditions are defined to an equivalent of physical model. The preprocessing section also includes discretization in which the solution region is divided into number of small triangular areas. The processing section is where the finite element objects are computed, boundary conditions are enforced and the system is solved to understand the flux density and flux intensity distribution. The results from the processing section are analyzed in the post-processing section. Post processor always operates in one of the three modes, depending on the task to be performed. These modes are point value modes, contour modes and block mode. Point value mode is used to calculate the field at any particular point inside the solution region. Contour Mode allows the user to define arbitrary contours in the solution region. Once a contour is defined, plots of field quantities can be produced along the contour and various line integrals can be evaluated along the contour. In Block Mode, a sub-domain in the solution region is defined. A variety of area and volume integrals can be taken over the defined sub-domain.

Integrals include stored energy (inductance), various kinds of losses, total current in the block, and so on [3]. The overall distribution of magnetic field in the solution region can be simulated in the post processor section. Some of the applications of FEMM can be found in literature [9–14]. This article deals only with the RMF produced due to the stator winding of an IM and hence modeling of rotor is not in the present scope. As number of turns and peak values of current does not alter the sinusoidal wave shape of MMF wave (no saturation conditions), and the RMF effect, these numerical values are of no greater significance for the present effort related to RMF demo.

Desired model is designed using plot, segment and arcs tools available in FEMM [3]. In the case of Induction motor considered, the stator slots (assumed to be circular in shape) are modeled with the help of 2D polar co-ordinate system. The center points of the circular slots can be found by using $x = r*\cos(\varphi)$ and $y = r*\sin(\varphi)$. Here, 'r' indicates the inner radius of stator and 'φ' indicates angle between the lines joining the center of the circular slots to the origin, whose value depends on the number of poles. In the present work a simplistic model with circular slots is adopted. In case of complex geometry, the model can be designed using CAD and then it can be imported in FEMM using.dxf file format [3].

Other general details of modeling in FEMM are given in Table 26.1. The following section explains about model details specific to 4-pole induction motor.

Table 26.1 General details of modelling in FEMM

Sr. no	General properties	Description
1	Type setting	Planar symmetry
2	Boundary conditions	Dirichlet condition
3	Mesh size	One
4	Circuits	Three circuits are created for 3-phases of the supply with current magnitude specified corresponding to each phase
5	Stator and rotor magnetic materials	Carpenter silicon core iron "A", 1066 C Anneal with $\mu_r = 7{,}000$
6	Windings	16 AWG copper magnet wire
7	No of turns	100

26.3 4-Pole Induction Motors

A three phase, 4-pole, 12 slots (the least possible number for 3-phase, 4-pole situation) induction motor is considered for simulation and modeled in FEMM. The number of slots per pole = 12/4 = 3; number of slots per pole per phase = 12/(4*3) = 1; Minimum 12 slots are required to house the three phase winding in a 4-pole induction motor and such a basic model is used for simulation purpose to demonstrate the RMF. Phase sequence followed is A–B–C where $A = 1 < 0°, B = 1 < 120°, C = 1 < 240°$ and the magnitude of current chosen are on per unit basis with peak value of unity. At a typical instant, with phase 'A' having its instantaneous value of zero (current), the phase 'B' and phase 'C' are assigned a value of 0.866 and −0.866 units, respectively.

Number of turns and peak values of current decides the magnitude of MMF and does not alter the sinusoidal wave shape, needed in demonstration of RMF effect; hence the numerical values are not attached any significance in the present simulation. The basic requirements are that the peak value of the sinusoidal current and the corresponding MMF should remain constant. Thus numerical values are chosen purely arbitrarily, only keeping in view the visual distinct display of magnetic field distribution, with colors mapped against the magnitudes of spatial magnetic field intensity distribution. After defining all the material properties as specified in Table 26.1, mesh is generated and Finite Element Analysis (FEA) is carried out [15]. Figure 26.1 shows the model of the 4-pole induction motor with 12 slots, 4 slots per phase. The mesh is generated with the mesh-size of one, as specified in Table 26.1. The snapshot of a quadrant of complete model with the mesh distribution as taken from FEMM is given in Fig. 26.2. After FEM analysis and processing of the data using the post processor gives the magnetic field distribution as shown in Fig. 26.3. FEMM gives the color coded (proportional to magnitude of the field intensity) spatial distribution which is useful to identify the

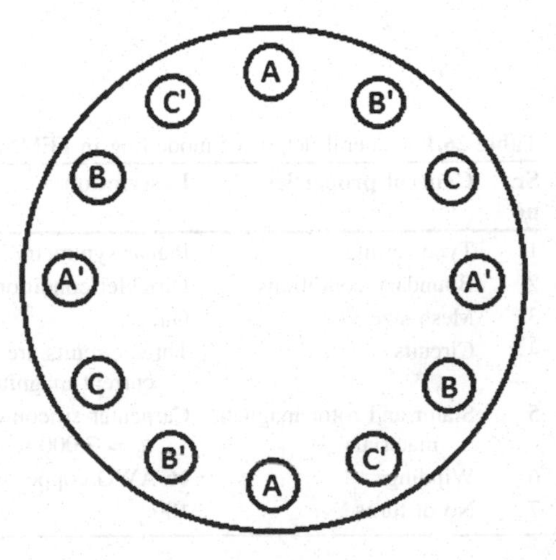

Fig. 26.1 Model of 4-pole induction motor showing positions of slots housing the conductors

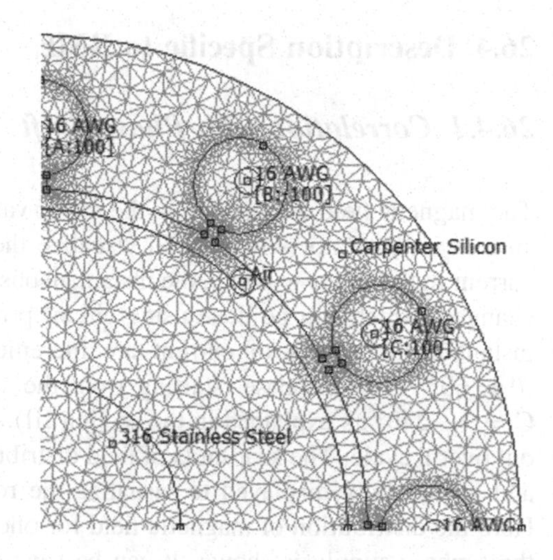

Fig. 26.2 A FEMM snapshot (a quadrant) of the model after generation of mesh. The materials chosen are from the library of FEMM

flux distribution. This can also be used to identify the pole positions at any given time instant on the electrical wave (of 50 Hz). Such a field distributions for a time instant when $I_a = 0$, $I_b = 0.866$, $I_c = -0.866$ is obtained (given in Fig. 26.3).

Such FEMM analysis is carried out at different instants of the time and these snapshots can be used to give a feel of RMF. This phenomenon of RMF and its correlation with the electrical wave and time instants is discussed in the sequel.

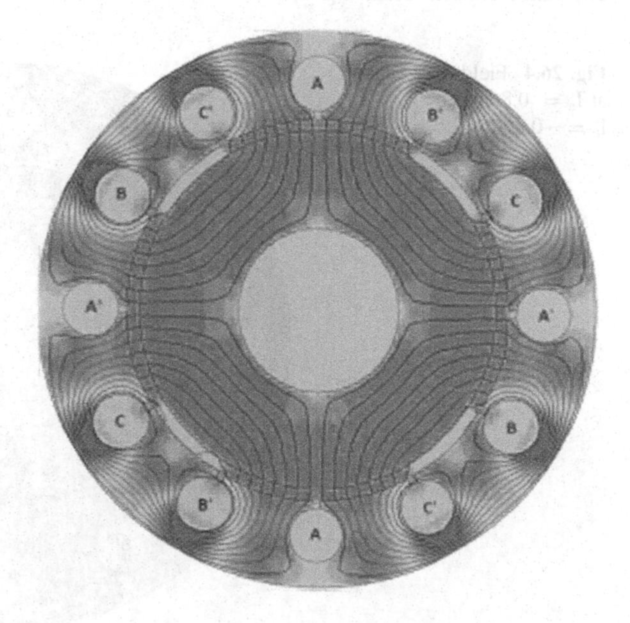

Fig. 26.3 Field distribution at $I_a = 0$, $I_b = 0.866$, $I_c = -0.866$

26.4 Description Specific to RMF

26.4.1 Correlation with Phase Shift

The magnetic field distribution in the motor varies with the change in magnitude of current in the stator winding. To visualize the rotating magnetic field due to the current in the stator winding, the instantaneous values of current in the winding are changed, which can be done using circuit property described in Table 26.1. The instantaneous values of current are incremented with the phase shift of 60° ($\theta = 60°$) in each phase (starting from the value of A = 1∠ 0°, B = 1∠ 120°, C = 1∠ 240° as explained in the section III). And FEMM simulations are carried out over a cycle. The snapshots of field distribution at different instances are saved and sequentially displayed to visualize the rotating field produced in the stator. Here the distribution of magnetic field for one half cycle (θ_e= 60°, 120°, 180°) of three phase supply is shown. It can be seen from the Figs. 26.3, 26.4, 26.5 and 26.6; the magnetic field attains its original position over the half cycle of three phase supply. As the field distribution is going to be same for another half cycle, the same snapshots of field distribution can be used to visualize the RMF over one full cycle.

From Fig. 26.6, it can be seen that RMF gets rotated by 90°(θ_m = 90°) from its original position (Fig. 26.3), for 180° increment in three phase supply (θ_e = 180°). The correlation between these electrical degree (θ_e) and mechanical degree (θ_m) is explained in the following section.

Fig. 26.4 Field distribution at I_a = 0.866, I_b = 0, I_c = −0.866

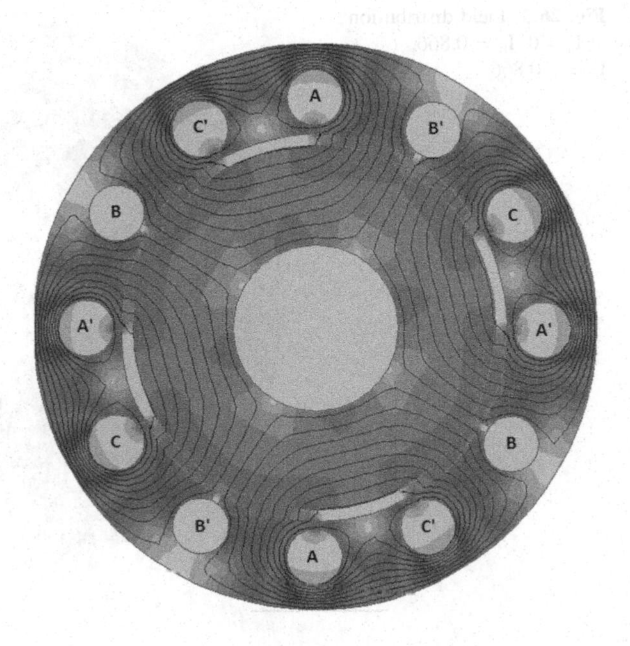

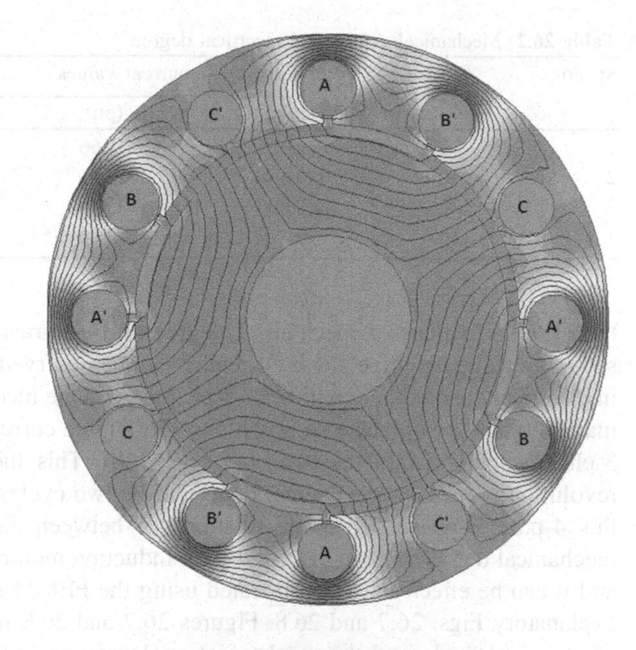

Fig. 26.5 Field distribution at $I_a = 0.866$, $I_b = -0.866$, $I_c = 0$

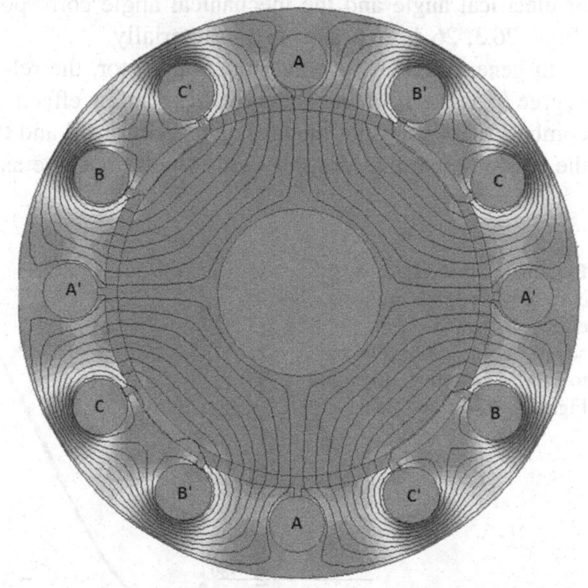

Fig. 26.6 Field distribution at $I_a = 0$, $I_b = -0.866$, $I_c = 0.866$

26.4.2 Mechanical Degree

From the Figs. 26.4, 25.5 and 26.6, the angle of rotation of RMF is calculated at each instances of current with respect to field distribution at the starting value ($I_a = 0$, $I_b = 0.866$, $I_c = -0.866$), corresponding to Fig. 26.3. Table 26.2 shows

Table 26.2 Mechanical degree and electrical degree

Sr. no	θ_e (deg)	Three phase current values			θ_m (deg)
		I_a (pu)	I_b (pu)	I_c (pu)	
1	0	0	0.866	−0.866	0
2	60	0.866	0	−0.866	30
3	120	0.866	−0.866	0	60
4	180	0	0.866	−0.866	90

the observed values of mechanical degree and electrical degree with the corresponding values of three phase current. It can be observed that for each 'θ' change in electrical degree, there will be a ($\theta/2$) change in the mechanical angle. When the magnetic field completes one revolution (360°), the current feeding each phase of 3-phase supply completes two cycles (2*360). This means that for every one revolution in mechanical degree, there will be two cycles in electrical degree (for this 4-pole, 3-phase, IM). The relationship between electrical degree 'θ_e' and mechanical degree 'θ_m' in case of 4-pole induction motor is given by $\theta_e = 2*(\theta_m)$ and it can be effectively demonstrated using the FEMM snapshots along with the explanatory Figs. 26.7 and 26.8. Figures 26.7 and 26.8 indicate the advancement of electrical angle and the mechanical angle corresponding to FEMM snapshots (Figs. 26.3, 26.4, 26.5 and 26.6), pictorially.

In general, for a 'P' pole induction motor, the relationship between electrical degree (θ_e) and mechanical degree (θ_m) can be effectively demonstrated through a combo FEEM snapshot pictures. This reinforced and thought as $\theta_e = (P/2)*\theta_m$ in the teaching learning process (with 4-pole machine as example).

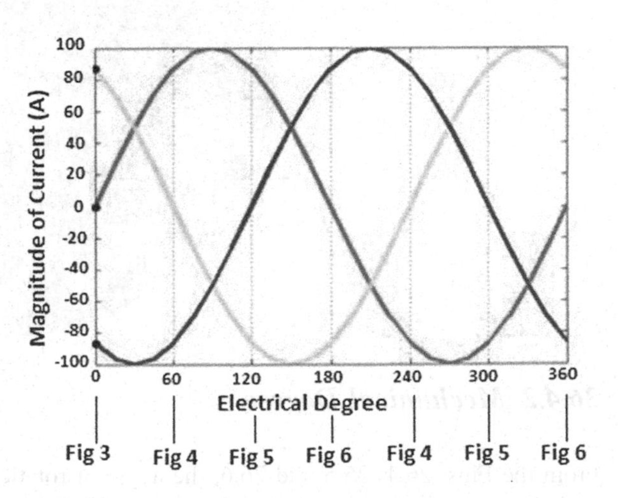

Fig. 26.7 Three phase current wave forms showing electrical angular advancements corresponding to FEMM snapshots given in Figs. 26.4, 26.5 and 26.6

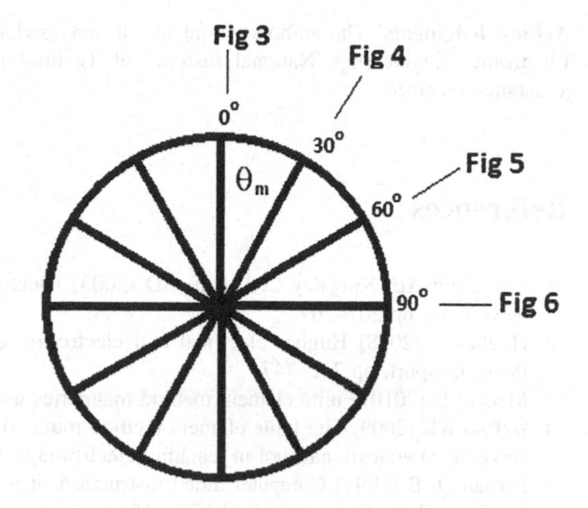

Fig. 26.8 Mechanical angular advancements with electrical wave advancements (Fig. 26.7), corresponding to FEMM snapshots given in Figs. 26.4, 26.5 and 26.6

26.4.3 Phase Reversal

The change in direction of rotation of motor can be obtained by changing the phase sequence of the three phase supply. To visualize this, the current in each phase is decreased by phase shift of 60^0, starting from the values A = 1∠ 0°, B = 1∠ 120°, C = 1∠ 240° as described in section III. As a result of decrement in phase shift, the phase sequence is modified. The simulation has been carried out for different instances of current and the snapshots of the FEMM simulation are saved. The earlier phase sequence followed was ABC which resulted in clockwise rotation of magnetic field of the induction motor. When the phase sequence is changed to ACB, the magnetic field of the motor will rotate in counter clockwise direction. From this, it will be seen that the direction of rotation of the resultant magnetic flux can be reversed by reversing the connections to any two of the three terminals of the motor.

The snapshot based demonstration of RMF in the power point projection form will be part of the presentation.

26.5 Conclusions

- The FEMM is an open source, simple to use software, which has ability to solve complex problems involving Laplace and Poisson's equation.
- One such application of FEMM is to demonstrate the rotating magnetic field in the stator of the induction motor, which has been successfully presented in this paper.
- Combining the field calculations using FEMM, open source software in demonstrating the concept of rotating magnetic field has been a unique attempt which aids in teaching–learning process.

Acknowledgments The authors would like to acknowledge the Department of Electrical and Electronics Engineering, National Institute of Technology Karnataka and MHRD India, for assistance received.

References

1. Fitzgerald AE, Kingsley C, Umans SD (2003) Electric machinery, 6th edn. McGraw-Hill, New York, pp 201–207
2. Hughes E (2008) Hughes electrical and electronic technology, 10th edn. Ashford Colour Press, Gosport, pp 742–747
3. Meeker D (2010) Finite element method magnetics user's manual, Version 4.1, 16 Oct 2010
4. Baltzis KB (2009) The finite element method magnetics (FEMM) freeware package: May it serve as an educational tool in teaching electromagnetic? Springer, Berlin
5. Fardanesh B (1992) Computer aided instruction of rotating electric machines via animated graphics. Trans Power syst 7(4):1579–1583
6. Bo M, Xiaofmg L, Xiqiang C, Jianhua Z, Yingli L (2001) Computer modeling of rotating magnetic fields for teaching purpose, vol 1. In: Proceedings of 5th international conference on electric machines and systems, pp 475–477
7. Panthala S (2004) Production of rotating magnetic fields in polyphase A.C. machines: a novel teaching approach. AU J Technol 7:105–110
8. Sadiku MNO (1989) A simple introduction to finite element analysis of electromagnetic problems. IEEE Trans Educ 32(2):85–93
9. Phuangmalai W, Konghirun M, Chayopitak N (2012) A design study of 4/2 switched reluctance motor using particle swarm optimization. In: 9th international conference on electrical engineering/electronics, computer telecommunications and information technology, pp 1–4
10. Punekar GS, Tesfaye N, Kishore NK (2011) On electric fields in belted cables and 3-phase gas insulated cables. In: International conference on power and energy systems, Dec 2011, pp 1 – 4
11. Hachicha MR, Hadj NB, Ghariani M, Neji R (2012) Finite element method for induction machine parameters identification. In: First international conference on renewable energy and vehicular technology
12. Zakaria Z, Mansor MSB, Jahidin AH, Shuhanaz M, Azlan Z, Rahim RA (2010) Simulation of magnetic flux leakage (MFL) analysis using FEMM software. In: IEEE symposium on industrialelectronics and applications
13. Litvinov BV, Davydenko OB (2009) A synchronous reluctance motor with a laminated rotor. Russ Electr Eng 80(1):29–32, ISSN 1068–3712
14. Saraiva E, Chaves MLR, Camacho JR (2008) Three-phase transformer representation using FEMM, and a methodology for air gap calculation. In: Proceedings of the international conference on electrical machines
15. Meeker D (2012) Induction Motor Example. FEMM website, available at http://www.femm.info/wiki/InductionMotorExample, July 2012

Chapter 27
Modelling of Integrating and Unstable Time Delay Processes

Bajarangbali and Somanath Majhi

Abstract Modelling of integrating and unstable time delay processes using describing function (DF) approximation of a relay with hysteresis is presented in this paper. Modelling of system parameters is necessary for tuning of controllers. In relay based identification methods describing function analysis is generally used to estimate the model parameters because of the general usefulness of the method and it is simple and straightforward. The effect of measurement noise in process modelling is an important issue as noise may change the actual amplitude of limit cycle output and also can fail the system identification test. To reduce the effect of measurement noise, a relay with hysteresis is considered in the proposed identification method. Simulation results are discussed to validate the proposed identification method.

27.1 Introduction

Process information is extracted from the limit cycle output and using this information process model parameters are obtained. This limit cycle output can be obtained from relay feedback test. Relay is a nonlinear device which can be approximated by an equivalent gain using describing function technique. In literature many authors have used this describing function analysis for modelling of integrating and unstable time delay processes. Åström and Hägglund [1] were the first to suggest the use of relay

Bajarangbali (✉) · S. Majhi
Department of Electronics and Electrical Engineering, Indian Institute of Technology, Guwahati 781039, India
e-mail: bajarangbali@iitg.ernet.in

S. Majhi
e-mail: smajhi@iitg.ernet.in

R. Malathi and J. Krishnan (eds.), *Recent Advancements in System Modelling Applications*, Lecture Notes in Electrical Engineering 188,
DOI: 10.1007/978-81-322-1035-1_27, © Springer India 2013

feedback technique with describing function technique to find the ultimate gain and frequency. Then Luyben [2] proposed the use of relay feedback with describing function analysis for process modelling. The measurement noise in process control systems is due to control valves, measuring devices or it can be from the process. Because of this measurement noise accurate process information cannot be obtained. More accurate estimation of process frequency can be done with the help of smooth limit cycle output. By using relay with hysteresis, the effect of measurement noise can be reduced [1]. Modelling of unstable processes with time delays is difficult. Since the condition for the limit cycle to exist is that the ratio of time delay to unstable time constant should be less than 0.693 [3, 4]. Using state space method and single relay feedback test Majhi and Atherton [5] suggested online tuning of process controllers for an unstable first order plus time delay systems (FOPDT). Li et al. [6] proposed modelling of stable and unstable processes by using two relay tests which is tiresome and also time consuming. Then few researchers used exact analysis methods like state space or Fourier analysis for improving relay autotuning. However, some practical constraints of relay based modelling remain unsolved. The advantage of conventional relay based offline identification method is that if the process information is not known then any controller cannot be used there, hence by using offline identification technique process can be modelled then controller can be designed. Majhi [7] suggested identification of integrating processes using single relay and state space method. Marchetti et al. [8] used two relay tests for open-loop unstable processes. Liu and Gao [9] used Newton–Raphson iteration method for obtaining exact expressions for integrating and unstable processes. Panda et al. [10] used single relay feedback test to estimate the process model parameters of integrating and time delay processes. Liu and Gao [11] proposed identification algorithms for obtaining FOPDT model for unstable processes. These exact methods take long time and not straightforward and also they do not give simple explicit expressions like describing function method for unstable and integrating process model parameters. In this paper, simple and straightforward identification methods are proposed by using relay with hysteresis using describing function analysis. Using describing function approach the authors in their recent publication [12] have proposed identification of stable FOPDT system using relay with hysteresis. The work has been extended here to model unstable and integrating time delay processes.

The contents of this paper are organized as follows. Section 27.2 illustrates the proposed identification methods and estimation of model parameters. Section 27.3 discusses about simulation examples. And conclusions are discussed in sect. 27.4.

27.2 Identification Methods

The procedure for estimation of integrating and unstable process model parameters is discussed in this section.

The conventional off-line identification scheme is shown in Fig. 27.1. Here the reference input r is considered to be zero.

27.2.1 Unstable FOPDT Process Model

The transfer function of the process $G_p(s)$ be represented by the unstable FOPDT process model transfer function as

$$G_m(s) = \frac{Ke^{-\theta s}}{Ls - 1} \tag{27.1}$$

The three unknown parameters to be estimated are K, L and θ which are the steady state gain, time constant and the time delay, respectively.

27.2.1.1 Estimation of L and θ

As shown in Fig. 27.1, relay induces sustained oscillations in the form of limit cycle output. Peak amplitude (A) and time period (P) of the limit cycle output are used to estimate the process model parameters L and θ. Here it should be noted that relay parameters h and ε are not zero. A stable limit cycle can be obtained for the following condition.

$$NG_m(j\omega) = -1 \tag{27.2}$$

where N is the equivalent gain of the relay with hysteresis and it can be obtained by the describing function analysis as

$$N = \frac{4h\left(\sqrt{A^2 - \varepsilon^2} - j\varepsilon\right)}{\pi A^2} \tag{27.3}$$

$G_m(j\omega)$ and N are substituted in (27.2) to result in (27.4) as

$$\frac{4h\left(\sqrt{A^2 - \varepsilon^2} - j\varepsilon\right)}{\pi A^2} \frac{Ke^{-j\omega\theta}}{j\omega L - 1} = -1 \tag{27.4}$$

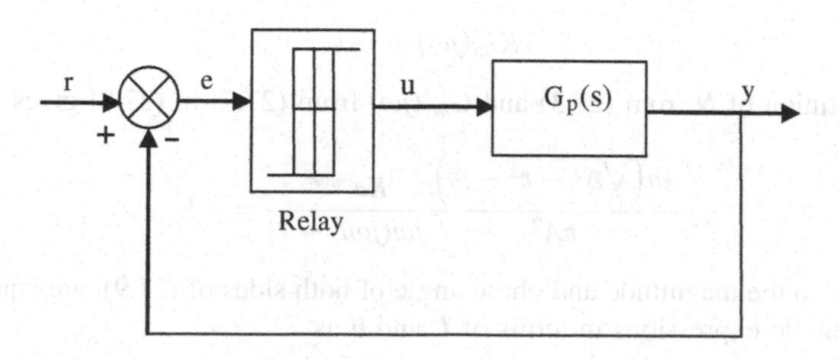

Fig. 27.1 Conventional off-line identification scheme

The magnitude and phase angle of both sides of (27.4), are equated to get the simple expressions in terms of L and θ as

$$L = \frac{\sqrt{\left(\frac{4hK}{\pi A}\right)^2 - 1}}{\omega} \tag{27.5}$$

$$\theta = \frac{\tan^{-1}(\omega L) - \tan^{-1}\left(\frac{\varepsilon}{\sqrt{A^2-\varepsilon^2}}\right)}{\omega} \tag{27.6}$$

where h and ε are relay parameters as relay height and hysteresis width respectively and $\omega = \frac{2\pi}{P}$

Hence the process model parameters L and θ are estimated from (27.5) and (27.6) respectively.

27.2.2 Integrating SOPDT Process Model

Now, in Fig. 27.1 let the transfer function of the process $G_p(s)$ be represented by integrating SOPDT process model transfer function as

$$G_m(s) = \frac{Ke^{-\theta s}}{s(Ls+1)} \tag{27.7}$$

Here also the process model transfer function has three unknown parameters as K, L and θ to be estimated.

27.2.2.1 Estimation of L and θ

Repeating the procedure of Sect. 27.2.1 to get sustained oscillations in the form of limit cycle output to estimate the process model parameters L and θ. Again to get a stable limit cycle the condition to be satisfied is

$$NG_m(j\omega) = -1 \tag{27.8}$$

Substitution of N from (27.3) and $G_m(j\omega)$ from (27.7) in (27.8) gives

$$\frac{4h\left(\sqrt{A^2-\varepsilon^2} - j\varepsilon\right)}{\pi A^2} \frac{Ke^{-j\omega\theta}}{j\omega(j\omega L+1)} = -1 \tag{27.9}$$

Here also the magnitude and phase angle of both sides of (27.9), are equated to get the simple expressions in terms of L and θ as

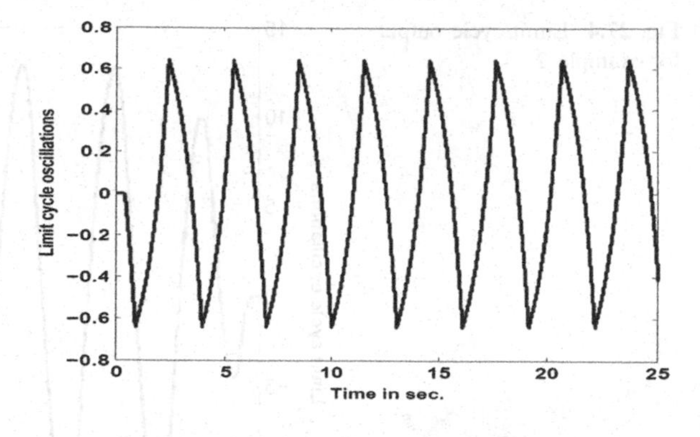

Fig. 27.2 Limit cycle output for example 1

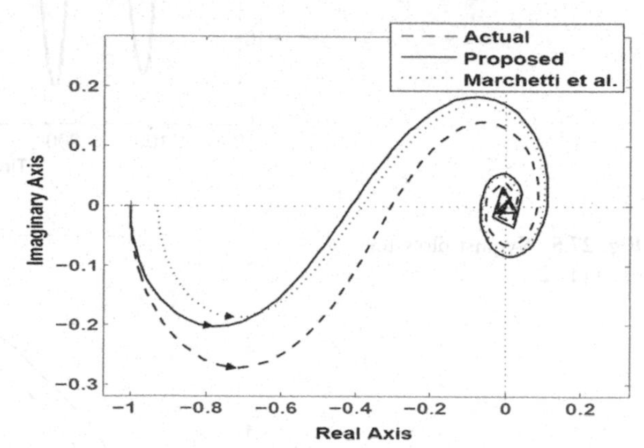

Fig. 27.3 Nyquist plots for example 1

$$L = \frac{\sqrt{\left(\frac{4hK}{\omega\pi A}\right)^2 - 1}}{\omega} \tag{27.10}$$

$$\theta = \frac{\frac{\pi}{2} - \tan^{-1}(\omega L) - \tan^{-1}\left(\frac{\varepsilon}{\sqrt{A^2 - \varepsilon^2}}\right)}{\omega} \tag{27.11}$$

Hence the process model parameters L and θ are estimated from (27.10) and (27.11) respectively.

27.3 Simulation Results

Two well-known examples are considered here to show the validity and general usefulness of the proposed method. Relay is connected as shown in Fig. 27.1 to induce limit cycle. From the limit cycle output, amplitude and time period are measured and substituted in corresponding expressions to get process model parameters L and θ. The steady state gain, K, is assumed to be known a priori.

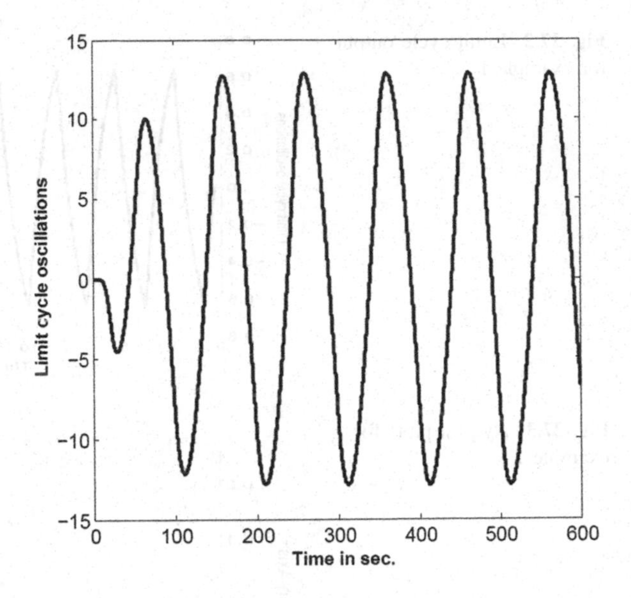

Fig. 27.4 Limit cycle output for example 2

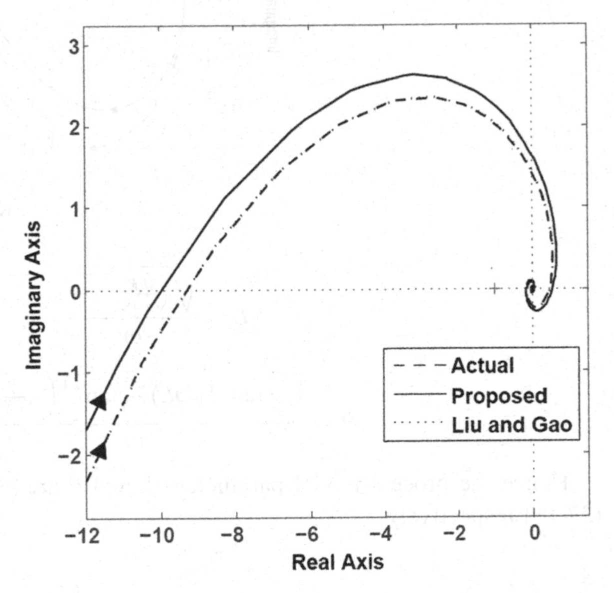

Fig. 27.5 Nyquist plots for example 2

Example 1
Let us consider the unstable FOPDT process transfer function [8]

$$G_p(s) = \frac{e^{-0.4s}}{s-1}$$

Marchetti et al. [8] have obtained the FOPDT process model parameters as $K = 0.928$, $L = 0.757$ and $\theta = 0.392$, by using two different relay tests. Using our proposed method with relay parameters $(h, \varepsilon) = (1, 0.1)$, the peak amplitude and time period are measured from the limit cycle output shown in Fig. 27.2. Substituting these parameters in (27.5) and (27.6), the model parameters L = 0.8305 and $\theta = 0.4287$ are estimated.

The Nyquist plots of actual process transfer function, identified process model and Marchetti et al.'s process model are compared as shown in Fig. 27.3. As the plots are close to each other, the proposed method can be applied to model unstable FOPDT processes.

Example 2

Consider the integrating SOPDT process transfer function [9]

$$G_P(s) = \frac{e^{-10s}}{s(20s+1)}$$

Liu and Gao [9] have identified the process model parameters as $K = 0.9983$, $L = 19.9443$ and $\theta = 10.027$, by using Newton–Raphson iteration method. As shown in Fig. 27.4, limit cycle output is obtained with relay parameters $(h, \varepsilon) = (1, 0.1)$. Substituting the peak amplitude and time period in (27.10) and (27.11) the process model parameters $L = 19.7229$ and $\theta = 10.8104$ are estimated.

Figure 27.5 shows the Nyquist plots of actual process transfer function, identified process model and Liu and Gao's process model. Since the plots are close to each other, the proposed method can be used to identify integrating SOPDT process model.

27.4 Conclusion

Modelling of integrating and unstable process with time delay is discussed in this paper by using relay with hysteresis. A simple and straightforward describing function technique is used to estimate the process model parameters. Two examples, one each for integrating and unstable processes, are illustrated to show the usefulness of the proposed method. These results are compared using Nyquist plots.

References

1. Åström KJ, Hägglund T (1984) Automatic tuning of simple regulators with specifications on phase and amplitude margins. Automatica 20:645–651
2. Luyben WL (1987) Derivations of transfer functions for highly non-linear distillation columns. Ind Eng Chem Res 26:2490–2495
3. Atherton DP (1997) Improving accuracy of autotuning parameter estimation. In: Proceedings of the IEEE international conference on control applications. Hartford, USA, pp 51–56

4. Majhi S, Atherton DP (1998) Autotuning and controller design for unstable time delay processes. In: Proceedings of IEEE conference, CONTROL'98, UKACC, 1998. Swansea, UK, pp 769–774
5. Majhi S, Atherton DP (2000) Online tuning of process controllers for an unstable FOPDT. In: IEE Proceedings control theory applications, vol 147(4), pp 421–427 July 2000
6. Li W, Eskinat E, Luyben WL (1991) An improved auto tune identification method. Ind Eng Chem Res 30:1530–1541
7. Majhi S (2007) Relay based identification of processes with time delay. J Process Control 17:93–101
8. Marchetti G, Scali C, Lewin DR (2001) Identification and control of open-loop unstable processes by relay methods. Automatica 37:2049–2055
9. Liu T, Gao F (2008) Identification of integrating and unstable processes from relay feedback. Comput Chem Eng 38:3038–3056
10. Panda RC, Vijayan V, Sujatha V, Deepa P, Manamali D, Mandal AB (2011) Parameter estimation of integrating and time delay processes using single relay feedback test. ISA Trans 50:529–537
11. Liu T, Gao F (2008) Alternative identification algorithms for obtaining a first-order stable/unstable process model from a single relay feedback test. Ind Eng Chem Res 47:1140–1149
12. Bajarangbali, Majhi S (2012) Relay based identification of systems. IJSER 3(6):ISSN 2229–5518

Chapter 28
Design and Implementation of Fractional-Order Controller for Fractional Order System

J. Prakash and S. R. Jayasurian

Abstract In this paper, the fractional-orders of integrator and differentiator in an optimal PID controller, for a first order plus dead time (FOPDT) model, are varied and a comparative study is made on the closed loop responses. In addition a new model based controller design approach is followed for designing fractional-order controllers for a class of fractional order systems.

Keywords Fractional-order systems · Fractional-order PID controllers · Model based control

28.1 Introduction

In recent years, fractional-calculus has gained an increasing attention in Control Engineering field. This is due to the fact that many real physical systems are well characterized by fractional order differential equations involving non integer order derivatives [1, 2]. In a view to improve control quality, new fractional order based control schemes are discussed in literatures, which includes fractional-order PID controllers, fractional-order lead lag compensators. The fractional-order PID controller is a generalized form of conventional PID controller with integral and derivative terms having non-integer orders instead of unity order [3, 4].

J. Prakash (✉) · S. R. Jayasurian
Department of Instrumentation Engineering, Madras Institute of Technology,
Chromepet, Chennai, India
e-mail: prakaiit@rediffmail.com

S. R. Jayasurian
e-mail: srjsurian@gmail.com

R. Malathi and J. Krishnan (eds.), *Recent Advancements in System Modelling Applications*, Lecture Notes in Electrical Engineering 188,
DOI: 10.1007/978-81-322-1035-1_28, © Springer India 2013

The structure of fractional-order PID controller is given by [5],

$$G_c(s) = k_p \cdot \frac{\tau_i s^\lambda + 1}{\tau_i s^\lambda} \cdot \frac{\tau_d s^\mu + 1}{\frac{\tau_d s}{N} + 1} \tag{28.1}$$

where, k_p is proportional gain, τ_i is integral time constant, τ_d is derivative time constant, λ is fractional-order of integrator, μ is fractional-order of differentiator and when $\lambda = \mu = 1$, the above structure reduces to classical PID controller.

In this work, the fractional-orders of integrator and differentiator, in an optimal PID controller for a FOPDT process model, are varied from 0 to 2 and closed loop responses were compared with performance index as integral absolute error (IAE). Also, a new model based controller design approach is followed to design fractional-order based controllers for a class of fractional-order systems.

This paper is organized as follows. In Sect. 28.2, the problem is formulated. In Sect. 28.3, simulation results are presented and comparative study is carried out. In Sect. 28.4, a new model based controller design for fractional order systems is discussed and corresponding simulation results are presented. Concluding remarks are presented in Sect. 28.5.

28.2 Problem Formulation

The structure of the FOPDT process model, considered in this work, is given as below.

$$G_p(s) = \frac{K}{Ts+1} e^{-Ls} \tag{28.2}$$

where, K is the process gain and T is the process time constant, and L is the dead time associated with the process. The fractional-order PID controller structure considered in this work is given by Eq. (28.1). The optimal values of controller parameters k_p, τ_i, τ_d were obtained from [5], in which genetic algorithm is used to obtain optimal values by minimizing IAE with constraints imposed on maximum sensitivity and they are given in table Closed loop responses for the process model $G_p(s) = \frac{1}{s+1} e^{-2.3s}$, were obtained for different values of λ, μ, while keeping the optimal values of k_p, τ_i, τ_d fixed. The fractional orders are varied in the following ways.

1. Keeping integral order fixed, derivative order has been varied.
2. Keeping derivative order fixed, integral order has been varied.
3. Both integral and derivative orders are increased.
4. Integral order increased and derivative order decreased.

For simulation purposes, the fractional order operators in the controller transfer function are approximated by using Oustaloup's approximation filter [6–9], which gives an equivalent integer order transfer function in poles-zeros form. The performance criterion IAE is evaluated for each case.

28.3 Simulation Results

The closed loop responses and the corresponding controller outputs for the system given in Eq. (28.2) with various values fractional orders of integrator λ and differentiator μ are presented in Figs. 28.1, 28.2, 28.3, 28.4, 28.5.

The performance criteria IAE for all the cases are reported in Tables 28.1, 28.2, 28.3.

28.4 Model Based Controller Design for Fractional-Order Systems

A new model based controller design approach has been proposed in [2], in which no constraint is imposed on model structure and no explicit model inversion is required. The process model $G_m(s)$ is factorized into two terms namely gain and dynamic elements as

$$\overline{G}_m(s) = k_p\overline{G}_m(s), \text{ where } \overline{G}_m(0) = 0. \tag{28.3}$$

Further,

$$\overline{G}_m(s) = \overline{G}_{m+}(s)\,\overline{G}_{m-}(s) \tag{28.4}$$

where $\overline{G}_{m+}(s)$ contains non-minimum phase elements and $\overline{G}_{m-}(s)$ contain stable elements.

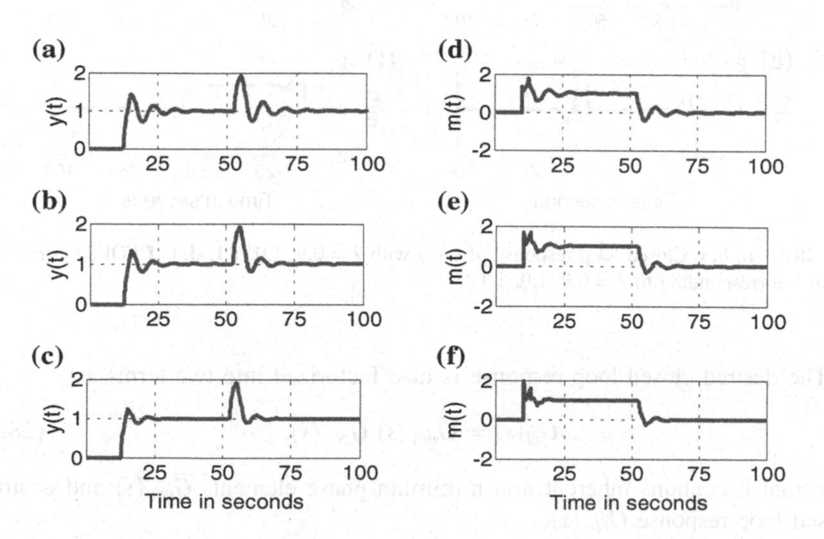

Fig. 28.1 **a, b, c** Closed loop response of G(s) with μ = 0.8, 0.9, 1.0, **d, e, f** FOPID controller outputs corresponding to μ = 0.8, 0.9, 1.0

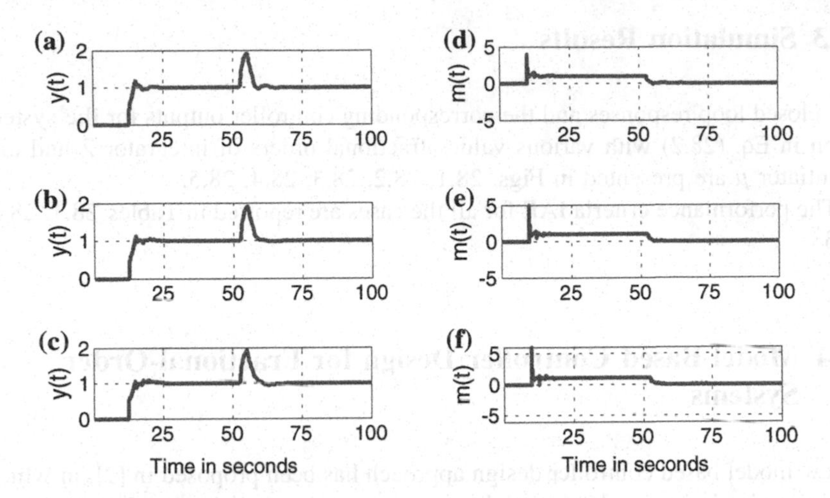

Fig. 28.2 **a, b, c** Closed loop response of G(s) with $\mu = 1.1, 1.2, 1.3$, **d, e, f** FOPID controller outputs corresponding to $\mu = 1.1, 1.2, 1.3$

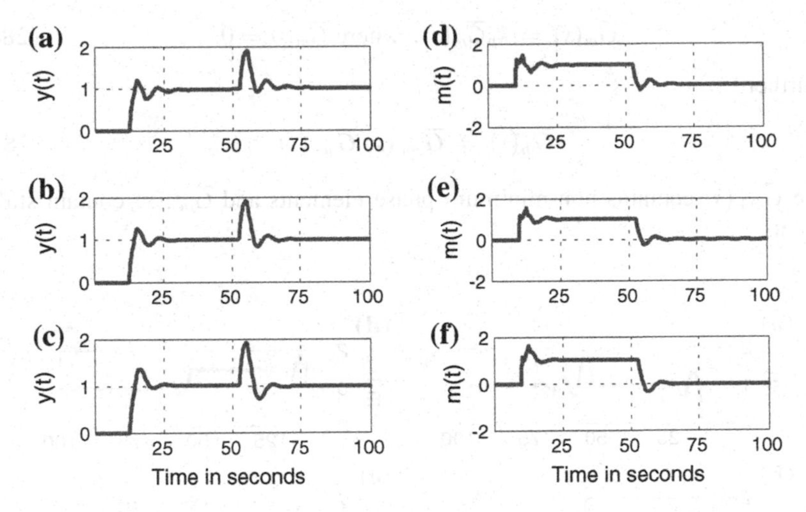

Fig. 28.3 **a, b, c** Closed loop response of G(s) with $\lambda = 0.9, 1.0, 1.1$, **d, e, f** FOPID controller outputs corresponding to $\lambda = 0.9, 1.0, 1.1$

The desired closed loop response is also factorized into two terms as:

$$G_{cl}(s) = \overline{G}_{m+}(s)\ G_{cl-}(s), \tag{28.5}$$

such that it contains inherent non minimum phase elements $\overline{G}_{m+}(s)$ and desired closed loop response $G_{cl-}(s)$.

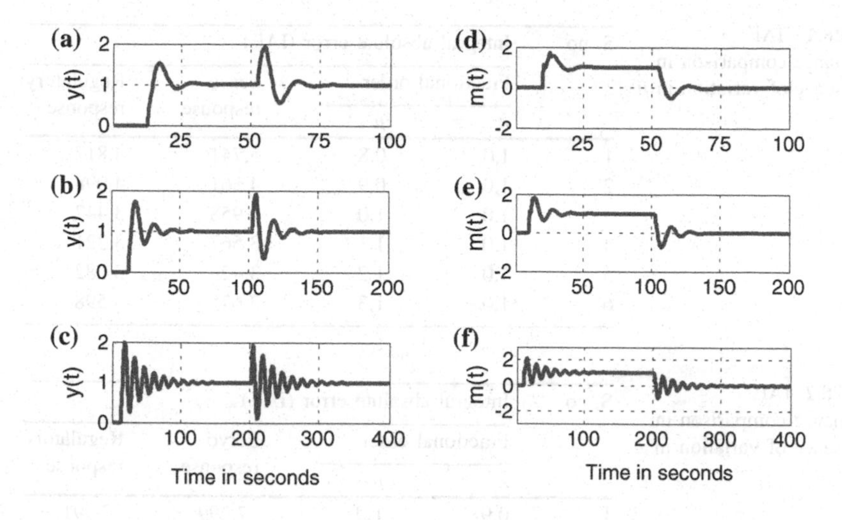

Fig. 28.4 **a**, **b**, **c** Closed loop response of G(s) with $\lambda = 1.2, 1.3, 1.4$, **d**, **e**, **f** FOPID controller outputs corresponding to $\lambda = 1.2, 1.3, 1.4$

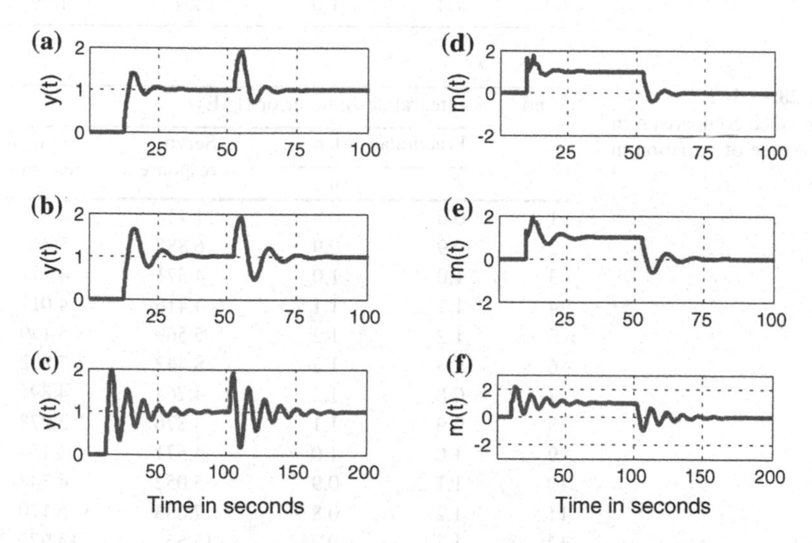

Fig. 28.5 **a**, **b**, **c** Closed loop response of G(s) with $\lambda = 1.1, 1.2, 1.3$ and $\mu = 0.9, 0.8, 0.7$, **d**, **e**, **f** FOPID controller outputs corresponding to $\lambda = 1.1, 1.2, 1.3$ and $\mu = 0.9, 0.8, 0.7$

Using synthesis approach, the controller transfer function is synthesized as

$$G_c(s) = k_m^{-1}\overline{G}_{m-}(s)^{-1}G_{cl-}(s)\frac{1}{1-\overline{G}_{m+}(s)^{-1}G_{cl-}(s)} \tag{28.6}$$

Table 28.1 IAE performance comparison in the presence of variation in μ

S. no	Integral absolute error (IAE)			
	Fractional order		Servo response	Regulatory response
	λ	μ		
1	1.0	0.8	5.741	4.817
2	1.0	0.9	4.561	4.098
3	1.0	1.0	3.958	3.447
4	1.0	1.1	3.567	3.321
5	1.0	1.2	3.423	3.282
6	1.0	1.3	3.675	3.598

Table 28.2 IAE performance comparison in the presence of variation in λ

S. no	Integral absolute error (IAE)			
	Fractional order		Servo response	Regulatory response
	λ	μ		
1	0.9	1.0	7.299	7.491
2	1.0	1.0	5.192	4.937
3	1.1	1.0	5.075	4.723
4	1.2	1.0	6.258	5.776
5	1.3	1.0	9.476	8.677
6	1.4	1.0	23.9	21.99

Table 28.3 IAE performance comparison in the presence of variation in λ and μ

S. no	Integral absolute error (IAE)			
	Fractional order		Servo response	Regulatory response
	λ	μ		
1	0.8	0.8	11.11	8.334
2	0.9	0.9	6.885	5.463
3	1.0	1.0	4.571	4.157
4	1.1	1.1	4.416	4.011
5	1.2	1.2	5.569	5.179
6	1.3	1.3	8.442	7.892
7	0.8	1.2	4.765	4.897
8	0.9	1.1	3.876	3.978
9	1.0	1.0	4.571	4.157
10	1.1	0.9	5.052	4.348
11	1.2	0.8	7.033	6.170
12	1.3	0.7	15.55	13.97

Assuming, $G_{cl-}(s) = \overline{G}_{m-}(s)$, the controller transfer function becomes,

$$G_c(s) = \frac{1}{1 - \overline{G}_m(s)} \tag{28.7}$$

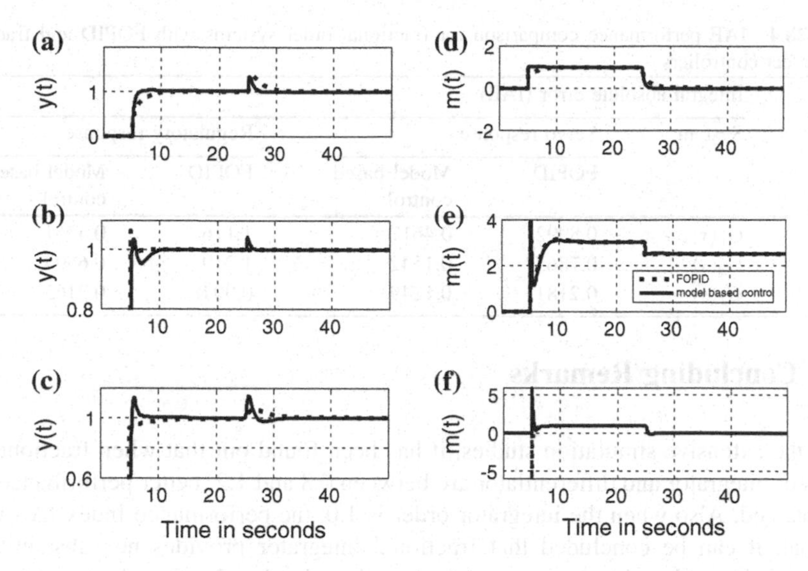

Fig. 28.6 **a, b, c** Closed loop responses of $G_1(s)$, $G_2(s)$, $G_3(s)$ with model based control (*solid line*) and FOPID control (*dotted line*), **d, e, f** FOPID controller outputs (*dotted line*) and model based controller outputs (*solid line*) corresponding to $G_1(s)$, $G_2(s)$, $G_3(s)$

This model based approach is followed to design controller for the following set of stable fractional-order systems [10]

$$\begin{aligned} G_1(s) &= \frac{3s^{0.8}+1}{s^{1.6}+s^{0.8}+1} \\ G_2(s) &= \frac{s^{1.2}+2}{s^2+s^{0.8}+3} \\ G_3(s) &= \frac{1}{s^{1.3}+1} \end{aligned} \tag{28.8}$$

Accordingly, the controller transfer functions for the above fractional order systems are given as below.

$$\begin{aligned} G_{c1}(s) &= \frac{s^{1.6}+s^{0.8}+1}{s^{1.6}-2s^{0.8}} \\ G_{c2} &= \frac{s^2+s^{0.8}+1}{s^2+s^{0.8}-s^{1.2}-1} \\ G_{c3}(s) &= \frac{s^{1.3}+1}{s^{1.3}} \end{aligned} \tag{28.9}$$

The closed loop responses for the fractional-order systems given in (28.8) using model based controllers given in Eq. (28.9) are obtained and they are compared with responses obtained using FOPID controller. The responses are shown in Fig. 28.6. The IAE performance criteria are compared in Table 28.4.

Table 28.4 IAE performance comparison for fractional order systems with FOPID and fractional order controllers

S. no	Integral absolute error (IAE)				
	System	Servo response		Regulatory response	
		FOPID	Model based control	FOPID	Model based control
1	$G_1(s)$	0.8992	0.4617	1.116	0.4361
2	$G_2(s)$	0.7865	0.1542	1.219	0.6604
3	$G_2(s)$	0.2181	0.1249	0.9978	0.2165

28.5 Concluding Remarks

From the extensive simulation studies, it has been found out that when fractional orders of integrator and differentiator are between 0.8 and 1.2, better performances are obtained. Also when the integrator order is 1.0, the performance Index IAE is minimal. It can be concluded that fractional integrator provides no substantial improvement in closed loop response, and on the other hand, when the derivative order is in the range of 0.8–1.2, better performances are obtained.

A new model based controller provides satisfactory control performance compared to conventional control scheme. This new approach can be extended to higher order processes, where a higher order process can be approximated using a fractional-order model and the same can be used to design controller, instead of implementing a higher order controller.

References

1. Podlubny I (1999) Fractional-order differential equations, 1st edn. Academic Press, San Diego, 340 pp
2. Pathiran A, Prakash J, Shah SL New model based control for single loop output feedback systems. ISA (submitted)
3. Hwang C, Leu JF, Tsay SY (2002) A note on time domain simulation of feedback fractional-order systems. IEEE Autom Control 47:625–631
4. Petras I (1999) The fractional-order controllers: methods for their synthesis and application. J Elect Eng 50(9–10):284–288
5. Padula F, Visioli A (2011) Tuning rules for optimal PID and fractional-order PID controller. J Process Control 21:69–81
6. Monje CA, Chen Y, Vinagre BM, Xue D, Feliu V (1999) Fractional-order systems and controls series: advances in industrial control. Springer, London
7. Zhao C, Xue D, Chen YQ (2005) A fractional-order PID tuning algorithm for a class of fractional-order systems. In: Proceedings of ICOMA'05
8. Dorcak L (1994) Numerical models for simulation the fractional-order control systems. The Academy of Science Institute of Experimental Physics, Kosice, 12 pp, UEF-04-94
9. Podlubny I (1999) Fractional-order systems and PIλ Dμ-controllers. IEEE Trans Autom Contr 44:208–214
10. Tavazoei MS (2011) Overshoot in the step response of fractional-order control systems. J Process Control 22:90–94

Chapter 29
Effect of Choice of Basis Functions in Neural Network for Capturing Unknown Function for Dynamic Inversion Control

Gandham Ramesh, P. N. Dwivedi, P. Naveen Kumar and R. Padhi

Abstract The basic requirement for an autopilot is fast response and minimum steady state error for better guidance performance. The highly nonlinear nature of the missile dynamics due to the severe kinematic and inertial coupling of the missile airframe as well as the aerodynamics has been a challenge for an autopilot that is required to have satisfactory performance for all flight conditions in probable engagements. Dynamic inversion is very popular nonlinear controller for this kind of scenario. But the drawback of this controller is that it is sensitive to parameter perturbation. To overcome this problem, neural network has been used to capture the parameter uncertainty on line. The choice of basis function plays the major role in capturing the unknown dynamics. Here in this paper, many basis function has been studied for approximation of unknown dynamics. Cosine basis function has yield the best response compared to any other basis function for capturing the unknown dynamics. Neural network with Cosine basis function has improved the autopilot performance as well as robustness compared to Dynamic inversion without Neural network.

G. Ramesh (✉) · P. N. Dwivedi
RCI, DRDO, Hyderabad, India
e-mail: gramesh.rci@gmail.com

P. N. Dwivedi
e-mail: prasiddha.dwivedi@gmail.com

P. Naveen Kumar
Department of ECE, Osmania University, Hyderabad, India

R. Padhi
Department of Aerospace Engineering, Indian Institute of Science, Bangalore, India

R. Malathi and J. Krishnan (eds.), *Recent Advancements in System Modelling Applications*, Lecture Notes in Electrical Engineering 188,
DOI: 10.1007/978-81-322-1035-1_29,

29.1 Introduction

Classically, missile autopilots are designed using linear control approaches by either in frequency domain or by applying linear quadratic regulators. In both the approaches, the plant is linearized around fixed operating point, which are suitably interpolated in the flight. The basic requirement for an autopilot is fast response and minimum steady state error for better guidance performance. Robustness against model uncertainties and decoupling between longitudinal and lateral motion in stressing engagement scenario are important for desired guidance performance. The highly nonlinear nature of the missile dynamics due to the severe kinematic and inertial coupling of the missile airframe as well as the aerodynamics has been a challenge for an autopilot that is required to have satisfactory performance for all flight conditions in probable engagements. Modern day missile are designed with high maneuverabilities to tackle highly agile and stealth target. High angle of attack operation region becomes imminent, resulting in high cross coupling of lateral and longitudinal plane. Linear autopilot (A/P) fails to address issue of cross coupling, since the fundamental assumption of design of linear A/P is decoupling of longitudinal and lateral motion.

A/P design using Dynamic Inversion/Time scale separated had been carried out by different authors like Menon [1, 2], Reiner [3], Buffington [4] for super maneuverable aircraft and missile. This class of A/P is also reported to be benchmark A/P for a tail control missile application Palumbo [5]. However it is also reported that DI A/P suffers from lack of robustness (Khargankar [6]). Many authors have tried to insert robustness in DI autopilot by inserting H_∞ loop which is designed using μ Synthesis approach [6–8]. Devaud et al. [9] has reported different traditional desired dynamics (with and without integrator) for design of DI A/P. Dwivedi et al. [10] has also utilized traditional 3 Loop architecture for realization of robust inner loop design. However all this methods achieve robustness by sacrificing the speed of response. To achieve robustness without sacrificing speed of response, model following adaptive techniques has to be implemented which will add an additional command in case of plant parameter perturbation. Padhi et al. [11–13] reported a new model-following adaptive control design technique for a class of non-affine and non-square nonlinear systems using neural networks. An appropriate nominal controller is assumed for a nominal system model. This nominal controller may not be able to guarantee stability/satisfactory performance in the presence of un modeled dynamics (neglected algebraic terms in the mathematical model) and/or parameter uncertainties present in the system model. In order to ensure stable behavior, an online control adaptation procedure is proposed. The controller design is carried out in two steps: (i) synthesis of a set of neural networks which capture matched un modeled (neglected) dynamics or model uncertainties because of parametric variations and (ii) synthesis of a controller that drives the state of the actual plant to that of a desired nominal model. Neuro adaptive nonlinear dynamic inversion controller has been proposed in this paper which will have enough robustness without sacrificing speed of response.

In Ref. [14], the author has shown the neuro adaptive controller with linear function approximation. The results shows that, it is able to capture the unknown function but with more oscillations. In order to reduce the oscillations, in this paper many basis functions has been studied. It has been found that for this kind of problem, cosine basis function gives the best result. It is also found that increasing the order of basis function is not giving any advantage.

29.2 Neuro Adaptive Controller Using Dynamic Inversion

In general there is presence of un modeled dynamics and parameter un-certainties in any system. Missile parameters are no exception. Due to this, nominal controller designed using Dynamic Inversion does not behave in the way it is required to. For worst combination of plant parameter perturbation, this may lead the system to become unstable. In 3 loop A/P architecture, an integrator is used in the inner loop to improve the robustness. Use of integrator impart accurate body rate tracking with less overshoot but at the cost of speed of response. Moreover gains are tuned to obtain adequate robustness for extreme combination of plant parameter [5]. To achieve this A/P need to be slowed down in most of the cases and hence performance is sacrificed. Hence there is always a need to impart required robustness in an adaptive way without slowing the A/P. Hence there arises a need for designing a controller which is more robust using adaptive approach[11–13], which is named as “Robust Adaptive Intelligent Design” (RAID). The design steps of RAID are given underneath:

(1) A nonlinear controller is designed based on the principle of dynamic inversion. In this work the nominal controller is available for us from the earlier design.
(2) The presence of off-nominal parameter values and unwanted inputs lead to performance degradation. To address this issue a model following function approximation is carried out.

As We pointed out earlier the first aim is to come up with a nominal controller, that meets the goals for the nominal model. The class of nonlinear system considered can be represented by the following equation

$$\dot{X}_d = f_d(X_d) + G_d(X_d)U_d \tag{29.1}$$

$$Y_d = h_d(X_d) \tag{29.2}$$

where $X_d \in R^n$ and $U_d \in R^m, Y_d \in R^p$ are the state and control variables of the nominal system respectively. The objective here is to design a controller U_d so that $Y_d \to Y^*$, where $Y^*\ (t)$ is the commanded signal, which is assumed to be bounded and smooth. The nominal controller U_d has been designed using dynamic inversion technique. The Eq. 29.3 may not truly represent the actual plant because of the presence of uncertainties in the model. Using the chain rule of derivative, the expression for $\dot{Y}$ can be derived as:

$$\dot{Y}_d = f_{yd}(X_d) + G_{yd}(X_d)U_d \tag{29.3}$$

where $f_{yd}[\partial h/\partial X_d]f_d(X_d)$ and $G_{yd}[\partial h/\partial X_d]G_d(X_d)$. Now the actual plant output is represented as

$$\dot{Y} = f_y(X) + G_y(X)U + d(X, X_d) \tag{29.4}$$

where $Y \in R^n$ is the output vector and $U \in R_m$ is the control vector of the actual system respectively. $d(X, X_d)$ is an unknown function that arises due to parameter uncertainties and unknown disturbances. The controller U needs to be designed online such that the states of the actual plant follow the respective states of the nominal model. In other words, the goal is to ensure that $Y \rightarrow Y_d$ as $t \rightarrow \infty$. To achieve this, the idea followed here is to first capture the unknown function $d(X, X_d)$, which is accomplished through a neural network approximation. For this purpose, an intermediate step is needed, which is to define an *approximate system* as follows

$$\dot{Y}_a = f_y(X) + G_y(X)U + \hat{d}(X, X_d) + K_a(Y - Y_a), \tag{29.5}$$

$$Y_a(0) = Y(0) \tag{29.6}$$

where K_a is selected as a positive definite gain matrix. A relatively easier way of doing this is to select k_{a_i} as a diagonal matrix with the ith element being $k_{a_i} > 0$. However, it is desirable to choose $k_{a_i} > 1$ because it leads to a smaller error bound. The reason for choosing an approximate system of the form in Eq. 29.5 is to facilitate meaningful bounds on the errors and weights. The approach followed here for ensuring $Y \rightarrow Y_d$ involves two steps. (i) $Y \rightarrow Y_a$ and (ii) $Y_a \rightarrow Y_d$ since $Y \rightarrow Y_d$ is desired but not possible, this objective is achieved using the above two steps. Figure 29.1 depicts the model following approach.

Step 1. Capturing $d(X, X_d)$ and ensuring $Y \rightarrow Y_a$. To capture the unknown function, first write $d_d(X, X_d)$ $[d_1(X, X_d) \ldots d_n(X, X_d)]$ where $d_i(X, X_d), i = 1, 2, \ldots n$ is the ith component of the $d(X, X_d)$. Each $d(X, X_d)$ is approximated as $\hat{d}(X, X_d)$ in a separate linear-in-the-weight neural network and write

$$\hat{d}(X, X_d) = \hat{W}^T{}_i \phi(X, X_d) \tag{29.7}$$

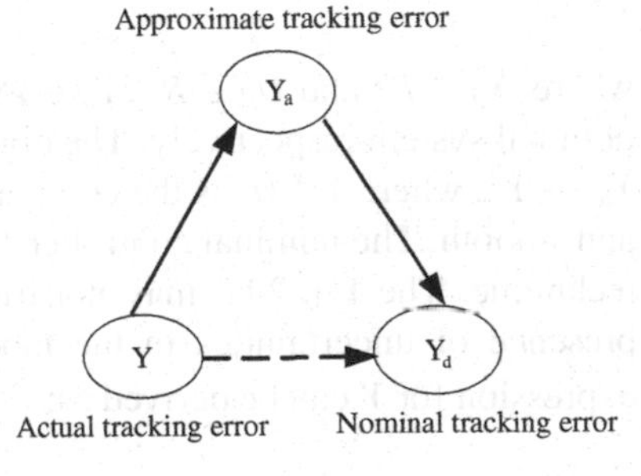

Fig. 29.1 Philosophy of model following approach

where $\hat{W}_i$ is the weight vector of the ith neural network and $\phi(X, X_d)$ is its basis function vector. At this point, it needs to be mentioned that even though generic radial basis functions can be used for this purpose; it is probably wiser to incorporate some prior knowledge about the system and judiciously select the basis functions, which will lead to faster learning of the unknown function. Note that the combination of n sub-networks can be interpreted to constitute a single neural network that represents $d(X, X_d)$. The idea of having n neural networks for n independent channels is to facilitate simpler mathematical analysis. More important, it leads to faster training because of reduced computational complexity, as none of the weights are linked to more than one output function. The next task is to update the weights of the neural network (i.e. to train them). Towards this end, the error between the actual state and the corresponding approximate state is defined as

$$e_{a_i} y_i - y_{a_i} \tag{29.8}$$

From Eqs. 4 and 5, the equations for the ith channel e_{a_i} is written as

$$\begin{aligned} \dot{e}_{a_i} d_i(X, X_d) - \hat{d}_i(X, X_d) - k_{a_i} e_{a_i} \\ = \tilde{W}_i^T \phi_i(X, X_d) + \in i - k_{a_i} e_{a_i} \end{aligned} \tag{29.9}$$

where $\tilde{W}\left(W_i - \hat{W}\right)$ is the error between the ideal weight and actual weight of the neural network. Next, define a series of Lyapunov function candidates L_i, $i = 1, 2, \ldots n$ such that

$$L_i = \frac{e_{a_i} p_i e_{a_i}}{2} + \frac{\tilde{W}_i^T \gamma_i^{-1} \tilde{W}_i}{2} \tag{29.10}$$

where $p_i > 0$ and $\gamma_i > 0$. Taking the time derivative of both sides of Eq. 29.10, using the fact that $\dot{\tilde{W}} = -\dot{\hat{W}}_i$ (Since W_i is a constant and on substituting for $\dot{e}_{a_i}$from Eq. 29.10

$$\begin{aligned} \dot{L}_i =& e_{a_i} p_i \dot{e}_{a_i} + \tilde{W}_i^T \gamma_i^{-1} \dot{\tilde{W}}_i \\ =& \tilde{W}_i^T \left(e_{a_i} p_i \phi(X, X_d) - \gamma_i^{-1} \dot{\hat{W}}_i \right) + e_{a_i} p_i \in_i - K_{a_i} e_{a_i}^2 p_i \end{aligned} \tag{29.11}$$

Note that our objective is to come up with a meaningful condition that will ensure $\dot{L}_i < 0$ which will ensure the stability of the error dynamics (of tracking error as well as weight error). However, the expression for $\dot{L}_i$ contains $\tilde{W}_i$ (which is unknown), and hence, nothing can be concluded about the sign of $\dot{L}_i$. To get rid of this difficulty, force the term multiplying it to zero and obtain the following weight update rule (training algorithm) for the ith neural network

$$\dot{\hat{W}}_i = \gamma_i e_{a_i} p_i \phi_i(X, X_d) \tag{29.12}$$

where γ_i can be interpreted as a learning rate for the ith network (its numerical value essentially dictates the rate of capturing the unknown function) $d_i(t, X, X_d)$.

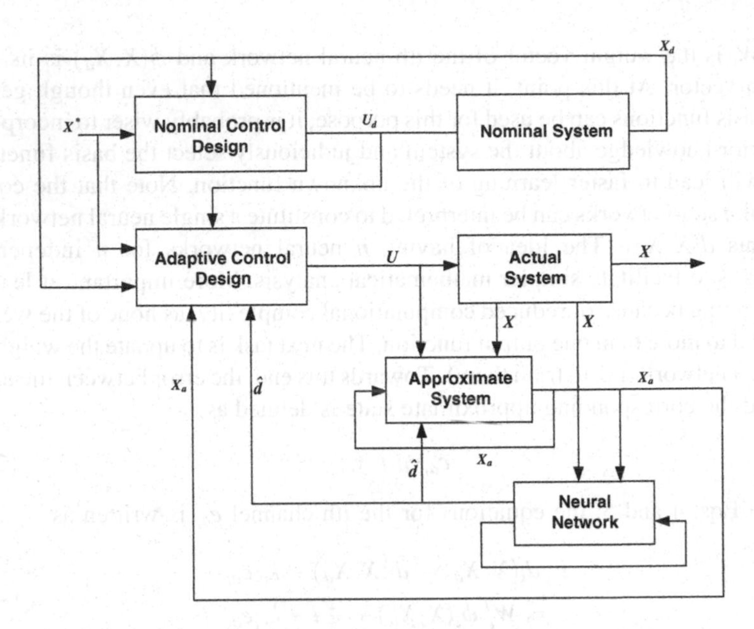

Fig. 29.2 Block diagram for adaptive control

Note that Eq. 29.12 is the weight update (learning) rule for $\tilde{W}_i$. Select the initial condition as $\tilde{W}_i(0) = 0$. This is compatible with the fact that if $d_i(t, X, X_d) = 0$ (i.e. there is no error in the model), then automatically $d_i(\hat{X}, X_d) = 0$. Substituting Eq. 29.12 in Eq. 29.10

$$\dot{L}_i = -k_{a_i} e_{a_i}^2 p_i + e_{a_i} p_i^2 \in_i \tag{29.13}$$

This eventually leads to the following (sufficient) condition $\dot{L}i < 0$ whenever $|e_{a_i}| > |\epsilon_i| / K_{a_i}$.

Step 2. Ensuring $Y_a \rightarrow Y_d$ and Computation of U.

As pointed out earlier, while ensuring $Y \rightarrow Y_a$ and capturing the unknown function $d(t, X, X_d)$ as a functional approximation $\hat{d}(t, X, X_d)$, it is simultaneously ensured that $Y_a \rightarrow Y_d$ as $t \rightarrow \infty$. To achieve this objective, the controller U is designed such that the following stable error dynamics is satisfied

$$\left(\dot{Y}_a - \dot{Y}_d\right) + K(Y_a - Y_d) = 0 \tag{29.14}$$

where K is chosen to be a positive definite gain matrix. A relatively easy way of choosing the gain matrix is to have $K = diag(1/\tau_1 \ldots 1/\tau_n)$, where τ_i can be interpreted as the desired time constant for the ith channel of the error dynamics in Eq. 29.14. The block diagram of adaptive control is shown in Fig. 29.2.

Table 29.1 Basis function used in simulation

Case	Basis function	Description
1	Linear	$\phi_1(X, X_d) = 1, \phi_2(X, X_d) = X_d$
2	Quadratic	$\phi_1(X, X_d) = 1, \phi_2(X, X_d) = X_d$ $\phi_3(X, X_d) = X_d^2$
3	Cubic	$\phi_1(X, X_d) = 1, \phi_2(X, X_d) = X_d$ $\phi_3(X, X_d) = X_d^2, \phi_4(X, X_d) = X_d^3$
4	Cosine	$\phi_1(X, X_d) = 1, \phi_2(X, X_d) = \cos(X_d)$
5	Quadratic cosine	$\phi_1(X, X_d) = 1, \phi_2(X, X_d) = \cos(X_d)$ $\phi_3(X, X_d) = \cos(X_d)^2$
6	Exponential	$\phi_1(X, X_d) = 1, \phi_2(X, X_d) = \exp(X_d)$

29.3 Choice of Basis Function

The choice of basis plays the major role in capturing the unknown function using neural network in Dynamic Inversion controller. In this paper we studied the effect of choice of basis function in capturing the unknown dynamics. The basis functions used here are linear, quadratic, cubic, cosine, quadratic cosine and exponential basis functions. The basis functions are shown in Table 29.1.

29.4 Simulation

For the simulation purpose the 3 DOF pitch dynamic of missile has been used as given below.

$$\dot{u} = \frac{-mg\sin\theta - D}{m} - qw \tag{29.15}$$

$$\dot{w} = \frac{-L - 0.5\rho V^2 S C_{N_\delta}\delta + mg\cos\theta}{m} + qu \tag{29.16}$$

$$\dot{q} = \frac{M}{I_y}; \quad \dot{\theta} = q \tag{29.17}$$

The autopilot design has been done in two part. First is outer loop design and second is inner loop. In outer loop first the ® command has been calculated as

$$\alpha_c = \frac{a_c * mass}{0.5\rho V^2 S C_{N_\alpha}} \tag{29.18}$$

where a_c is the guidance demand. Then desire dynamic is chosen as

$$\dot{\alpha}_c = \omega_o(\alpha_c - \alpha) \tag{29.19}$$

where ω_o is outer loop gain chosen as 10. Once the $\dot{\alpha}_c$ is calculated the q^* has been calculated by as

$$q^* = \dot{\alpha}_c \tag{29.20}$$

In inner loop, the robustness study is carried out where the states and output are pitch body rates. The objective as explained before is to come up with a nominal controller, which will meet the goals of nominal system i.e. the nominal states $q \rightarrow q^*$ as $t \rightarrow \infty$ and hence the adaptive control will meet the goals of the actual system. The governing equations used for designing the nominal controller are

$$\dot{q} = \frac{M}{I_y} \tag{29.21}$$

where M is aerodynamic moment given by

$$M = 0.5\rho V^2 S_d (C_{m_\alpha}\alpha + C_{m_\delta}\delta) \tag{29.22}$$

The technique used for designing nominal controller is Dynamic inversion. The system is of the form $\dot{Y}_d = f_{yd}(X_d) + G_{yd}(X_d)U_d$ where $F(X)$ and $G(X)$ are defined as

$$f_{yd}(X_d) = \frac{0.5\rho V^2 S_d C_{m_\alpha}\alpha}{I_y} \tag{29.23}$$

$$G_{yd}(X_d) = \frac{0.5\rho V^2 S_d C_{m_\delta}}{I_y} \tag{29.24}$$

Enforcing a first order error dynamics, one can solve for nominal controller as

$$U_d = \left[G_{yd}(X_d)\right]^{-1}\left[f_{yd}(X_d) + K_g\left(Y_d - Y_d^*\right)\right] \tag{29.25}$$

where K_g is autopilot gain. The actual dynamics differs from the nominal dynamics because the actual dynamics contains nominal parameters along with their corresponding uncertainties and disturbance. The actual output dynamics equation looks like

$$\dot{Y} = f_y(X) + G_y(X)U + \hat{d}(t, X, X_d) \tag{29.26}$$

$d(t, Y, Y_d)$ crops up because the states are seeing now the perturbed coefficients. Finally the adaptive controller is designed using

$$\dot{Y}_a = f_y(X) + G_y(X)U + \hat{d}(t, X, X_d) + K_a(Y - Y_a) \tag{29.27}$$

First order error dynamics are imposed to ensure that $Y_a \rightarrow Y_d$. The error dynamics are written as $(\dot{Y}_a \rightarrow \dot{Y}_d) + K(Y_a - Y_d) = 0$. The values of K and K_a selected are as $K = 30$ and $K_a = 5$. Next we discuss the neural network selection and its training. The Lyapunov function is as mentioned in the previous section and the values of the scalars 'p' and the learning rate γ are as follows: $p = 1$ and $\gamma = 200$.

29.5 Result and Discussion

To show the effect of choice of basis function for capturing the unknown dynamics, 6 cases has been chosen. In all cases the missile pitching moment coefficient is perturbed by +50 %. In case 1 Unknown function is approximated as linear function. In case 2 Unknown function is approximated as quadratic function. In case 3 Unknown function is approximated as cubic function. In case 4, 5, 6, Unknown function is approximated as nonlinear function approximation with cosine basis functions, quadratic cosine basis functions and exponential basis functions. All the 6 cases are discussed in the following section.

29.5.1 Case 1

In case 1 the missile pitching moment coefficient is perturbed by +50 % with linear function approximation. Figure 29.3 shows the body rate demanded and achieved by controller with and without adaptive control. This shows that the adaptive controller make the controller fast as well as steady state error is less. The above plot also show that controller without adaptation has more steady state error and settling time. Figure 29.4 shows the unknown function and its neural network approximation. Initially some oscillation has been observed in $(d(\hat{x}))$. It is because it is taking some time to train the network. Once the network is trained, the Neural network approximation $(d(\hat{x}))$ is following the unknown perturbation. This shows the importance of the capturing the unknown function in case of unknown perturbation in aero parameter. Even though Linear function approximation is able to capture the unknown function but has more oscillation and settling time. Therefore

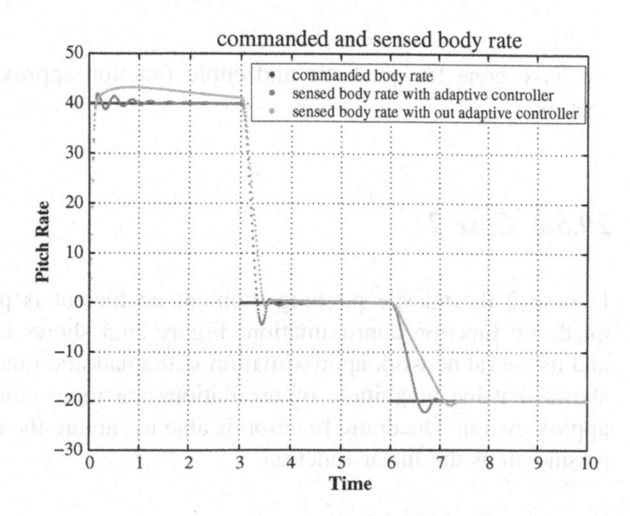

Fig. 29.3 Demanded and achieved body rate case 1

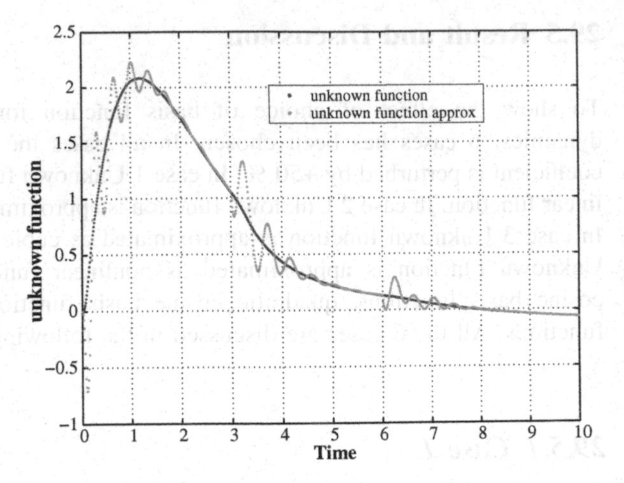

Fig. 29.4 $(d(x))$ and $(d(\hat{x}))$ of case 1

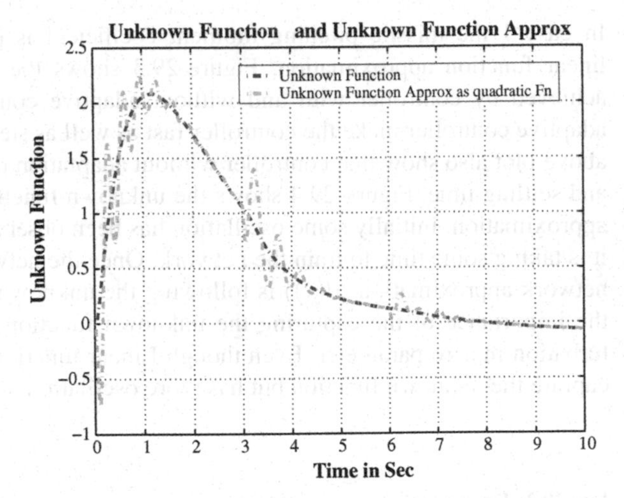

Fig. 29.5 $(d(x))$ and $(d(\hat{x}))$ of case 2

we have gone for quadratic and cubic function approximation in the following section.

29.5.2 Case 2

In case 2 the missile pitching moment coefficient is perturbed by +50 % with quadratic function approximation. Figure 29.5 shows the unknown perturbation and its neural network approximation with quadratic function approximation. This shows that the magnitude of oscillations are more compared to linear function approximation. Quadratic function is able to capture the unknown function but not as smooth as the linear function.

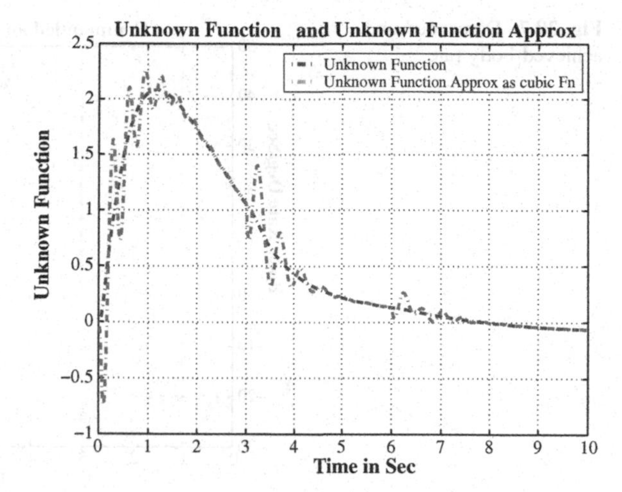

Fig. 29.6 $(d(x))$ and $(d(\hat{x}))$ of case 3

29.5.3 Case 3

In case 3 the missile pitching moment coefficient is perturbed by +50 % with cubic function approximation. Figure 29.6 shows the unknown perturbation and its neural network approximation. The transient response with cubic function approximation is almost same as that of quadratic function approximation. It is clear that even with increase in order of the approximating function we are not going to achieve required performance. This is the reason we have gone for different basis function from case 4 onward.

29.5.4 Case 4

In case 4 the missile pitching moment coefficient is perturbed by +50 % with cosine basis function. Figure 29.7 shows the body rate demanded and achieved by controller with adaptive control. In this case controller is able to track the command with less oscillation. The percentage overshoot and also the steady state error and settling time are less compared to linear, quadratic and cubic function approximation. Figure 29.8 shows the unknown perturbation and its neural network approximation. Here it can be observed that the choice of basis function is important for particular kind of problem. It is able to track the unknown function without sacrificing the speed of response. To see whether it improves the performance compared to cosine basis function we have gone for quadratic cosine function in the following section.

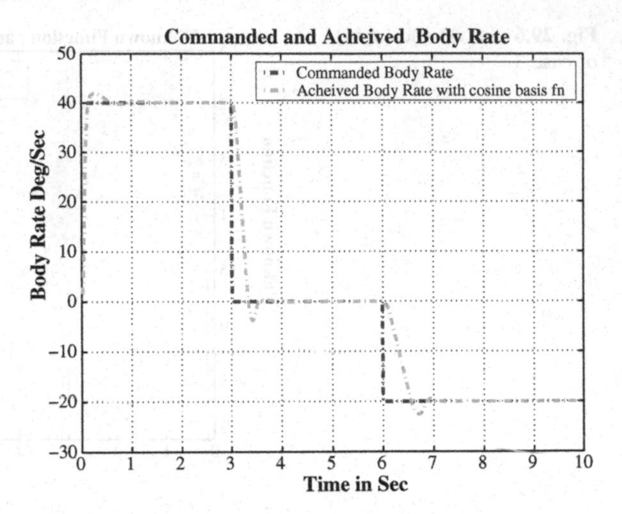

Fig. 29.7 Demanded and achieved body rate

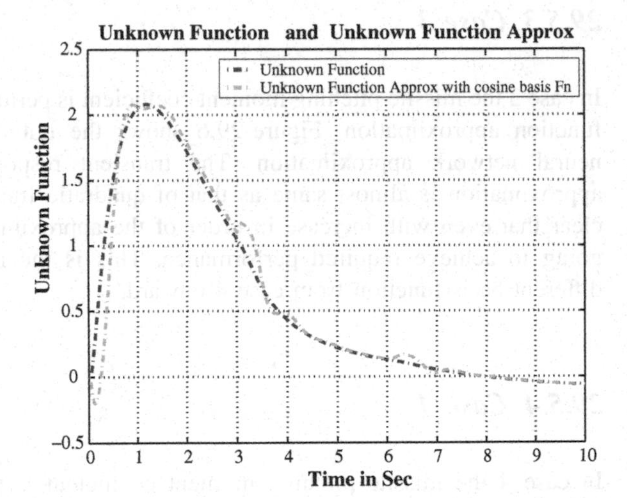

Fig. 29.8 $(d(x))$ and $(d(\hat{x}))$ of case 4

29.5.5 Case 5

In case 5 the missile pitching moment coefficient is perturbed by +50 % with quadratic cosine basis function. Figure 29.9 shows the unknown perturbation and its neural network approximation. In this case neural network is able to track the command with less oscillation compared to linear function approximation. It is evident that quadratic cosine basis function approximation has more settling time compared to cosine basis function approximation. It is clear that increase in order of the approximating function is not improving the required performance.

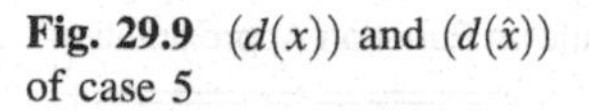

Fig. 29.9 $(d(x))$ and $(d(\hat{x}))$ of case 5

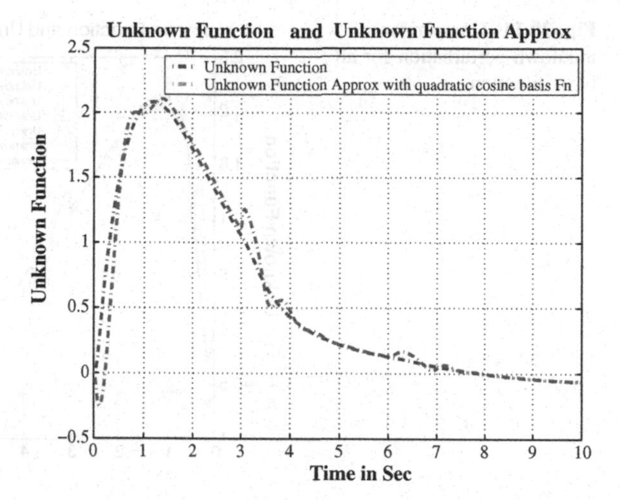

Fig. 29.10 $(d(x))$ and $(d(\hat{x}))$ of case 6

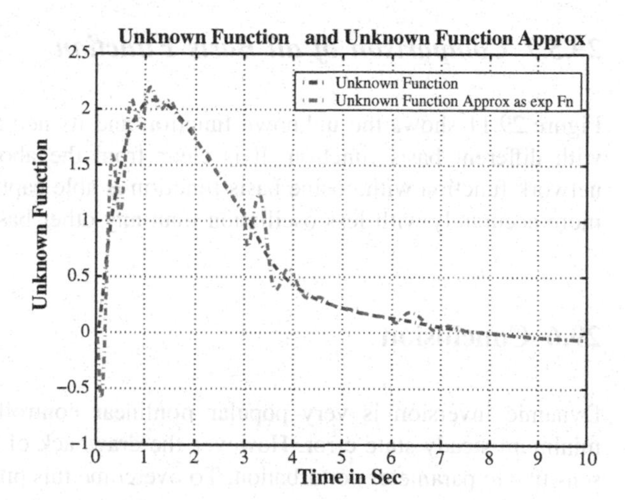

29.5.6 Case 6

In case 6 the missile pitching moment coefficient is perturbed by +50 % with exponential basis function. Figure 29.10 shows the unknown perturbation and its neural network approximation. From the above figure it is clear that it is able capture the unknown function more accurately than linear, quadratic and cubic function approximation. But the magnitude of oscillations are more compared to cosine basis function approximation.

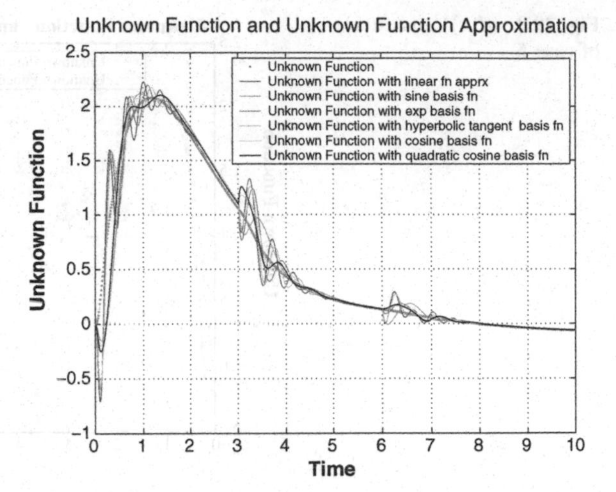

Fig. 29.11 Comparison of unknown perturbation for all basis function

29.5.7 Comparison of all Basis Function

Figure 29.11 shows the unknown function and its neural network approximation with different basis function. It is clear from the above figure that the neural network function with cosine basis function is able capture the unknown function more accurately with less oscillation than any other basis function.

29.6 Conclusion

Dynamic inversion is very popular nonlinear controller for fast response and minimum steady state error. However the drawback of this controller is that it is sensitive to parameter perturbation. To overcome this problem, neural network has been used to capture the parameter uncertainty on line. The choice of basis function plays the major role in capturing the unknown dynamics. Here in this paper, many basis function has been studied for approximation of unknown dynamics. Cosine basis function has yield the best response compared to any other basis function for capturing the unknown dynamics. Neural network with Cosine basis function has improved the autopilot performance as well as robustness compared to Dynamic inversion without Neural network. The future work in this direction is to apply this method in complete Six DOF model with real constrains and robustness of complete A/P has to be assessed.

Acknowledgments The authors gratefully acknowledge the Dr. S.K. Chaudhuri, Outstanding Scientist and Director, RCI for his guideline and constant encouragement. Author is thank full to Abhijit Bhattacharya, PD, AD(M), PGAD for his suggestion and support towards this work. Authors are also grateful to SP Rao, Technology Director PGCT, RCI, Hyderabad for his constant encouragement during the course of this work. Author is also grateful to Dr. P. Chandra shekhar, Head of ECE Dept, Osmania University for his support towards this work. The authors are grateful to Prasanth Bhale, Scientist 'D', DRDL for his timely help during the course of this work.

References

1. Menon PK, Badget ME, Walker RA, Duke EL (1987) Nonlinear flight test trajectory controllers for aircraft. J Guid Control Dyn 10(1):67–72
2. Menon PK, Iragavarapu VR, Ohlmeyer EJ (1997) Nonlinear missile autopilot using time scale separation. AIAA Paper 96-3765
3. Reiner J, Balas G, Garrard WL (1994) Design of a flight control systems for a highly maneuverable aircraft using robust dynamic inversion. AIAA Paper 94-3682-CP
4. Buffington J, AdAms R, Banda SS (1993) Robust, nonlinear, high AOA control design for a supermaneuvring vehicle. AIAA Paper 93-3774-CP
5. Palumbo NF, Reardon BE, Blauwkamp RA (2004) Integrated guidance and control for homing missiles. Johns Hopkins APL Tech Dig 25(2):121–139 (2004)
6. Schumacher C, Khargankar PP (1998) Missile autopilot design using *H1* control with gain scheduling and dynamic inversion. J GCD 21(2):234–243
7. Hogui S, Balas G, Garrard WL (1995) Design of a robust dynamics inversion lateral flight controller. AIAA 95-32534-CP
8. Mafarland MB, D'Souza CN (1994) Missile flight control with DI and structured singular value synthesis. AIAA 94-3604-CP
9. Devaud E, Harcaut J-P, Siguerdidjane H (2001) Three-axes missile autopilot design: from linear to nonlinear control strategies. J. Guid Control Dyn 24(1):64–71. ISSN:0731-5090
10. Bhattacharyaa A, Dwivedi PN, Kumar P, Prashant, Bhale G, Bhattacharjee RN (2007) A practical approach for robust scheduling of nonlinear time-scale separated autopilot. In: ACCORDS 2007. IISC, Banglore
11. Padhi R, Chunodhkar A (2007) Precision attitude maneuver of spacecraft using model following neuro adaptive control. J Syst Sci Eng 16:1
12. Padhi R, Unnikrishnan N, Balakrishnan SN (2007) Model following neuro adaptive control design for non-square, non-affine systems. IET, Control Theory Appl 1(6):1650–1661
13. Padhi R, Kothari M (2007) An optimal dynamic inversion based neuro adaptive approach for treatment of chronic mylogenous leukemia. Comput Methods Prog Biomed 87:208–228
14. Dwivedi PN, Bhattacharyaa A, Padhi R (2009) Robust dynamic inversion missile autopilot using neural network. In: Proceedings of ICEAE 2009

Chapter 30
Design, Development and Evaluation of Longitudinal Autopilot for An Unmanned Aerial Vehicle Using X-Plane/Simulink

A. Kaviyarasu, P. Sivaprakash and K. Senthilkumar

Abstract Presently there is a vast interest in unmanned aerial vehicle (UAV) development given its civilian and military applications. One of the main UAV components is autopilot system. Its development invariably demands several lab simulations and field tests. Generally after an UAV crash few parts remain unused. Thus, before embedding an autopilot system, it has to be exhaustively lab tested. With educational and research purposes in autopilot control systems development area, a test platform is herein proposed. It employs Matlab/Simulink to run the autopilot controller under test. The autopilot controller designed on Matlab/Simulink is tested by controlling an aircraft on X-Plane. The inputs given to the aircraft flight control surfaces in the X-Plane are simultaneously sent to the microcontroller which translates these commands into effective servo movement control. Through this platform, designed autopilot systems can be applied into models similar to real aircraft minimizing risks and increasing flexibility for design changes. As study case, tests results from a pitch attitude autopilot system are presented.

Keywords Autopilot · X-plane

A. Kaviyarasu (✉) · P. Sivaprakash · K. Senthilkumar
Department of Aerospace Engineering MIT, Anna University, Chennai, India
e-mail: isrokavi@gmail.com

P. Sivaprakash
e-mail: psivaprak07@gmail.com

R. Malathi and J. Krishnan (eds.), *Recent Advancements in System Modelling Applications*, Lecture Notes in Electrical Engineering 188,
DOI: 10.1007/978-81-322-1035-1_30, © Springer India 2013

30.1 Introduction

The advancement of modern aircraft design requires the development of many technologies such as aerodynamics, structures, materials, propulsion and flight controls. Currently, the aircraft design strongly relies on automatic control system to monitor and control many of the aircraft subsystems. Those control systems can also provide artificial stability to improve the flying qualities of an aircraft.

To stabilize the aircraft after being disturbed from its wing-level equilibrium flight attitude, the autopilot system was primarily conceived [1]. The modern autopilot systems are much more complex and are essential to flight aiding crew in navigation, flight management, stability augmentation and landing operations.

With the development of modern aircraft new problems arose due to their dynamic stability. For instance, some oscillations are inherent to the aircraft can be adequately damped or controlled by the pilot if the period is around 10 s or more. On the other hand, if the period of oscillation is 4 s or less, the pilot's reaction time is very short. Thus, such oscillations have to be artificially well damped [2].

The oscillations named "Short period" in pitch for longitudinal motion, and "Dutch roll" for lateral-directional motion fall into the category of a 4 s oscillation. In more conventional aircraft that kind of oscillation may be damped by handling their constructive characteristics. However, for modern jets artificial damping is required. It is provided by an automatic system [2].

The above concepts are also applicable to unmanned aerial vehicle (UAVs) whose development are presently given a vast interest in increasing possibilities of its civilian and military applications. Some of the applications are reconnaissance, surveillance, search and rescue, remote sensing, traffic monitoring missions, etc. [3].

One of the main components of UAVs is the autopilot system. An autopilot is a mechanical, electrical, or hydraulic system used to guide vehicle without any assistance from human being. The autopilot systems development invariably demands several lab simulations and field tests. The later ones are of high risk particularly when conducted on UAVs. Generally after an UAV crash few parts remain usable. Thus, before embedding an autopilot system, it has to be exhaustively lab tested.

Based on this necessity, with educational and research purpose in autopilot control systems development area, a test platform is here in proposed and various control laws are verified before making it to the final integration process.

30.2 Objective

This article intends to describe a test platform developed to aid in autopilot design process.

The test platform herein proposed provides an environment where thc designed autopilot system can be applied into models similar to real aircraft. Several flight situations can be simulated. Parameters of flight as well as aircraft responses can be monitored and easily analyzed. These features increase the flexibility to implement changes and immediately check out the results.

Basically, this test platform employs the Matlab/Simulink to run the autopilot controller to be tested, the flight simulator X-plane with the aircraft to be commanded, a microcontroller to command model aircraft flight control surfaces and a servo to drive these control surfaces. All this resources are interconnected through data buses in order to exchange information.

As study case, this article presents the results obtained from tests conducted over an autopilot control system designed for longitudinal movement, specifically for pitch mode control. This system was conceived for Cessna aircraft.

30.3 Aircraft Translational Motions

An airplane in free flight has three translational motions (vertical, horizontal and transverse), three rotational motions (pitch, yaw and roll) and numerous elastic degrees of freedom.

In order to reduce the complexity of this mathematical modeling problem, some simplifying assumptions may be applied. First, it is assumed that the aircraft motion consists of small deviations from its equilibrium flight condition. Second, it is assumed that the motion of the airplane can be analyzed by separating the equations into two groups [2]:

- Longitudinal Equations: composed by X-force, Z-force and pitching moment equations.
- Lateral Equations: composed by Y-force, yawing and rolling moment's equations.

Figure 30.1 shows the above mentioned aircraft reference axes, forces and moments.

30.3.1 Longitudinal: Directional Motion

The longitudinal-directional motion of an aircraft disturbed from its equilibrium state is pitching.

The longitudinal-directional motion, specifically the pitch motion, is the movement intended to be simulated on the proposed test platform as demonstration example. To simplify the studied case, it was chosen to address specifically the pitch motion in this work. Even though this simplifies the dynamic model equations, the platform functionality and characteristics, are very well demonstrated.

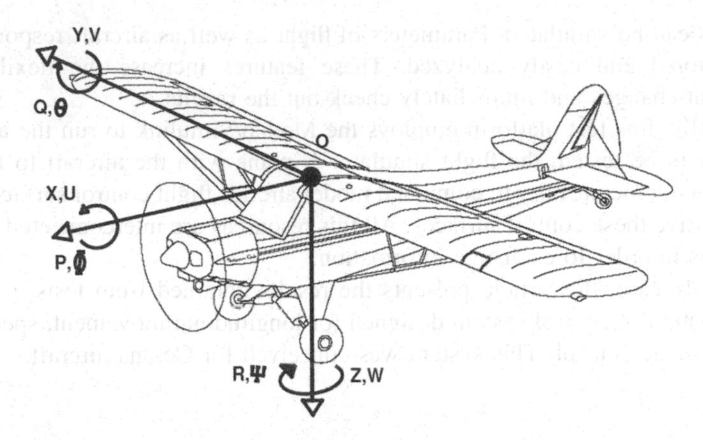

Fig. 30.1 Aircraft axes, forces and moments

30.3.2 Pitch Motion Closed-Loop Control System

The aircraft pitch motion can be controlled by an autopilot system block diagram as shown in the Fig. 30.2.

Through this simplified system, the pitch angle (θ) can be controlled by a reference angle applied as an input. Primarily, the pitch attitude autopilot is designed to maintain the aircraft with leveled wings, or with $\theta = 0$. Usually, this simplified system does not fulfill the control design performance requirements in relation to damping ratio, overshoot and undamped natural frequency.

So that, in the case study herein presented, a more efficient control system is employed. It is shown in Fig. 30.3.

In this case, an accurate aircraft pitch to elevator transfer function for Cessna 172SP aircraft is employed in the pitch autopilot system. This transfer function was calculated based on aircraft physical characteristics and stability derivatives as detailed in [4]. Moreover, a typical elevator actuator transfer function is employed.

The pitch attitude autopilot system gains *KG, Ka* and *KRG* are calculated through the root locus method considering the above rationale.

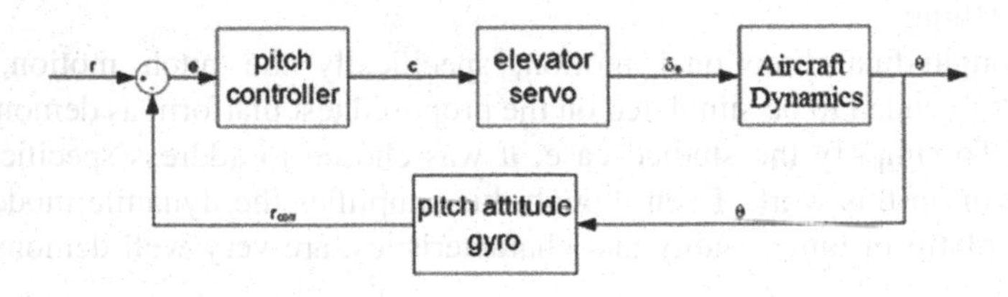

Fig. 30.2 A pitch attitude autopilot system

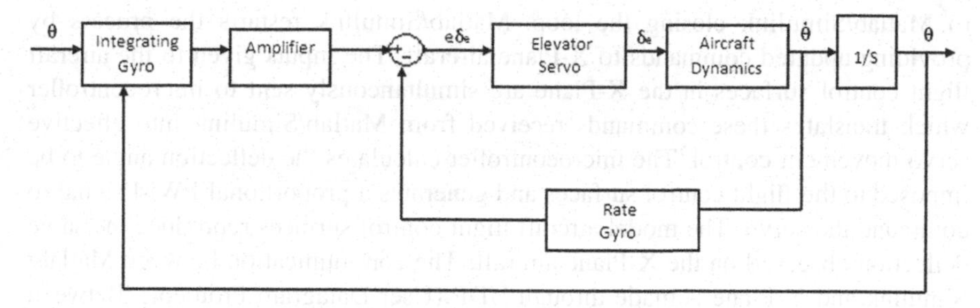

Fig. 30.3 Pitch attitude autopilot system

30.4 The Test Platform

The main components of the test platforms are as follows.

- Matlab/Simulink containing the autopilot control system.
- Flight Simulator X-Plane containing the aircraft model is to be controlled.
- Microcontroller to command the aircraft flight control surfaces.

The test platform concept is based on the block diagram presented in Fig. 30.4. In the test platform, the block "Controller" is replaced by the designed autopilot system model. This model runs into the Matlab/Simulink environment.

Similarly, the block "Aircraft Dynamics" is replaced by the flight simulator X-Plane with the aircraft model that is to be controlled.

Thus, the basic principle of the test platform consists of establishing the communication between Matlab/Simulink, X-Plane, microcontroller and servo in the following mode: The parameters calculated by the autopilot control system are sent to X-Plane in order to command the aircraft flight control surfaces. The X-Plane calculates the new aircraft attitude according to the inputs received from Matlab/Simulink. The X-Plane sends those new aircraft attitude parameters back

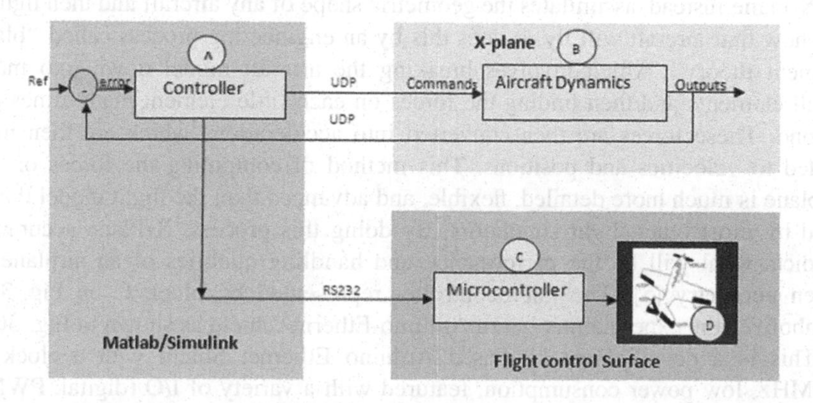

Fig. 30.4 Test platform block diagram

to Matlab/Simulink closing the loop. Matlab/Simulink restarts the process by providing updated commands to X-Plane aircraft. The inputs given to the aircraft flight control surfaces in the X-Plane are simultaneously sent to microcontroller which translates these commands received from Matlab/Simulink into effective servo movement control. The microcontroller calculates the deflection angle to be imposed to the flight control surfaces and generates a proportional PWM signal to command the servo. The model aircraft flight control surfaces reproduce the same deflection observed on the X-Plane aircraft. The communication between Matlab/Simulink and X-Plane is made through UDP (User Datagram Protocol). Between Matlab/Simulink and microcontroller a RS-232 serial communication is used. Block diagram shown by Fig. 30.4 summarizes the platform concept.

The flight simulator X-Plane provides very accurate aircraft models and has a very important feature. That is the possibility to exchange data with external systems [5]. Giving its realistic simulations capability, X-Plane is also federal aviation administration (FAA) approved for pilot training. The aircraft models simulated in X-Plane are built based on their real physical dimensions, power and weight among other characteristics. X-Plane is not considered a game, but can be categorized as an engineering tool that can be used to predict the flying qualities of fixed and rotary wing aircraft [5].

Most other flight simulators use stability derivatives method to compute how an airplane flies. This technique involves forcing the aircraft nose to return to a centered position along the flight path with certain acceleration for each degree of offset from straight-ahead flight of the airplane. This is too simplistic to be used across the entire flight envelope of the airplane. Stability derivatives will not normally take into proper account the asymmetric effects of engine failures, the chaotic effects of turbulence, stalls, spins and the myriad of dynamic effects that airplane flight generates. In other words, the commonly used stability derivatives are gross over simplifications of how an airplane flies. In summary, those simulators can not predict how the airplane will fly. Basically, the airplane designer teaches the simulator how the airplane should fly, and the simulator reproduces that information right back to the user [5].

X-Plane instead, assimilates the geometric shape of any aircraft and then figures out how that aircraft will fly. It does this by an engineering process called "blade element theory", which involves breaking the aircraft model down into many small elements and then finding the forces on each little element many times per second. These forces are then converted into accelerations which are then integrated to velocities and positions. This method of computing the forces on the airplane is much more detailed, flexible, and advanced than the flight model that is used by most other flight simulators. By doing this process, X-Plane accurately predicts what will be the performance and handling qualities of an airplane of given geometry [5]. The microcontroller represented by block C on Fig. 30.4 symbolizes the experimenter board Arduino-Ethernet shield as shown in Fig. 30.5.

This is a development kit based Arduino Ethernet Shield with a clock of 16 MHz, low power consumption, featured with a variety of I/O (digital, PWM), besides RAM and flash memory space. The choice for this device is justified by its

Fig. 30.5 Microcontroller unit

Fig. 30.6 Model aircraft digital servo

inbuilt Ethernet microcontroller chip, serial port (communicate between I/O devices in the external world) as well as easy integration with digital servos through PWM I/O ports. On block D of Fig. 30.4 it represents the digital servo that commands model aircraft flight control surfaces. A typical servo is shown in Fig. 30.6.

It receives PWM signals from the microcontroller unit and converts them into proportional axis movement.

30.5 X-Plane Data Interface

Flight Simulator X-Plane has an important feature that is essential to this test platform development. Its capacity of sending and receiving data to and from X-plane software and MATLAB. One way to implement this communication is by employing protocol UDP. Data packets are sent and received through the computer Ethernet port. Figure 30.4 shows UDP protocol being used to establish the communication between Matlab/Simulink (Controller) and X-Plane (Aircraft Dynamics).

UDP uses a simple transmission model without implicit hand-shaking dialogues for guaranteeing reliability, ordering, or data integrity. Thus, UDP provides an unreliable service and data packets may arrive out of order, appear duplicated or go missing without any notice. Error checking and correction are considered either not necessary or performed in the application. This way, UDP avoids the overhead of such processing at the network interface level being extremely fast [6]. Time-sensitive applications such as this test platform often use UDP. Dropping packets

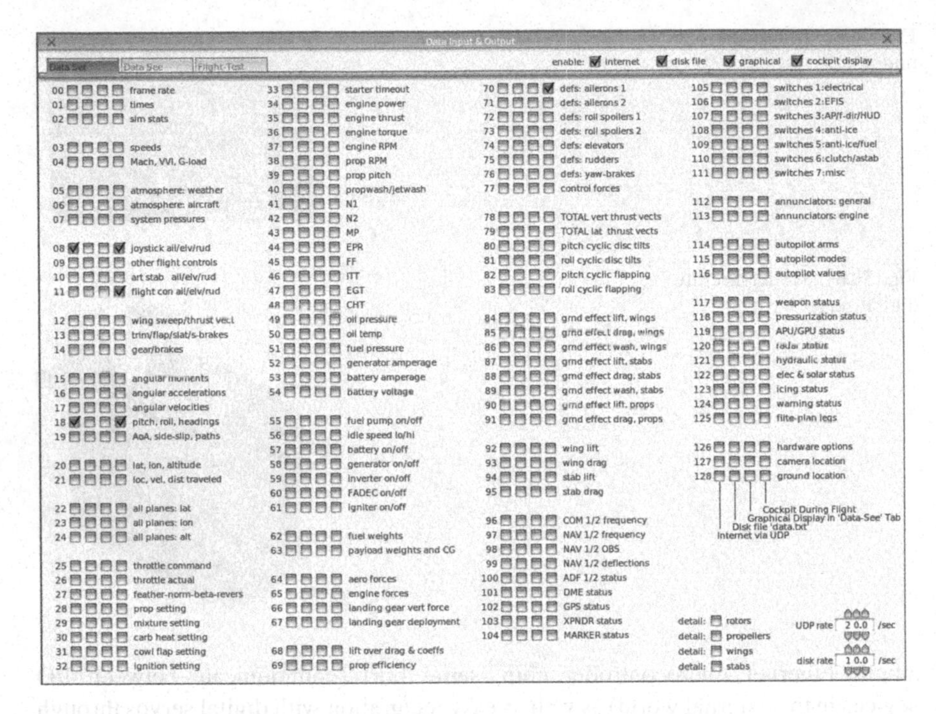

Fig. 30.7 X-plane parameters user interface

is preferable to waiting for delayed packets, which it is not an option in real-time systems.

Hence UDP speed constitutes a key point in the test platform, once the communication between Matlab/Simulink and X-Plane must be fast enough to synchronize commands, data processing and system responses.

X-Plane is able to send or receive up to 99.9 data packets per second via UDP. Each data package may be configured to carry aircraft parameters data that are selected on X-Plane Data Input and Output interface.

For example, in the case of lateral motion, parameters such as roll, yaw, pitch, altitude and speed shall be selected for transmission. Each parameter receives a numeric label for proper identification.

Figure 30.7 shows the interface that serves to select data for input and output from X-Plane.

Besides parameters selection for UDP data transmission or receiving, it also serves for selecting parameters to be shown during aircraft flight simulation, file recording and graphs construction.

Once the parameters are selected on X-Plane, through UDP it is possible to receive them at Matlab/Simulink environment and after processing, to send commands back to X-Plane aircraft. In the case of pitch attitude autopilot proposed example, Matlab/Simulink generates the pitch angle reference and control signal sending them to X-Plane. In its turn, X-Plane commands the simulated aircraft

1	2	3	4	5	6	7	8	9	10	11	12	13	14	15	16	17		38	39	40	41	Bytes
"D"	"A"	"T"	"A"	I	L_1	L_2	L_3	L_4	B_{11}	B_{12}	B_{13}	B_{14}	B_{21}	B_{22}	B_{23}	B_{24}	...	B_{81}	B_{82}	B_{83}	B_{84}	

Fig. 30.8 X-plane data package pattern

according to the inputs received. The new aircraft pitch angle position is sent from X-Plane back to Matlab/Simulink, restarting the process.

The X-Plane data package follows a pattern shown on Fig. 30.8.

Basically it is composed by a sequence of bytes that need to be properly interpreted. X-Plane uses what are known as single precision floating point variables for just about everything sent over the network. This means that the numbers can be stored using four bytes. The first four bytes of the packet shown on Fig. 30.7 represents the characters "DATA" used to indicate that this is a data packet. The fifth byte is an internal code (I). The next four bytes represent the parameter label (L1, L2, L3, and L4). Following there are eight sets of four bytes (B11, B12, B13, B14 to B81, B82, B83, B84) representing the data itself in single precision floating point. Taking each set of four bytes, the first bit is the sign bit. It tells whether the number is positive or negative. The next eight bits are the biased exponent. The remaining 23 bits represents the mantissa. The next seven sets of four bytes completes the data. So, the Matlab/Simulink model has to decode this data packet accordingly. It has to properly encode the data to be sent to X-Plane. Most of the blocks of the test platform implemented on Matlab/Simulink environment are dedicated to decode and encode data packets from and to X-Plane.

30.6 Implementation

Figure 30.9 illustrates the test platform implemented.

In one computer Matlab/Simulink runs a model containing the autopilot control system and other blocks responsible for UDP data packets decode and encode process as well as serial communication with microcontroller. In the other computer X-Plane simulates the aircraft to be controlled. Both computers communicate to each other through their Ethernet port using UDP protocol. Through serial bus the microcontroller receives the same commands Matlab/Simulink sent to X-Plane. It decodes this data converting them into PWM signal. An interface amplifies this signal to apply it to the servo that commands the model aircraft flight control surfaces.

30.7 Application Example: Test Results

In order to demonstrate how the test platform could be applied in laboratory classes and also help autopilot systems development, the roll attitude autopilot system was designed and integrated into the platform. The dynamic system used as

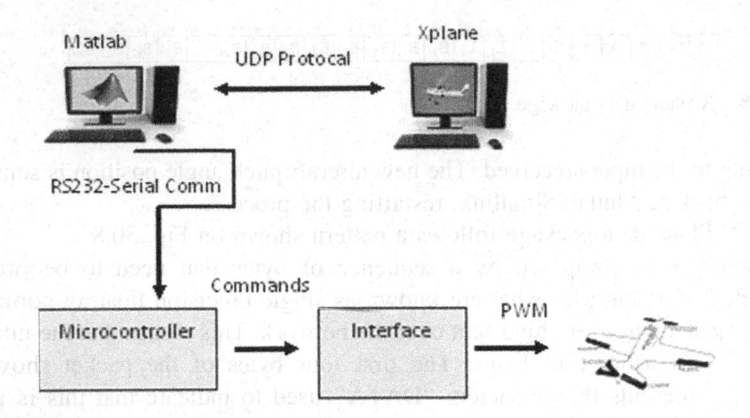

Fig. 30.9 Autopilot test platform implementation

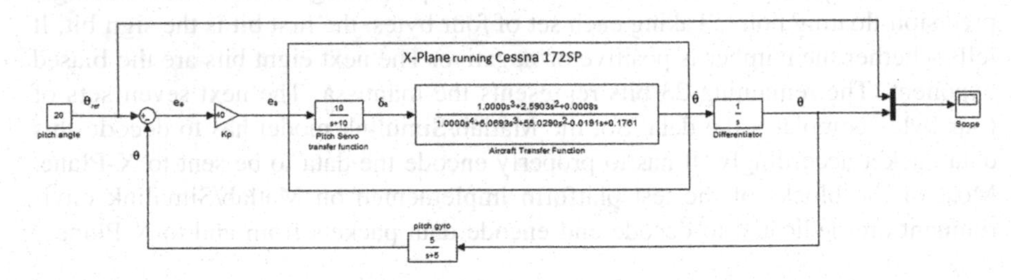

Fig. 30.10 Pitch attitude autopilot system for Cessna 172SP

the case to be studied and tested using the proposed platform can be represented by Fig. 30.3. The gains *KG, Ka* and *KRG* of that system were calculated through the root locus method using the systems dynamics as in (2) and (3), which are the Cessna pitch to elevator and typical Pitch servo transfer functions.

Pitch Dynamics:

$$\frac{N^{\theta}_{\delta_e}}{\Delta_{long}} = \frac{1.0000s^3 + 2.5903s^2 + 0.0008s}{1.0000s^4 + 6.0693s^3 + 55.0290s^2 - 0.0191s + 0.1761}$$

Pitch Servo:

$$S(s) = \frac{10}{s + 10}$$

The designed system result is shown on Fig. 30.10. It is also shown which blocks will be simulated by X-Plane and which ones will be integrated into the test platform model at Matlab/Simulink.

The Pitch to elevator and servo transfer functions are simulated by the Cessna 172 SP at X-Plane software. The other autopilot blocks are distributed into the test

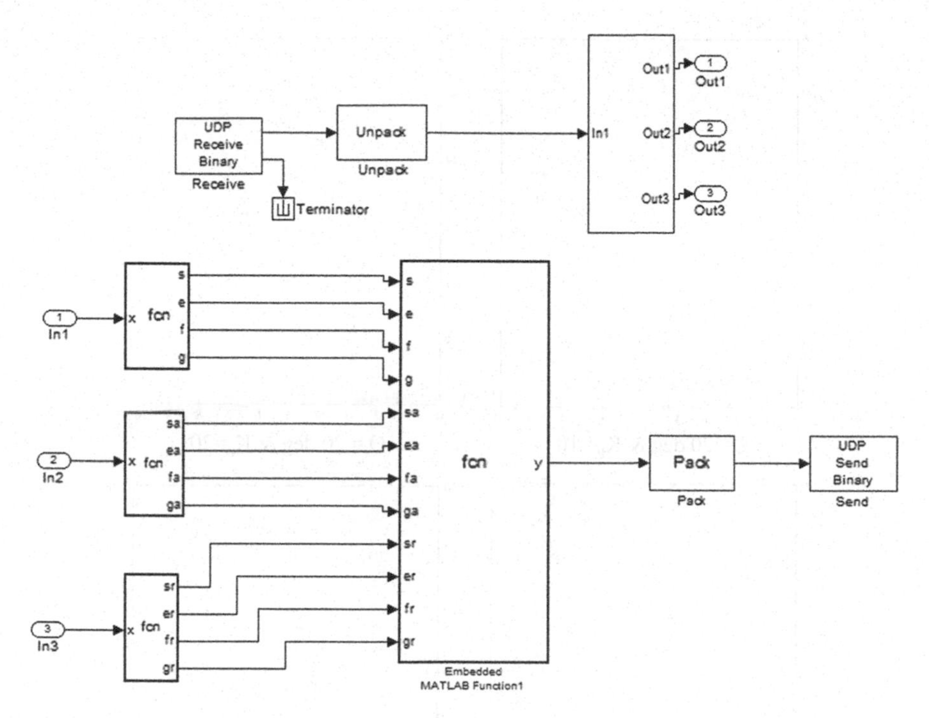

Fig. 30.11 Autopilot test platform model at matlab/simulink

platform model. Figure 30.11 shows the entire test platform model implemented at Matlab/Simulink environment.

In a typical lab class, the student should find the test platform ready to be used with the communication interfaces appropriately configured. The gains *KG, Ka* and *KRG* shall then be fed into the controller implemented at Matlab/Simulink. The gains calculation may be performed at the lab or can be previously done, being the lab classes preferably used for test at the platform.

To initialize the simulation, X-Plane is loaded with aircraft Cessna 172 SP in a cruise flight at 40,000 ft altitude. As soon as the test platform model shown on Fig. 30.11 runs, the designed autopilot system takes the aircraft control. In principle, for any pitch angle applied as reference into the autopilot system, the aircraft at X-Plane shall respond following that reference.

The reference signals applied as system input, the system response and also the commands sent to the aircraft flight control surface actuators can be monitored through real time graphs at Matlab/Simulink. In the proposed study case, the input is the reference pitch angle, the system response is the new aircraft Pitch angle attitude and the commands are the deflections to be imposed to the elevators. Figure 30.12 shows the results for the proposed pitch attitude autopilot system implemented at the test platform.

The graph on Fig. 30.12 presents the autopilot Pitch angle reference 20° applied into the system. During simulation it is possible to see the Cessna 172SP make a

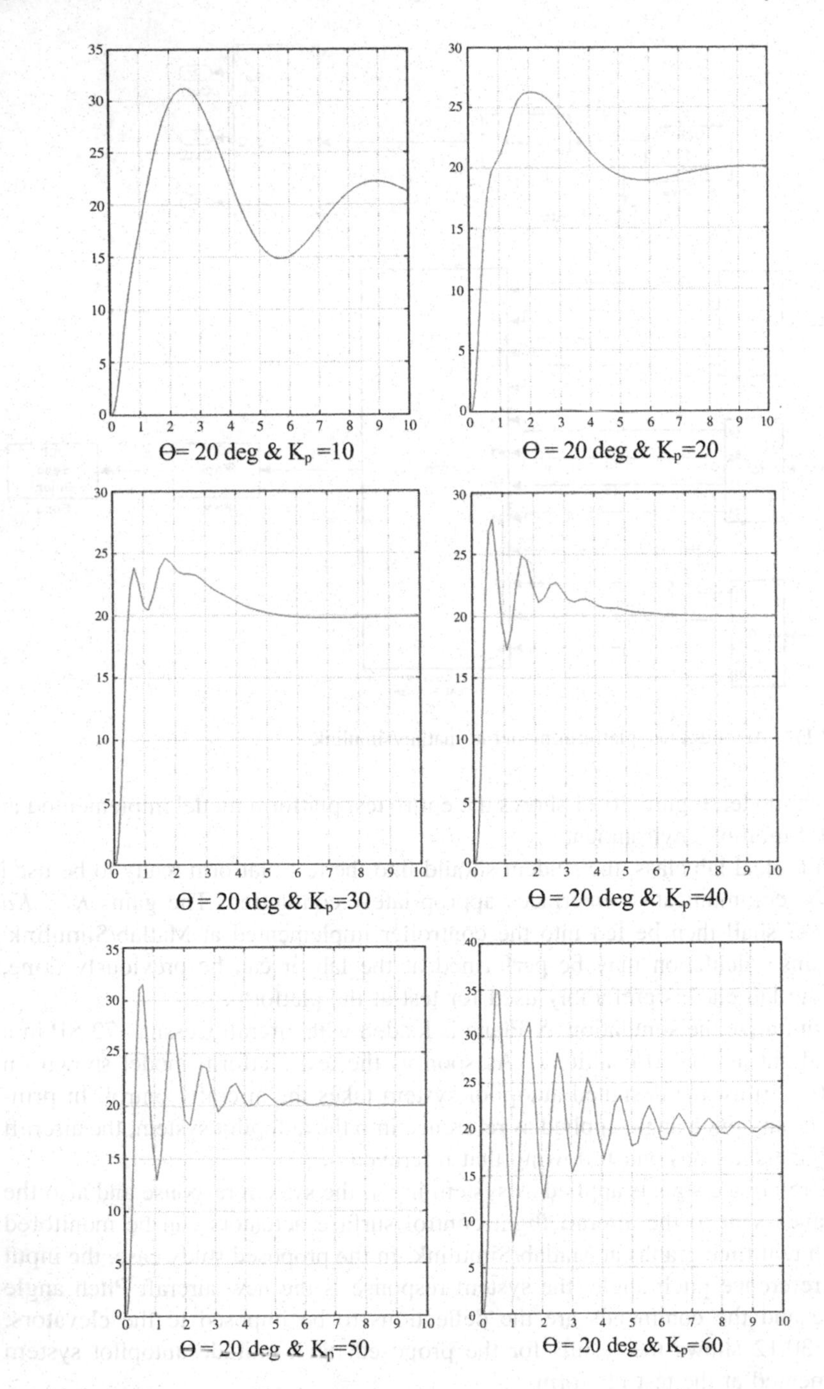

Fig. 30.12 Pitch reference and response

20° pitch angle attitude and performing a slight pitch up. In parallel it is possible to verify the servo movements in response to these commands in the aircraft model. Interesting to notice that even with the aircraft leveled and keeping a certain pitch attitude there are small commands to the flight control surfaces. This is something that only was possible to observe due to the realism provided by X-Plane simulation which introduces small perturbations to aircraft flight reproducing a real atmosphere with wind, turbulences, etc. One exercise that can also be performed on the designed autopilot system is to vary the loop gains and check out the aircraft responses. This can be used to properly tune the loop gains to optimize system performance.

30.8 Conclusion

The development of this test platform resulted in a valuable tool for the students and professionals to aid autopilot system study and design. It allows monitoring the aircraft responses for a designed autopilot system with high degree of realism. It is possible to change control system parameters very easily and check the results out in a friendly environment. This possibility easies the design task as well minimizes risks of embedding the system for field tests.

The test platform also permits the study of longitudinal and lateral-directional movement of different fixed and rotary wing unmanned aerial vehicle platform for which the autopilot system is being designed. A possible extension of this study is to test the platform autopilot design with the Hardware in loop system (HILS) to simulate the UAV in the Real-time.

References

1. Nelson RC (1998) Flight stability and automatic control, 2nd edn. McGraw-Hill, New York
2. Blacklock JH (1991) Automatic control of aircraft and missiles. Wiley Inter science Publication, New York
3. Suwandi Ahmad A, Sembiring J (2007) Hardware in the loop simulation for simple low cost autonomous UAV (unmanned aerial vehicle) autopilot system research and development. Institute Technology Bandung, Indonesia
4. Laminar research (2009) X-plane description, X-plane manual
5. AeroSim-Aeronautical Simulation block set (2005) Version 1.1, User's Guide, unmanned dynamics, USA
6. Roskam J (1982) Airplane flight dynamics and automatic flight controls, part I roskam aviation and engineering corporation

Chapter 31
Comparison of State Estimation Algorithms on the Tennessee Eastman Process

J. Vinoth Upendra and J. Prakash

Abstract Design and Implementation of State estimators for nonlinear, large dimensional system have gain widespread attention in the field of advanced process control. In this paper an Extended Kalman Filter (EKF) and Unscented Kalman Filter (UKF) based state estimation schemes have been applied to estimate the state variables of Tennessee Eastman Process. The efficacy of the derivative based and derivative free state estimation schemes on the Tennessee Eastman Process have been assessed through simulation studies.

Keywords Tennessee Eastman process · EKF · UT · UKF

31.1 Introduction

State estimation involves computing all state variables of the dynamic system using available measurements and this becomes necessary for model predictive control and other advanced process control schemes. Solutions for the state estimation of linear system under Gaussian noise are provided by a conventional Kalman Filter [1]. But most of the dynamic systems are not linear in nature, so the conventional Kalman filter cannot be used for estimation problem. In such instants, a natural extension of Kalman filter namely extended Kalman filter (EKF) [1, 2] is

---

J. Vinoth Upendra (✉) · J. Prakash
Department of Instrumentation Engineering, Madras Institute of Technology Campus, Anna University, Chennai, India
e-mail: jvinothupendra@yahoo.co.in

J. Prakash
e-mail: prakaiit@rediffmail.com

R. Malathi and J. Krishnan (eds.), *Recent Advancements in System Modelling Applications*, Lecture Notes in Electrical Engineering 188,
DOI: 10.1007/978-81-322-1035-1_31,

used for estimation of nonlinear dynamic system by linearization of nonlinear system dynamics around the mean of a Gaussian distribution.

Although EKF seems to be a good estimator for wide industrial problems, it relies on linearization to propagate the mean and covariance of the state by jacobian computation at every time instants. To avoid the computation of Jacobian, derivative free estimators such as unscented Kalman Filters (UKF) proposed by Julier and Uhlmann [5] has been widely used. In UKF, sample points are chosen deterministically and are propagated through the nonlinear function, and the mean of the transformed variable is estimated from the transformed sample points and their associated weights.

In this work, a large dimensional Tennessee Eastman process is chosen to analyze the performance of EKF and UKF state estimation schemes and the performance of both the filters has been assessed based on performance measure namely Root Mean Squared Error (RMSE) criteria.

The organization of the paper is as follows. Section 31.1 discusses different nonlinear state estimation algorithm formulation. Description of the Tennessee Eastman process is presented in Sect. 31.2. Simulation results are presented in Sect. 31.3 followed by conclusions drawn from the simulation studies are presented in Sect. 31.4.

31.2 State Estimation Algorithms

31.2.1 Extended Kalman Filter

The Extended Kalman filter (EKF) is the most widely applied state estimation algorithm for nonlinear systems.

Consider the nonlinear system and measurement equations:

$$x_k = f(x_{k-1}, u_{k-1}) + w_{k-1} \tag{31.1}$$

$$y_k = h(x_k) + v_k \tag{31.2}$$

where $x_k \in R^n$ is the state, $y_k \in R^m$ is the measurement, f is the state evolution function and, h is the measurement model function. The process noise and measurement noise are denoted by w_k and v_k respectively. Both are assumed to be white, zero mean and, uncorrelated with covariance matrices Q_k and R_k respectively which is represented as:

$$w_k \sim N(0, Q_k) v_k \sim N(0, R_k)$$
$$E\left[w_k w_k^T\right] = Q_k E\left[v_k v_k^T\right] = R_k$$

Also, two random vectors w_k and v_k are uncorrelated.

$E\left[w_k v_k^T\right] = 0$; For all k and j.

The EKF equations are expressed in two steps, prediction step, where information about the process is used and update step, where measurements are used to update the states

Prediction Step:

$$\hat{x}_k^- = f\left(\hat{x}_{k-1}^+, u_{k-1}\right) \tag{31.3}$$

$$P_k^- = F_{k-1}P_{k-1}^+F_{k-1}^T + Q_{k-1} \tag{31.4}$$

Update Step:

$$S_k = H_kP_k^-H_k^T + R_k \tag{31.5}$$

$$K_k = P_k^-H_k^TS_k^{-1} \tag{31.6}$$

$$\hat{x}_k^+ = \hat{x}_k^- + K_k\left[y_k - h\left(\hat{x}_k^-\right)\right] \tag{31.7}$$

$$P_k^+ = (I - K_kH_k)P_k^{-1} \tag{31.8}$$

where the error covariance $P_k \in \Re^{n \times n}$ is defined as $P_k = E\left[e_ke_k^T\right]$ and here $e_k = x_k - \hat{x}_k$.

The EKF algorithm uses a Taylor series based transformation to approximate the nonlinear function. However, for large dimension systems jacobian computation of nonlinear system functions and measurement functions become computationally intensive and error prone.

Hence, in order to eliminate such approximations and jacobian computations, derivative free methods such as Unscented Kalman filters [1, 3] has been developed.

31.2.2 Unscented Kalman Filter

The state distribution is represented by Gaussian random variables which are formed using Unscented Transform (UT).

UT is a method for calculating statistics of a random variable, which undergoes a nonlinear transformation. In UT, a minimal set of sigma points are carefully chosen which capture the desired moments namely mean and covariance of original distribution of x exactly and, when propagated through the true nonlinear system, capture the posterior mean and covariance accurately until the third order for any nonlinearity.

The nonlinear system and measurement equations are shown in Eqs. (31.1) and (31.2). The Unscented transformation is carried out as follows:

Here, n-dimensional random variable x and mean $\bar{x}$ and covariance P is approximated by 2n+1 sigma points. A set of 2n+1 sigma points are first computed from the columns of the matrix $\sqrt{(n + \lambda)P}$.

$$
\begin{aligned}
x^{(0)} &= \bar{x} \\
x^{(i)} &= \bar{x} + \left(\sqrt{(n+\lambda)P}\right), \quad i = 1,\ldots,n \\
x^{(i)} &= \bar{x} - \left(\sqrt{(n+\lambda)P}\right), \quad i = n+1,\ldots,2n
\end{aligned}
\tag{31.9}
$$

and the associated weighting coefficients are given by:

$$
\begin{aligned}
W_m^{(0)} &= \frac{\lambda}{(n+\lambda)} \\
W_c^{(0)} &= \frac{\lambda}{(n+\lambda)} + (1 - \alpha^2 + \beta) \\
W_m^{(i)} = W_c^{(i)} &= \frac{1}{2(n+\lambda)} \quad ,i = 1,\ldots,2n
\end{aligned}
\tag{31.10}
$$

where, $\lambda = \alpha^2(n+\kappa) - n$

α Determines the spread of the sigma points

N No. of states in the process

κ Secondary scaling factor.

Now, each sigma points are propagated through the process model given by

$$\hat{x}_k^{(i)} = f\left[\hat{x}_{k-1}^{(i)}, u_k\right] \tag{31.11}$$

The predicted mean is now given by:

$$\hat{x}_k^- = \sum_{i=0}^{2n} W_m^i \hat{x}_k^{(i)} \tag{31.12}$$

Priori error covariance is computed as follows:

$$P_k^- = \sum_{i=0}^{2n} W_m^i (\hat{x}_k^{(i)} - \hat{x}_k^-)(\hat{x}_k^{(i)} - \hat{x}_k^-)^T \tag{31.13}$$

After the prediction step is completed each prediction points are passed through the measurement model given by:

$$\hat{y}_k^{(i)} = h\left(\hat{x}_k^{(i)}\right) \tag{31.14}$$

Now the predicted measurement is calculated as follows:

$$\hat{y}_k = \sum_{i=0}^{2n} W_m^{(i)} \hat{y}_k^{(i)} \tag{31.15}$$

Covariance of the predicted measurement is now estimated as follows:

$$P_y = \sum_{i=0}^{2n} W_m^i \left(\hat{y}_k^{(i)} - \hat{y}_k\right)\left(\hat{y}_k^{(i)} - \hat{y}_k\right)^T + R_k \tag{31.16}$$

Cross-covariance of the state and measurement is given by:

$$P_{xy} = \sum_{i=0}^{2n} W_m^i \left(\hat{x}_k^{(i)} - \hat{x}_k^-\right)\left(\hat{y}_k^{(i)} - \hat{y}_k\right)^T \tag{31.17}$$

Now, the filter gain, measurement update of the state estimate and covariance is computed as follows:

$$K_k = P_{xy} P_y^- \tag{31.18}$$

$$\hat{x}_k^+ = \hat{x}_k^- + K_k(y_k - \hat{y}_k) \tag{31.19}$$

$$P_k^+ = P_k^- - K_k P_y K_k^T \tag{31.20}$$

31.3 Process Description

31.3.1 Process Overview

The Tennessee Eastman Process, first introduced by Downs and Vogel [4], a challenging process in Chemical Engineering Community is suited for wide variety of studies such as plant-wide control, model predictive control, Estimation etc. It has been studied by many authors in the field of model predictive control [5], decentralized control schemes [6, 7], plant wide control [8] and as an optimization problem.

This process produces two products (G and H) from four reactants (A, C, D, and E). A further inert trace component (B) and one byproduct (F) is also present.

All reactions are irreversible and exothermic with rates that depend on temperature and on the reactor gas phase concentration of the reactants.

$$\begin{array}{ll} A(g) + C(g) + D(g) \rightarrow G(liq) & \text{(Product 1)} \\ A(g) + C(g) + E(g) \rightarrow H(liq) & \text{(Product 2)} \\ A(g) + E(g) \rightarrow F(liq) & \text{(Byproduct)} \\ 3D(g) \rightarrow 2F(liq) & \text{(Byproduct)} \end{array}$$

All feed streams and recycle streams in gaseous phase are fed into the reactor. In the reactor every 3 mol of gas is converted to 1 mol of liquid and there is a hold up of G and H components in the reactor but no liquid effluent. Product and unconverted reactant leave the reactor as vapor. Liquefaction of vapor takes place in the condenser and this vapor–liquid mixture is separated at Flash drum.

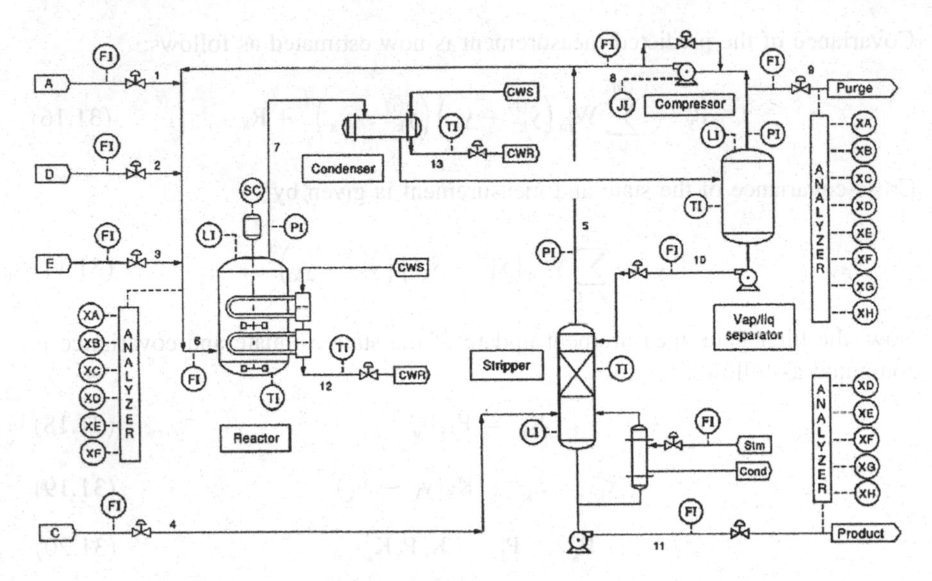

Fig. 31.1 Flow sheet of the Tennessee Eastman process

Vapor and liquid get separated at flash drum and liquid is withdrawn from the bottom of flash drum and sent to the stripper column. Non-condensed components from the top of the Flash drum are sent back to the reactor. Stripper column is basically used to obtain the desired purity in the product on the basis of mass transfer. To control the accumulation of inert product B, purge is used. Purge is the only outlet for B, which is non-condensable like A and C. The flow sheet of the process is shown in Fig. 31.1.

The mathematical model is based on a number of typical assumptions: all vapors behave as ideal gases, the vapor/liquid equilibrium follows Raoult's law with the vapor pressure calculated using the Antoine equation, and all vessels are assumed perfectly mixed.

31.3.2 Process Model and Estimation Details

The process model equations for the TE process consists of energy balance in addition to that of mass balance equations as discussed in Jockenhovel et al. [9], where the model equations and initial state values for the mixing zone, the reactor, the separator and the stripper is discussed in detail.

The model contains 30 ordinary differential equations and 149 explicit algebraic equations. It is modeled of 30 dependent variables and 123 model parameters. The state vector contains 30 state variables as a result of integration of 30 ODE's and there are 11 manipulated input variables with eight stream flows, two cooling water flow rates and one steam flow rate.

These 30 state variables are: temperature in the mixing zone, reactor temperature, separator temperature, stripper temperature, and the molar holdup of all the reactants in the mixing zone, reactor, separator and stripper. Except temperature all other states are difficult to measure directly but they play an important role in monitoring, control and process model accuracy. Hence, the purpose of estimation of these states is justified for the above mentioned reasons.

$$Xs = [Tm, Tr, Ts, Tstr, Ni,m, Nir, Nis, NGstr, NHstr]; \; i = A, \ldots, H$$

A total of 30 process variables are measured in the process: temperatures in the reactor, separator, and stripper, two cooling water flows, three liquid levels of reactor, separator and stripper, reactor and separator pressure, the reactor feed rate and concentrations of the reactor feed stream, of the purge stream and of the final product stream.

The base case values have been used in the simulation studies. The levels of reactor, separator and stripper are controlled by the input streams A, D and E using a PI controller. Now the variations in these levels are made to perturb the process and estimation is performed. Here few values are maintained within the safe margin of the shutdown limit to operate at different load conditions. For example,

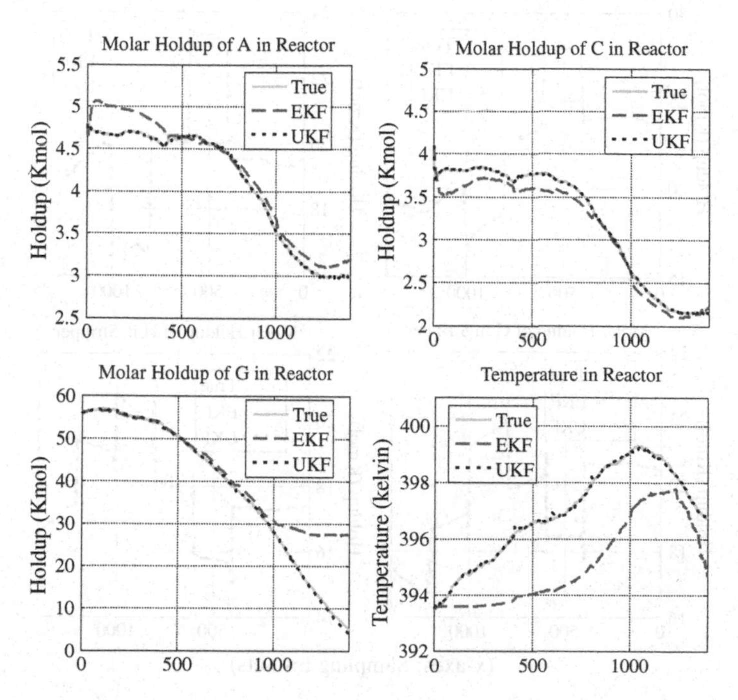

Fig. 31.2 Evolution of true and estimated states (molar holdup of component *A*, *C*, and *G* and *temperature*) in the reactor

the upper limit of the reactor pressure is selected to be 2800 kPa in order to maintain the 200 kPa margin from the shut down limit in spite of its increased operating costs Ricker [10].

The state variables in the update step are updated by using 22 process variables measured from the process at each sampling instants with the same rate and without delay.

31.4 Simulation Results

The performance of EKF and UKF state estimation schemes is quantified based on the root mean squared error (RMSE) performance criterion. This criterion will facilitate a quantitative comparison of the performance of these two state estimators. The state estimation results of EKF and UKF are shown in Figs. 31.2 and 31.3. The graphs show the various true and estimated states of reactor, separator, and stripper. Table 31.1 represents the comparison of RMSE between EKF and UKF for the states illustrated. This data indicates the inability of EKF to estimate

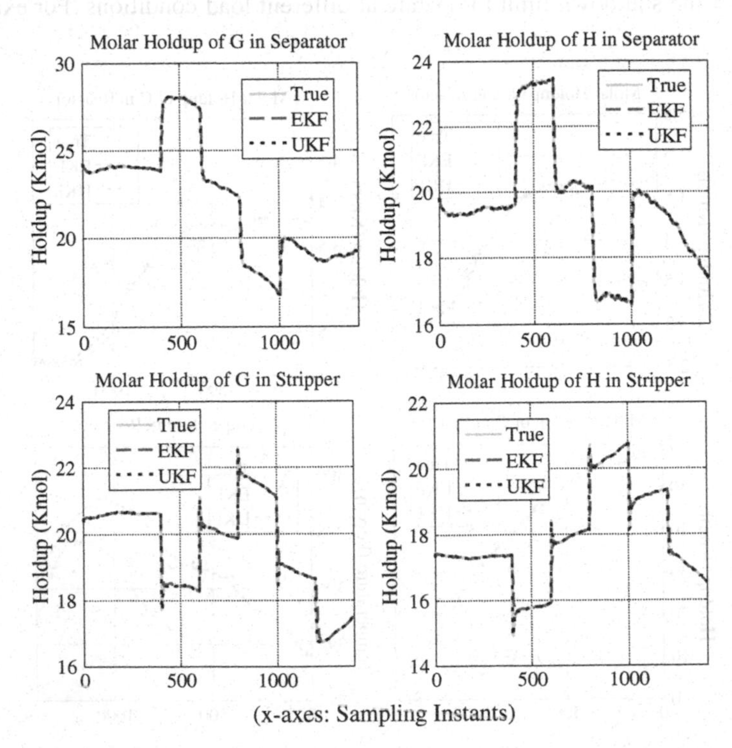

Fig. 31.3 Evolution of true and estimated states (molar holdup of component *G* and *H*) in the separator and stripper

Table 31.1 RMSE values of EKF and UKF for the unmeasured state variable

Unmeasured state variables	EKF	UKF
Molar holdup of A in reactor	4.6241	0.0177
Molar holdup of C in reactor	3.9637	0.0187
Molar holdup of G in reactor	147.0714	3.3318
Temperature in reactor	68.3601	0.0312
Molar holdup of G in separator	0.1665	0.0149
Molar holdup of H in separator	0.7679	0.0027
Molar holdup of G in stripper	0.3766	0.2488
Molar holdup of H in stripper	0.2980	0.1563

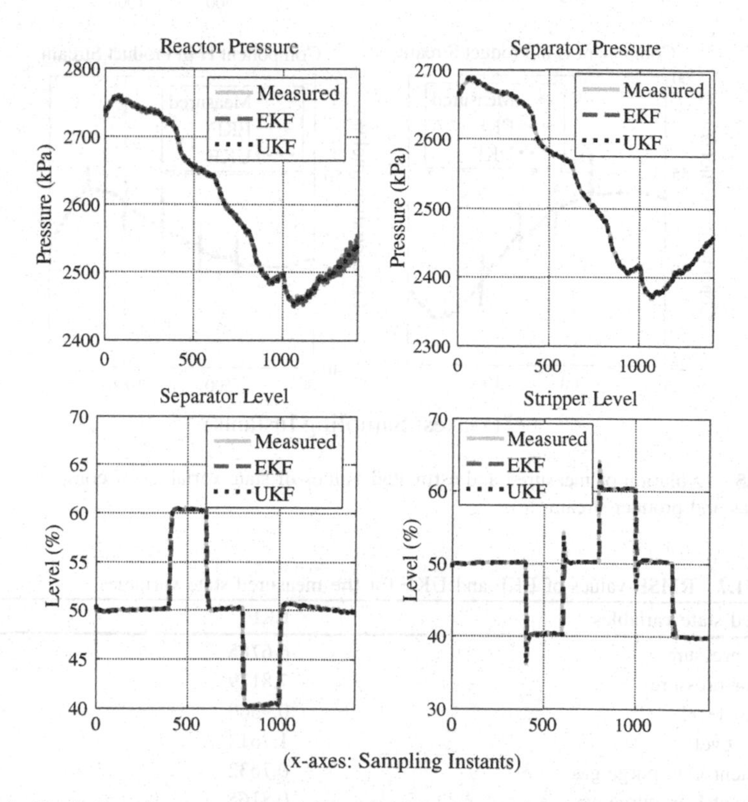

Fig. 31.4 Evolution of measured and estimated values of state variables namely pressure in the reactor, pressure and liquid level in the separator and the stripper

the states of the process accurately. A large deviation between true and estimated values is seen in EKF due to poor linearization approximation.

Figures 31.4 and 31.5 show the evolution of measured and estimated values of the state variables of Tennessee Eastman process.

The RMSE values of EKF and UKF for various unmeasured process state variables are shown in Table 31.1.

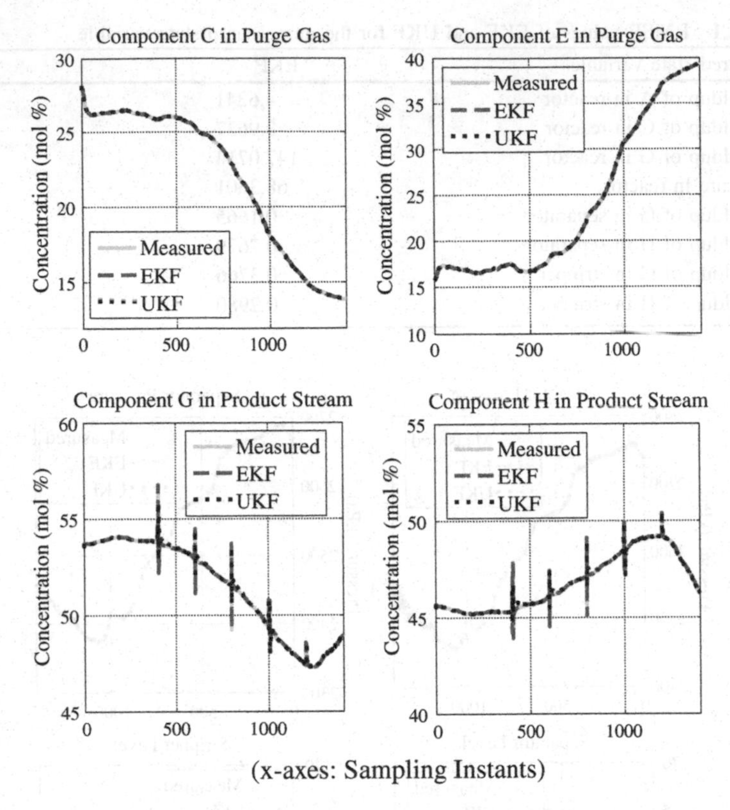

Fig. 31.5 Evolution of measured and estimated values of state variables of components in the purge gas and product stream

Table 31.2 RMSE values of EKF and UKF for the measured state variables

Measured state variables	EKF	UKF
Reactor pressure	6.6785	0.1204
Separator pressure	7.8179	1.5846
Separator level	0.9300	0.0653
Stripper level	1.7613	1.2489
Component C in purge gas	0.7632	0.0085
Component E in purge gas	0.8165	0.0126
Component G in product stream	0.6210	1.0668
Component H in product stream	0.3453	1.1551

The RMSE values of EKF and UKF for various measured states variables are shown in Table 31.2.

The covariance matrices of state noise and measurement noise are assumed as

$$\begin{aligned} Q = \text{diag}\ (&0.0005, 0.00005, 0.0005, 0.00005, 0.005,\\ &0.0001, 0.00001, 0.00001, 0.003, 0.0001,\\ &0.00005, 0.0001, 0.00001, 0.01, 0.0001,\\ &0.05, 0.05, 0.003, 0.0005, 0.00005, 0.0005,\\ &0.00001, 0.001, 0.0001, 0.005, 0.005, 0.003,\\ &0.0002, 0.0002, 0.003). \end{aligned}$$

$$\begin{aligned} R = \text{diag}\ (&0.001, 0.007, 0.0001, 0.005, 0.005, 0.0001,\\ &0.0001, 0.0001, 0.0001, 0.0001, 0.0001,\\ &0.0001, 0.0001, 0.0001, 0.0001, 0.0001,\\ &0.0001, 0.0001, 0.0001, 0.0001, 0.0001,\\ &0.0001, 0.0001). \end{aligned}$$

31.5 Conclusions and Future Work

In this paper, the performances of the EKF and UKF state estimation schemes have been compared on the Tennessee Eastman process, a benchmark problem in Chemical Engineering. Due to the linearization errors, a large deviation between true and estimated values is seen in the case of EKF. On the other hand, UKF perform much better compared to EKF and RMSE for some of the states is nearly 10 % lesser when compared to that of EKF. Implementation of other state estimation schemes in combination with nonlinear model predictive control may be the scope for the future work.

References

1. Simon D (2006) Optimal state estimation. Wiley, New York
2. Gelb A, Kasper J, Nash RA, Price CF (1974) Applied optimal estimation. MIT Press, Cambridge
3. Julier SJ, Uhlmann JK (2004) Unscented filtering and nonlinear estimation. Proc IEEE 92(3):401–422
4. Downs JJ, Vogel EF (1993) A plant-wide industrial process control problem. Comput Chem Eng 17:245–255
5. Ricker NL, Lee JH (1995) Nonlinear model predictive control of the Tennessee Eastman challenge process. Comput Chem Eng 19:961–981
6. McAvoy TJ, Ye N (1994) Base control for the Tennessee Eastman problem. Comput Chem Eng 18:383–413

7. Luyben WL, Tyreus BD, Luyben ML (1999) Plant wide process control. McGraw-Hill, New York
8. Palavajjhala S, Motard R, Joseph B (1993) Plant wide control of the Eastman Tennessee process. In: Annual AIchE Meeting, paper 148, St. Louis
9. Jockenhövel T, Biegler LT, Wächter A (2003) Dynamic optimization of the Tennessee Eastman process using the opt control centre. Comput Chem Eng 27:1513–1531
10. Ricker NL (1995) Optimal steady state operation of the Tennessee Eastman challenge process. Comput Chem Eng 19:949–989

Chapter 32
Modeling of ECG Signal and Validation by Neural Networks

N. Sathya and R. Malathi

Abstract ECG signal classification is widely used for diagnosing many cardiac diseases, which is the main cause of mortality in developed countries. Since most of the clinically useful information in the ECG signal is found in the intervals and amplitudes. The development of accurate and robust methods for automatic ECG signal classification is a subject of major importance. Modeling techniques like Least Square Estimation (LSQ) and Autoregressive (AR) modeling have been performed on the ECG signal. The model coefficients extracted using autoregressive modeling technique was found to be resourceful, so it has been taken for further validation. The ECG data is taken from standard MIT-BIH Arrhythmia database. AR coefficients obtained from the AR modeling are fed to the back-propagation neural network which classifies the ECG signal. In order to train the modeling coefficients with the back-propagation neural network the architecture implemented with 2 input neurons, 2 hidden neurons and 2 output neurons. In this work all neurons uses sigmoid activation function.

Keywords Least square estimation · Autoregressive modeling · Back-propagation neural network

---

N. Sathya (✉)
Department of Electronics and Instrumentation Engineering, M.A.M. College of Engineering, Tiruchirappalli, TamilNadu, India
e-mail: sathyaeie@gmail.com

R. Malathi
Department of Electronics and Instrumentation Engineering, Annamalai University, Chidambaram, TamilNadu, India
e-mail: vsmalu@gmail.com

R. Malathi and J. Krishnan (eds.), *Recent Advancements in System Modelling Applications*, Lecture Notes in Electrical Engineering 188,
DOI: 10.1007/978-81-322-1035-1_32,

32.1 Introduction

Electrocardiograph deals with the electrical activity of the heart. Monitored by placing sensors at the limb extremities of the subject, the electrocardiogram (ECG) is a record of the origin and propagation of electrical potential through cardiac muscles. It is consider a representative signal of cardiac physiology, useful in diagnosing cardiac disorders.

The medical state of the heart is determined by the shape of the Electrocardiogram, which contains important pointer to different types of diseases afflicting the heart. However, the electrocardiogram signals are irregular in nature and occur randomly at different time intervals during a day. Thus arises the need for continues monitoring of the ECG signal, which by nature are complex to comprehend and hence there is a possibility of the analyst missing vital information which can be crucial in determining the nature of the disease. Thus computer based automated analysis is recommended for early and accurate diagnosis.

ECG signals consist of P wave, QRS complex, and T wave. They are designated by capital letters P, Q, R, S, and T. In the normal beat phase of a heart, the main parameters inspected include the shape, the duration, and the relationship with each other of P wave, QRS complex, and T wave components and R–R interval. The changes in these parameters indicate an illness of the heart that may occur by [1] any reason. All of the irregular beat phases are generally called arrhythmia and some arrhythmias are very dangerous for patient life.

The biggest challenge faced by the models for automated heart beat classification is the variability of the ECG waveform from one patient to another patient even within the same person. However, different types of arrhythmias have certain characteristics which are common among all the patients. Thus the objective of a heart beat classifier is to identify those characteristics so that the diagnosis can be general and as reliable as possible. Modeling and simulation of ECG under various conditions are very important in understanding the functioning of the cardiovascular system as well as in the diagnosis of heart diseases. One of such methods which can be reliably used for ECG classification is the use of neural networks. Neural networks are one of the most efficient pattern recognition [2] tools because of their high nonlinear structure and tendency to minimize error in test inputs by adapting itself to the input output pattern and thus establishing a nonlinear relationship between the input and output.

However, the performance of the neural network is highly dependent on the number of input elements in the computational [3] layers. A large number of elements would lead to a number of multiplication and additions and the network would become expensive on computing resources. Thus to reduce the number of inputs a pre-processing layer is used.

This pre-processor uses white noise to "smooth" out the ECG waveform and reduce the number of samples while preserving all the distinct signal features. Also the use of modeling techniques makes use of the model to be implemented easily in real time processing. The model so obtained was implemented in a real-time

Fig. 32.1 General processing architecture

strategy is shown below. Each block represents a processing milestone. The first block is the pre-processor which was described previously and the second block is the neural network block which does the actual processing. Refer Fig. 32.1.

32.2 Electrophysiology and Electrical Activity of Cardio Vascular System

Cardiac electrophysiology is the science of elucidating, diagnosing, and treating the electrical activities of the heart. The term is usually used to describe studies of such phenomena by invasive (intracardiac) catheter recording of spontaneous activity as well as of cardiac responses to programmed electrical stimulation (PES). These studies are performed to assess arrhythmias, elucidate symptoms, evaluate abnormal electrocardiograms, assess risk of developing arrhythmias in the future, and design treatment. Cardiac Electrophysiology (also referred to as clinical cardiac electrophysiology, Arrhythmia [4] Services, or electrophysiology), is a branch of the medical specialty of clinical cardiology [5] and is concerned with the study and treatment of rhythm disorders of the heart (Fig. 32.2).

The normal electrical conduction of the heart allows electrical propagation to be transmitted from the Sinoatrial Node through both atria and forward to the Atrioventricular Node. Normal/baseline physiology allows further propagation from the AV node to the Ventricle or Purkinje Fibers and respective bundle branches and subdivisions/fascicles. Both the SA and AV nodes stimulate the

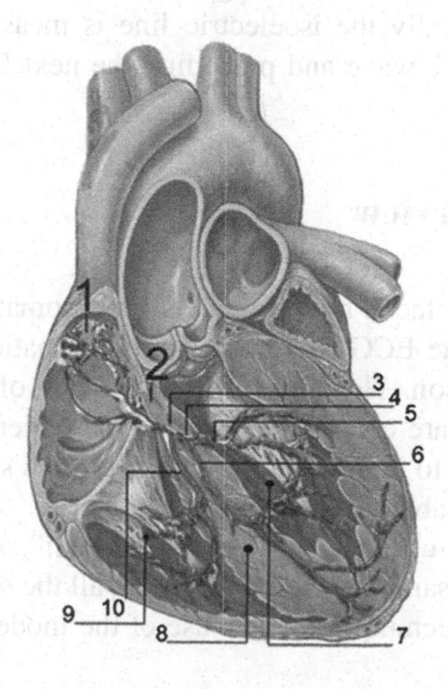

Fig. 32.2 Electrical activity of heart

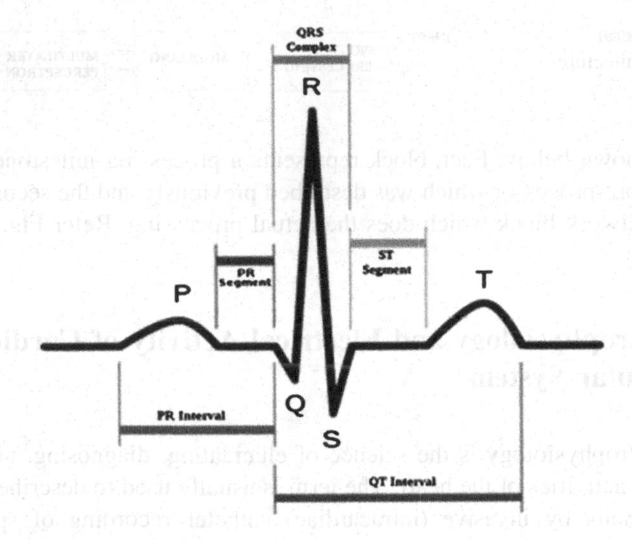

Fig. 32.3 Electrocardiograph signal

Myocardium. Time ordered stimulation of the myocardium allows efficient contraction of all four chambers of the heart, thereby allowing selective blood perfusion through both the lungs and systemic circulation.

A single beat consists of four sub waves P waves, QRS complex wave, T wave and less frequently observed T wave. Below we explain the relationship between each of this wave with corresponding cardiac activity. Note that generation of waves is purely due to electric activity within heart and not due to mechanical movements. A typical ECG tracing of the cardiac cycle (heartbeat) consists of a P wave, a QRS complex, a T wave, and a U wave which is normally visible in 50–75 % of ECGs. The baseline voltage of the electrocardiogram is known as the *isoelectric line*. Typically the isoelectric line is measured as the portion of the tracing following the T wave and preceding the next P wave (Fig. 32.3).

32.3 System Overview

The biggest challenge faced by the models for automated heart beat classification is the variability of the ECG waveform from one patient to another patient even within the same person. However, different types of arrhythmias have certain characteristics which are common among all the patients. Thus the objective of a heart beat classifier is to identify those characteristics so that the diagnosis [6] can be general and as reliable as possible.

This pre-processor uses white noise to "smooth" out the ECG waveform and reduce the number of samples while preserving all the distinct signal features. Also the use of modeling techniques makes use of the model to be implemented easily

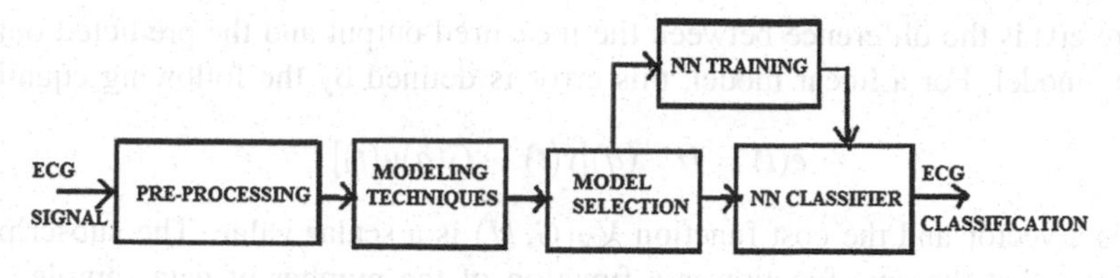

Fig. 32.4 Block diagram of the ECG classifier

in real time processing. The model so obtained was implemented in a real-time strategy is shown below. Each block represents a processing milestone. The first block is the pre-processor which was described previously and the second block is the neural network block [7] which does the actual processing. Refer Fig. 32.1.

Modeling and simulation of ECG under various conditions are very important in understanding the functioning of the cardiovascular system as well as in the diagnosis of heart diseases. One of such methods which can be reliably used for ECG classification is the use of neural networks. Neural networks are one of the most efficient pattern recognition [8] tools because of their high nonlinear structure and tendency to minimize error in test inputs by adapting itself to the input output pattern and thus establishing a nonlinear relationship between the input and output (Fig. 32.4).

However the performance of the neural network is highly dependent on the number of input elements in the computational layers. A large number of elements would lead to a number of multiplication and additions and the network would become expensive on computing resources. Thus to reduce the number of inputs a pre-processing layer is used.

32.4 Modeling Techniques

The ECG signal parameters are extracted from the preprocessed signal by using modeling techniques like Least Square Estimation and Autoregressive Modeling.

32.4.1 Least Square Estimation Modeling

Estimate model parameters using iterative prediction-error minimization method. PEM uses optimization to minimize the cost function, defined as follows for scalar outputs:

$$V_N(G,H) = \sum_{t=1}^{N} e^2(t)$$

where e(t) is the difference between the measured output and the predicted output of the model. For a linear model, this error is defined by the following equation:

$$e(t) = H^{-1}(q)[y(t) - G(q)u(t)]$$

e(t) is a vector and the cost function $V_N(G, H)$ is a scalar value. The subscript N indicates that the cost function is a function of the number of data samples and becomes more accurate for larger values of N.

32.4.2 Autoregressive Modeling

AR analysis models the ECG signal as the output [9] of a linear system driven by white noise of zero mean and unknown variance. AR models have the form

$$V[K] = \sum_{i=2}^{P+1} a_k v[k - i + 1] + n[k]$$

where $v[k]$ is the ECG time series, $n[k]$ is zero mean white noise, a_i's are the AR coefficients, and P is the AR order.

A critical issue in AR modeling [10] is the AR order used to model a signal. It is necessary to select an appropriate AR order so that the signal is modeled with sufficient accuracy so as to be useful for classification. Various model orders were used to estimate the accuracy of the reconstructed signals.

ECG data for Normal Sinus Rhythm and Arrhythmia were taken from the MIT-BIH database and simulated with least square estimation and autoregressive modeling using [11] MATLAB software. The model order, model coefficients and simulated output waveform has been obtained.

The preprocessed Normal Sinus Rhythm ECG signal waveform is shown in Fig. 32.5a.

The model order extraction using modeling technique is shown in the Fig. 32.5b. Here the model order is 12.

The Normal Sinus Rhythm Simulated output for model order 12 is represented in the Fig. 32.5c.

The Arrhythmia waveform is shown in Fig. 32.6a.

The model order is shown in the Fig. 32.6b. Here the model order is 6.

The Simulated Arrhythmia output for model order 6 is represented in the Fig. 32.6c.

The model coefficients obtained using autoregressive modeling was found to be resourceful, so it is taken for further validation with neural network.

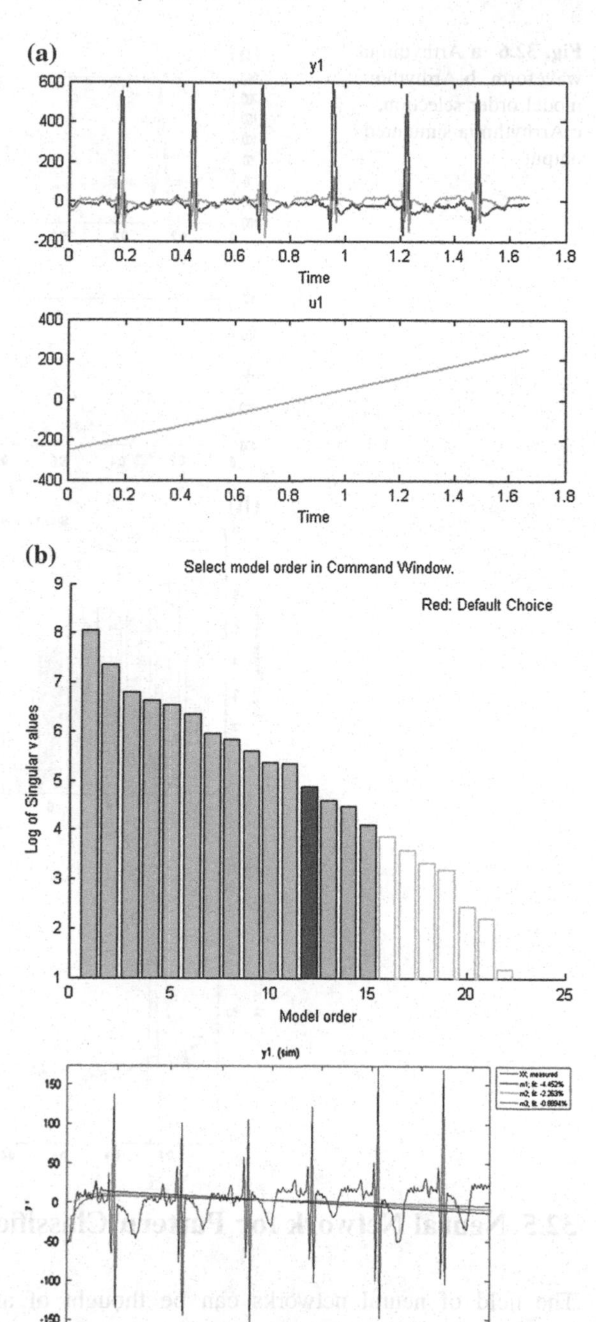

Fig. 32.5 **a** Normal sinus rhythm waveform. **b** Normal sinus rhythm model order selection. **c** Normal sinus rhythm simulated output

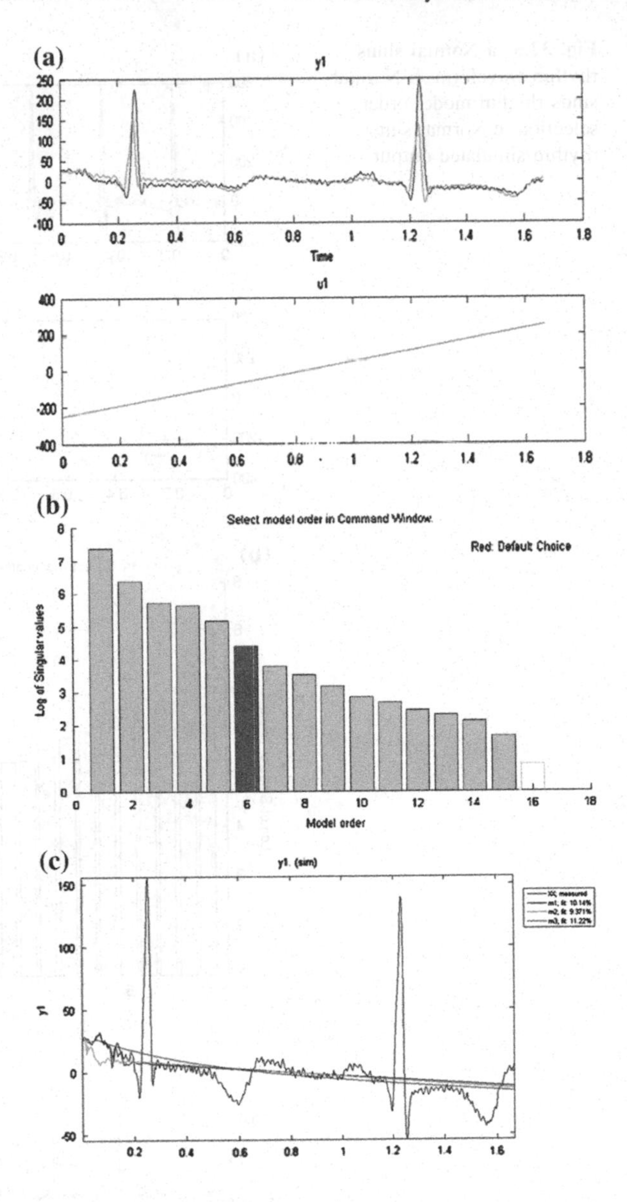

Fig. 32.6 **a** Arrhythmia waveform. **b** Arrhythmia model order selection. **c** Arrhythmia simulated output

32.5 Neural Network for Pattern Classification

The field of neural networks can be thought of as being related to artificial intelligence, machine learning, parallel processing, statistics, and other fields. The attraction of neural networks (NN) is that they are best suited in solving the problems that are the most difficult to solve by traditional [12] computational methods.

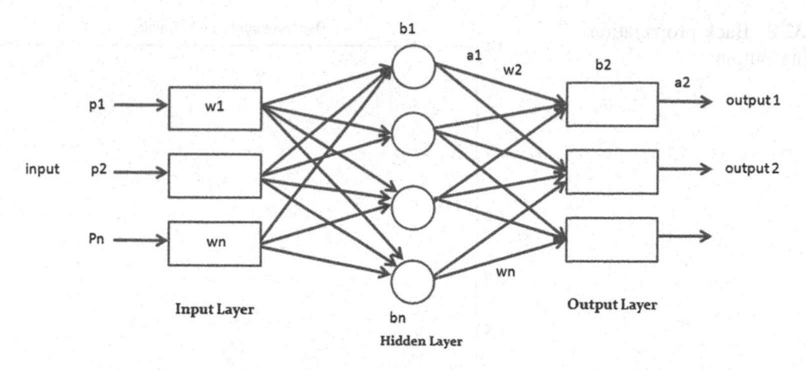

Fig. 32.7 Architecture of back propagation neural network

32.5.1 Neural Network Classifier

Neural networks are one of the most efficient pattern recognition tools because of their high nonlinear structure and tendency to minimize error in test inputs by adapting itself to the input output pattern and thus establishing a nonlinear relationship between the input and output. It is a supervised learning method. It requires a teacher that knows, or can calculate, the desired output for any given input. Back propagation requires that the activation function used by the artificial neurons (or "nodes") is differentiable (Fig. 32.7).

Classification of arrhythmias is a complicated problem. To solve this in order to train the modeling coefficients with the back propagation neural network [13] the architecture implemented with 2 input 1 hidden (2 neurons) and 2 output neurons. In this work all neurons uses sigmoid activation function. Random weights and biases are initialized by using back propagation algorithm the network has been trained with moderate values of learning rate (lr) and momentum *(w)*. The weights are updated for every training vector, and the termination [14] condition is that the sum square error reaches a minimum value.

The network is trained based on the several input vector sets of modeling coefficient values of normal ECG. After extensive training the best weights are selected. Based on these weights the abnormal ECG coefficients are validated.

32.6 Experimental Result and Conclusion

The ECG signals are preprocessed by adding white noise for removing the unwanted signal. The pre-processed ECG output is subjected to modeling techniques and required coefficients are extracted, which contain the information about the ECG. The model coefficients [15] extracted using autoregressive modeling was found to be useful and these coefficients are taken as input to neural network, for further pattern classification.

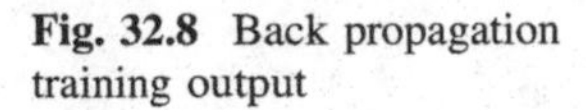
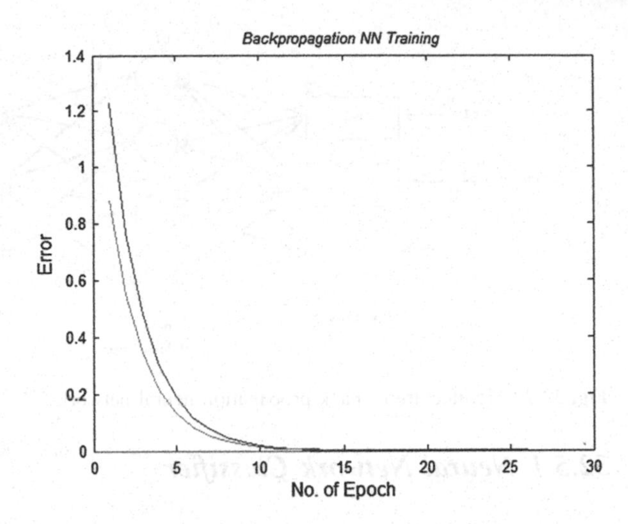

Fig. 32.8 Back propagation training output

Table 32.1 The neural network ECG classification data based on normal and abnormal ECG

	Target	Output	Error
Normal ECG	0–0.25	0.1756–0.2112	−0.0506 to −0.0862
	0.8–1.0	0.9065–0.9681	−0.0065 to −0.0681
Abnormal ECG	0.6–1	0.09228–0.9672	−0.0228 to −0.0672
	0.1–0.4	0.1823–0.2013	−0.0573 to −0.0763

The Back propagation neural network is trained with sigmoid activation function for a set of normal ECG coefficients and the weights obtained are evaluated with abnormal ECG coefficients, hence the difference can be distinguished whether the person is normal or abnormal i.e. suffering from disease or not. The corresponding training algorithm of the network show's plot between the error and the no. of epochs (Fig. 32.8).

32.6.1 Inference

The network is trained to input sample (modeling coefficients) the error is minimizing to zero, in order to achieve the target (Table 32.1).

Hence based on the neural network output data we can deduce whether the subject ECG is having the abnormality or not and the performance of Neural Network can be improved by enhancing in clinical applications.

Acknowledgments I wish to acknowledge my Brother N. Rajesh, Parents and family members for their valuable support, which helped me to prepare this paper.

References

1. Ge D, Srinivasan N, Krishnan SM (2002) Cardiac arrhythmia classification using autoregressive modeling. IEEE Trans Biomed Eng
2. Melo SL, Caloba LP, Nadal J (2000) Arrhythmia analysis using artificial neural network and decimated electrocardiographic data. Comp Cardiol 27:73–76
3. Sun Y (2001) Arrhythmia recognition from electrocardiogram using non-linear analysis and unsupervised clustering techniques. Ph.d. dissertation, Nanyang Technological University
4. Coast DA, Stren RM, Cano GG, Briller SA (1990) An approach to cardiac arrhythmia analysis using hidden Markov models. IEEE Trans Biomed Eng
5. Goldschlager N, Goldman MJ (1989) Principles of clinical electrocardiography. Appleton and Lange, California
6. Jain M, Chaturvedi S, Mithal V Detection of abnormalities and diseases in ECG data
7. Coast DA, Stren RM, Cano GG, Briller SA (1990) An approach to cardiac arrhythmia analysis using hidden Markov models. IEEE Trans Biomed Eng 37:826–836
8. Connor JT, Martin RD, Atlas LE (1994) Recurrent neural networks and robust time series prediction. IEEE Trans Neural Networks 5(2)
9. Anderson CW, Stolz EA, Shamssunder S (1998) Multivariate autoregressive models for classification of spontaneous electroencephalographic signals during mental tasks. IEEE Trans Biomed Eng 45:277–286
10. Arnold M, Miltner WHR, Witte H (1998) Adaptive AR modeling of nonstationary time series by means of Kalman filtering. IEEE Trans Biomed Eng 45:553–562
11. Math Works (2000) MATLAB user's guide. Math Works Inc
12. Bishop CM (2001) Neural networks for pattern recognition. Oxford University Press, New York
13. Demuth H, Beale M (1992) Neural networks toolbox manual. Math Works Inc
14. Kohonen T (1990) The self-organizing map. Proc IEEE 78:1464–1480
15. Haykin S (2000) Neural networks. Second Edition, Addison Wesely Long-man

Chapter 33
Detection of Obstructive Sleep Apnoea Using ECG Signal

A. Aishwarya, N. S. Bharath, M. Swathi and Ravi Prabha

Abstract Sleep Apnoea is a disorder characterized by abnormal pauses in breathing during sleep. Obstructive Sleep Apnoea is a condition where breathing is interrupted by a physical block to airflow despite respiratory effort. It is based on the fact that heart rate dynamics of a healthy person differs from that of a person suffering from OSA. The heart rate typically shows cyclic increases and decreases associated with the Apnoea phase and resumption of breathing. Identification of the oscillatory dynamics is done using the RR inter beat interval series. Hilbert transformation is applied to the sinus inter beat interval time series to derive the instantaneous amplitudes and frequencies of the series thus monitoring the presence or absence of OSA.

Keywords Obstructive sleep apnoea · Polysomnography · ECG signal · QRS complex · Hilbert transform · Amplitude and frequency threshold

A. Aishwarya (✉) · N. S. Bharath · M. Swathi · R. Prabha
Department of Medical Electronics, M. S. Ramaiah Institute of Technology, Bangalore, India
e-mail: aishwaryaanand2003@gmail.com

N. S. Bharath
e-mail: bharath.somashekar.1990@gmail.com

M. Swathi
e-mail: swathi.makam@yahoo.in

R. Prabha
e-mail: prabharavi90@gmail.com

R. Malathi and J. Krishnan (eds.), *Recent Advancements in System Modelling Applications*, Lecture Notes in Electrical Engineering 188,
DOI: 10.1007/978-81-322-1035-1_33, © Springer India 2013

33.1 Introduction

Apnoea is a Greek term for suspension of breathing. Sleep Apnoea is a disorder characterized by abnormal pauses in breathing during sleep. Each pause in breathing can last from a few seconds to minutes, and may occur 5–30 times or more an hour. An estimate of the severity of Apnoea is calculated by dividing the number of Apnoea instances by the number of hours of sleep, giving an Apnoea index (AI in Apnoea instances per hour); the greater the AI, the more severe the Apnoea. Similarly, a Hypopnea Index (HI) can be calculated by dividing the number of Hypopneas by the number of hours of sleep. The Apnoea-Hypopnea index (AHI) is an index of severity that combines Apnoeas and Hypopneas. Combining them gives an overall severity of Sleep Apnoea including sleep disruptions and de-saturations.

There are three forms of Sleep Apnoea:

1. *Central Sleep Apnoea* (*CSA*)—breathing is interrupted by a lack of respiratory effort; in OSA.
2. *Obstructive Sleep Apnoea* (*OSA*)—breathing is interrupted by a physical block to airflow despite respiratory effort.
3. *Mixed Sleep Apnoea*—It is a combination of CSA and OSA.

This disorder is characterized by numerous effects like daytime fatigue and sleepiness, insomnia, poor concentration and attention, memory problems, anxiety, irritability and headaches. High blood pressure, stroke, congestive heart failure and mood disorders can be caused or worsened by Sleep Apnoea [1].

33.2 Traditional Treatment

33.2.1 Common Treatment Using CPAP

For moderate to severe Sleep Apnoea, the most common treatment is the use of a continuous positive airway pressure (CPAP) or automatic positive airway pressure (APAP) device, which 'splints' the patient's airway open during sleep by means of a flow of pressurized air into the throat. Several levels of obstruction may be addressed in physical treatment, including the nasal passage, throat (pharynx), base of tongue, and facial skeleton. Surgical treatment for obstructive Sleep Apnoea needs to be individualized in order to address all anatomical areas of obstruction.

Often, correction of the nasal passages needs to be performed in addition to correction of the oropharynx passage. Septoplasty and turbinate surgery may improve the nasal airway [2] (Fig. 33.1).

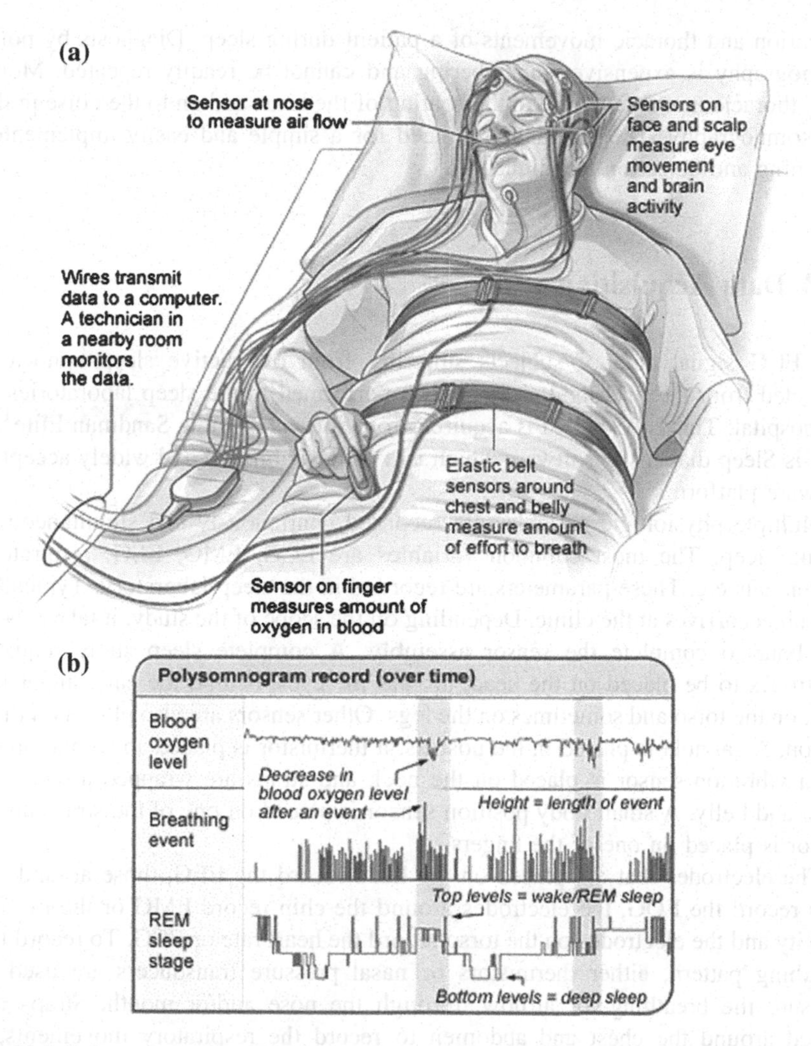

Fig. 33.1 Polysomnography **a** Placement of various sensors on the patient's body. **b** Polysomnogram record

33.2.2 Polysomnography

Studies of respiration during sleep demonstrate apnoeic episodes in the presence of respiratory muscle effort. Apnoeic episodes greater than 10 s in duration are considered clinically significant. The apnoeic episodes, as monitored by nasal and oral airflow, are typically 20–30 s in duration; rarely, episodes up to several minutes in duration can occur [3].

The traditional method of diagnosing Sleep Apnoea is based on the conjoint evaluation of clinical symptoms and the result of a formal sleep study called polysomnography. This method includes the monitoring of nasal airflow, oxygen saturation and thoracic movements of a patient during sleep. Diagnosis by polysomnography is expensive, encumbering and cannot be readily repeated. Moreover, thoracic movements due to the beating of the heart add on to the noise in the polysomnography. This demands the need for a simple and easily implemented screening and detection technique [4].

33.3 Data Acquisition

The ECG signal data of subjects suffering from obstructive sleep apnoea is extracted from the overnight sleep studies performed in the sleep laboratories in the hospital. This ECG signal is acquired from software Embla Sandman Elite™. This is Sleep diagnostic software which is a well established and widely accepted software platform.

Multiple physiologic variables are recorded continuously and simultaneously during sleep. The most common variables are ECG, EMG, EEG, respiratory movements etc. These parameters are recorded in the sleep laboratory. Typically, the subject arrives at the clinic. Depending on the scope of the study, it takes about one hour to complete the sensor assembly. A complete sleep study requires electrodes to be placed on the head, around the eyes, behind the ears, under the chin, on the torso and sometimes on the legs. Other sensors are also placed on their person. A cannula is placed in the nostrils, a thermistor is placed above the upper lip, a vibration sensor is placed on the neck and straps are wrapped around the chest and belly. A small body position sensor is placed on one of the straps and a sensor is placed on one of the fingers.

The electrodes that are placed on the head record the EEG, those around the eyes record the EOG, the electrodes around the chin record EMG or the muscle activity and the electrodes on the torso record the heart rate or ECG. To record the breathing pattern, either thermistors or nasal pressure transducers are used to measure the breathing, or airflow, through the nose and/or mouth. Straps are placed around the chest and abdomen to record the respiratory movements, a vibration sensor is placed on the neck to detect snoring and a position sensor is also used to signify major sleeping positions of the person, because people may breathe differently in different positions. Saturation levels are also recorded to detect the level of oxygen in the blood. The level of oxygen decreases in the blood when apnoea instances take place. The decrease is called de-saturation. An infrared sensor, usually placed on the finger, detects the colour of the blood through the fingernail. Once the electrode attachment and hook-up is complete, the person goes to sleep. A technician makes sure that all the electrodes and sensors stay in place and that the recording equipment works properly during the sleep study.

Once the study is complete, the technician removes all the sensors and electrodes. The data is sorted on a computer and the clinician uses a diagnostic software (Embla Sandman Elite™) to analyze the signals.

To achieve our objective, the EKG2 signal which is the lead II ECG signal of the patient is exported from the Embla Sandman Elite™ software as a text file. This text file is later imported into MATLAB for further processing [5].

33.4 Block Diagram

See Fig. 33.2.

33.5 Methodology

33.5.1 Baseline Wander Removal

Baseline wanders is considered as an artifactual data when measuring the ECG parameters, especially the ST segment measures are strongly affected by this wandering. It is mainly caused due to respiration, electrode impedance change due to perspiration and increased body movements. It is removed using linear filtering [6].

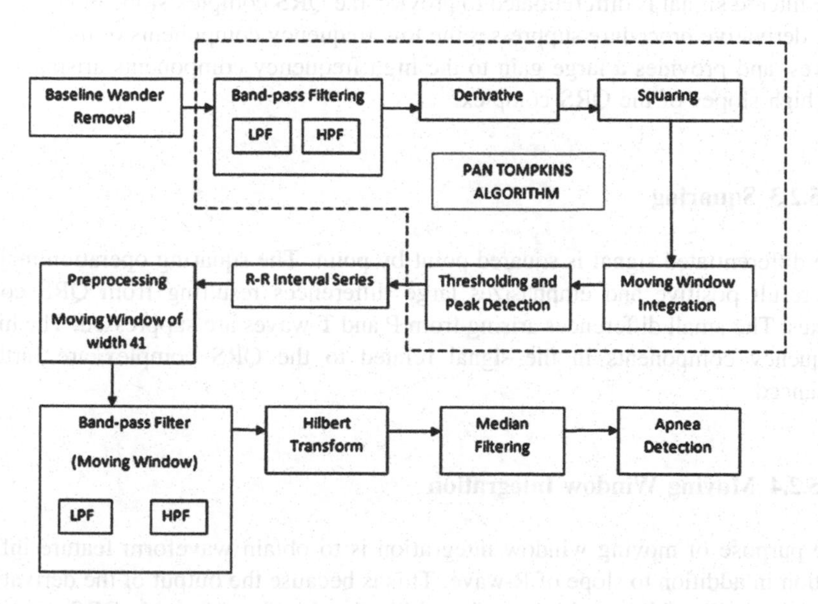

Fig. 33.2 Block diagram for OSA detection

33.5.2 Pan Tompkins Algorithm

J Pan and WJ Tompkins together formulated an algorithm for real time QRS complex detection from an ECG signal. The algorithm includes a series of filters and methods that perform low pass, high pass, derivative, squaring, integration and thresholding [7].

33.5.2.1 Band Pass Filtering

The band pass filter reduces the influence of muscle noise, 50 Hz power line interference, baseline wander and T-wave interference. The desirable pass band to maximize the QRS energy is approximately 5–15 Hz. Since the chosen sampling rate is 200 samples/s, a band pass filter cannot be designed directly for the desired pass band of 5–15 Hz. Therefore, we cascade the low pass and high pass filters to achieve a pass band from about 5–12 Hz, reasonably close to the design goal.

A butterworth low pass filter of order 1 is used whose cut off frequency is 11 Hz on the original ECG signal. A butterworth high pass filter of order 1 is used whose cut off frequency is 5 Hz on the low pass filtered ECG signal.

33.5.2.2 Derivative

The filtered signal is differentiated to provide the QRS complex slope information. The derivative procedure suppresses the low frequency components of the P and T waves, and provides a large gain to the high frequency components arising from the high slopes of the QRS complex.

33.5.2.3 Squaring

The differentiated signal is squared point by point. The squaring operation makes the result positive and emphasizes large differences resulting from QRS complexes. The small differences arising from P and T waves are suppressed. The high frequency components in the signal related to the QRS complex are further enhanced.

33.5.2.4 Moving Window Integration

The purpose of moving window integration is to obtain waveform feature information in addition to slope of R-wave. This is because the output of the derivative operation will exhibit multiple peaks within the duration of single QRS complex (Fig. 33.3).

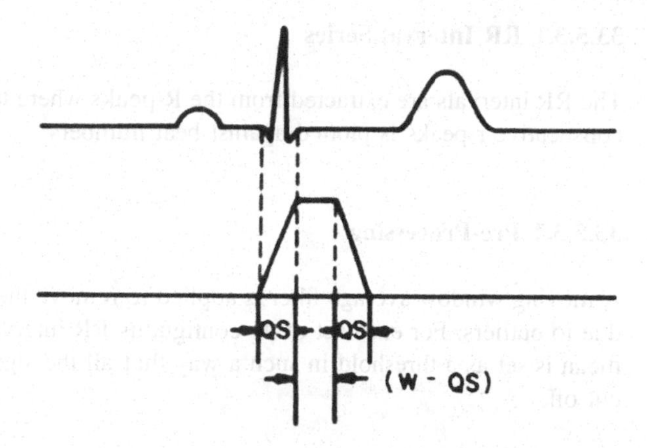

Fig. 33.3 Relationship of QRS complex to moving window integration waveform

The number of samples in the moving window is important. Generally the width of the window should be approximately the same as the widest possible QRS complex. If the window is too wide, the integration waveform will merge the QRS and the T complexes together. If it is too narrow, some QRS complexes will produce several peaks in the integration waveform. These can cause difficulty in subsequent QRS detection processes. For our sample rate 200 samples/s, the window is 30 samples wide.

33.5.2.5 Threshold and Peak Detection

When analyzing the amplitude of the moving window integrator output, the algorithm uses two threshold values. The first pass through the signal uses these thresholds to classify each non zero sample as either signal or noise. If the current peak value is greater than the threshold, the location is identified as a QRS complex and the signal level is updated as 1. If the current peak value is less than the threshold, then that location is identified as a noise peak and the noise level is updated as 0.

The QRS complex has to be identified in the adaptive threshold output. This can be done by recognizing the 1 levels of the adaptive threshold output and then stemming the average of the QRS complex [7].

33.5.3 RR Interval Processing Using Hilbert Transform

The heart rate oscillations accompanying prolonged OSA are transient with varying amplitudes and frequencies. These properties cause standard spectral techniques such as the Fourier Transform and autoregressive methods to be ineffective in reliably detecting and localizing episodes of OSA. Therefore, the Hilbert Transform which is an analytical technique is best suited for processing ECG signals to detect OSA [8].

33.5.3.1 RR Interval Series

The RR intervals are extracted from the R-peaks where the difference between the consecutive r peaks is plotted against beat numbers

33.5.3.2 Pre-Processing

A moving window average filter is applied to remove the remaining impulse noise due to outliers. For each set of 41 contiguous RR intervals, a local mean and this mean is set as a threshold in such a way that all the signals beyond the mean are cut off.

33.5.3.3 Band Pass Filtering

Here, the frequency range of the band pass filter is selected to ensure that OSA oscillations will not be filtered out. A low pass filter and a high pass filter are cascaded.

A moving window boxcar filter of width 5, giving a 3 dB cut-off at 0.09 Hz where the value of each point in the series is replaced by the average value over the window focused at that point.

The high pass filter is a moving window of 81 data points, giving a high pass 3 dB cut-off at 0.01 Hz. At each point in the series, the slope of regression line over a window focused at that point is computed, and the value of this fit at that point is subtracted from the actual value at this point [9].

33.5.3.4 Hilbert Transform

A real function and its Hilbert transform are related to each other in such a way that they together create a so called strong analytic signal. The strong analytic signal can be written with amplitude and a phase where the derivative of the phase can be identified as the instantaneous frequency. Hilbert transform is applied to the filtered RR interval series and the instantaneous amplitudes and frequencies at each point are computed [10].

33.5.3.5 Median Filtering

Since the amplitudes and frequencies of the transformed data exhibit large fluctuations around their mean values, to eliminate this noise they are each median filtered using a moving window of 60 points.

33.5.3.6 Apnoea Detection

Amplitude and frequency are plotted and threshold conditions are checked. OSA is said to be detected if the frequency goes below 0.02 Hz and the amplitude goes above 0.4 units at approximately the same time.

33.6 Results and Discussions

The ECG signal from the sleep study data is extracted and processed using Hilbert transform. Processing is done for signals obtained from a normal subject and signals obtained from subjects suffering from Obstructive Sleep Apnoea. The results are graphically represented and conclusions regarding the presence of Obstructive Sleep Apnoea are made based on the threshold conditions (Figs. 33.4, 33.5).

It is observed in Fig. 33.6 that the frequency has fallen below 0.02 Hz and the amplitude has risen above 0.4 units at approximately the same time. This satisfies the threshold conditions. Therefore, OSA is detected.

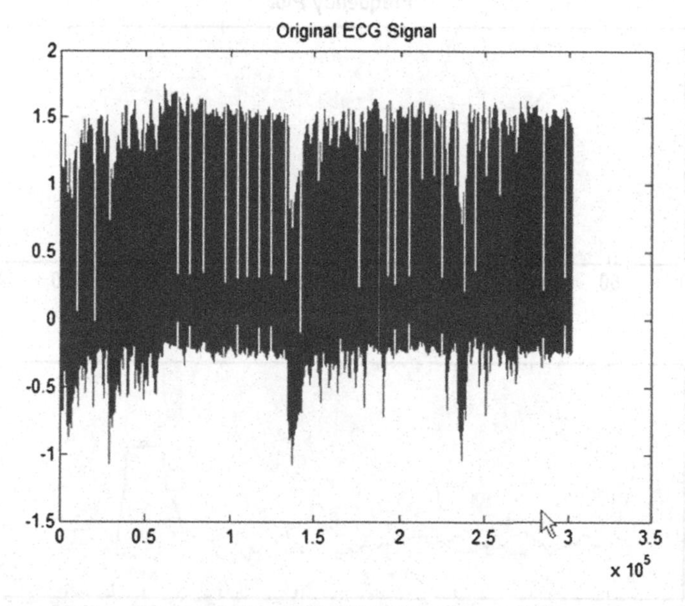

Fig. 33.4 A screenshot of figure window in MATLAB showing ECG signal of the subject having OSA

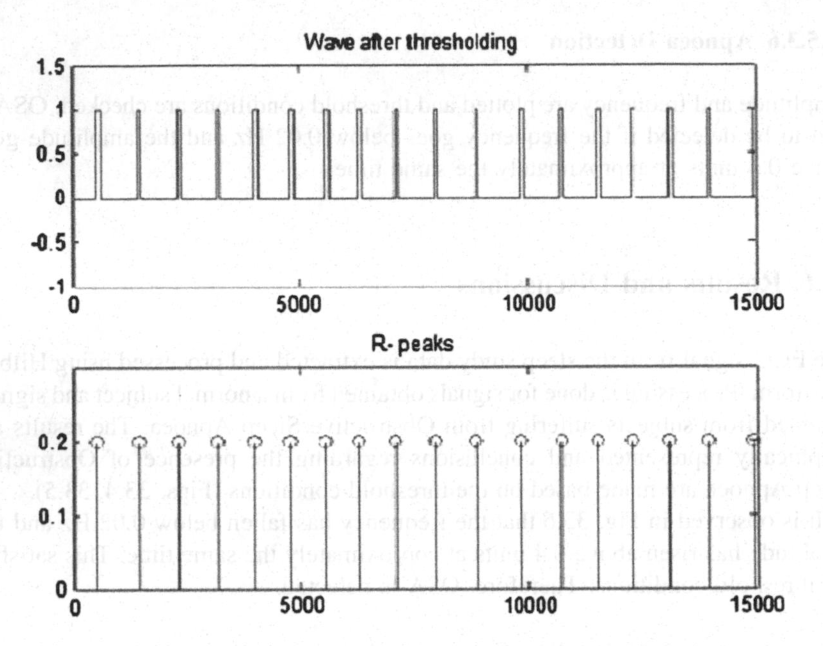

Fig. 33.5 A screenshot of figure window in MATLAB showing plot of R peaks

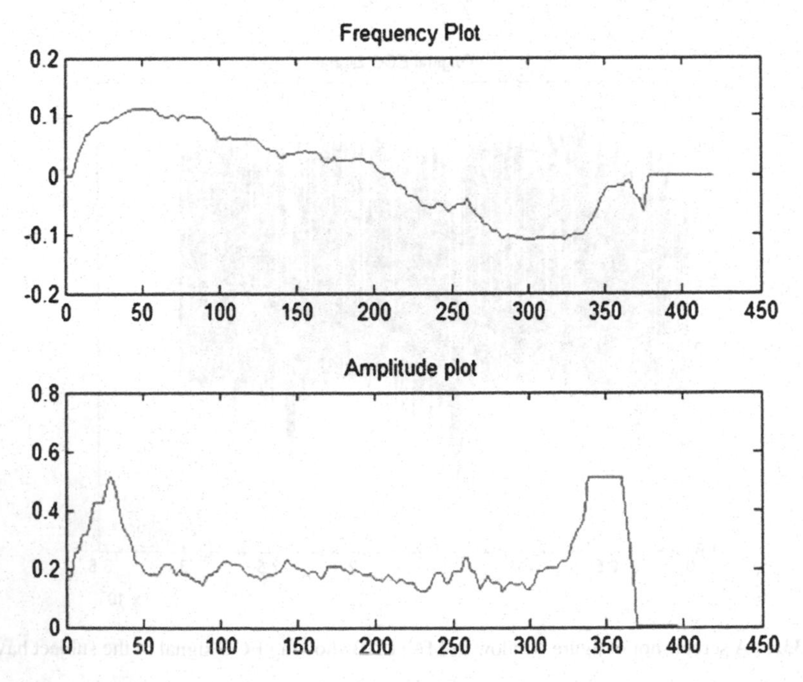

Fig. 33.6 A screenshot of figure window in MATLAB showing median filtered output which shows the presence of OSA in the subject

33.7 Summary and Conclusions

OSA is detected using the ECG signal and the discrimination of cessation of breathing and its resumption is shown graphically.

The frequency and amplitude of a person suffering from OSA is shown to exceed threshold conditions. This proves that the subject is suffering from OSA. The threshold conditions are defined as follows:

1. Frequency value decreases below 0.02 Hz.
2. Amplitude value increases above 0.4 units.
3. Both the frequency and amplitude threshold conditions are met at approximately the same instances of time.

ECG signals from a normal subject (not suffering from OSA) are also processed. The processed results of normal subjects prove that they are not suffering from OSA as the threshold conditions are not met.

33.8 Scope for Future Work

This method of detecting OSA from ECG signal is comparatively less expensive and easier compared to the traditional method of polysomnographic study. But more work on this technique has to be carried out to successfully implement it in the hospitals. If the following points are considered and implemented, this method can replace the traditional method for diagnosis of OSA.

1. Improvements in the filtering techniques and parameter adjustments may further increase the efficacy of this technique.
2. The algorithm can be extended for quantitative measurement of Apnoea by displaying AI.
3. This method can be implemented in hardware where the ECG monitored in the patient monitor can be processed in an IC programmed to detect Apnoea. The output will possess information about Apnoea instances in the ECG. Thus, a separate slot can be included in the patient monitor to detect and measure Apnoea.
4. OSA can be detected using various algorithms and various other signals like SpO_2, EEG etc. and a comparison can be made.
5. Detection of OSA from the ECG signal of a person having significant arrhythmias can also be attempted.

Acknowledgments We would like to offer our gratitude to Mr P. G. Kumaravelu, HOD of Department of Medical Electronics, MSRIT who gave us the support and encouragement during the course of the project. We are extremely grateful to Dr. Uma Maheshwari, M. S Ramaiah Memorial Hospital who helped us in Data acquisition. We would also like to thank Mr. Sanjay H. S, for helping and guiding us through this project. Most importantly we would like to thank Mr. Anand M, Research engineer, C Dot, Bangalore and Dr. Sethu Selvi, HOD, Department of Electronics and Communication, MSRIT whose help was provided during the signal processing part of the project.

References

1. http://en.wikipedia.org
2. http://www.medicinenet.com/sleep_apnea/article.htm
3. Harwick JD (2007) (ed) Otolaryngologic clinics of North America sleep disorders
4. Chaudari SK (2008) Concise medical physiology, 5th edn. New Central Book Agency Pvt Ltd
5. http://www.embla.com/index.cfm/id/17/Sleep-Studies/
6. Mozaffary B, Tinati MA (2006) ECG baseline wander elimination using wavelet packets
7. (1985) IEEE Transactions on Biomedical Engineering, vol BME-32, no 3, March 1985 http://www.engr.wisc.edu/bme/faculty/tompkins_willis/Pan.pdf
8. Mietus JE, Peng CK, Ivanov PC, Goldberger AL (2000) Detection of obstructive sleep apnoea from cardiac interbeat interval time series. Beth Israel Deaconess Medical Center and Harvard Medical School, Boston
9. http://www.mathworks.in/products/signal/
10. Physiobank (2000) Physiobank chanllege 2000. Physiobank. http://www.physionet.org/challenge/2000/

Chapter 34
Automated Detection and Classification of Blood Diseases

A. N. Nithyaa, R. Premkumar, D. Kanchana and Nalini A. Krishnan

Abstract The aim of this paper is to automate the classification of various blood diseases using digital image processing technique in MATLAB software. The analysis of blood smear is a powerful diagnostic tool for the prediction of diseases like Malaria, Elephantiasis, Trypanosomiasis, Sickle cell anaemia and Polycythemia. As they are life threatening diseases and an enormous global health problem, rapid and precise differentiation is necessary in clinical settings. Automation of disease detection in life science laboratories can be done by extracting the statistical features of the blood smear images taken by the digital microscopes and processing it using Digital Image Processing technique in MATLAB software.

Keywords Blood smear · Malaria · Elephantiasis · Trypanosomiasis · Sickle cell anaemia · Polycythemia · Statistical features

A. N. Nithyaa (✉) · R. Premkumar · D. Kanchana · N. A. Krishnan
Department of Biomedical Engineering, Rajalakshmi Engineering College, Anna University, Thandalam, Chennai 602105, India
e-mail: nithyaavasan@gmail.com

R. Premkumar
e-mail: indoprem@gmail.com

D. Kanchana
e-mail: pepsi.ruby26@gmail.com

N. A. Krishnan
e-mail: nalinikrishnan19@gmail.com

R. Malathi and J. Krishnan (eds.), *Recent Advancements in System Modelling Applications*, Lecture Notes in Electrical Engineering 188,
DOI: 10.1007/978-81-322-1035-1_34, © Springer India 2013

34.1 Introduction

In life science laboratories, Blood smear analysis is done by manual observation using microscopes to predict the various blood borne diseases. Visual inspection of microscopic images is the most widely used technique for determination of diseases and it is a labor-intensive repetitive and time consuming task [1]. This motivation behind this project is to produce quick quantitative results by automated Blood smear analysis with minimal human intervention.

Microscopic examination of stained thick and thin blood films remains the 'gold standard' for routine malaria diagnosis [2]. For slide preparation working solutions of Giemsa were made by adding 100 μl stock solution to each milliliter of distilled water. Dried thin blood films were fixed with methanol for 30 s, poured off and stained with Giemsa for 20 min. The stain was rinsed off with tap water for 10 s. Upon drying, slides were used immediately or stored for future use. Image was captured by connecting high resolution Digital camera to microscope. By adjusting microscope magnification image is captured [3]. The recognition accuracy largely depends on subjective factors like experience and fatigue due to human tiredness. With the increasing demand for more number of such examinations along with the need for quality results, there arose a necessity for the automation of the whole process. This not only reduced the burden on haematologists but also yielded accurate results in significantly short period of time [4]. Automation of this task is very helpful for improving the haematological procedure and accelerating diagnosis of many diseases [5]. The statistical features use gray level histogram and saturation histogram of the pixels in the image and based on such analysis, the mean value is calculated [6]. The extracted features are processed in MATLAB using digital image processing technique.

34.2 Materials and Methods

34.2.1 Data

For this project 20 known microscopic image samples were taken for feature extraction program and 20 unknown microscopic image samples were taken for the classification program (Figs. 34.1, 34.2, 34.3, 34.4, 34.5, 34.6).

34.2.2 Method in Existence

The most popular method of evaluating the clinical status which is in existence is by means of visual inspection of blood smear using a compound or light microscope. The blood smear is prepared and then observed under microscope. The field of view (FOV) is varied by adjusting the knob and the observation is done for

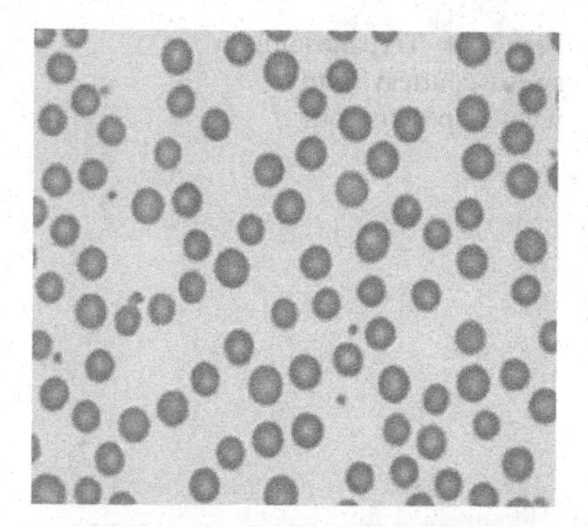

Fig. 34.1 Normal blood smear courtesy: http://www.serc.carleton.edu

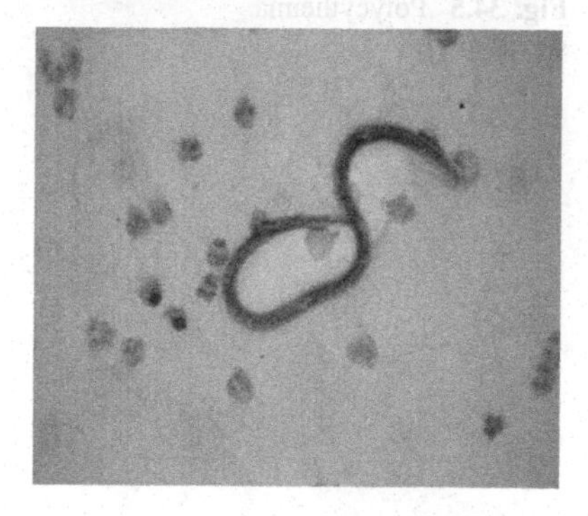

Fig. 34.2 Elephantiasis smear courtesy: http://www.flickr.com

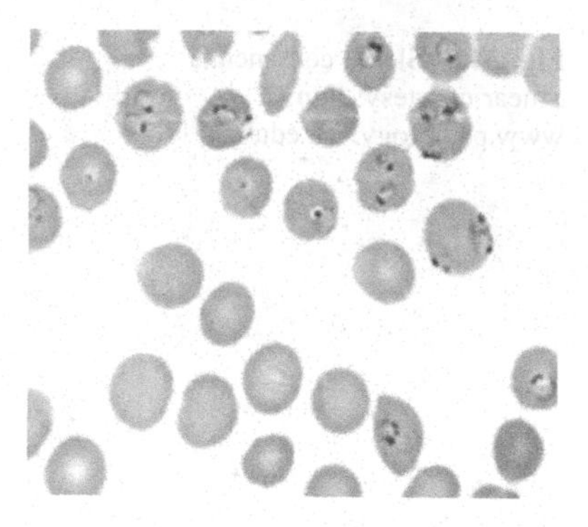

Fig. 34.3 Malaria smear courtesy: http://www.southsudanmedicaljournal.com

disease prediction. The Magnification of microscope is varied to have a better view of the parasites and cell structure.

34.2.3 Proposed Method

The image of microscopic slide is taken using a digital microscope connected to a laptop/Personal computer. Once the image is captured it is loaded into the MATLAB for processing. In this project an algorithm for detecting Normal blood

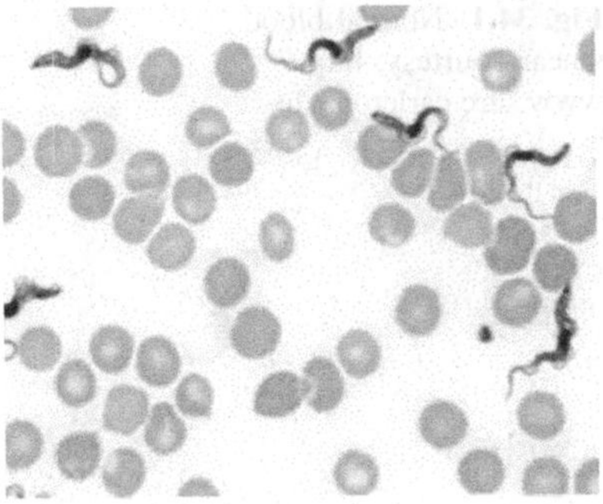

Fig. 34.4 Trypanosomiasis smear courtesy: http://www.allposters.com

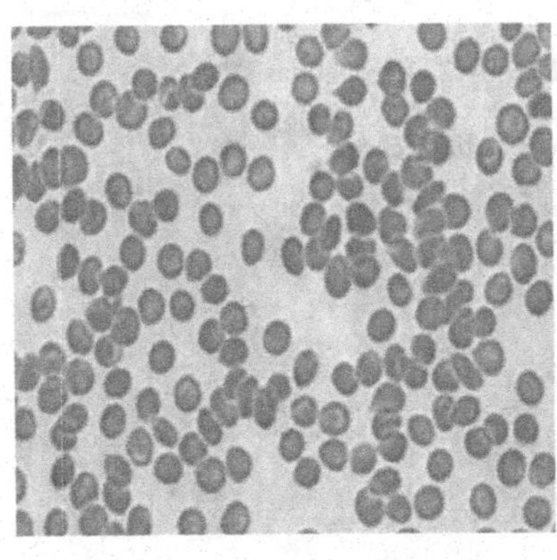

Fig. 34.5 Polycythemia smear

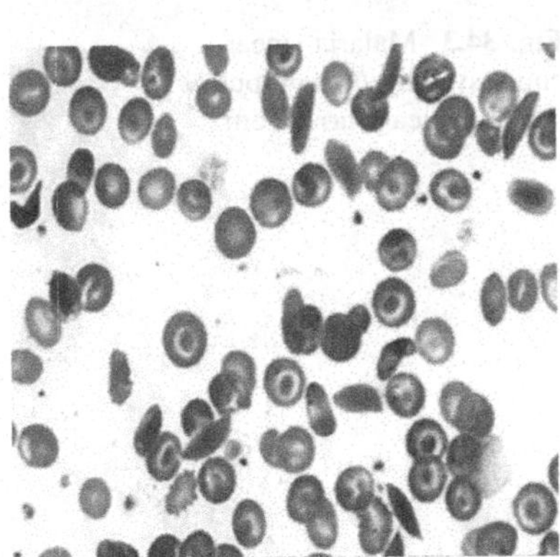

Fig. 34.6 Sickle cell anemia Smear courtesy: http://www.pathology.vcu.edu

smear, Malaria, Elephantiasis, Trypanosomiasis, Polycythemia, and Sickle Cell Anaemia is coded in MATLAB (Fig. 34.7).

The present work was carried out with the following objectives:

(1) Feature extraction of known samples
(2) Master feature set creation (Mean, Variance, Moments, Entropy) of known samples
(3) Feature set creation of test sample

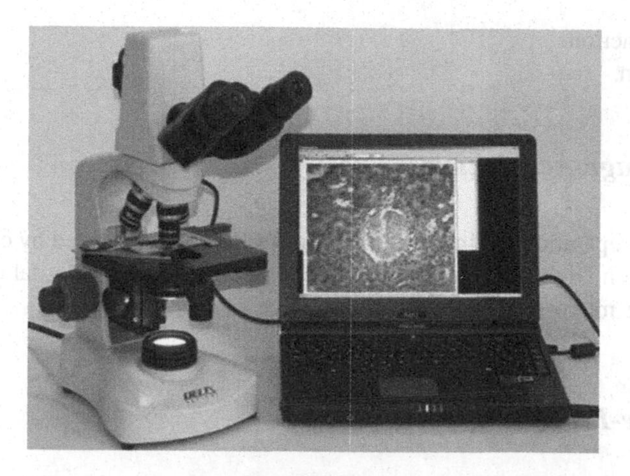

Fig. 34.7 Image acquistion system courtesy: http://www.365astronomy.com

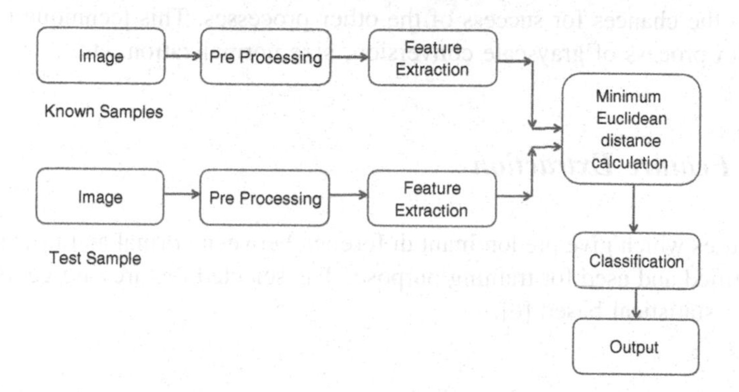

Fig. 34.8 General block diagram

(4) To classify the disease based on the minimum Euclidean distance calculation of the tested sample with the known master feature set values as Normal, Malaria, Elephantiasis, Trypanosomiasis, Polycythemia, Sickle cell anaemia.

34.3 Implementation

34.3.1 Block Diagram

Microscopic Image classification involves the following processes (Fig. 34.8):

Image input
Pre-processing

Feature extraction
Classification.

34.3.2 Image Acquisition

After slide preparation the image of the blood smear was captured by connecting a high resolution Digital camera to microscope or by means of a digital microscope. By adjusting microscope magnification, image is captured [3].

34.3.3 Pre-Processing

After acquiring a digital image, the next step deals with the preprocessing of that image. The key function of preprocessing is to improve the image in way that increases the chances for success of the other processes. This technique typically deals with process of grayscale conversion, size normalization, etc.

34.3.4 Feature Extraction

The features which give predominant difference between normal and infected cells are identified and used for training purpose. The selected features are geometrical, color and statistical based [6].

34.3.5 Statistical Approach

A statistical approach sees an image texture as a quantitative measure of the arrangement of intensities in a region. In general this approach is easier to compute and is more widely used, since natural textures are made of patterns of irregular sub elements. Statistical features (mean, variance, 3rd order moment, 4th order moment, entropy) are taken for classification. Texture is generated from the grayscale image matrices of the red, green and blue components, as well as the intensity component from the hue-saturation-intensity image space. First order features, based on the image histograms are used. The following features are extracted from the microscopic images.

1. Average gray level or mean
2. Variance
3. Third moment or skewness
4. Fourth order moment or Kurtosis
5. Entropy or Randomness.

Fig. 34.9 Algorithm-feature extraction

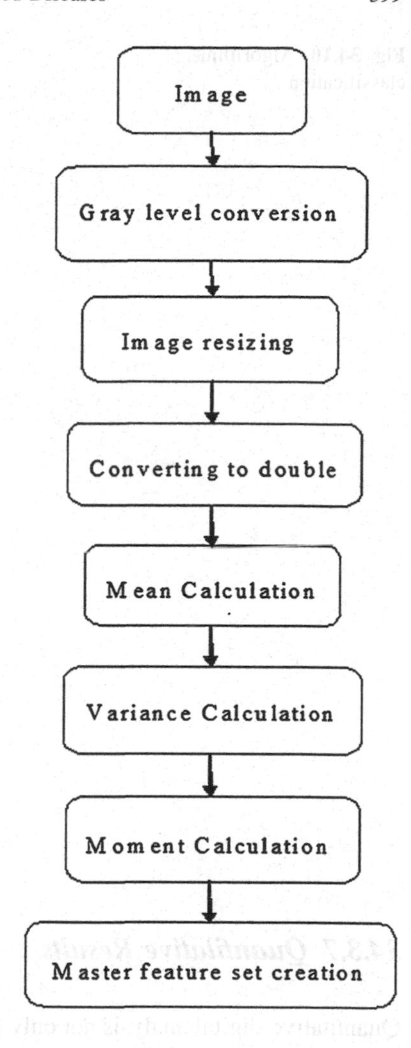

34.3.6 Methodology Involved

The known sample images are taken and their statistical features are extracted using the feature extraction program. Using the extracted features a master feature set is created for each disease taken for classification. Then the unknown test sample images are taken as input and the minimum Euclidean distance between the known sample's master feature set and the test sample is calculated. The output is displayed based on the minimum value of the Euclidean distance between the known and test sample images. Thus the disease classification is automated (Figs. 34.9, 34.10).

Fig. 34.10 Algorithm-classification

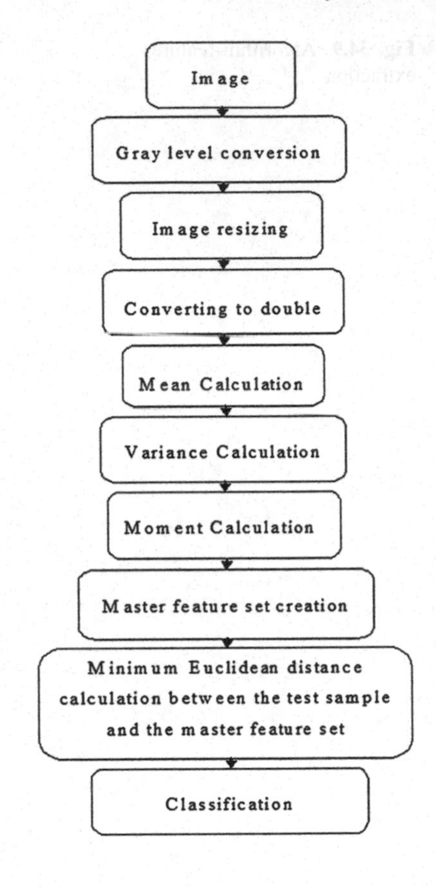

34.3.7 Quantitative Results

Quantitative digital analysis not only provides classification results approximate to that of the human experts but furnishes reproducibility in classification not available from experts. The master feature set is calculated by taking the average of the extracted features of the normal blood smear images. Similarly the features are extracted for Malaria, Elephantiasis, Trypanosomiasis, Sickle cell anemia and Polycythemia.

When the Normal blood smear images are taken and loaded into Feature extraction program written in the MATLAB, the output is generated as follows:

***** FEATURE EXTRACTION—NORMAL *****

For IMAGE 1 : NORMAL 1

Mean	Variance	Moment-3	Moment-4	Entropy
0.8487	836.1780	−0.0008	0.0002	4.8937

For IMAGE 2 : NORMAL 2

Mean	Variance	Moment-3	Moment-4	Entropy
0.8491	811.4493	−0.0008	0.0002	4.9385

For IMAGE 3 : NORMAL 3

Mean	Variance	Moment-3	Moment-4	Entropy
0.8494	811.4828	−0.0008	0.0002	4.7959

For IMAGE 4 : NORMAL 4

Mean	Variance	Moment-3	Moment-4	Entropy
0.8537	803.0929	−0.0009	0.0002	4.7941

***** Averaged Feature values—Normal *****

Mean	Variance	Moment-3	Moment-4	Entropy
0.8502	815.5507	−0.0008	0.0002	4.8556

The following table represents the Master feature set Creation for the normal blood smear images (Table 34.1).

Once the Master feature set is created for all known image samples, the test samples are taken and loaded into the Classification code written in MATLAB. The following output is generated when an unknown test sample is taken as input.

Table 34.1 Feature value extraction

Sample	Mean	Variance	Moment 3rd order	Moment 4th order	Entropy
Normal 1	0.8487	836.1780	−0.0008	0.0002	4.8937
Normal 2	0.8491	811.4493	−0.0008	0.0002	4.9385
Normal 3	0.8494	811.4828	−0.0008	0.0002	4.7959
Normal 4	0.8537	803.0929	−0.0009	0.0002	4.7941
Master feature set	0.8502	815.5507	−0.0008	0.0002	4.8556

***** FEATURE EXTRACTION—TEST SAMPLE *****

For TEST SAMPLE:				
Mean	Variance	Moment-3	Moment-4	Entropy
0.7791	388.4736	−0.0005	0.0002	5.8678

***** NORMALISED VALUES *****
Normalised Feature values—TRAINING DATA SET

For Normal:				
Mean	Variance	Moment-3	Moment-4	Entropy
0.8502	0.2106	−0.0008	0.0002	0.8026

For ELEPHANTIASIS:				
Mean	Variance	Moment-3	Moment-4	Entropy
0.7591	0.1370	−0.0013	0.0005	0.9429

For MALARIA:				
Mean	Variance	Moment-3	Moment-4	Entropy
0.8681	0.2574	−0.0010	0.0004	0.8522

For TRYPANOSOMIASIS:				
Mean	Variance	Moment-3	Moment-4	Entropy
0.9320	0.1842	−0.0017	0.0009	0.8194

For POLYCYTHEMIA:				
Mean	Variance	Moment-3	Moment-4	Entropy
0.8170	0.2644	−0.0003	0.0002	1.0000

For SICKLE:				
Mean	Variance	Moment-3	Moment-4	Entropy
0.8316	1.0000	−0.0071	0.0037	0.9533

For TEST SAMPLE:				
Mean	Variance	Moment-3	Moment-4	Entropy
0.7791	0.1003	−0.0005	0.0002	0.9699

***** MINIMUM DISTANCE *****

Minimum distance between normal and test sample
0.0452

Minimum distance between elephantiasis and test sample
0.0025

Minimum distance between malaria and test sample
0.0465

Minimum distance between trypanosomiasis and test sample
0.0531

Minimum distance between polycythemia and test sample
0.0293

Minimum distance between sickle cell anemia and test sample
0.8125

******* OUTPUT *******
The tested sample is elephantiasis ≫

34.4 Conclusion

In laboratories, manual inspection is prone to errors and accuracy is not assured. There are problems of observation errors, reading mistake, fatigue which greatly tend to alter the results. In order to overcome this difficulty, the computers are programmed to recognize and classify the cells. While several languages can be used for computation, MATLAB was found to be an efficient tool for obtaining the desired results. The combination of MATLAB and image processing help us to classify the cells into their categories. The results were found to be more accurate and in synchronization with the results of interpretation from human experts. It was found to be more advantageous than manual inspection and proved to be better and effective. Thus this project can be a valuable asset in the life science laboratories.

In future, this project can be further extended for the detection and classification of other blood cell disorders like cancer and tumors.

Acknowledgments The authors wish to thank the Chairman and Chairperson of Rajalakshmi Engineering College, Chennai for providing all the facilities required for carrying out this work. They also thank the Principal, Rajalakshmi Engineering College, Chennai and Mr. S. Mahesh Anand, Director, Scientific Computing solutions, Chennai for his constant encouragement and guidance.

References

1. Hirimutugoda YM, Wijayarathna G (2010) Image analysis system for detection of red cell disorders using artificial neural networks. Sri Lanka J BioMed Inf 1(1):35–42
2. Frean J (2010) Microscopic determination of malaria parasite load: role of image analysis 862:866
3. Jigyasha S (2011) Advanced image analysis based system for automatic detection of malaria parasite in blood images using SUSAN approach. Int J Eng Sci Technol (IJEST) 3(6) ISSN: 0975-5462
4. Bandyopadhyay SK, Roy S (2012) Detection of sharp contour of the element of the WBC and segmentation of two leading elements like nucleus and cytoplasm. Int J Eng Res Appl (IJERA) 2(1):545–551 ISSN: 2248-9622
5. Hiremath PS, Bannigidad P, Geeta S (2010) Automated identification and classification of white blood cells (Leukocytes) in digital microscopic images. IJCA special issue on "recent trends in image processing and pattern recognition" RTIPPR
6. Savkare SS, Narote SP (2011) Automatic detection of malaria parasites for estimating parasitemia. Int J Comput Sci Secur (IJCSS) 5(3)

Chapter 35
An Automated Breathing Device for Critically Ill Patients

N. Padmasini and J. Archana

Abstract In this paper the project titled an automated breathing device for critically ill patients has been discussed. Ambu bag (AB) is a flexible reservoir bag used for artificial ventilation connected by tubing and non-rebreathing valve to a face mask or endo-tracheal tube. It is a hand held device used to provide positive pressure ventilation to a patient who is not breathing or who is breathing inadequately [1]. In ambulances and in emergency wards of hospitals manually operated ambu bag is used. Bagging is necessarily regular in medical emergencies when the patient's breathing is insufficient or has ceased completely and also to provide mechanical ventilation in preference to mouth to mouth resuscitation. The problems involved in this type of manual ambu bag is that sometimes due to the negligence of the caretakers required quantity of oxygen is not carried over to the lungs, secondly even if the caretakers are alert one cannot assure that they expel a constant quantity of oxygen into the patient's lungs, thirdly this type of mechanism basically is a stress to the caretakers. To overcome this, an automatic mechanism for administering oxygen into the patient's body has been developed and tested successfully. This automated ambu bag is economically cheap and gives much comfort to the care takers and the required amount of oxygen can be delivered appropriately according to the need of the patient.

Keywords Ambu bag · DC motor · PIC microcontroller · Proximity sensor

N. Padmasini (✉) · J. Archana
Department of BME, Rajalakshmi Engineering College, Chennai, India
e-mail: archana91jayaram@gmail.com

R. Malathi and J. Krishnan (eds.), *Recent Advancements in System Modelling Applications*, Lecture Notes in Electrical Engineering 188,
DOI: 10.1007/978-81-322-1035-1_35, © Springer India 2013

35.1 Introduction

Manual resuscitator (Ambu bag) is one of the most common devices used to ventilate and oxygenate patients in medical practice [1]. Even though most health providers has a general idea about the way these devices function, only few of them know the exact mechanisms of it. This device has an advantage over the manual resuscitator used in the hospitals, and is shown to be superior to the anesthesia machine ventilator in the situation of sub atmospheric pressure and big leak in the circuit. On the other side, to make anesthesia providers aware of the serious complications of endotracheal nasogastric tube placement [2], this type of arrangement is proposed. In this project we have the mechanical setup which has metal plates over which the ambu bag is been kept and the bagging operation is done with the help of linear movement of the shaft which is attached with the cam arrangement and is operated using DC shunt motor, the shaft end is attached with plastic ball to avoid the wear and tear of the bag. The speed of the DC motor is controlled using microcontroller unit [3]. Three varying speeds of DC motor has been set, in order to suit adults, children and, infants. By varying the DC motor speed the delivery of the output i.e. air is controlled. Adults need 16–20 bagging operation per minute, children need 25–30, and infants need 30–35 bagging operation per minute. The size of the ambu bag also varies according to the ages [4].

35.2 Materials and Methods

Figure 35.1 shows the block diagram of the system and the various materials used are discussed below:

35.2.1 Ambu bag

The ambu bag also known as BVM (Bag Valve Mask) consists of a flexible air chamber (the "bag", about the size of an American football), attached to a face mask via a shutter valve. When the face mask is properly applied and the "bag" is squeezed, the device forces air through into the patient's lungs; when the bag is released, it self-inflates from its other end, drawing in either ambient air or a low pressure oxygen flow supplied by a regulated cylinder, while also allowing the patient's lungs deflate to the ambient environment (not the bag) past the one way valve. Bag and valve combinations can also be attached to an alternate airway adjunct, instead of the mask. For example, it can be attached to an endotracheal tube laryngeal mask airway. Often a small Heat and Moisture exchanger, or humidifying/bacterial filter (HME) is used [5]. A bag valve mask can be used without being attached to an oxygen tank to provide air to the patient, often called

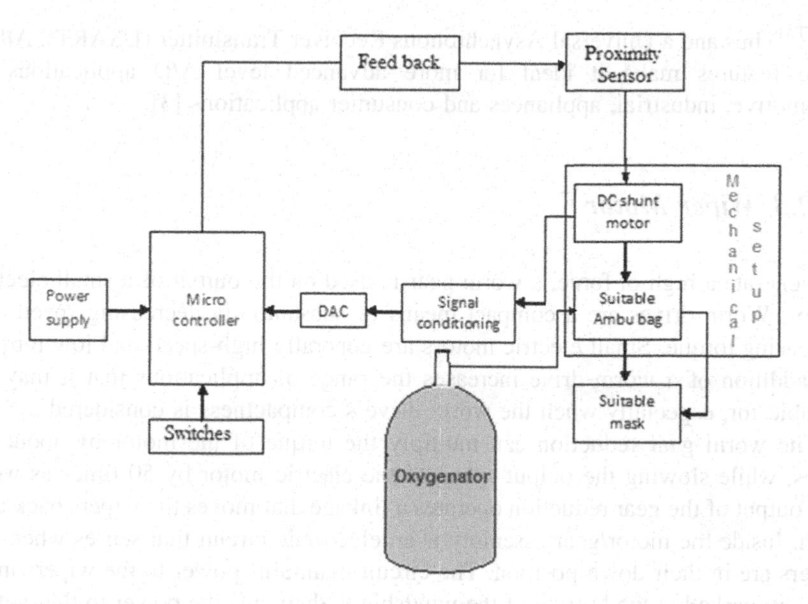

Fig. 35.1 Block diagram of the system

"room air" in the U.S. Supplemental oxygen increases the partial pressure of oxygen inhaled, helping to increase perfusion in the patient [6]. Most devices also have a reservoir which can fill from an oxygen source during the patient's expiratory phase (a process which happens passively for patients in respiratory arrest), in order to increase the amount of oxygen that can be delivered to the patient to nearly 100 %. Bag valve masks come in different sizes to fit infants, children, and adults. The mask size may be independent of the bag size; for example, a single pediatric-sized bag might be used with different masks for multiple face sizes, or a pediatric mask might be used with an adult bag for patients with small faces [7, 8]. Most types of the device are disposable and therefore designed for single use, while others are designed to be cleaned and reused. In our project we have used the reusable one [9].

35.2.2 PIC16F877A

This powerful (200 ns instruction execution) yet easy-to-program (only 35 single word instructions) CMOS FLASH-based 8-bit microcontroller packs Microchip's powerful PIC® architecture into an 40- or 44-pin package and is upwards compatible with the PIC16C5X, PIC12CXXX and PIC16C7X devices. The PIC16F877A features 256 bytes of EEPROM data memory, self programming, 2 Comparators, 8 channels of 10-bit Analog-to-Digital (A/D) converter, 2 capture/compare/PWM functions, the synchronous serial port can be configured as either 3-wire Serial Peripheral Interface (SPI™) or the 2-wire Inter-Integrated Circuit

(I^2C^{TM}) bus and a Universal Asynchronous Receiver Transmitter (USART). All of these features make it ideal for more advanced level A/D applications in automotive, industrial, appliances and consumer applications [3].

35.2.3 Wiper Motor

To generate a high of force, a worm gear is used on the output of a small electric motor. Worm drives are a compact means of substantially decreasing speed and increasing torque. Small electric motors are generally high-speed and low-torque; the addition of a worm drive increases the range of applications that it may be suitable for, especially when the worm drive's compactness is considered.

The worm gear reduction can multiply the torque of the motor by about 50 times, while slowing the output speed of the electric motor by 50 times as well. The output of the gear reduction operates a linkage that moves the wipers back and forth. Inside the motor/gear assembly is an electronic circuit that senses when the wipers are in their down position. The circuit maintains power to the wipers until they are parked at the bottom of the windshield, then cuts the power to the motor. This circuit also parks the wipers between wipes when they are on their intermittent setting [10].

35.2.4 NPN Proximity Sensors

When no metal is approaching the sensor its output will be high. When a metal approaches the sensor its output will be low.

35.2.5 Working of Mechanical Set Up and Microcontroller Unit

Cam and shaft arrangement, DC shunt motor, Ambu bag, Plastic/metallic ball, Tools are required for mechanical set up, O_2 Cylinder and mask are the components used along with it.

As shown in Fig. 35.2 Cam mechanism is used and at the end of cam, shaft is connected to rod. When the motor starts rotating in clockwise direction, causes linear motion in the shaft, so that the rod moves in the up and down direction. To the end of the rod, a plastic ball is attached, which prevents damage to the ambu bag [10].

In this system we have the mechanical setup which has metal plates over which the ambu bag is been kept and the bagging operation is done with the help linear movement of the shaft which is attached with the cam arrangement which is operated using DC shunt motor the shaft end is attached with plastic ball to avoid

Fig. 35.2 Front view of the developed project

the wear and tear of the bag. The speed of the DC motor is controlled using microcontroller programming [3]. Because for adults, children, infants the delivery of the output i.e. air is different for adults they need 16–20 bagging operation per minute, for children they need 25–30, for infants 35–30 bagging operation per minute and the size of the ambu bag varies according to the ages.

The speed of the motor is controlled using microcontroller. The PIC Microcontroller board consists of circuits necessary to operate a Microcontroller with PC interface. The board contains provisions for interfacing 8 analog inputs and 23 Digital level signals [11].

35.3 Results and Discussion

In the developed system of our project we have the mechanical setup which has metal plates over which the ambu bag has been kept and the bagging operation is done with the help linear movement of the shaft which is attached with the cam arrangement which is operated using DC shunt motor the shaft end is attached with metallic ball to avoid the wear and tear of the bag. The speed of the DC motor is controlled using microcontroller programming. Because for adults, children, infants the delivery of the output i.e. air is different. For adults 16–20 bagging operation per minute, for children 25–30, and for infants 35–30 bagging operation per minute is needed and hence the size of the ambu bag varies according to the ages [8]. The fabricated automated ambu bag has been successfully tested on various patients who are in emergency and in need of oxygen.

35.3.1 Future Work

At the inlet of the oxygen delivery system to the patient, an air pressure sensor can be used to monitor the pressure of the air delivered to the patient and hence depending on the patient condition automatic control of flow rate can be done by using the microcontroller unit.

35.4 Conclusion

The fabricated automated breathing device will have a significant role in every ambulance and in hospitals. It could be used to save the life of the patients who is suffering from breathing disorders in a better manner when compared with conventional type. This automated Ambu bag system can supply oxygen to needy patients who are suffering during an emergency situation. It will reduce the distress and strain for the caretakers. In the manual operation of AB (currently existing method), there is a risk of over-inflating the lungs due to high pressure which can cause gastric distension [12]. The patient may tend to vomit due to additional airway problems [9, 13–15]. This project provides automated assist device for breathing such that the oxygen is supplied at constant amount for the patients with pulmonary failure. Hence vomiting or other complications do not occur using this automated type.

References

1. Elsharydah A (2003) Manual resuscitators (ambu bags) can ventilate the lungs adequately despite big subatmospheric pressure in the breathing circuit. Internet J Anesthesiol 7(2):6
2. (1983) A complication of nasogastric tube intubation: pulmonary hemorrhage. Anesthesiology 59(4):356–368
3. Capelli J, Ferlo N, Foley A, Smith V, Atlas G (2011) Bioengineering conference (NEBEC), IEEE 37th Annual North East, 1–3
4. Clementsen HJ (1963) The E. M. O. inhaler and the ambu bag. Br Med J 2(5369):1409
5. Finkelstein JA MD (1996) Oral ambu-bag insufflation to remove unilateral nasal foreign bodies. Am J Emerg Med 14(1)
6. Kannan S, Morrow B, Furness G (1999) Tension pneumothorax and pneumomediastinum after nasogastric tube insertion. Anaesthesia 54(10):1012–1013
7. Mazidi MA, McKinlay RD, Danny causey, Pic microcontroller and embedded systems using assembly and C, Chapter 9
8. Thio M, Bhatia R, Dawson JA Arch dis child fetal neonatal ed doi:10.1136/adc.2009.166462,Oxygen delivery using neonatal self-inflating resuscitation bags without a reservoir
9. Zwagill LW, Metzeroff KO (1973) Mechanical ventilation comprised by misplacement of nasogastric tube. Respir Care 18:60–61

10. Ramesh G, Narasimharao I, Robinson A (2010) Intelligent engine with micro controller valve actuation and eliminating the cam linkage arrangement, frontiers in automobile and mechanical engineering (FAME), Sathyabama University, Chennai, pp 151–155
11. http://www.faadooengineers.com/threads/4373-8051-microcontroller-Ebook-Mazidi-amp-Mazidi-PDF
12. Saini S, Taxak S (1997) Malpositioned nasogastric tube: a cause of ventilatory volume loss. Trop Doct 27(2):125
13. Soroker D, Ezri T, Szmuk P (1994) An unusual case of failure to ventilate the lungs. Anaesthesia 49(12):1105
14. Stirt JA, Lewenstein LN (1981) Circle system failure induced by gastric suction. Anesth Intensive Care 9:161–162
15. Wood G, Milne B, Spjeda V, Lewis J (1990) Ventilatory failure due to an improperly placed nasogastric tube. J Anaesthesia 37(5):587–588

Chapter 36
Evaluation of Hypolipidemic Effect of Various Extracts of Whole Plant of *Bauhinia purpurea* in Rat Fed with High Fat Diet

C. D. Shajiselvin and A. Kottai Muthu

Abstract The objective of the present study was to investigate the hypolipidemic effect of various extracts of whole plant of *Bauhinia purpurea* in rat fed with high fat diet. The elevated levels of total cholesterol, ester and free cholesterol, phospholipids, triglycerides, low-density lipoprotein, and very low-density lipoprotein due to high fat diet (HFD). The group receiving ethyl acetate extract of *Bauhinia purpurea* at the dose of 250 mg/kg (Group IV) was significantly ($P < 0.001$) reduced the lipid profile and lipoprotein levels. A significant ($p < 0.001$) reduction in HDL-cholesterol was noticed in HFD fed groups (II); however, a significant increased the HDL level was produced by the administration of ethyl acetate extract of *Bauhinia purpurea*. There was a noticed increase in the body weight in HFD fed group (II), which was significantly ($p < 0.001$) reduced by the administration of ethyl acetate extract of *Bauhinia purpurea*. Therefore, it was concluded that the ethyl acetate extract of whole plant of *Bauhinia purpurea* as definite cardio protective effect against hyperlipidemia.

Keywords *Bauhinia purpurea* · Ethyl acetate extract · HFD rats · Hypolipidemia

C. D. Shajiselvin
Karpagam University, Coimbatore, India

A. Kottai Muthu (✉)
Annamalai University, Annamalai Nagar, Chidambaram, India
e-mail: arthik03@yahoo.com

R. Malathi and J. Krishnan (eds.), *Recent Advancements in System Modelling Applications*, Lecture Notes in Electrical Engineering 188,
DOI: 10.1007/978-81-322-1035-1_36, © Springer India 2013

36.1 Introduction

Hyperlipidemia contributes significantly in the manifestation and development of atherosclerosis and coronary heart diseases (CHD). Atherosclerosis, are the most common cause of mortality and morbidity worldwide. Although several factors, such as diet high in saturated fats and cholesterol, age, family history, hypertension and life style play a significant role in causing heart failure, the high levels of cholesterol particularly TC, TG and LDL-cholesterol is mainly responsible for the onset of CHDs. A 20 % reduction of blood cholesterol level can decrease about 31 % of CHD incidence, and 33 % of its mortality rate [1]. In addition hyperlipidemia is induced by secondary effect of diabetes, therefore, the agent having some antioxidant and anti-diabetic effect also showed favorable effect to hyperlipidemia. HMG Co A reductase inhibitor has been used in the treatment of hyperlipidemia, and simvastatin is one of the most prevalently used HMG CO A reductase inhibitors [2].

Coronary heart disease resulting from progressive atherosclerosis remains the most common cause of morbidity and mortality all over the world [3]. In developing countries, the incidence of cardiovascular disease is increasing alarmingly. India is on the verge of a cardiovascular epidemics [4, 5]. The circulatory system disorders are going to be the greatest killer in India by the end of year 2015 [6]. A large number of allopathic hypolipidemic drugs are currently available in the market but these lag behind the desired properties such as efficacy and safety on long term use, cost and simplicity of administration. These factors do not fulfill conditions for patient's compliance [7]. Plants and herbs are mines of large number of bioactive phytochemicals that might serve as lead for the development of effective, safe, cheap novel drugs. A number of medicinal plants have shown their beneficial effect on the cardiovascular disease (CVD) by virtue of their lipid lowering, antianginal, antioxidant and cardioprotective effects [8, 9].

Bauhinia purpurea Linn. (Leguminosae) is a medium sized deciduous tree, sparingly grown in India. This plant is used traditionally in dropsy, pain, rheumatism, convulsions, delirium, and septicaemia [10]. The bark of the plant is used as an astringent in the treatment of diarrhea. Its decoctions are recommended for ulcers as a useful was solution [11]. They are reported to exhibit various pharmacological activities such as CNS activity, cardiotonic activity, lipid-lowering activity, anti-oxidant activity, hepatoprotective activity, hypoglycemic activity, etc. [12]. Even through, traditionally, leaves of *Bauhinia purpurea* (Linn) were extensively used for the treatment of variety of wounds [13], and no scientific data in its support is available. There is no earlier report regarding the antihyperlipidemic effect of various extract of whole plant of *Bauhinia purpurea* (Linn). Hence, the objective of the present study was to investigate the antihyperlipidemic effect of various extracts of whole plant of *Bauhinia purpurea* in rat fed with high fat diet.

36.2 Materials and Methods

36.2.1 Plant Materials

The whole plant of *Bauhinia purpurea* (Linn), were collected from Nagercoil, Kanyakumari District of Tamil Nadu, India. Taxonomic identification was made from Botanical Survey of Medical Plants Unit Siddha, Government of India, Palayamkottai. The whole plant of *Bauhinia purpurea* (Linn), were dried under shade, segregated, pulverized by a mechanical grinder and passed through a 40 mesh sieve.

36.2.2 Preparation of Extracts

The above powered materials were successively extracted with Petroleum ether (40–60 °C) by hot continuous percolation method in Soxhlet apparatus [14] for 24 h. Then the marc was subjected to ethyl acetate (76–78 °C) for 24 h and then marc was subjected to methanol for 24 h. The extracts were concentrated by using a rotary evaporator and subjected to freeze drying in a lyophilizer till dry powder was obtained. The extracts were suspended in 2 % tween 80 [15].

36.2.3 Animals and Experimental Design

Male Wistar rats of 16–19 weeks age, weighing 150–175 g were procured from the Central Animal House, Sankaralingam Bhuvaneswari College of Pharmacy, Sivakasi. The rats were kept in cages, 2 per cage, with 12:12 h light and dark cycle at 25 ± 2 °C. The rats were maintained on their respective diets and water ad libitum. Animal Ethical Committee's clearance was obtained for the study. Rats were divided into following 6 groups of 6 rats each:

Group I Standard chow diet (Control)
Group II High fat diet
Group III High fat diet + Petroleum ether extract of *Bauhinia pupurea* (250 mg/kg B.wt)
Group IV High fat diet + Ethyl acetate extract of *Bauhinia pupurea* (250 mg/kg B.wt)
Group V High fat diet + Methanol extract of *Bauhinia pupurea* (250 mg/kg B.wt)
Group VI High fat diet + standard drug atorvastatin (1.2 mg/kg B.wt)

At the end of 9 weeks all the rats were sacrificed by cervical dislocation after overnight fasting. Just before sacrifice, blood was collected from the retro-orbital sinus plexus under mild ether anaesthesia and blood sample collected in heparinised tubes and plasma was separated. Liver, heart and aorta were cleared of adhering fat, weighed accurately and used for the preparation of homogenate. Animals were given enough care as per the Animal Ethical Committee's recommendations.

36.2.4 Animal Diet

The compositions of the two diets were as follows [16]. *Control diet:* Wheat flour 22.5 %, roasted bengal gram powder 60 %, skimmed milk powder 5 %, casein 4 %, refined oil 4 %, salt mixture with starch 4 % and vitamin and choline mixture 0.5 %. *High fat diet:* Wheat flour 20.5 %, roasted bengal gram 52.6 %, skimmed milk powder 5 %, casein 4 %, refined oil 4 %, coconut oil 9 %, salt mixture with starch 4 % and vitamin and choline mixture 0.5 %, cholesterol 0.4 %.

36.2.5 Biochemical Estimation

Plasma samples were analyzed for total cholesterol, HDL-cholesterol and triglycerides were estimated using Boehringer Mannheim kits by Erba Smart Lab analyzer, USA. LDL-cholesterol and VLDL-cholesterol were calculated by using Friedwald et al. [17] method. Ester cholesterol [18] and free cholesterol [18] were analyzed by using digitonin. Portions of liver, heart and aorta tissues were blotted, weighed and homogenized with methanol (3 volumes) and the lipid extracts were obtained by the method of Folch et al. [19]. Extracts were used for the estimation of ester cholesterol and free cholesterol, triglycerides [20] and phospholipids [21]. Plasma total cholesterol: HDL-cholesterol ratio and LDL-cholesterol: HDL-cholesterol ratio was also calculated to access the atherogenic risk.

36.2.6 Statistical Analysis

Results were expressed as mean ±SE of 6 rats in each group. The statistical significance between the groups was analyzed by using one way analysis of variance (ANOVA), followed by Dunnet's multiple comparison test. Significance level was fixed at 0.05.

36.3 Results

The average body weight changes in control and experimental rats were presented in Table 36.1. The body weight of High fat fed rats (group II) were increased significantly ($p < 0.001$) in comparison with normal control rats (group I). The average body weight was reduced significantly ($p < 0.001$) by the administration of ethyl acetate extract of *Bauhinia purpurea* at the dose of 250 mg/kg body weight as well as atorvastatin 1.2 mg/kg in comparison with HFD rats (group II).

Effect of various extract of *Bauhinia purpurea* on plasma lipid profiles are summarized in Table 36.2. There was a significant ($p < 0.001$) increased in the level of plasma lipid profile in the group II rats fed with high fat diet in comparison with the control rats (group I). Administration of ethyl acetate extract of *Bauhinia purpurea* to rat fed with HFD significantly ($p < 0.001$) decreased in the level of total cholesterol, free cholesterol, ester cholesterol, triglyceride and phospholipids as compared to HFD rats (group II). The similar result was not observed in the other two extract treatment group. However, the administration of ethyl acetate extract of *Bauhinia purpurea* treated rats with HFD showed that the plasma cholesterol was restored to near normal as that of atorvastatin (group VI). The Atherogenic Index (AI) is used as a marker to assess the susceptibility of atherogenesis. It was significantly ($p < 0.001$) increased on feeding high fat diet to rats (group II) as compared to control rats (group I).The ethyl acetate extract of *Bauhinia purpurea* significantly reduced the atherogenic index when compared with HFD fed rats (group II).

As shown in Table 36.3. The HDL-cholesterol levels were reduced in HFD rats (Group II) as compared to control rats (group I). After administration of ethyl

Table 36.1 Average body weight changes in control and experimental rats

Groups	Initial weight (g)	Final weight (g)	Average body weight gain (g)
Group I	145 ± 1.29^{bNS}	$205.83 \pm 3.51^{b*}$	$60.00 \pm 3.87^{b*}$
Group II	131.73 ± 0.66^{aNS}	$276.04 \pm 0.68^{a**}$	$144.31 \pm 0.98^{a**}$
Group III	$152.33 \pm 0.95^{aNS,\ bNS}$	$262.50 \pm 10.78^{aNS,\ b*}$	$119.33 \pm 5.93^{aNS,\ b*}$
Group IV	$156.40 \pm 0.31^{aNS},\ ^{bNS}$	$232.37 \pm 0.92^{aNS,\ b**}$	$76.29 \pm 1.17^{aNS,\ b*}$
Group V	$152.33 \pm 0.95^{aNS,\ bNS}$	$269.17 \pm 7.35^{aNS,\ b*}$	$111.83 \pm 3.05^{aNS,\ b*}$
Group VI	$189.67 \pm 0.19^{aNS,\ bNS}$	$255.44 \pm 0.60^{aNS,\ b**}$	$65.76 \pm 0.58^{aNS,\ b**}$

Values are mean ±SE of 6 rats
P values: * < 0.001, ** < 0.05
NS: Non significant
a → group I compared with groups II, III, IV, V, VI
b → group II compared with groups III, IV, V, VI
Group I: Standard chow diet (Control)
Group II: High fat diet
Group III: High fat diet + Petroleum ether extract of *Bauhinia purpurea* (250 mg/kg B.wt)
Group IV: High fat diet + Ethyl acetate extract of *Bauhinia purpurea* (250 mg/kg B.wt)
Group V: High fat diet + Methanol extract of *Bauhinia purpurea* (250 mg/kg B.wt)
Group VI: High fat diet + standard drug atorvastatin (1.2 mg/kg B.wt)

Table 36.2 Effect of various extracts of *Bauhinia purpurea* on plasma lipid profile in control and experimental rat

Groups	Total cholesterol (mg/dl	Free cholesterol (mg/dl)	Ester cholesterol (mg/dl)	Phospholipid (mg/dl)	Triglyceride (mg/dl)	Athrogenic index
Group I	$122.38 \pm 0.83^{b*}$	$28.63 \pm 0.18^{b*}$	$94.91 \pm 1.51^{b*}$	$99.56 \pm 0.12^{b*}$	$59.02 \pm 0.90^{b*}$	$2.05 \pm 0.03^{b*}$
Group II	$178.13 \pm 2.59^{a*}$	$47.76 \pm 0.18^{a*}$	$132.49 \pm 0.92^{a*}$	$138.44 \pm 0.26^{a*}$	$72.42 \pm 0.69^{a*}$	$4.57 \pm 0.03^{a*}$
Group III	$142.20 \pm 1.88^{a**,\ b**}$	$40.62 \pm 0.51^{a*,\ b*}$	$102.64 \pm 1.21^{a**,\ b*}$	$134.49 \pm 0.11^{a*,\ b**}$	$69.53 \pm 0.27^{a*,\ b**}$	$3.35 \pm 0.05^{a**,\ b**}$
Group IV	$106.74 \pm 1.63^{a*,\ b*}$	$22.41 \pm 0.49^{a*,\ b*}$	$84.45 \pm 1.82^{a*,\ b*}$	$100.19 \pm 0.27^{a*,\ b*}$	$53.96 \pm 0.54^{a*,\ b*}$	$1.87 \pm 0.03^{a*,\ b*}$
Group V	$128.80 \pm 1.87^{a*,\ b*}$	$24.83 \pm 0.12^{a*,\ b*}$	$104.84 \pm 2.07^{a*,\ b*}$	$123.34 \pm 0.21^{a*,\ b**}$	$58.36 \pm 0.61^{a*,\ b**}$	$2.45 \pm 0.04^{a*,\ b*}$
Group VI	$107.84 \pm 2.61^{a*,\ b*}$	$29.00 \pm 0.23^{a*,\ b*}$	$78.60 \pm 1.75^{a*,\ b*}$	$97.78 \pm 0.31^{a*,\ b*}$	$51.74 \pm 0.34^{a*,\ b*}$	$1.80 \pm 0.02^{a*,\ b*}$

Values are expressed as mean ±SE (n = 6 rats)
P values: * < 0.001, ** < 0.05
NS: Non significant

Table 36.3 Effect of various extracts of *Bauhinia purpurea* on plasma lipoprotein in control and experimental rats

Groups	HDL-cholesterol (mg/dl)	LDL-cholesterol (mg/dl)	VLDL-cholesterol (mg/dl)	LDL-c/HDL-c ratio	HDL-c/TC ratio
Group I	$59.03 \pm 0.35^{b*}$	$24.84 \pm 0.43^{b*}$	$11.94 \pm 0.13^{b*}$	$0.41 \pm 0.01^{b*}$	$0.47 \pm 0.02^{b*}$
Group II	$38.91 \pm 0.43^{a*}$	$42.68 \pm 0.23^{a*}$	$14.47 \pm 0.17^{a*}$	$1.08 \pm 0.09^{a*}$	$0.21 \pm 0.01^{a*}$
Group III	$42.50 \pm 0.31^{a**, b*}$	$34.03 \pm 0.33^{a*, b*}$	$13.90 \pm 0.06^{a*, b*}$	$0.79 \pm 0.01^{a**, b*}$	$0.29 \pm 0.01^{a*, b**}$
Group IV	$56.93 \pm 0.46^{a*, b*}$	$21.91 \pm 0.38^{a*, b*}$	$10.86 \pm 0.09^{a*, b*}$	$0.38 \pm 0.01^{a*, b**}$	$0.52 \pm 0.01^{a*, b*}$
Group V	$52.54 \pm 0.35^{a*, b*}$	$27.01 \pm 0.46^{a*, b*}$	$11.57 \pm 0.15^{a*, b**}$	$0.50 \pm 0.01^{a*, b**}$	$0.40 \pm 0.01^{a**, b*}$
Group VI	$59.72 \pm 0.37^{a*, b*}$	$19.96 \pm 0.36^{a*, b*}$	$10.35 \pm 0.08^{a*, b*}$	$0.33 \pm 0.02^{a*, b*}$	$0.55 \pm 0.03^{a*, b*}$

Values are expressed as mean ±SE (n = 6 rats)
P values: * < 0.001, ** < 0.05
NS: Non significant

acetate extract of *Bauhinia purpurea* significantly ($p < 0.001$) increased the beneficial HDL-cholesterol concentration in rats fed with high fat diet as compared to HFD rats (Group II). HFD fed rats (group II) are elevated levels of LDL and VLDL-cholesterol when compared with the control rats (group I). Administration of ethyl acetate extract of *Bauhinia purpurea* were significantly reduced the levels of LDL and VLDL-cholesterol in plasma when compared with HFD rats (group II). The ratios of total cholesterol: HDL-cholesterol and LDL-cholesterol: HDL-cholesterol was presented in Table 36.3. High fat diet rats caused significant ($P < 0.001$) increase in the ratios of total cholesterol: HDL-cholesterol and LDL-cholesterol: HDL-cholesterol. Administration of ethyl acetate extract of *Bauhinia purpurea* along with HFD was found significantly reduced the ratios of total cholesterol: HDL-cholesterol and LDL-cholesterol: HDL-cholesterol when compared to HFD rats (Group II). But the of ethyl acetate extract of *Bauhinia purpurea* along with HFD (Group IV) was showed similar result to standard group rats (VI).

Effect of free and ester cholesterol in plasma and tissue were presented in Tables 36.4 and 36.5. The significant ($P < 0.001$) increase in the levels of both free and ester cholesterol were observed in plasma and tissue of rats fed with high

Table 36.4 Effect of various extracts of *Bauhinia purpurea* on tissues ester cholesterol profile in control and experimental rats

Groups	Ester cholesterol (mg/g tissue)		
	Liver	Heart	Aorta
Group I	$1.94 \pm 0.05^{b*}$	$2.73 \pm 0.09^{b*}$	$2.02 \pm 0.42^{b*}$
Group II	$3.55 \pm 0.02^{a*}$	$7.07 \pm 0.16^{a*}$	$6.81 \pm 0.23^{a*}$
Group III	$2.88 \pm 0.09^{a*}, {}^{b**}$	$6.96 \pm 0.02^{a*}, {}^{b**}$	$6.34 \pm 0.15^{a*}, {}^{b**}$
Group IV	$1.86 \pm 0.01^{a*}, {}^{b*}$	$3.17 \pm 0.02^{a*}, {}^{b**}$	$2.58 \pm 0.01^{a*}, {}^{b*}$
Group V	$2.59 \pm 0.03^{a*}, {}^{b*}$	$5.57 \pm 0.01^{a*}, {}^{b**}$	$4.98 \pm 0.24^{a*}, {}^{b**}$
Group VI	$1.98 \pm 0.09^{a*}, {}^{b*}$	$2.98 \pm 0.01^{a*}, {}^{b*}$	$2.83 \pm 0.11^{a*}, {}^{b*}$

Values are expressed as mean ±SE (n = 6 rats)
P values: * < 0.001, ** < 0.05
NS: Non significant

Table 36.5 Effect of various extracts of *Bauhinia purpurea* on tissues free cholesterol profile in control and experimental rats

Groups	Free cholesterol (mg/g tissue)		
	Liver	Heart	Aorta
Group I	$0.79 \pm 0.01^{b*}$	$0.73 \pm 0.01^{b*}$	$0.40 \pm 0.01^{b*}$
Group II	$1.28 \pm 0.01^{a**}$	$1.03 \pm 0.03^{a*}$	$2.30 \pm 0.02^{a*}$
Group III	$1.26 \pm 0.05^{a**, b**}$	$0.92 \pm 0.02^{a*, b**}$	$1.42 \pm 0.03^{a*, b**}$
Group IV	$0.85 \pm 0.02^{a*, b*}$	$0.63 \pm 0.03^{a*, b*}$	$0.74 \pm 0.01^{a*, b*}$
Group V	$1.02 \pm 0.03^{a*, b*}$	$0.81 \pm 0.03^{a*, b**}$	$1.06 \pm 0.05^{a*, b**}$
Group VI	$0.86 \pm 0.04^{a*, b*}$	$0.64 \pm 0.04^{a*, b*}$	$0.63 \pm 0.04^{a*, b*}$

Values are expressed as mean ±SE (n = 6 rats)
P values: * < 0.001, ** < 0.05
NS: Non significant

Table 36.6 Effect of various extract of Bauhinia purpurea on plasma and tissue triglyceride level in control and experimental rats

Groups	Triglyceride (mg/g tissue)		
	Liver	Heart	Aorta
Group I	8.25 ± 0.01[b*]	10.78 ± 0.11[b*]	10.02 ± 0.03[b*]
Group II	29.47 ± 0.09[a*]	48.24 ± 0.17[a*]	22.14 ± 0.19[a*]
Group III	25.08 ± 0.27[a*, b**]	40.03 ± 0.13[a**, b**]	18.68 ± 0.09[a*, b**]
Group IV	15.25 ± 0.06[a*, b*]	18.35 ± 0.08[a*, b*]	14.46 ± 0.14[a*, b*]
Group V	18.84 ± 0.15[a*, b*]	26.32 ± 0.15[a**, b*]	16.52 ± 0.11[a*, b*]
Group VI	10.65 ± 0.09[a*, b*]	18.78 ± 0.16[a*, b*]	13.15 ± 0.09[a*, b*]

Values are expressed as mean ±SE (n = 6 rats)
P values: * < 0.001, ** < 0.05
NS: Non significant

Table 36.7 Effect of various extracts of *Bauhinia purpurea* on tissues phospholipids level in control and experimental rats

Groups	Phospholipids (mg/g tissue)		
	Liver	Heart	Aorta
Group I	19.40 ± 0.15[b*]	23.48 ± 0.07[b*]	9.32 ± 0.05[b*]
Group II	29.63 ± 0.09[a*]	37.41 ± 0.12[a*]	16.31 ± 0.09[a*]
Group III	25.85 ± 0.05[a**, b*]	34.19 ± 0.15[a**, b**]	15.24 ± 0.12[a*, b**]
Group IV	19.21 ± 0.06[a*, b*]	26.32 ± 0.09[a*, b*]	10.35 ± 0.11[a*, b*]
Group V	23.63 ± 0.11[a*, b**]	32.27 ± 0.07[a*, b**]	12.51 ± 0.13[a*, b*]
Group VI	18.72 ± 0.17[a*, b*]	25.39 ± 0.18[a*, b*]	10.67 ± 0.10[a*, b*]

Values are expressed as mean ±SE (n = 6 rats)
P values: * < 0.001, ** < 0.05
NS: Non significant

(group II) when compared to control rats (group I). Both plasma and tissue free and ester cholesterol reduced remarkably on treating the HFD rats with ethyl acetate extract of *Bauhinia purpurea* (group IV) than that of other two extract treatment group.

Effect of various extract of whole plant of *Bauhinia purpurea* on plasma and tissue triglyceride were depicted in Table 36.6. The concentrations of plasma and tissue triglyceride were elevated in rats fed high fat diet (group II) as compared to control rats (group I). Both plasma and tissue triglyceride levels were significantly reduced in rats treated with ethyl acetate extract of *Bauhinia purpurea* at the dose of 250 mg/kg and as well as standard drug atorvastatin along with high fat diet in comparison with HFD rats (group II).

Effects of various extract of whole plant of *Bauhinia purpurea* on plasma and tissue phospholipids are presented in Table 36.7. The concentration of plasma phospholipids was significantly increased in rats fed HFD (group II) as compared to control rats (group I). After administration of ethyl acetate extract of *Bauhinia purpurea* along with HFD were shown significantly ($p < 0.001$) reduced level of phospholipids in comparison with HFD fed rats (group II). Similar result was not found in other two extract treatment groups.

36.4 Conclusion

The plasma lipid profile were elevated in rats fed with high fat diet (Group II), Earlier studies reveal significant elevation of lipid parameters in plasma and tissue response to atherogenic diet or high fat diet [22–24]. This high cholesterol concentration in circulation may be damage the endothelial cells lining the large arteries and aorta and this may be an initial event in the etiology of atherosclerosis [25]. Administration of ethyl acetate extract of *Bauhinia purpurea* at the dose of 250 mg/kg was significantly reduced the level of lipid profile when compare with high fat diet rats.

In rats fed with high fat diet was showed the elevated level of LDL-C and VLDL-C as compared to control rats. High levels of LDL- and VLDL-cholesterol are major risk factor for coronary heart disease [26]. Studies show that both LDL and VLDL have a positive role in atherogenesis [27]. The reduced level of HDL-C was found in HFD rats, it has been demonstrated that an increase in the concentration of HDL-C correlates inversely with coronary heart disease [28, 29]. After administration of ethyl acetate extract of *Bauhinia purpurea* at the dose of 250 mg/kg was showed significantly ($p < 0.001$) reduced LDL-C, VLDL-C and increased the HDL-C, since HDL-C removes cholesterol to the liver for excretion, the increase in HDL-C will be appropriate for the increased total cholesterol and thus reduce the risk of coronary artery disease. High fat diet rats caused significant ($P < 0.001$) increase in the ratios of total cholesterol: HDL-cholesterol and LDL-cholesterol: HDL-cholesterol. These results are consistent with earlier reports [24, 30]. Administration of ethyl acetate extract of *Bauhinia purpurea* along with HFD was found significantly reduced the ratios of total cholesterol: HDL-cholesterol and LDL-cholesterol: HDL-cholesterol when compared to HFD group (II).

The concentration of plasma and tissue triglyceride was elevated in rats fed with high fat diet. HFD rats significant increase in the level of plasma triglyceride due to decrease in the activity of lipoprotein lipase [31, 32]. Similarly the concentration of plasma and tissue phospholipids also significantly increased in rats fed with HFD, this may be due to decreased phospolipase activity [33, 34]. Administration of ethyl acetate extract of *Bauhinia purpurea* significantly reduced the level of triglyceride and phospholipids when compare with high fat diet rats. The plant extract may have stimulation of lipoprotein.

Lipase activities resulting in decrease of plasma triglyceride and might increase the uptake of triglyceride from plasma by skeletal muscle and adipose tissues [35].

From these result it can be concluded that the ethyl acetate extract of whole plant *Bauhinia purpurea* which decreases plasma lipid and lipoprotein profile and lowers the risk of atherosclerosis in high fat diet fed rats. Therefore, further investigations need to be carried out to isolate and identify the active compounds present in the plant extract.

Acknowledgments The authors are thankful to the Karpagam University, Coimbatore, India, for providing laboratory and technical support for the present investigation.

References

1. Zamani M, Rahimi AO, Mahdavi R, Nikbakhsh M, Jabbari MV, Rezazadeh H, Delazar A, Nahar L, Sarker SD (2007) Assessment of anti-hyperlipidemic effect of *Citrullus colocynthis*. Braz J Pharmacogn 17(4): 492–496
2. Ku SK, Ahn HC, Lee HS (2006) Hypolipidemic effect of water extract of Picrorhiza in PX407 Induced hyperlipidemia ICR mouse model with hepatoprotective effects: a prevention study. J Ethnopharmacol 105:380–386
3. Yusuf S, Reddy S, Ounpuu S, Anand S (2001) Global burden of cardiovascular disease. Part I: General consideration, the epidemiologic transition, risk factor and impact of urbanization. Circulation 104:2746–2753
4. Grover A, Bahl A, Vijayvergiya R, Kumar RM, Thingan ST (2003) Changing trends of burden of cardiovascular disease in India. Implication for prevention based on 2001 census. Indian J Cardiol 6:59–63
5. Okrainec K, Banerjee DK, Eisenberg MJ (2004) Coronary artery disease in the developing world. Am Heart J 148:7–15
6. Kaul U, Sapra R, Ghose T (1998) In: Sethi KK (ed) Coronary artery disease in Indians: a global perspective, pp 11–25
7. Davidson MH, Tooth PP (2004) Comparative effect of lipid lowering therapies. Prog Cardiovasc Dis 47:173–204
8. Dwivedi S (2004) Atherosclerosis revisited. Indian J Cardiol 7:6–12
9. Wang HX, Ng TB (1999) Natural products with hypoglycemic, hypotensive, hypocholesterolemic, antiathero-sclerotic and antithrombotic activities. Life Sci 65: 2663–2677
10. Asolker LV, Kakkar KK, Chakre OJ (2000) Supplement to glossary of Indian medicinal plants. Part-I (A–K). National Institute of Science Communication, New Delhi, pp 116–117
11. Kirthikar KR, Basu DB (2000) Indian medicinal plants, vol 4, 2nd edn. Oriental Enterprises, Dehradun, pp 1255–1257
12. Rajanarayana K, Reddy MS, Chaluvadi MR, Krishna DR (2001) Bioflavonoids classification, pharmacological, biochemical effects and therapeutic potential. Indian J Pharmacol 33:2–16
13. Chopda MZ, Mahajan RT (2009) Wound healing plants of Jalgaon district of Maharashtra state, India. Ethnobot Leafl 13:1–32
14. Harborne JB (1984). Phytochemical methods, 11 edn. Chapman and Hall, New York, pp 4–5
15. Waynforth BH (1980) Injection techniques: experimental and surgical techniques in the rats. Academic Press, London, p 3
16. Kottai Muthu A, Sethupathy S, Manavalan R, Karar PK (2005) Hypolipidemic effect of methanolic extract of *Dolichos biflorus* Linn in high fat diet fed rats. Indian J Exp Biol 43:522–525
17. Freidewald WT, Levy RI, Fredericks0n DS (1972) Estimation of the concentration of low density lipoprotein cholesterol in plasma without use of the preparative ultracentrifuge. Clin Chem 18:499–502
18. Sperry WM, Webb M (1950) Revision of cholesterol determination. J Biol Chem 187:97
19. Folch J, Lees M, Sloane GH (1957) A simple method for the isolation and purification of total lipids from animal's tissues. J Biol Chem 226:497
20. Foster CS, Dunn O (1973) Stable reagents for determination of serum triglyceride by colorimetric Hantzsch condensation methods. Clin Chem 19:338
21. Zilversmit B, Davis AK (1950) Microdetermination of plasma phospholipids by trichloroacetic acid precipitation method. J Lab Clin Inv 35:155
22. Chandar R, Khanna AK, Kapoor NK (1996). Lipid lowering activity of gugulsterone from Commiphora mukul in hyperlipidemic rats. Phyt Res 10:508
23. Guido S, Joseph T (1997) Effect of chemically different antagonists on lipid profile in rats fed on a high fat diet. Indian J Exp Biol 30:292

24. Prasad K (2005) Hypocholesterolemic and antiatherosclerotic effect of flax lignin complex isolated from flax seed. Atherosclerosis 179:269–275
25. Hennig B, Chow CK (1998) Lipid peroxidation and endothelial cell injury: implication in atherosclerosis. Free Radic Biol Med 4(2):99
26. Temme EH, Van HPG, Schouten EG, Kesteloot H (2002) Effect of a plant sterol-enriched spread on serum lipids and lipoproteins in mildly hypercholesterolaemic subjects. Acta card 57:111–115
27. Parthasarathy S, Quinin MT, Schwenke DC, Carew TE, Steinberg B (1989) Oxidative modification of beta very low density lipoprotein. Potential role in monocyte recruitment and foam cell formation. Atherosclerosis 9:398
28. Mayes PA, Murray RK, Granner DK and Rodwell VW (1996) Lipid transport and storage. In: Harper's biochemistry, 24th edn. Prentice Hall International Inc, USA, pp 254–255
29. Philip DM (1995) Plasama enzymes in diagnosis. In: Clinical chemistry in diagnosis and treatment, 6th edn. Arnold Publishers, London, pp 303–307
30. Vijaimohan K, Jainu M, Sabitha KE, Subramaniyam S, Anandhan C, Shyamala Devi CS (2006) Beneficial effects of alpha linoleic acid rich flax seed oil on growth performance and hepatic cholesterol metabolism in high fat diet fed rat. Life Sci 79:448–454
31. Kavitha R, Nalini N (2001) Hypolipidemic effect of green and red chilli extract in rats fed high fat diet. Med Sci Res pp 17–21
32. Van Heek M, Zilversmith DB (1998) Evidence for an inverse relation between plasma triglyceride and aortic cholesterol in the coconut oil cholesterol fed rabbit. Atherosclerosis 71:185–192
33. Mirhadi SA, Singh S (1991) Effect of garlic supplementation to cholesterol rich diet on development of atherosclerosis in rabbit. Indian J Exp Biol 29:162
34. Whereat AF, Robinowitz JL (1975) Aortic mitochondrial synthesis of lipid and its response to cholesterol feeding. Am J Cardiol 55:567
35. El-Hazmi MA, Warsy AS (2001) Evaluation of serum cholesterol and triglyceride levels in 1–6 year-old Saudi children. J Trop Pediatr 47:181–185

Chapter 37
Immunomodulatory Activity of Aqueous Leaf Extract of *Ocimum sanctum*

V. V. Venkatachalam and B. Rajinikanth

Abstract *Objective*: To evaluate the immunomodulatory activity of aqueous leaf extract of *Ocimum sanctum* (ALEOS) by in-vitro and in-vivo methods. *Materials and Methods*: Extracts and standard drug were administered orally for 14 days. The immunomodulatory activity were studied in Wistar strain rats by the following parameters like Delayed Type Hypersensitivity (DTH), Humoral Antibody Titre (HAT), Total Leucocyte Count (TLC) and Differential Leucocyte Count (DLC). *Results*: The effect of ALEOS 200 and 400 mg/kg on DTH response is significantly ($P < 0.05$) increased when compared with the control group. In HAT the ALEOS 200 mg/kg showed significant ($P < 0.05$) effect and ALEOS 400 mg/kg, Standard treatment produced highly significant ($P < 0.01$) when compared with control group. ALEOS 200, 400 mg/kg and standard treatment shows highly significant ($P < 0.01$) in TLC when compared with control group. In DLC count ALEOS 200 mg/kg showed significant effect ($P < 0.05$) only but in ALEOS 400 mg/kg and standard treatment shows highly significant ($P < 0.01$) compared with the control group. Physiochemical examination of the extract showed the presence of carbohydrates, glycosides, saponins, phenolics, tannins and alkaloids. *Conclusion*: Based on the results it can be concluded that the plant *Ocimum sanctum* was found to be a better herb for immunomodulatory activity. Further studies are required to understand the mechanism of action at the molecule level to support these findings.

V. V. Venkatachalam (✉) · B. Rajinikanth
Department of Pharmacy, Annamalai University, Annamalai Nagar,
Chidambaram, 608 002 Tamil Nadu, India
e-mail: Venkat_alamelu@yahoo.co.in

B. Rajinikanth
e-mail: rajini_pharm@yahoo.co.in

R. Malathi and J. Krishnan (eds.), *Recent Advancements in System Modelling Applications*, Lecture Notes in Electrical Engineering 188,
DOI: 10.1007/978-81-322-1035-1_37, © Springer India 2013

Keywords *Ocimum sanctum* · Immunomodulatory · Delayed type hypersensitivity · Humoral antibody titre

Abbreviations

ALEOS	Aqueous leaf extract of *Ocimum sanctum*
OS	*Ocimum sanctum*
DTH	Delayed type hypersensitivity
HAT	Humoral antibody titre
TLC	Total leukocyte count
DLC	Differential leukocyte count
CAS	Centre of Advanced Studies at Marine Biology Department of Annamalai University
WBC	White blood cells

37.1 Introduction

Millions of people around the world use traditional systems of medicine for developing immunity, resistance against infections/diseases, to prevent or alleviate the symptoms of the disease or cure it. The main factors that make natural products more attractive because due to their availability, cost effectiveness and safety [1]. A large number of medicinal plants included in rasayanas have been claimed to possess immunomodulatory activities. Medicinal plants with immunomodulatory effect provide alternative potential to conventional chemotherapy for a variety of diseases, especially in relation to host defense mechanism [2]. Traditionally, several ancient medicinal systems, including Ayurveda, Greek, Roman, Siddha and Unani medicinal system have mentioned various therapeutic properties of OS [3]. The aqueous and alcoholic extract from the leaves of this plant have been investigated extensively for various pharmacological activities including their activity against cancer [4]. OS leaves contains 0.7 % volatile oil comprising about 71 % eugenol and 20 % methyl eugenol. Additional components are carvacrol sesquiterpene hydrocarbon, caryophyllene, apigenin, luteolin, apigenin-7-O glucoronide, orientin, molludistin, and urosilic acid [3]. Mechanism of immunomodulation activity occur mainly via phagocytosis stimulation, macrophages activation, immunostimulatory effect on peritoneal macrophages, lymphoid cell stimulation, cellular immune function enhancement and nonspecific cellular immune system effect, antigen-specific immunoglobulin production increase, increased nonspecific immunity mediators and natural killer cell numbers, reducing chemotherapy induced leukopenia and increasing circulating total white cell counts and interleukin-2 levels. Modulation of immune responses through the stimulatory or suppressive activity of a phyto-extract may help maintain a disease-free state in normal or unhealthy people [5]. The present work is to investigate the immunomodulatory properties of ALEOS in wistar rats by DTH, HAT, TLC and DLC.

37.2 Materials and Methods

37.2.1 Plant Materials Collection and Identification

Fresh leaves were collected from Annamalai Nagar, Tamil Nadu. The specimen was authenticated at the Botany Department of Annamalai University. The shade, dried powdered leaves were stored in air tight container.

37.2.2 Extraction of Plant Materials

The powdered plant material (500 g) was weighed and soaked in chloroform at room temperature for 3 days to remove the chlorophyll. Then washed with distilled water 3–4 times, extracted with distilled water for 24 h at room temperature. The drug and water were kept in the ratio of 1:5. Then it was filtered through a thin muslin cloth. The resultant extract was freeze dried at CAS. The yield was 9 g.

37.2.3 Experimental Animals

Wistar strain rats of either sex weighing between 150 and 250 g were maintained under standard laboratory conditions and provided with standard diet (Pranav Agro Industries Ltd. India) and water *ad libitum*. The experimental protocol has been approved by Institutional Animal Ethics Committee Rajah Muthaiah Medical College, Annamalai University, Reg No. 160/1999/CPCSEA, Proposal number-498.

37.2.4 Preparation of Antigen

Fresh blood was collected from sacrificed sheep in the local slaughter house. Sheep red blood cells (SRBCs) were washed three times in normal saline and adjusted to a concentration of 0.1 ml containing 1×10^8 cells for immunization and challenge [6].

37.2.5 Immunomodulatory Activity

To assess the immunomodulatory activity 24 rats were divided into four groups of six animals each. Group 1-served as control which received 10 ml of potable water, Group 2 and 3 served as extract treatment which received *Ocimum sanctum.,*Linn aqueous extract at a dose 200 and 400 mg/kg/day. Group-4 served as standard treatment Levamisole 50 mg/kg/day [7]. Extracts and standard drug were administered orally for 14 days [8]. Immunomodulatory effect of the extract was evaluated based on the following parameters.

37.2.5.1 Delayed Type Hypersensitivity Response

The animals were immunized by injecting 0.1 ml of SRBC suspension, containing 1×10^8 cells, intraperitoneally on day 0. On day 8, after immunization the thickness of the right hind footpad was measured using a vernier caliper. The rats were then challenged by injection of 1×10^8 cells SRBCs in the left hind footpad. The footpad thickness was measured again after 24 h of challenge. The percentage difference between the pre and post challenge footpad thickness was taken as a measure of DTH response which was calculated as follows [9].

$$\text{DTH} = \frac{\text{Left footpad challenged with antigen} - \text{right footpad control}}{\text{Left footpad challenged with antigen}} \times 100$$

37.2.5.2 Humoral Antibody Titre

The rats were immunized by injecting 0.1 ml of SRBCs suspension containing 1×10^8 cells, on the first day intraperitoneally. Blood samples were collected in centrifuge tubes from individual rats of all the groups by retro orbital vein puncture on tenth day. The blood samples were centrifuged and the serum was separated. Antibody levels were determined by the hemagglutination technique [9].

37.2.5.3 Total Leukocyte Count

The sample blood is diluted with Turk's fluid in WBC pipette. The WBC was counted using Neubauer's chamber [10].

37.2.5.4 Differential Leukocyte Count

A drop of blood was added on the centre line of glass slide about 1 cm from one end and blood diluted with leishmans stain for 30 min and washed with distilled water and dried at room temperature. For counting of DLC the slide was examined under microscope at 100× using cedar wood oil. Finally total number of Neutrophils, Lymphocytes, Eosinophils and Monocytes in the 100 cells were counted and results were expressed in percentage [10].

37.3 Statistical Analysis

The data are expressed as mean ± SEM and subjected to one way ANOVA, followed by a post hoc comparison using Dunnett's *t* test. The probability level more than 95 % was considered to be statistical significance.

37.4 Results

The result obtained in DTH indicates (Table 37.1) that the control animals did not show any characteristic increase in paw edema. The animals treated with both ALEOS 200 and 400 mg/kg showed a significant increase ($P < 0.05$) in paw edema as compared with the control. The standard drug Levamisole showed highly significant ($P < 0.01$) increase in paw edema when compared with control.

Similar to DTH response the control animals the control animals did not show significant humoral antibody titre (Table 37.2). Animals treated with ALEOS 200 mg/kg showed significant increase, but in ALEOS 400 mg/kg as well as standard drug Levamisole at a dose of 50 mg/kg showed highly significant ($P < 0.01$) titre value compared to control.

The results obtained from TLC (Table 37.3) indicates that the group received ALEOS 200, 400 mg/kg and the group received standard drug levamisole exhibited highly significant ($P < 0.01$) increase the mean TLC as compared to control. Further there is increase in the count as the dose is increased indicating that there may be dose dependent increased in the activity. Oral administration of ALEOS for 14 days showed the following differential count in rats.

Table 37.1 DTH response in rats using sheep's RBC's as antigen

Group	Treatment	Dose	DTH response in (mm) mean paw edema ± SEM (n = 6)
1	Control	10 ml/kg of potable water	2.133 ± 0.033
2	ALEOS	200 mg/kg	2.267 ± 0.041*
3	ALEOS	400 mg/kg	2.367 ± 0.036*
4	Standard levamisole	50 mg/kg	3.667 ± 0.033**

DUNNET *t* test and P values as significant * if $p < 0.05$. Highly significant ** if $p < 0.01$ as compared to control

Table 37.2 Humoral antibody titre

Group	Treatment	Dose	Antibody titre mean ± SEM (n = 6)
1	Control	10 ml/kg of potable water	8 ± 0.78
2	ALEOS	200 mg/kg	298.26 ± 21.33*
3	ALEOS	400 mg/kg	414.53 ± 63.12**
4	Standard levamisole	50 mg/kg	426.67 ± 85.33**

DUNNETT *t* test and P values as significant * if $p < 0.05$. Highly significant ** if $p < 0.01$ as compared to control

Table 37.3 Total leukocyte count

Group	Treatment	Dose	Mean leukocyte count ± SEM (n = 6)
1	Control	10 ml/kg of potable water	4736 ± 31.80
2	ALEOS	200 mg/kg	7816 ± 60.99**
3	ALEOS	400 mg/kg	8906 ± 12.02**
4	Standard levamisole	50 mg/kg	9173 ± 14.53**

DUNNETT *t* test and P values as highly significant ** if $p < 0.01$ as compared to control

Table 37.4 Differential leukocyte count

Group	Treatment	Dose	Mean % of lymphocyte ± SEM (n = 6)	Mean % of eosinophils ± SEM (n = 6)	Mean % of neutrophils ± SEM (n = 6)
1	Control	10 ml/kg of potable water	16.00 ± 0.33	3.67 ± 0.67	64.71 ± 0.432
2	ALEOS	200 mg/kg	23.67 ± 1.20**	2.67 ± 0.33	65.83 ± 0.67*
3	ALEOS	400 mg/kg	24.00 ± 1.15**	4.12 ± 0.33**	66.51 ± 1.3**
4	Standard levamisole	50 mg/kg	25.33 ± 0.33**	4.67 ± 0.23**	67.83 ± 0.60**

DUNNETT t test and P values as significant * if $p < 0.05$. Highly significant ** if $p < 0.01$ as compared to control

The results obtained in Table 37.4 indicates that the animal treated with ALEOS 200 mg/kg showed a significant increase in mean percentage of neutrophils and lymphocytes. Mean percentage of eosinopils did not show any significant increase at the ALEOS 200 mg/kg, when compared with control. ALEOS 400 mg/kg and standard drug treatment showed highly significant (P < 0.01) increase in eosinophils, lymphocytes and neutrophil count. This is also dose dependent.

37.5 Discussion

There are few reports on immunomodulatory activity of the extracts from oil of the leaves of OS. Hence it was thought that it will be appropriate to investigate the immunomodulatory activity of the leaves of OS. In the present study the ALEOS was used to evaluate immunomodulatory activity, since most of the herbal drugs are used only as aqueous extract in indigenous system of medicine. Physio-chemical examination of the extract showed the presence of carbohydrates gly-cosides, saponins, phenolics, tannins and alkaloids. In DTH the ALEOS in lower and higher doses increases the paw edema volume similar to the standard drug. It is due to stimulatory effect of the standard drug on chemotaxis dependent leu-kocyte migration, which is known to induce local inflammation with increased vascular permeability, edema and infiltration of PMN leukocytes. These results indicated that the cell mediated immunostimulatory activity of OS. HAT: The antigen antibody reaction results in agglutination. The relative strength of an antibody titre is defined as the reciprocal of the highest dilution which is still capable of causing visible agglutination. The antibody titre is useful to measure the changes in the amount of the antibody in the course of an immune response. The present study indicated that the ALEOS has increased the amount of antibody produced against SRBC i.e., there is a dose dependent humoral immunostimula-tory activity of OS. TLC: There is an increase in the count as the dose is increased there is increase in the activity also. DLC: The higher dose of ALEOS increase in mean percentage of lymphocytes, eosinophils and neutrophils. These results also support the cell mediated immunostimulatory activity. Thus, the ALEOS shows both cell mediated and humoral immunity and dose dependent. The activity may be due to eugenol, carvacrol or apigenin. Further studies are required to establish the molecular basis for mechanism of action.

References

1. Mediratta PK, Sharma KK, Singh S (2002) Evaluation of immunomodulatory potential of *Ocimum sanctum* seed oil and it's possible mechanism of action. J Ethnopharmacol 80:15–20
2. Kumar SV, Kumar SP, Rupesh D (2011) Immunomodulatory effects of some traditional medicinal plants. J Chem Pharm Res 3(1):675–684
3. Dharamani P, Kumar V, Maurya R (2004) Evaluation of anti-ulcerogenic and ulcer healing properties of *Ocimum sanctum* Linn. J Ethnopharmacol 93:197–206
4. Prakash J, Gupta SK (2000) Chemopreventive activity of *Ocimum sanctum* seed oil. J Ethnopharmacol 72:29–34
5. Kumar D, Arya V, Kaur R (2012) A review of immunomodulators in the Indian traditional health care system. J Microbiol Immunol Infect 45(3):165–184
6. Venkatachalam VV, Kannan K, Suresh V (2009) Study of immunomodulatory activity of trikattu churanam in mice. JCPS 2(2):88–90
7. Dashputre NL, Naikwade NS (2010) Immunomodulatory activity of *Abutilon indicum* Linn on Albino mice. IJPSR 1(3):178–184

8. Govinda HV, Asdaq SMB (2011) Immunomodulatory potential of methanol extract of *Aegle marmelos* in animals. Indian J Pharm Sci 73(2):235–240
9. Venkatacham VV, Kannan K, Ganesh S (2009) Preliminary immunomodulatory activities of aqueous extract of *Morus alba* Linn. Int J Chem Sci 7(4):2233–2238
10. Yadav Y, Mohanty PK, Kasture SB (2011) Evaluation of immunomodulatory activity of hydroalcoholic extract of *Quisqualis indica* Linn. Flower in wistar rats. IJPLS 2(4):687–694

Chapter 38
Isolation and Characterization of Active Components Derived from Whole Plant of *Mucuna pruriens* (Linn.)

D. Satheesh Kumar, A. Kottai Muthu and K. Kannan

Abstract The aim of the present investigation was isolation and characterization of active components derived from whole plant of *Mucuna pruriens*. The plant were extracted with various solvents (pet. ether, ethyl acetate and methanol), methanol was found to be more active among them. The preliminary phytochemical results revealed that phytosterols, flavonoids and amino acids as active constituents in methanolic extract of *Mucuna pruriens*. The methanolic extract of *Mucuna pruriens* was undergone column chromatography with different solvent fractions. Despite, three compounds were isolated from methanolic extract of *Mucuna pruriens* with the compound 1 was eluted with benzene: Chloroform 90:10, v/v and compound 2 were eluted with eluted with ethyl acetate: methanol 80:20, v/v and then compound 3 were eluted with ethyl acetate: methanol, 70:30, v/v. The structures of the two isolated compounds were characterized by using FT-IR, NMR and Mass spectrophotometric methods. Thus, the compound 1 was characterized as acetate of 3*β*-Hydroxy-5α-Cholanic acid ($C_{26}H_{38}O_3$), the compound 2 was characterized as 3, 5, 7, 4′-Tetrahydroxy-6-methoxyflavone ($C_{16}H_{12}O_7$) and the compound 3 was characterized as Ethyl 2-amino-5-hydroxy-3, 6, 6-trimethyl heptonate ($C_{12}H_{25}NO_3$). Furthermore, pharmacological studies required for the isolated compounds.

Keywords *Mucuna pruriens* · Isolation · Column chromatography · FT-IR · NMR

D. Satheesh Kumar · A. Kottai Muthu (✉) · K. Kannan
Department of Pharmacy, Annamalai University, Annamalai Nagar,
Chidambaram 608 002, India
e-mail: arthik03@yahoo.com

R. Malathi and J. Krishnan (eds.), *Recent Advancements in System Modelling Applications*, Lecture Notes in Electrical Engineering 188,
DOI: 10.1007/978-81-322-1035-1_38, © Springer India 2013

38.1 Introduction

Mucuna pruriens Linn belongs to the family fabaceae, commonly known as cowhage plant or kapikacho or kevach in Hindi, is the most popular drug in Ayurvedic system of medicine [1]. Traditionally, in India, the seeds of *Mucuna pruriens* are used as a tonic and aphrodisiac for male virility. It has been reported to be antidiabetic [2]. Its different preparations (from seeds) are used for the management of several free radical-mediated diseases such as ageing, rheumatoid arthritis, diabetes, atherosclerosis, male infertility and nervous disorders. It is also used as an aphrodisiac and in the management of Parkinsonism, as it is a good source of L-dopa [3]. The anti-epileptic and anti-neoplastic activity of methanol extract of *Mucuna pruriens* has been reported [4]. It had been reported analgesic and anti-inflammatory [5]. It is also used as a fertility agent in men [6]. The plant is rich in alkaloids such as prurienine, prurieninine and prurienidine [7]. Triterpenes and sterols (β-sitosterol, ursolic acid, etc.,) were found in the root and seeds of *Mucuna pruriens*. The seeds contain proteins, amino acids such as L-DOPA [8]. Therefore, the objective of the present investigation was isolation and characterization of active components derived from whole plant of *Mucuna pruriens* by using FT-IR, NMR and mass spectrophotometric methods.

38.2 Experimental Section

38.2.1 Plant Material

The whole plant of *Mucuna pruriens* (Linn), were collected from Neiyur dam, Kanyakumari District of Tamil Nadu, India. Taxonomic identification was made from Botanical Survey of Medicinal Plants Unit Siddha, Government of India, Palayamkottai. The whole plant of *Mucuna pruriens* (Linn), were dried under shade, segregated, pulverized by a mechanical grinder and passed through a 40 mesh sieve.

38.2.2 Extraction

The powdered plant materials were successively extracted with Petroleum ether (40–60 °C) by hot continuous percolation method in Soxhlet apparatus [9] for 24 h. Then the marc was subjected to Ethyl acetate (76–78 °C) for 24 h and then marc was subjected to Methanol for 24 h. The extracts were concentrated by using a rotary evaporator and subjected to freeze drying in a lyophilizer till dry powder was obtained. All the three extract were stored in screw cap vial at 4 °C until further use.

38.2.3 Preliminary Phytochemical Screening

The extract was subjected to preliminary phytochemical screening for the detection of various plant constituents present. The various extracts of *Mucuna pruriens* was subjected to the following chemical tests such as tests for Alkaloids [10], test for Carbohydrates [10], tests of Glycosides [10], tests for Phytosterol [11], test for Coumarins [11], test for Flavonoids [12, 13], test for Tannins and Phenolic compounds [14], tests for Proteins and Amino Acids [10], test for Saponins [10], test for Fixed Oils [10].

38.2.4 TLC Characterization of Methanolic Extract of Mucuna pruriens

The principle of separation is either partition or adsorption. The constituent which is having more affinity for mobile phase moves with it, while the constituent which is having more affinity for stationary phase gets adsorbed on it. This way various compounds appear as a band on the TLC plate, having different R_f values. The methanolic extract of *Mucuna pruriens* was subjected to thin layer and high performance thin layer chromatographic studies for the separation and identification of their components.

38.2.5 Preparation of Plates

100 g of silica gel G was weighed and made into a homogenous suspension with 200 ml of distilled water to form slurry. The slurry was poured into a TLC applicator, which was adjusted to 0.25 mm thickness on flat glass plate of different dimensions (10 × 2, 10 × 5, 30 × 5, 20 × 10 cm etc.,). The coated plates were allowed to dry in air, followed by heating at 100–105 °C for 1 h, cooled and protected from moisture. Before using, the plates were activated at 110 °C for 10 min.

38.2.6 Separation of Components

The methanolic extracts of *Mucuna pruriens* was dissolved in methanol separately and spotted using a capillary tube on TLC plates 2 cm above from the bottom of the plate. The selection of solvent systems was based on increasing the order of polarity. The different spots developed in each system were detected by means of iodine staining.

38.2.7 *Isolation of Methanolic Extract of* Mucuna Pruriens *by Using Column Chromatography*

The 20 g of methanolic extract of *Mucuna pruriens* was admixed with 20 g silica gel (60/120 meshes) to get uniform mixing. 200 g of silica gel (70/325 meshes) was taken in a suitable column and packed very carefully without air bubbles using hexane as filling solvent. The column was kept aside for 1 h and allowed for close packing. Admixture was then added at the top of the stationary phase and started separation of compounds by the eluting with various solvent mixtures with increasing order of polarity. All the column fractions were collected separately and concentrated under reduced pressure. Finally the column was washed with ethyl acetate and methanol.

38.2.8 *Characterization of Isolated Compounds*

38.2.8.1 FT-IR

IR spectra of the compounds isolated from the methanolic extracts of *Mucuna pruriens* were recorded using a Nicolet 170SX. The spectral resolution for the Nicolet 170SX was 0.25 cm^{-1}, and the spectral data were stored in the database at intervals of 0.5 cm^{-1} at 4000–2000 cm^{-1}, and of 0.25 cm^{-1} at 2000–400 cm^{-1}. Liquid samples were measured with liquid film method, and solid samples were measured by using KBr disc methods.

38.2.8.2 ^{1}HNMR

^{1}HNMR spectra of the compounds isolated from the methanolic extracts of *Mucuna pruriens* was recorded using a JEOL AL-400 (399.65 MHz). The measuring conditions for the most of the spectra were as follows: flip angle of 22.5–30.0°, pulse repetition time of 30 s. The long pulse repetition time and small flip angle is used to ensure precise relative intensities. The ^{1}HNMR chemical shifts were referred to TMS in organic solvents and TSP in D_2O.

38.2.8.3 ^{13}CNMR

^{13}CNMR spectra of the compounds isolated from the methanolic extracts of *Mucuna pruriens* was recorded with a Bruker AC-200 (50.323 MHz). The measuring conditions for the most of the spectra were as follows: a pulse flips angle of 22.45–45°, a pulse repetition time of 4–7 s, and a resolution of 0.025–0.045 ppm. The spectra whose spectral codes started with "CDS" were reconstructed from peak positions, intensities, and line widths by assuming all resonance peaks were Lorenz lines. The chemical shift was referred to a TMS for all solvents.

38.2.8.4 Mass Spectrum

Mass spectra of the compounds isolated from the methanolic extracts of *Mucuna pruriens* was recorded with JEOL JMS-700 by the electron impact method where an electron is accelerating voltage 75 eV and an ion accelerating voltage of 8–10 nV. The reservoir inlet systems were used. The dynamic range for the peak intensities were three digits and the accuracy of the mass number was 0.5.

38.3 Results and Discussion

The various extracts of *Mucuna pruriens* (Linn.) were subjected to screening for its phytochemical constituents. The phytochemical screening results are shown in Table 38.1. The petroleum ether extract of *Mucuna pruriens* was contains phytosterols and fixed oils and fats. Ethyl acetate extracts containing Alkaloids, Carbohydrates, protein and amino acid compounds and fixed oils and fats. The Methanolic extracts containing Alkaloids, Carbohydrates, glycoside, Phenolic compounds, Saponins, Tannins, Protein and amino acids, coumarins and flavonoids.

Petroleum ether, ethyl acetate and methanol were used individually as solvent for the extraction of *Mucuna pruriens*. The methanolic extract of *Mucuna pruriens* was found more active among them. Therefore, the methanolic extract of *Mucuna pruriens* was subjected to the TLC chromatographic profile and column chromatographic separation. The methanolic extract of *Mucuna pruriens* dissolved in their mother solvent was taken in a capillary tube and spotted on TLC plates 2 cm above its bottom. Most of the sample for application were between 0.1 and 1 %.

Table 38.1 Phytochemical analysis of various extracts of whole plant of *Mucuna pruriens* (Linn.)

S.No.	Test	Petroleum ether	Ethyl acetate	Methanol
I	Alkaloids	–	+	+
II	Carbohydrates and glycosides	–	+	+
III	Phytosterols	+	–	+
IV	Fixed oil and fats	+	+	–
V	Saponins	–	–	+
VI	Phenolic compounds and tannins	–	–	+
VII	Protein and amino acid	–	+	+
VIII	Coumarins	–	–	+
IX	Flavonoids	–	–	+

+ Positive; – Negative

Table 38.2 TLC profiles of methanolic extracts of *Mucuna pruriens*

S.No	Solvent system	No. of Spot	Rf Value
1.	Benzene : Chloroform (90:10)	2	0.72, 0.35
2.	Benzene : Chloroform (80:20)	2	0.72, 0.26
3.	Benzene : Chloroform (70:30)	2	0.57, 0.37
4.	Ethyl acetate: Methanol (70:30)	2	0.88, 0.65
5.	Ethyl acetate: Methanol (50:50)	2	0.68, 0.47, 0.34

The applied spots were of equal size as far as possible and diameter ranging from 2 to 3 mm. The solvent system for methanolic extracts was developed by trial and error method using various solvents which were differing in polarities (Table 38.2).

The methanolic extract of *Mucuna pruriens* was subjected to column chromatographic separation using normal phase silica gel column. The dark brown solid (20 g methanolic extract of *Mucuna pruriens*) was adsorbed on silica gel (20 g) and transferred to a column of silica gel (200 g equilibrated with benzene). Despite, three compounds were isolated in column chromatography with different solvents. Obviously, the compound 1 (140 mg) was eluted with benzene: Chloroform 90:10, v/v, compound 2 (165 mg) was eluted with ethyl acetate: methanol, 80:20 v/v and compound 3 was eluted with ethyl acetate: methanol, 70:30, v/v. This active fraction was used to identify the chemical tests showed the presence of phytosterols, flavonoids and amino acid as active compounds. The actual compounds were isolated from column chromatography as mentioned in the experimental section. The spectra (IR, ^{1}H and ^{13}CNMR and Mass) of these compounds as mentioned in the experimental section.

38.3.1 Characterization of Compound 1

The spectral data IR, ^{1}HNMR and ^{13}CNMR and Mass of the compound 1 are good in agreement with the structure proposed for the compound. The melting point of the compound 1 was found as 149 °C. The IR spectrum of the compound 1 was analysed from the IR data. The absorption at 3,427 cm^{-1} indicates the presence of –OH group. The presence of –C=O group in this compound is revealed by the strong absorption at 1692 cm^{-1}. The unsaturation such as –C=C is known from the absorption at 1458 cm^{-1}. The strong absorption at 1027 cm–1 indicates the presence of –C–O-C link.

The ^{1}HNMR chemical shift values the chemical shift values at δ, 11.94 and 2.50 ppm show the presence of –OH group (–COOH). The singlet at δ, 5.13 and 4.29 ppm are due to proton at C_6 and C_{22} carbon atom respectively. The chemical shifts as multiplets at δ, 2.11–0.86 ppm are due to the presence of methylene and methyl groups. The ^{13}CNMR chemical shift at δ, 178.58 and 178.28 ppm are due to the carbonyl carbon of acid group (–COOH) and acetyl group(–CO–O-CH_3). The

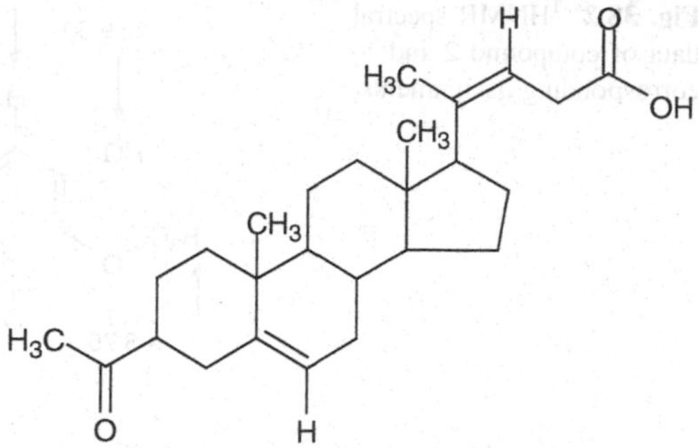

Fig. 38.1 Structure of compound 1(acetate of 3*β*-hydroxy-5*α*-cholanic acid)

chemical shift at *δ*, 138, 124 and 121 ppm are due to C_{22}, C_6 and C_3 respectively. The mass spectral analysis of compound 1 led to the molecular peak *m/z* 404(M + 4), which indicated the molecular formula $C_{26}H_{38}O_3$. Thus, the compound 1 was characterized as acetate of 3*β*-Hydroxy-5*α*-Cholanic acid was given in Fig. 38.1. The Molecular Formula of the compound 1 was deduced as $C_{26}H_{38}O_3$.

38.3.2 Characterization of Compound 2

The spectral data IR, ^{1}HNMR and ^{13}CNMR and Mass of the compound 2 are good in agreement with the structure proposed for the compound. The melting point of the compound 2 was found as 235 °C. The IR spectrum of the compound 2 was analysed from the IR data. The presence of –OH group known from the absorption at the range 3330–3463 cm^{-1}. A strong band at the range 1655–1617 cm^{-1} is due to the presence of –C=O group. The presence of –C–O–C– indicates in the absorption at 1060 cm^{-1}. The ^{1}HNMR and ^{13}CNMR chemical shift values of the compound 2 was found to be 3, 5, 7, 4′-Tetrahydroxy-6-methoxyflavone (Figs. 38.2 and 38.3). The mass spectral analysis of compound 2 led to the molecular peak *m/z* 316, which indicated the molecular formula $C_{16}H_{12}O_7$. Thus, the compound 2 was characterized as 3, 5, 7, 4′-Tetrahydroxy-6-methoxyflavone was given in Fig. 38.4. The Molecular Formula of the compound 2 was deduced as $C_{16}H_{12}O_7$. This is the first report of occurrence of this compound in nature as well as the alkaloids in this plant.

38.3.3 Characterization of Compound 3

The spectral data IR, ^{1}HNMR and ^{13}CNMR and Mass of the compound 3 are good in agreement with the structure proposed for the compound. The melting point of the compound 3 was found as 235 °C. The IR spectrum of the compound 3 was analysed from the IR data. Absorption at 3412 cm^{-1} shows the presence of –OH

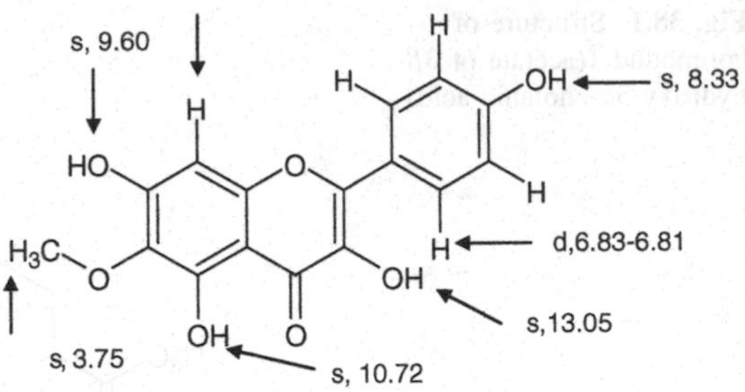

Fig. 38.2 ^{1}HNMR spectral data of compound 2 and corresponding assignments

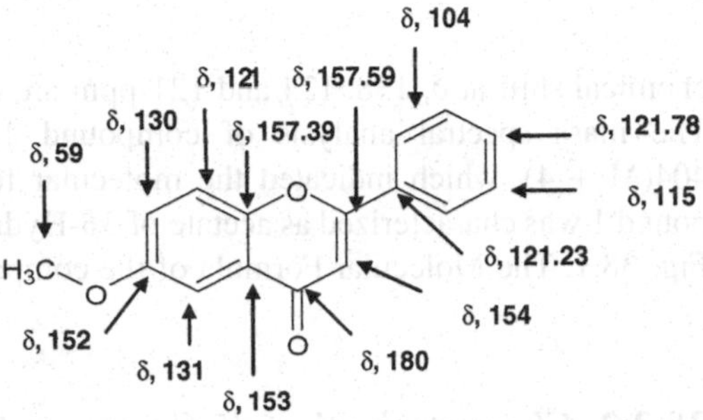

Fig. 38.3 ^{13}CNMR spectral data of compound 2 and corresponding assignments

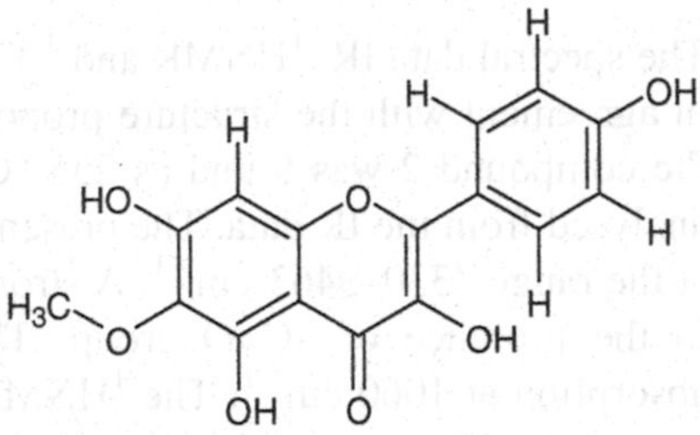

Fig. 38.4 Structure of Compound 2(3, 5, 7, 4′-tetrahydroxy-6-methoxyflavone)

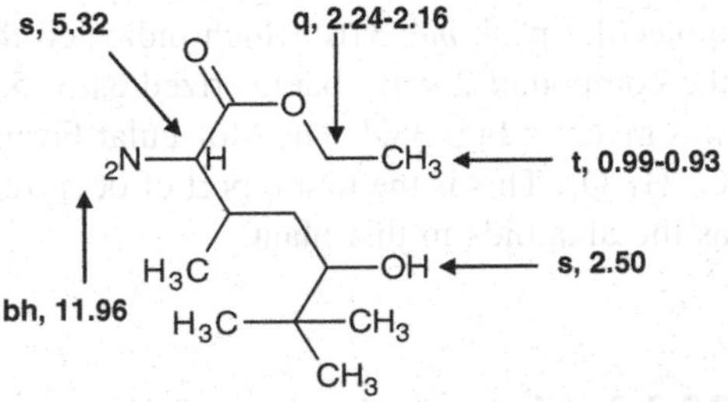

Fig. 38.5 ^{1}HNMR spectral data of compound 2 and corresponding assignments

group, whereas the strong band at 1734 cm^{-1} indicates the presence of carbonyl group of aliphatic compound. The absorption at 1173 cm^{-1} reveals the presence of –C–N stretching. The ^{1}HNMR and ^{13}CNMR chemical shift values of the compound 2 was found to be ethyl 2-amino-5-hydroxy-3, 6, 6-trimethyl heptonate (Figs. 38.5 and 38.6). The mass spectral analysis of compound 3 led to the

Fig. 38.6 ^{13}CNMR spectral data of compound 2 and corresponding assignments

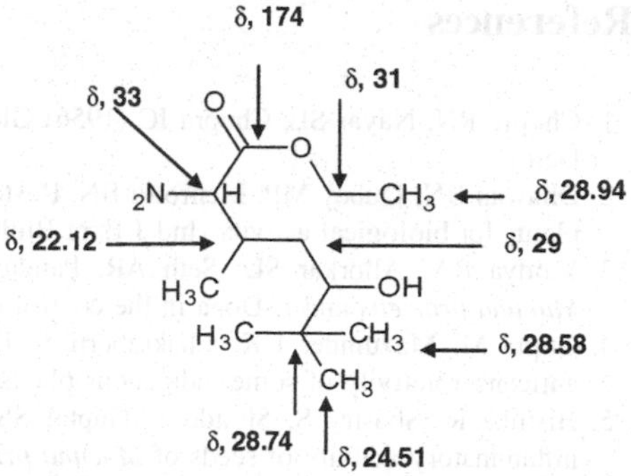

Fig. 38.7 Structure of Compound 3 (ethyl 2-amino-5-hydroxy-3, 6, 6-trimethyl heptonate)

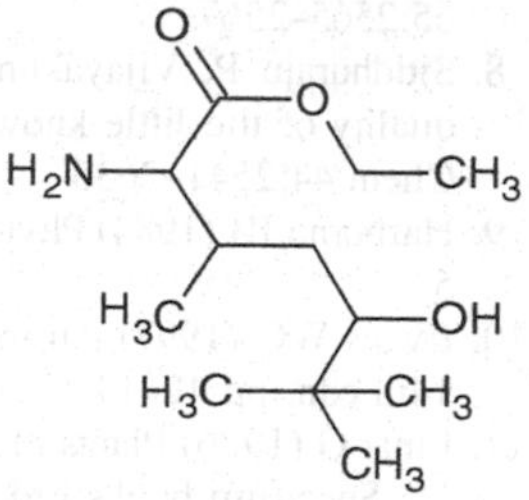

molecular peak *m/z* 434, which indicated the molecular formula $C_{16}H_{12}O_7$. Thus, the compound 2 was characterized as ethyl 2-amino-5-hydroxy-3,6,6-trimethyl heptonate was given in Fig. 38.7. The Molecular Formula of the compound 3 was deduced as $C_{12}H_{25}O_3$.

38.4 Conclusion

From the above reports, three compounds were isolated from methanolic extract of *Mucuna pruriens* (Linn.) such as 3β-Hydroxy-5α-Cholanic acid ($C_{26}H_{38}O_3$), 3, 5, 7, 4′-Tetrahydroxy-6-methoxyflavone ($C_{16}H_{12}O_7$) and ethyl 2-amino-5-hydroxy-3,6,6-trimethyl heptonate ($C_{12}H_{25}NO_3$). However, this is the first report of occurrence of 3, 5, 7, 4′-Tetrahydroxy-6-methoxyflavone ($C_{16}H_{12}O_7$) in nature as well as the alkaloid in this plant. Therefore, further in-depth biological investigations need for the isolated compounds.

Acknowledgments The authors are grateful to University Grants Commission, Govt. of India, New Delhi for providing financial assistance and motivation for the present investigation.

References

1. Chopra RN, Nayar SL, Chopra IC (1956) Glossary of Indian medicinal plants. CSIR, New Delhi
2. Dhawan BN, Dubey MP, Mehrotra BN, Rastogi RP, Tandon JS (1980) Screening of Indian plants for biological activity. Ind J Expt Biol 18:594–606
3. Vaidya RA, Allorkar SD, Seth AR, Panday SK (1978) Activity of bromoergocryptine, *Mucuna pruriens* and L-Dopa in the control of hyperprolactinemia. Neurology 26:179–186
4. Gupta M, Mazumder UK, Chakraborti S, Rath N, Bhawal SR (1997) Antiepileptic and anticancer activity of some indigenous plants. Indian J Physiol Allied Sci 51(2):53–56
5. Hishika R, Shastry S, Shinde S, Guptal SS (1981) Prliminary, phytochemical and anti-inflammatory activity of seeds of *Mucuna pruriens*. Indian J Pharmacol 13(1):97–98
6. Buckles D (1995) Velvet bean: a ‘new’ plant with a history. Econ Bot 40:13–25
7. Misra L, Wagner H (2004) Alkaloidal constituents of mucuna pruriens seeds. Phytochemistry 65:2565–2567
8. Siddhuraju P, Vijayakumari K, Janardhanan K (1996) Chemical composition and protein quality of the little known legume, velvet bean (*Mucuna pruriens* (L.) DC.). J Agric Food Chem 44:2541–2636
9. Harborne JB (1984) Phytochemical methods 11th edn. In: Chapman & Hall, New York, pp 4–5
10. Evans WC (1997) An index of medicinal plants. A text book of pharmacognosy, vol 7(5), 14th edn, pp 12–14
11. Finar G (1986) Plants of economic importance. Medicinal plants and medicine in Africa, vol 78. Spectrum books Ltd, Ibadan, pp 150–153
12. Dey PM, Harborne JB (1987) Methods in plant biochemistry. Academic Press, London
13. Evans WC (1989) Pharmacognosy, 13th edn. Balliere-Tindall, London
14. Mace Gorbach SL (1963) Anaerobic bacteriology for clinical laboratories. Pharmacognosy 23:89–91

Chapter 39
Despeckling of Polycystic Ovary Ultrasound Images by Fuzzy Filter

H. Prasanna Kumar and S. Srinivasan

Abstract Ultrasound imaging is a widely used and safe medical diagnostic technique, due to its noninvasive nature, low cost, real time imaging. However the usefulness of ultrasound imaging is degraded by the presence of signal depended noise known as speckle. Removing speckle noise from the original image is still a challenging research in image processing. Nonlinear techniques have recently assumed significance as they are able to suppress speckle noise which is also called as multiplicative noise to preserve important signal elements such as edges and fine details. Among nonlinear techniques, the fuzzy logic based approaches are important as they are capable of reasoning with vague and uncertain information. This paper presents a fuzzy filter for suppressing noise in polycystic ovary image to show the feasibility of the proposed noise reduction using Fuzzy filter. The better performance of the filter is demonstrated on the basis of MSE and RMSE values calculated from the original and restored images respectively.

Keywords Ultrasound · Speckle noise · Fuzzy logic · MSE RMSE

39.1 Introduction

In medical image processing, it is very important to obtain precise images to facilitate accurate observations for the given application. Low image quality is an obstacle for effective segmentation, feature extraction, analysis, recognition and

H. P. Kumar (✉) · S. Srinivasan
Department of Instrumentation, MIT Campus, Anna University, Chennai 600044, India
e-mail: Uvcehpk@gmail.com

S. Srinivasan
e-mail: srini@mitindia.edu

R. Malathi and J. Krishnan (eds.), *Recent Advancements in System Modelling Applications*, Lecture Notes in Electrical Engineering 188,
DOI: 10.1007/978-81-322-1035-1_39, © Springer India 2013

quantitative measurements. Therefore, there is a fundamental need of noise reduction from medical images. There are currently a number of imaging modalities that are used for study of medical image processing. Among the Ultrasound imaging [1] are believed to be very potential for accurate measurement of organ anatomy in a minimally invasive way. It is used for Imaging soft tissues in organs like liver, kidney, Spleen, uterus heart, brain etc. The common problem in ultrasound image is speckle noise [2] which is caused by the imaging technique used that may be based on coherent waves such as acoustic to laser imaging [3, 4]. For these kinds of noises, de-noising should be performed to improve the image quality for more accurate diagnosis. The main objective of image-de-noising techniques is to remove such noises while retain as much as possible the important signal features. There is no unique technique for noise removing from affected image. Different algorithms are used depending on the noise model. The averaging filtering technique [5] can successfully remove noise from the distorted image but in this case the filtered image suffers the blurring effect. For the mean filtering [5, 6] techniques each pixel is considered to calculate the mean and also every pixel is replaced by that calculated mean. So affected pixels are considered to calculate the mean and unaffected pixels are also replaced by this calculated mean. The median filter was once the most popular nonlinear filter for removing noise, because of its good de-noising power [6] and computational efficiency. Their main drawback is that the noisy pixels are replaced by some median Value in their vicinity without taking into account local features such as the possible presence of edges. Hence details and edges are not recovered satisfactorily, especially when the noise level is high. This paper presents a despeckling method, based on fuzzy weighted mean filter [7] is an extension of the adaptive weighted mean filter. The idea behind the filter is that the weights should take values in [0, 1], instead of only the crisp values 0, 1, and that the weights should not depend on a threshold value, but should be determined by means of fuzzy rules. For the image quality performance measure we used mean square error (MSE) and Root mean square error (RMSE) they are better measurements for speckle noise.

39.2 Speckle Noise

Speckle noise affects all coherent [2, 4] imaging systems including laser, Synthetic Aperture Radar (SAR) imagery, and ultrasound. Speckle may appear distinct in different imaging systems, but it is always manifested in a granular pattern due to image formation under coherent waves. Goodman has described the basic properties of speckle [8]. A general model for speckle noise proposed by Jain [6]. In the following, we will formulate the ultrasound speckle model. Denote by $\mathbf{N(x, y)}$ a noisy recorded ultrasound image, $\mathbf{M(x, y)}$ the noise-free image that has to be recovered and by $\mathbf{S_m (x, y)}$ and $\mathbf{S_a(x, y)}$, the corrupting multiplicative and additive speckle noise components, respectively one can write:

$$\mathbf{N(x,y) = M(x,y)S_m(x,y) + S_a(x,y)} \tag{39.1}$$

Generally, the effect of the additive component of the speckle in ultrasound images is less significant than the effect of the multiplicative component. Thus, ignoring the term, one can rewrite Eq. (39.1) as

$$\mathbf{N(x, y) = M(x, y)S_m(x, y)} \tag{39.2}$$

39.3 Fuzzy Filter

The adaptive weighted mean filter [7] replaces the gray value of a pixel **(i, j)** by a weighted average of the gray values in a neighborhood of that pixel. The choice of the weights is based on the gray value differences $|\mathbf{x(a, b)-x(a-c, b-d)}|$: if this difference exceeds a certain threshold, one defines $\mathbf{w_{ij}(a, b) = 0}$; in the other case $\mathbf{w_{ij}(a, b) = 1}$. The weights take values in the range **[0, 1]**, and they do not depend on a threshold value, but they are determined by means of fuzzy rules. The filter approach has two stages; each of them uses a fuzzy rule based [9, 10] system. In the first stage, try to determine whether a pixel is a noisy pixel or not. For this purpose, we use a fuzzy rule based system [8, 9] to determine a degree for each pixel of the image. The degree is a real number in the range **[0, 1]**. If the degree of a pixel is equals to 1, we'll assume that the pixel is not corrupted, and if it is less than 1, we will assume the pixel is noisy. The nearer the degree of a pixel to zero, the more it is considered as a noisy pixel. After finishing this stage, we'll have a degree matrix as the same size as the corrupted image. We use this matrix in the next stage which performs fuzzy smoothing by weighting the contributions of neighborhood pixel values.

39.3.1 Noise Estimation Fuzzy

In this part we want to determine whether a pixel is corrupted or not. For this, the following criteria are considered:

1. If a pixel is severely noisy, there isn't any similar gray level value in its neighborhood pixels, so the minimum gray value difference of that pixel and its 8-neighborhood pixels is large. Reversely, if minimum gray level difference of a pixel and its neighborhood pixels is small, one assumes that the pixel is not categorized as a noisy pixel. Hence we use minimum gray level differences as the first parameter of our fuzzy rule based system:

$$\mathbf{Diff} = \mathbf{min}|x(a,b) - x(a',b')| \tag{39.3}$$

where $\mathbf{x(a', b)}$ is an 8-neighborhood pixel of **(a, b)**

2. If a pixel has many similar pixels in its neighborhood, one assumes that it is uncorrupted, so we can use number of similar pixels to an assumed pixel in its 8-neighborhood as an important parameter to realize whether the pixel is corrupted or not. For this, we determine the number of pixels in the 8-neighborhood of a given pixel that their gray level differences with central pixel is less than a predefined threshold. We exploit this number as the second parameter of our fuzzy. Rule based system: with central pixel is less than a predefined threshold. We exploit this number as the second parameter of our fuzzy rule based system:

$$\mathbf{Similar\ pixel} = (\mathbf{Number\ Of}(\mathbf{a}',\mathbf{b}') \in \mathbf{N_8}(\mathbf{a},\mathbf{b})\ \mathbf{and}\ |\boldsymbol{x}(\boldsymbol{a},\boldsymbol{b}) - \boldsymbol{x}(\boldsymbol{a}',\boldsymbol{b}')| < \mathbf{Threshold}) \quad (39.4)$$

Here we set threshold is equal to five, but it may vary for each image dynamically to get the good result. The output of the fuzzy system is a degree associated to each pixel that is a real number between 0 and 1. It denotes the degree which a pixel is considered as an uncorrupted pixel. Fuzzy membership functions are illustrated in Fig. 39.1 fuzzy rules are as follows.

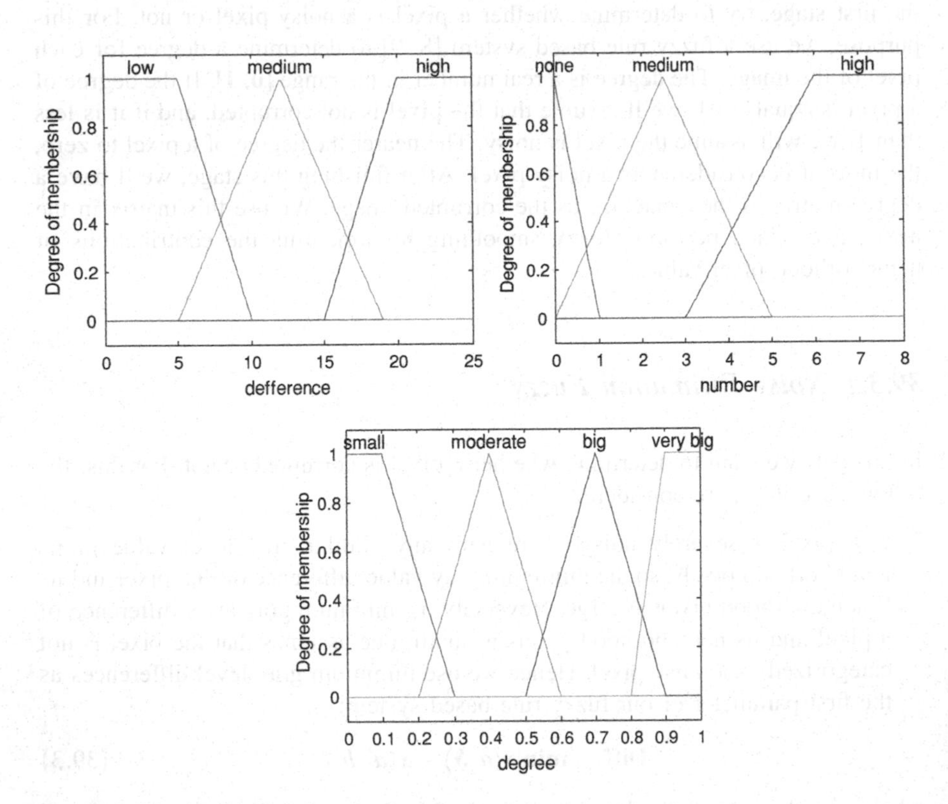

Fig. 39.1 Membership functions for fuzzy noise estimation

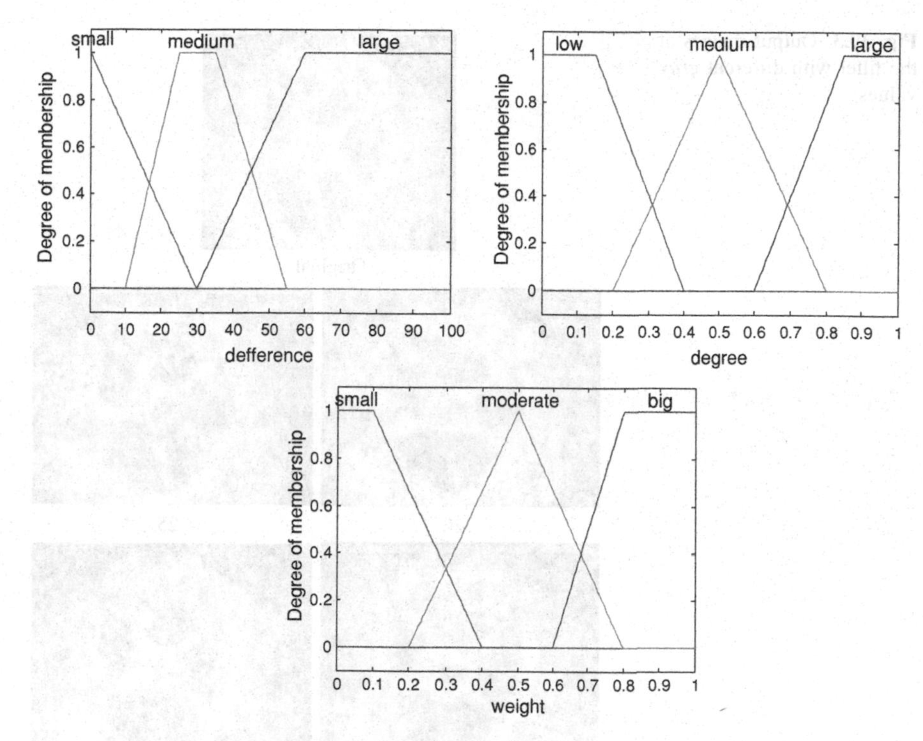

Fig. 39.2 Membership functions for fuzzy smoothing

1. If (difference is small) and (number is none) then (degree is moderate)
2. If (difference is small) and (number is few) then (degree is big)
3. If (difference is small) and (number is many) then (degree is very big)
4. If (difference is med) and (number is none) then (degree is small)
5. If (difference is med) and (number is few) then (degree is moderate)
6. If (difference is med) and (number is many) then (degree is big)
7. If (difference is high) and (number is none) then (degree is small)
8. If (difference is high) and (number is few) then (degree is moderate)
9. If (difference is high) and (number is many) then (degree is moderate)

39.3.2 Fuzzy Smoothing

The fuzzy weighted averaging is accomplished in this stage. The pixels weights of each neighborhood are obtained using a fuzzy smoothing technique. For this, we consider the following criterion: if the

Difference $|\mathbf{x}(\mathbf{a}, \mathbf{b}) - \mathbf{x}(\mathbf{a}-\mathbf{c}, \mathbf{b}-\mathbf{d})|$) is large; a small weight $\mathbf{w}_{ij}(\mathbf{c}, \mathbf{d})$. Must be applied to the pixel to reduce its contribution in averaging process. Hence we use

$$|\mathbf{x}(\mathbf{a}, \mathbf{b}) - \mathbf{x}(\mathbf{a}-\mathbf{c}, \mathbf{b}-\mathbf{d})|$$

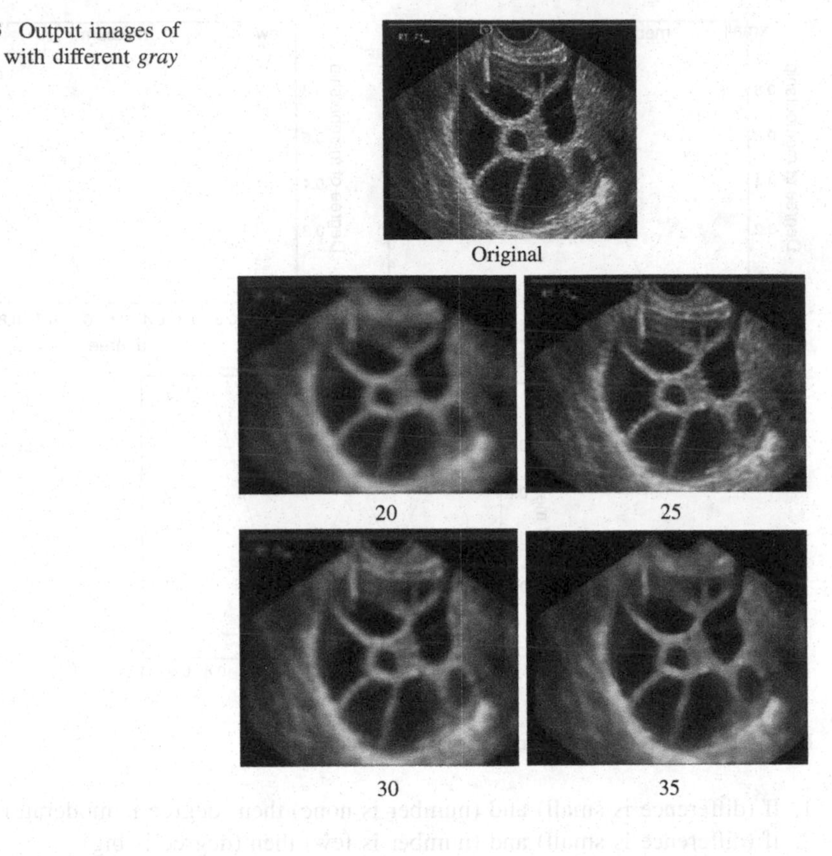

Fig. 39.3 Output images of the filter with different *gray* values

Table 39.1 Comparison of different gray values

Difference	MSE	RMSE
20	36.456	6.037
25	27.774	5.270
30	30.667	5.537
35	33.286	5.769

As first parameter in Fuzzy smoothing process

$$\textbf{Difference} = |x(a,b) - x(a-c,b-d)| \quad (39.5)$$

The second parameter is the output of previous fuzzy rule based system, which denotes the degree that a pixel is not considered as a noisy pixel. The output of fuzzy smoothing is the weight of pixel in the averaging process. Fuzzy membership functions are shown in Fig. 39.2. Fuzzy rules are as follows.

1. If (diff is small) and (degree is low) then (weight is medium)
2. If (difference is small) and (degree is medium) then (weight is medium)
3. If (difference is small) and (degree is large) then (weight is large)

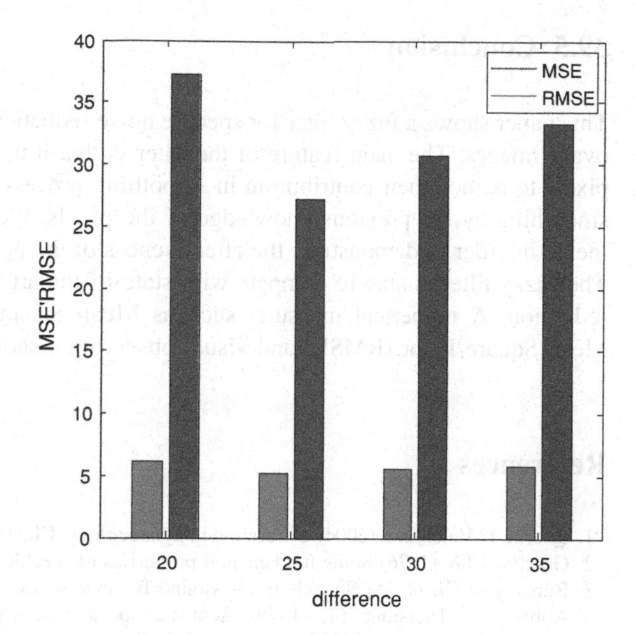

Fig. 39.4 Performance analysis of the different *gray* values

4. If (difference is medium) and (degree is low) then (weight is low)
5. If (difference is medium) and (degree is medium) then (weight is medium)
6. If (difference is medium) and (degree is large) then (weight is large)
7. If (difference is large) and (degree is low) then (weight is low)
8. If (difference is large) and (degree is medium) then (weight is low)
9. If (difference is large) and (degree is large) then (weight is medium)

39.4 Results

The performance of various spatial enhancement approaches are analyzed and discussed. The statistical measurement could be used to measure enhancement of the image. The Mean Square Error (MSE) and Root Mean Square Error (RMSE) are used to evaluate the enhancement performance [11]. Figure 39.3 original polycystic ultrasound image and filtered image of the polycystic image obtained by fuzzy filtering techniques are shown in figure. If the value of MSE and RMSE is low then the enhancement approach is better. As shown in the Table 39.1 the different window statistics values. In this paper we also tried with different gray scale value difference and gray scale value 25 gives the good result. Because the polycystic ovaries occur between the gray value of 25–30. Figure 39.4 show the performance analysis of the different gray scale values.

39.5 Conclusion

This paper shows a fuzzy filter for speckle noise reduction in ultrasound polycystic ovary images. The main feature of the filter is that it tries to determine corrupted pixels to reduce their contribution in smoothing process. Hence it performs fuzzy smoothing by the previous knowledge of the pixels. We performed some experiments in order to demonstrate the effectiveness of the proposed filtering approach. The fuzzy filter is able to compete with state-of-the-art filter techniques for noise reduction. A numerical measure, such as Mean Square Error (MSE) and Root Mean Square Error (RMSE) and visual observation show convincing results.

References

1. Bourne T, Valentin L (2004) Ultrasound in gynaecology. Clin Obstet Gynaecol 18(1):A1–A20
2. Goodman JW (1976) Some fundamental properties of speckle. J Opt Soc Am 66:1145–1150
3. Burckhardt CB (1978) Speckle in ultrasound B-mode scans. IEEE Trans Ultrason 25:1–6
4. Abbott JG, Thurstone FL (1979) Acoustic speckle: theory and experimental analysis. Ultrason Imaging 1:303–324
5. Gonzalez RC, Woods RE (1992) Digital image processing. Addison-Wesley Publishing Company, Reading
6. Jain AN (1989) Fundamentals of digital image processing. Prentice-Hall, Englewood Cliffs
7. Arakawa K (2000) Fuzzy rule-based image processing with optimization, fuzzy techniques in image processing. Springer, Berlin
8. Yongjian Y, Acton ST (2002) Speckle reducing anisotropic diffusion. IEEE Trans Image Process 11(11):1260–1270
9. Schulte S, Nachtegael M, Witte VD, Van der Weaken D, Kerre EE (2006) A fuzzy impulse noise detection and reduction method. IEEE Trans Image Process 15(5):1153–1162
10. Van De Ville D, Nachtegael M, Van der Weken D, Kerre E, Philips W, Lemahieu I (2003) Noise reduction by fuzzy image filtering. IEEE Trans Fuzzy Syst 11(4):429–436
11. Gagnon L, Smaili FD (1996) Speckle noise reduction of airborne SAR images with symmetric daubechies wavelets. In: SPIE Proceedings #2759, p 1424

Chapter 40
Characterisation of Streptomycetes from Lignite Mines and their Antagonistic Activity Against Bacteria and Fungi

V. Parthasarathy, G. S. Prasad and R. Manavalan

Abstract It has been well established that microorganisms are virtually an unlimited source of natural products, many of which have potential therapeutic applications. Among the various organisms the filamentous soil bacteria of the genus *Streptomyces* are remarkable and are considered as a potential source of important bioactive compounds. In the present study we have isolated actinomycetes from six locations in Neyveli lignite mine area, Tamilnadu, India and tested for their antagonistic activity. Seven isolates, which possess good antagonistic activity against bacterial and fungal pathogens, were selected and all the isolates represented the genus *Streptomyces*. The isolates have been identified up to the species level as per ISP procedures. The results of the study prove that the lignite mines are very promising zone for potential actinomycetes.

Keywords Actinomycetes · Streptomycetes · Antibiotics · Antifungal antibiotic · Soil bacteria · Actinomycetes in coal mine

40.1 Introduction

Among the microorganisms actinomycetes, especially Streptomycetes gain special importance and they comprise of over 500 mainly soil-dwelling saprophytic species [1–4]. This gram positive bacterial genus is the most potent source for production of antibiotics [5] and other bioactive secondary metabolites [6, 7].

V. Parthasarathy (✉) · G. S. Prasad · R. Manavalan
Department of Pharmacy, Annamalai University, Annamalai Nagar,
Chidambaram 608002, Tamil Nadu, India
e-mail: vapartha@yahoo.com

R. Malathi and J. Krishnan (eds.), *Recent Advancements in System Modelling Applications*, Lecture Notes in Electrical Engineering 188,
DOI: 10.1007/978-81-322-1035-1_40, © Springer India 2013

Streptomycetes have been isolated from almost every kind of substrate viz., soil, volcanic soil [8] and compost, aquatic, marine, indoor environments, air, building material and dust samples [9–11].

The actinobacteria are characterized by their gram-positive nature and high guanine-plus-cytosine (G + C) content in their genomes. Ecologically, actinobacteria and, particularly, the *Streptomyces* sp. are saprophytic, soil-dwelling organisms that spend majority of their life cycle as semidormant spores [12–15]. The streptomycetes are believed to play a major role in organic matter degradation especially that of polymers such as lignocellulose, starch, chitin in soil [16–19] and involved in humification processes in composts [20]. While their role in other microbially mediated soil processes has not been so extensively studied, evidence indicates that actinomycetes are quantitatively and qualitatively important in the rhizosphere [21–25].

Recent studies have well characterised the actinobacteria for their economic importance as producers of two-thirds of the microbially derived secondary metabolites (antibiotics) known today [12, 14], which have different biological activities, such as antibacterial [26], antifungal [27, 28], anti-parasitic and anti-tumour actions [29], enzyme production [30]. Very recently major compound meroparamycin was isolated from Streptomycetes, which is found to have antagonistic activity against gram-positive, gram-negative representatives and *Candida albicans* [31]. Streptomycetes are not particularly pathogenic, although a few species may cause infections [32]. They are potent inducers of inflammatory responses in vitro and in vivo [10]. A few actinomycetes produce herbicidal and insecticidal compounds [28, 33, 34]. Actinomycetes in general and streptomycetes in particular occur in a wide range of environment. However, the studies on actinomycetes from the unusual habitats such as lignite mines are scanty. Hence, the present study was carried out for the first time to isolate potential antagonistic actinomycetes from very unusual environment i.e. lignite mines of Neyveli, Tamilnadu, India.

40.2 Materials and Methods

40.2.1 Media

All isolation media were prepared in distilled water and were adjusted to pH between 7.0–7.2 with 0.1 N NaOH prior to autoclaving.

(i) **Trace salt solution**
Ferrous sulphate (hydrate) 1 g l^{-1}, Manganese chloride (hydrate) 1 g l^{-1}, Zinc Sulphate (hydrate) 1 g l^{-1}. The salts were dissolved in distilled water and sterilised by autoclaving.

(ii) **Glycerol-Czapek's agar**
Glycerol 30.0 g l^{-1}, Sodium nitrate 2.0 g l^{-1}, Dipotassium hydrogen phosphate 1.0 g l^{-1}, Magnesium Sulphate (hepta hydrate) 0.5 g l^{-1}, Potassium Chloride 0.5 g l^{-1}, Ferrous Sulphate (hepta hydrate) 0.01 g l^{-1}, Agar–Agar (Oxoid) 20.0 g l^{-1}. The pH of the medium was adjusted to 7.2.

(iii) **Modified Nutrient Agar**
Glucose 5.0 g l^{-1}, Peptone (peptone) 5.0 g l^{-1}, Beef extract (Hi media) 3.0 g l^{-1}, Sodium chloride 5.0 g l^{-1}, Agar agar 20.0 g l^{-1}. The pH of the medium was adjusted to 7.0 (±0.2).

(iv) **Oat meal Agar**
Oatmeal 20.0 g, Agar–Agar 18.0 g. The oatmeal was suspended in 1,000 ml of distilled water. The slurry was steamed for 20 min. It was filtered through cheese cloth. The volume of the filtrate was restored to 1,000 ml with water. 1 ml of trace salt solution was added. pH was adjusted to 7.2. Agar was added and the mixture was steamed at 100 °C for 10–15 min to liquefy the medium. The medium was sterilised in flasks.

(V) **ISP medium**

Solution I
Soluble starch 20.0 g l^{-1} of distilled water.
Solution II
Dipotassium hydrogen phosphate 2.0 g l^{-1}, Magnesium Sulphate 2.0 g l^{-1}, Sodium Chloride 2.0 g l^{-1}, Ammonium Sulphate 4.0 g l^{-1}, Calcium Carbonate 4.0 g l^{-1}, Trace salt solution 2 ml l^{-1}. pH of the solution adjusted between 7.0 and 7.4. The solution I and solution II were mixed. Agar (40 gm) was added. The agar was dissolved by steaming the final product at 100 °C for 10–15 min. The medium was distributed into flasks for pouring into Petri dishes. It was sterilised by autoclaving.

40.2.2 Carbon Sources and Method of Sterilisation

Chemically pure carbon sources, free of admixture with other carbohydrates or carbohydrate containing materials were used. Carbon sources for this test were negative control (no carbon source), D-glucose (positive control), Arabinose, D-fructose, Sucrose, Rhamnose, D-Xylose, Raffinose, L-Inositol, D-Mannitol. The sugar was sterilized without heat by one of the following methods.

40.2.2.1 Membrane Filtration

10 % w/v aqueous solution of the carbon source (with the exception of inositol which was not soluble in water) was sterilized by filtration through membrane filter.

40.2.2.2 Ether Sterilization

An appropriate amount of the dry carbon source was weighed and spread as a shallow layer in a pre-sterilized Erlenmeyer flask, fitted with a loose cotton plug. Sufficient acetone free diethyl ether $[(C_2H_5)_2O]$ was added to cover the carbon source. Ether was allowed to evaporate at room temperature under ventilated fume cup-board (fume hood) overnight or longer. When all the ether had evaporated, sterile distilled water was added aseptically to make a 10 % w/v solution or suspension of the carbon source.

40.2.2.3 Ethylene Oxide Sterilization

The carbon source, as a 10 % w/v solution or suspension was chilled to 3–5 °C and 1 % cold ethylene oxide liquid was added, mixed well and left aside in the fume hood at room temperature for 1 h. It was slightly warmed (45 °C) to expel residual ethylene oxide.

Carbon source sterilized by one of these three methods was added to the basal mineral salts agar to give a final concentration of 1 % w/v.

40.2.3 Isolation of Actinomycetes from Soil

The soil specimens were collected from different locations in the mining area of Neyveli Lignite mines at a depth of 200–300 m from the surface with due permission of the authorities concerned. While sampling, the topsoil was removed to a depth of 1 cm and discarded. The sample beneath topsoil was scooped with a metal spatula into a pre-sterilised boiling tube and was immediately plugged with sterile non-absorbent cotton wool. The samples were transported to the laboratory without any delay. The soil samples that were in a relatively dry condition were stored in a refrigerator until the plating stage. The samples were placed in an incubator at 30 °C overnight prior to dilution. Tenfold serial dilutions of the samples were made viz. 1:10, $1{:}10^2$, $1{:}10^3$, $1{:}10^4$, $1{:}10^5$ and $1{:}10^6$ by using sterile water containing $1{:}10^6$ dilution of sodium lauryl sulphate. 0.5 ml of the serially diluted samples was plated into Glycerol-Czapek's agar. The tubes were incubated at 28–30 °C for 5 days. The colonies of actinomycetes, with characteristic and distinct morphological features, that appeared from the fifth day onwards were marked and sub-cultured into culture tubes of Glycerol-Czapek's agar medium. The tubes were incubated at 28 °C for not less than 7 days. The medium did not contain peptone or meat extract as either of them permits the overgrowth of bacteria and prevent the development of actinomycetes which are slow growing in nature.

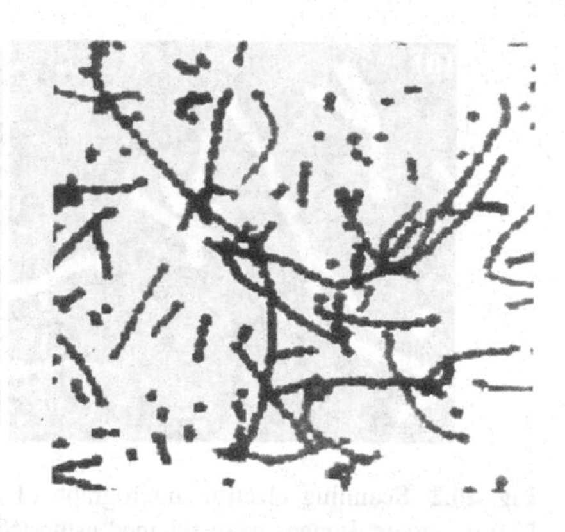

Fig. 40.1 Transmission Electron Microscope image of Streptomycete (NM1) from Neyveli Lignite mines (×2,000). The Images were scanned using Transmission Electron Microscope—Hitachi H-7000 [38]

40.2.4 Morphological Characterization

Aerial mass colour was determined by growing the culture in oatmeal agar [32]. Production of melanoid pigment was determined by using the medium ISP-4 [11, 32]. Microbial species must be characterized by multiple, readily recognizable and reasonably stable properties [35]. In few cases, the production of melanoid pigments was delayed or weak, and therefore, it was not distinguishable. This was indicated as variable (V). Spore morphological characters of the strains were studied by inoculating a loopful of one-week old cultures into 1.5 % agar medium contained in culture tubes at 37 °C. The culture was suspended in the semisolid agar medium and a drop of the medium was spread well on the slide and allowed to solidify into a thin film to facilitate direct observation under microscope at 100–700x magnifications. The cultures were incubated at 28 ± 2 °C and examined periodically for the formation of aerial mycelium, sporophore structure and spore morphology.

40.2.5 Spore Surface

Spore morphology and its surface ornamentation were observed under the transmission (TEM) (Hitachi H-7000 TEM, Hitachi, Co, Tokyo, Japan) (Figure 40.1) and scanning electron microscope (SEM) (Model S-450, Hitachi, Co, Tokyo, Japan) (Figures 40.2a, b). Scanning electron microscopy was performed with cells grown for 7 days on Glycerol-Czapek's agar. Cell preparation dehydration and staining procedures were carried out [36, 37].

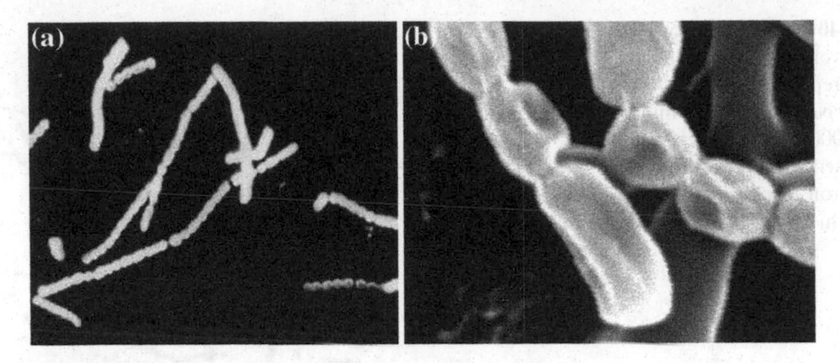

Fig. 40.2 Scanning electron micrograph of spore surface of the strain (NM1) from Neyveli Lignite mines. Images were scanned using (SEM) (Model S-450, Hitachi, Co, Tokyo, Japan) at ×2,000) (**a**). The image scanned at ×30,000) (**b**)

40.2.6 Assimilation of Carbon Source

Chemically pure carbon source, certified to be free of admixture with other carbohydrates or contaminating materials, were used for this purpose. Carbon sources for this test were glucose, xylose, arabinose, rhamnose, fructose, galactose, raffinose, mannitol, inositol, salicin and sucrose. These carbon sources were subjected to ether sterilization without heating. The media and plates were prepared and inoculated according to the convention of ISP [32].

40.2.7 Assay for Antagonistic Activity

The purified isolates of actinomycetes from lignite mine were tested for their antagonistic activity [5] against the growth of *Bacillus subtilis, Escherichia coli and Candida albicans* by an in vitro plate assay by employing the cross streak method [5]. A single streak of the actinomycetes was inoculated on the surface of modified nutrient agar plate to one side of centre. The culture was incubated at 28 ± 2 °C for about 5–8 days. After observing a good ribbon like sporulation on the Petri plates, the test microbes such as *Bacillus subtilis*, *Escherichia coli and Candida albicans* were streaked at right angle to the original streak of actinomycete and incubated at 28 ± 2 °C for 24–48 h (Fig. 40.3). The antagonistic activity was indicated when pathogenic microbial growth in the direction of the actinomycete colony was retarded or prevented. Single streak of each of the actinomycete was made on the surface of modified nutrient agar and incubated at room temperature (28 ± 2 °C). The test performed in duplicate to confirm that the results are consistent. The standard cultures of pathogenic strains were obtained from National culture collection centre, Pune, India.

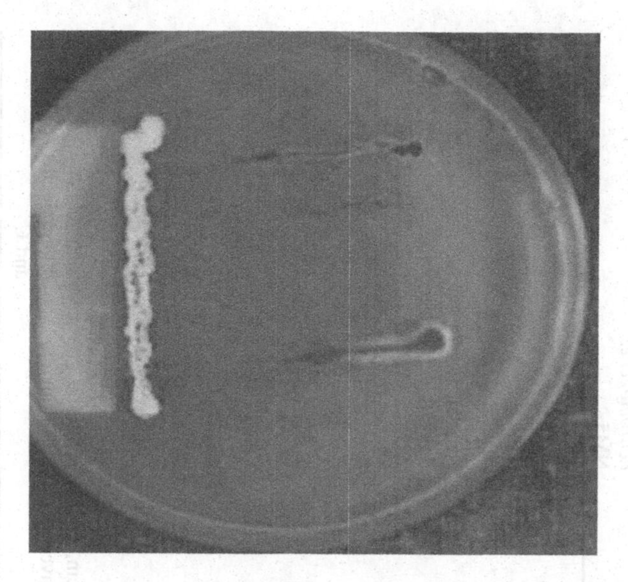

Fig. 40.3 Antagonistic activity of Streptomycete isolated from Neyveli lignite mines tested against *Bacillus subtilis*, *Escherichia coli* and *Candida albicans* by an in vitro plate assay by employing the cross streak method [5]. When growth along the streak developed fully in the form of a ribbon in modified nutrient agar plate, the test organisms were streaked at right angles to the original streak of the actinomycete and incubated at 30 ± 2 °C for not less than 24 h. The extent of inhibition was recorded

40.3 Results

All the seven isolates of antagonistic actinomycetes belonged to the genus *Streptomyces* as this has been ascertained on the basis of their morphological characters (Table 40.2). The results were compared with those of the *Streptomyces* species described in the Bergey's Manual of Determinative Bacteriology. These results clearly indicate that the samples collected from the Neyveli lignite mine area contained the actinomycetes of antibiotic producing nature. The actinomycetes were identified as six different species of *Streptomycetes*. All these species except *Streptomyces baarnensis* have been reported to produce antibiotics.

40.3.1 Spore Surface Ornamentation

The spore chain morphology of the strains was examined using direct observation under light microscope at 100–700x magnifications. The sporophore structure of actinomycete was studied with Transmission Electron Microscope (Hitachi H-7000 TEM microscope [38] followed by the scanning electron microscopy (SEM) (Model S-450, Hitachi, Co, Tokyo, Japan) [36, 39]. Spore chain morphology was

Table 40.1 Comparison between the isolates (NM1 to NM7) and reported[a] species of Streptomycetes

Character studied	*S. xanthocidicus*/ NM1	*S. baarnensis*/ NM2, NM7	*S. nitrosporeus*/ NM2	*S. collinus*/NM4	*S. achromogenes*/ NM5	*S. flavovirens*/NM6
Aerial Mycelium	Grey	White	Grey	Grey	Grey	Grey
Melanoid pigment	–	–	–	+	+	–
Spore chain	RF	RF	RF	RF	RF	RF
Spore surface	Smooth	Smooth	Smooth	Smooth	Smooth	Smooth
Glucose	+	+	+	+	+	+
Xylose	+	+	+	+	+	+
Arabinose	+	+	+	+	+	+
Rhamnose	–	+	+	+	+	+
Fructose	+	+	– (+)	+	+	+
Galactose						
Raffinose		– (+)				
Mannitol	– (+)					
Inosital				– (+)		
Salicin	–(ND)	+(ND)	–(ND)	(ND)	+(ND)	–(ND)
Sucrose	+	(ND)	–	+	–	– (+)
Antibiotic[a] production	Xanthocidicin	Nil	Nitrosporin	Cholrampheni-col, Collimycin, napthomycin and rubromycin.	Achromo-viromycin	Actinomycin and pillaromycins A, B_I, B_{II} and C

[a] As given in Bergey's Manual of Determinative Bacteriology [43–45]

Table 40.2 The colonial morphology of the isolates of Actinomycetes from lignite mines

Isolate number	Form	Profile	Margin	Colour	Soluble pigment	Size (mm)
NM1	Circular	Raised, convex	Entire	Grey	–	3
NM2	Circular	Raised, flat	Entire	Grey	–	2
NM3	Punctiform	Convex	Entire	Grey	–	3
NM4	Circular	Convex	Entire	Grey	Light brown	4
NM5	Circular	Raised, flat	Entire	Grey	Light brown	4
NM6	Circular	Raised, flat	Entire	Grey	–	2

Table 40.3 The Actinomycetes isolated from Neyveli lignite mines were tested for their antagonistic activity [5] against three different test micro organisms viz. *Bacillus subtilis, Escherichia coli and Candida albicans* by cross streak method

Isolate Number	*Bacillus subtilis* NCIM 2063	*Escherichia coli* NCIM 2256	*Candida albicans* ATCC 10231
NM1, NM3	+	–	–
NM2,NM5, NM7	+	+	+
NM4	+	+	+
NM6	–	–	–

The extent of inhibition was recorded
NM Neyveli Mine, + antagonism, – no antagonism

analysed using the methods of the International *Streptomyces* Project (ISP) [40], and Bergey's Manual of Systematic Bacteriology. The spore silhouettes were characterised as smooth, spiny, hairy or warty (Tables 40.1 and 40.2).

40.3.2 Antagonistic Activity

The isolates NM1, NM2, NM3, NM4, NM5 and NM7 showed good antagonistic activity against *Bacillus subtilis* NCIM 2063, *Escherichia coli* NCIM 2256, *Candida albicans* ATCC 10231 (Table 40.3).

40.4 Discussion

The isolates of actinomycetes from lignite mines originated from a non-rhizosphere associated substrate namely the lignite mine environment and there was not much diversity with regard to their micro morphological characters such as aerial mycelium, spore chain, spore surface and pigmentation. This is attributed to partial depletion of organic nutrients in the mine environment as well as the deep location of sampling from the terrestrial surface. The lack of diversity is attributed to the very special nature of the soil in the mine environment [15, 16, 28].

The in vitro antagonism of the isolates showed that majority of them inhibited the growth of gram +ve bacterium while the activity against gram −ve bacterium and the yeast was limited. This implies that the antagonists were not enriched in the non-rhizosphere substrate. The pattern of antagonism against *Bacillus subtilis*, *Escherichia coli*, and *Candida albicans* was more or less uniform. Those actinomycetes shown to be antagonists by agar plate diffusion studies may or may not be active antagonist in soil, while those shown to be non antagonist by plate assay are generally inactive in soil [41].

The actinomycetes play a vital role in protecting the plant population growing in organic rich soil against plant pathogens. A similar kind of role of actinomycetes in very rare environment like the lignite mine environment is not well understood. They may probably help in enrichment of the mine with lignite material with simultaneous depletion of available nutrients. This effect is essentially short lived and as such the predominance of microflora in mine environment is observed to be relatively low as compared to organic rich rhizosphere environment where the protective role of microbes in general and actinomycetes in particular is very well understood [42].

40.5 Conclusion

The search for new bioactive compounds is unending. Streptomycetes continue to be a fascinating and potential source of valuable metabolites. The chemotaxonomic approaches are capable of revealing the identity of these microbial consortia. It is also expected that screening of natural substrates of rare type such as the mine area is likely yield clues regarding the microbial diversity in this substrate and establish their ecological relationship. Secondly the probability of isolating producers of therapeutic agents from this substrate is very promising. The present work threw light on the occurrence of the industrially most important group of micro-organisms i.e. Streptomycetes in the lignite mine area.

References

1. Chater KF (1993) Genetics of differentiation in streptomyces. Annu Rev Microbiol 47:685–713
2. Clark CA, Chen C, Ward-Rainey N, Pettis GS (1998) Diversity within Streptomyces ipomoeae based on inhibitory interactions, rep-PCR, and plasmid profiles. Phytopathology 88:1179–1186
3. Lanoot B, Vancanneyt M, Cleenwerck I, Wang L, Li W, Liu Z, Swings J (2002) The search for synonyms among streptomycetes by using SDS-PAGE of whole-cell proteins. Emendation of the species Streptomyces aurantiacus, Streptomyces cacaoi subsp. cacaoi, Streptomyces caeruleus, and Streptomyces violaceus. Int J Syst Evol Microbiol 52:823–829

4. Zhang X, Clark CA, Pettis G (2003) Interstrain Inhibition in the Sweet Potato Pathogen Streptomyces ipomoeae: purification andcCharacterization of a highly specific bacteriocin and cloning of Its structural gene. Appl Environ Microbiol 69(4):2201–2208
5. Crawford DL, Lynch JM, Whipps JM, Ousley MA (1993) Isolation and characterisation of actinomycetes antagonists of a fungal root pathogen. Appl Environ Microbiol 59:3899–3905
6. Balagurunathan R (1997) A source of valuable products. Seshauyana Actinomycetes 5:17
7. El-Naggar MY, Hassan MA, Said WY, El-Aassar SA (2003) Effect of support materials on antibiotic MSW2000 production by immobilized Streptomyces violatus. J Gen Appl Microbiol 49:235–243
8. Ding L, Hirose T, Yokota A (2007) Amycolatopsis echigonensis sp. nov. and Amycolatopsis niigatensis sp. nov., novel actinomycetes isolated from a filtration substrate. Int J Syst Evol Microbiol 57(Pt 8):1747–1751
9. Anderson AS, Wellington EMH (2001) The taxonomy of streptomyces and related genera. Int J Syst Evol Microbiol 51:797–814
10. Suutari M, Lignell U, Hirvonen MR Nevalainen A (2000) Growth pH ranges of Streptomyces spp. ASM News 66(10):588
11. Suutari M, Lignell U, Hyvärinen A, Nevalainen A (2002) Media for cultivation of indoor streptomycetes. J Microbiol Methods 51:411–416
12. Berdy J (1989) The discovery of new bioactive microbial metabolites: screening and identification. Prog Indust Microbiol 27:3–25
13. Coombs JT, Franco CMM (2003) Isolation and Identification of Actinobacteria from surface-sterilized wheat roots. Appl Environ Microbiol 69(9):5603–5608
14. Kieser T, Bibb MJ, Buttner MJ, Chater KF, Hopwood DA (eds) (2000) Practical streptomyces genetics. John Innes Centre, Norwich
15. Mayfield CI, Williams ST, Ruddick SM, Hatfield HL (1972) Studies on the ecology of actinomycetes in soil. IV. Observations on the form and growth of streptomycetes in soil. Soil Biol Biochem 4:79–91
16. Lechevalier MP (1989) Actinomycetes in agriculture and forestry. In: Goodfellow M, Williams ST, Mordarski M (eds) Actinomycetes in biotechnology. Academic Press, New York, pp 327–358
17. Nishimura M, Ooai O, Davies J (2006) Isolation and characterization of Streptomycetes sp. NL.15-2Kcapable of degrading Ligin- related aromatic compounds. J Biosci Bioeng 102(2):124–127
18. Srinivasan MC, Laxman RS, Despharde MV (1991) Physiology and nutritional aspects of actinomycetes: an overview. World J Microbiol Biotechnol 7:171–184
19. Zimmerman W (1990) Degradation of lignin by bacteria. J Biotechnol 13:119–130
20. Steger K, Jarvis A, Vasara T, Romantschuk M, Sundh I (2007) Effects of differing temperature management on development of Actinobacteria populations during composting. Res Microbiol 158 (7):617–24
21. Miller HJ, Henken G, van Veen JA (1989) Variation and composition of bacterial populations in the rhizospheres of maize, wheat, and grass cultivars. Can J Microbiol 35:656–660
22. Miller JJ, Liljeroth E, Henken G, van Veen JA (1990) Fluctuations in the fluorescent pseudomonad and actinomycete populations of rhizosphere and rhizoplane during the growth of spring wheat. Can J Microbiol 36:254–258
23. Miller JJ, Liljeroth E, Willemsen-de Klein MJEIM, van Veen JA (1990) The dynamics of actinomycetes and fluorescent pseudomonas in wheat rhizoplane and rhizosphere. Symbiosis 9:389–391
24. Mohamed ZK (1982) Physiological and antagonistic activities of streptomycetes in rhizosphere of some plants. Egypt J Phytopathol 14:121–128
25. SardiP, Saracchi M, Quaroni S, Petrolini B, Borgonovi GE, Nesli S (1992) Isolation of endophytic Streptomyces strains from surface-sterilized roots. Appl Environ Microbiol 58:2691–2698
26. Parvateesam M, Bulchandani BD (2003) Screening of actinomycetes isolated from soil samples of Ajmer, Rajasthan for antimicrobial activity. Hindustan Antibiot Bull 45–46 (1–4):22–28

27. Jain PK, Jain PC (2007) Isolation, characterization and antifungal activity of Streptomyces sampsonii GS 1322. Indian J Exp Biol 45(2):203–206
28. Lukic A, Welty RE, Lucas GB (1972) Antifungal spectra of actinomycetes isolated from Tobacco. Antimicrobial Agents Chemother 1(4):363–365
29. Demain AL (1999) Pharmaceutically active secondary metabolites of microorganisms. Appl Microbiol Biotechnol 52:455–463
30. Jain PK, Jain R, Jain PC (2003) Production of industrially important enzymes by some actinomycetes producing antifungal compounds. Hindustan Antibiot Bull 45–46(1–4):29–33
31. El-Naggar MY, El-Assar SA, Abdul-Gawad SM (2006) Meroparamycin production by Newly isolated *Streptomyces* sp. Strain MAR01: taxonomy, fermentation, purification and structural elucidation. J Microbiol 44(4):432–438
32. Mishra SK, Gordon RE, Barnett DA (1980) Identification of nocardiae and streptomycetes of medical importance. J Clin Microbiol 11:728–736
33. DeFrank J, Putnam AR (1985) Screening procedures to identify soil-borne actinomycetes that can produce herbicidal compounds. Weed Sci 33:271–274
34. Hayashi H, Takiuchi K, Murao S, Araj M (1991) Identification of the microorganism which produces insecticidal indole alkaloids, okaramines A and B. Agric Biol Chem 55:2177–2178
35. Burkholder PR, Sun SH, Ehrlich J, Anderson LE (1954) Criteria of speciation in the genus Streptomyces. Ann N Y Acad Sci 60:102
36. Rheims H, Schumann P, Rohde M, Stackebrandt E (1998) Verrucosispora gifhornensis gen. nov. sp. nov, a new member of the actinobacterial family Micromonosporaceae. Int J Syst Bacteriol 48:1119–1127
37. Williams ST, Davies FL (1967) Use of a scanning electron microscope for the examination of actinomycetes. J Gen Microbiol 48:171–177
38. Curry A, Appleton H, Dowsett B (1996) Application of transmission microscopy to the clinical study of viral and bacterial infections: present and future. Micron 37(2):91–106
39. Rhee KH (2002) Isolation and characterization of *Streptomyces* sp. KH-614 producing anti-VRE (vancomycin-resistant enterococci) antibiotics. J Gen Appl Microbiol 48:321–327
40. Shirling EB, Gottlieb D (1966) Methods for characterisation of Streptomycetes species. Int J Syst Bacteriol 16:313
41. Broadbent P, Baker KF, Waterworth Y (1971) Bacteria and actinomycetes antagonistic to fungal root pathogens in Australian soils. Aust Biol Sci 24:925–944
42. Reddi GS, Rao AS (1971) Antagonism of soil actinomycetes to some soil-borne plant pathogenic fungi. Indian Phytopathol 24:649–657
43. Arai T, Kuroda S, Mikami Y (1976) Classification of actinomycetes antibiotics production. In: Arai T (ed) The actinomycetes—the boundary microorganisms. Toppan Company Ltd., Tokyo, p 543
44. Buchanan RE, Gibbons NE (1974) Bergey's manual of determinative bacteriology, 8th edn. The Williams and Wilkins Co., Baltimore, p 747
45. Waksman SA (1957) Species concept among the actinomycetes with special reference to the genus *Streptomyces*. Bacteriol Rev 21:1

Chapter 41
Biogeography Based Optimization Technique for Economic Emission Dispatch

S. Rajasomashekar and P. Aravindhababu

Abstract The economic emission dispatch (EED) assumes a lot of significance to meet the clean energy requirements of the society and simultaneously minimizes the cost of generation. The biogeography based optimization (BBO), inspired from the geographical distribution of biological species, has some features that are common to genetic algorithm and particle swam optimization; and searches for optimal solution through the migration and mutation operators. This paper presents an effective BBO strategy for obtaining the robust solution of EED problem. The feasibility of the proposed approach is evaluated through three test systems and the results are presented to highlight its suitability for practical applications.

Keywords Biogeography based optimization · Economic load dispatch · Economic emission

List of Symbols

a_i b_i c_i	Fuel cost coefficients of the *i*th generator
B B_o B_{oo}	Loss coefficients
BBO	Biogeography based optimization
d_i e_i	Coefficients of valve point effects of the *i*th generator
E	Maximum possible emigration rate
EED	Economic emission dispatch
ELD	Economic load dispatch
$E_i(P_{Gi})$	Emission cost function of the *i*th generator in ton/h
$F_i(P_{Gi})$	Fuel cost function of the *i*th generator in \$/h
h_i	Price penalty factor of the *i*th generator in \$/ton

S. Rajasomashekar (✉) · P. Aravindhababu
Department of Electrical Engineering, Annamalai University,
Annamalainagar, 608002 Chidambaram, Tamil Nadu, India
e-mail: rajasomashekar@yahoo.in

R. Malathi and J. Krishnan (eds.), *Recent Advancements in System Modelling Applications*, Lecture Notes in Electrical Engineering 188,
DOI: 10.1007/978-81-322-1035-1_41, © Springer India 2013

I	Maximum possible immigration rate
$Iter^{max}$	Maximum number of iterations for convergence check
k	Number of species in kth island
K_1 K_2	Weight constants
MED	Minimum emission dispatch
n	Maximum number of species
nd	Number of decision variables
ng	Number of generators
ni	Number of islands
nei	Number of elite islands
PM	Proposed method
P_{Gi}	Real power generation at ith generator
P_{Gi}^{min} & P_{Gi}^{max}	Minimum and maximum generation limits of ith generator respectively
P_D	Total power demand
P_L	Net transmission loss
P^{mod}	Island modification probability
P_m	Mutation probability
S^{max}	Maximum species count
SIV	Suitability index variable
SI	Suitability index
w	Trade-off parameter in the range of [0, 1] α_i β_i γ_i ξ_i and δ_i emission coefficients of ith generator
λ	Immigration rate
μ	Emigration rate
Φ	Objective function to be minimized
Ψ	Augmented objective function to be minimized

41.1 Introduction

Economic Load Dispatch (ELD) plays an important role in maintaining a high degree of economy and reliability in power system operational planning. It is a computational process of allocating the total required generation among the available generating units subject to load and operational constraints such that the cost of operation is minimum [1].

Operating at absolute minimum cost can no longer be the only criterion for dispatching electric power due to increasing concern over the environmental considerations. The generation of electricity from fossil fuels such as coal, oil and gas, releases several contaminants such as sulphur oxides (SOx), nitrogen oxides (NOx) and carbon dioxide into the atmosphere. SO_2 is dependent on the amount of fuel burned. The sulphur enters the boiler as a part of the fuel. During combustion process, some of sulphur from the fuel reacts with oxygen in the combustion air to form SO_2 and the remaining part settles at the bottom as ash in the boiler. If stack gas clean up equipment is present, most of the SO_2 is removed and the remaining SO_2 exits through the stack as emission.

There are two sources of nitrogen that combine with oxygen from the fuel and combustion air to produce NOx. The second source is nitrogen in the fuel that produces an emission called fuel NOx. In coal, there is no apparent correlation between the amount of fuel-bound nitrogen and the fuel NOx produced. Besides the particulate matter pollutes the whole atmosphere when it exceeds the limit. Hence, it becomes necessary to supply power with minimum emission as well as minimum fuel cost.

The nature and quantity of these atmospheric emissions depend upon the fuel type and quality. Recently the problem, which continues to attract attention, is to minimize the pollution due to the pressing public demand for clean air. The enactment of the 'Clean Air Act Amendment of 1990' and its acceptance by all nations forces the utilities to modify their operating strategies to meet rigorous environmental standards set by this legislation.

Various strategies like installing post combustion cleaning system, switching to low sulphur content coal and Minimum Emission Dispatch (MED) have been proposed for reducing the emissions, which minimizes only the emissions that result in high operating cost; and efforts initiated by researchers to develop algorithms for Economic Emission Dispatch (EED) that minimizes the cost of generation and emission levels simultaneously. Several researchers have considered emissions either in the objective function or treated emissions as additional constraints [2].

Traditional mathematical programming techniques such as lambda iteration, gradient search, linear programming and Lagrangian relaxation [1] and modern heuristic optimization techniques such as genetic algorithms [3, 4], evolutionary programming [5, 6], particle swarm optimization [7, 8] and harmony search [9] have been widely applied in solving the EED problems.

Recently, a Biogeography-Based Optimization (BBO) modeled on the theory of biogeography, which involves the study of the geographical distribution of biological organisms, has been proposed for solving optimization problems by Simon [10]. Like other evolutionary algorithms, BBO, a population based stochastic optimization technique sharing information between candidate solutions based on their fitness values with a view of obtaining the global best solution, has been suggested. Since its introduction, it has been applied to a variety of power system optimization problems [11, 12].

The evolutionary search algorithms such as SA, GA, EP, PSO and ACO have been widely applied in solving EED problems. The SA based approaches have been used to find the optimal solution using point-by-point iteration rather than a search over a population of individuals. The approaches are simple to formulate and require lower memory requirement than that of GA. However, they suffer from huge computational burden and end up with consuming exhaustively large execution times. In addition most of the GA based approaches involve binary representation of variables and the solution process involves complex search process. The effort in this article is to solve the EED problem using BBO with a view of obtaining the global best solution and explore its applicability for present day power systems. The paper is divided into six sections. Section 41.1 gives the introduction, Sect. 41.2 explains the BBO technique, Sect. 41.3 outlines the EED problem, Sect. 41.4 describes the proposed method, Sect. 41.5 discusses the simulation results and Sect. 41.6 concludes the article.

41.2 Biogeography Based Optimization

BBO, suggested by Dan Simon in 2008, is a stochastic optimization technique for solving multimodal optimization problems [10]. It is based on the concept of biogeography, which deals with the distribution of species that depend on different factors such as rain fall, diversity of vegetation, diversity of topographic features, land area, temperature, etc. In the science of biogeography, an island/habitat is defined as the ecological area that is inhabited by a particular plant or animal species and geographically isolated from other habitats.

Over evolving periods of time, some islands may tend to accumulate more species than others as they posses certain environmental features that are more favorable than islands with fewer species. Many species like plants and animals on islands with large population emigrate into neighboring islands with less number of species for their survival and better living and share their characteristics with those islands enabling islands with less population to inherit a high species immigration rate. The immigration and emigration processes help the species of less favorable area to acquire good features from the species in the favorable islands and strengthen the weak elements. The rate of immigration (λ) and the emigration (μ) are the functions of the number of species in the islands. Figure 41.1 shows the immigration and emigration curves indicating the movement of species in an island.

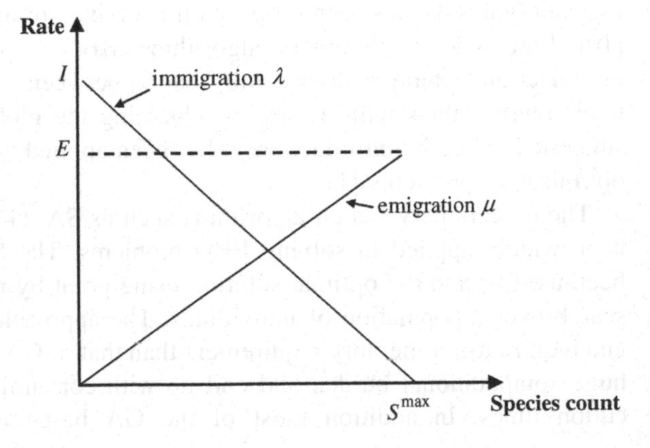

Fig. 41.1 Species model of an island

In BBO, a solution is represented by an island consisting of solution features named Suitability Index Variables (*SIV*), which are represented by real numbers. It is represented for a problem with *nd* decision variables as

$$Island = [SIV_1, SIV_2, SIV_3, \ldots, SIV_{nd}] \tag{41.1}$$

The suitability of sustaining larger number of species of an island can be modeled as a fitness measure referred to Suitability Index (*SI*) in BBO as

$$SI = f(island) = f(SIV_1, SIV_2, SIV_3, \ldots, SIV_{nd}) \tag{41.2}$$

High *SI* represents a better quality solution and low *SI* denotes an inferior solution. The aim is to find optimal solution in terms of *SIV* that maximizes the *SI*.

Each island, representing a solution point, is characterized by its own immigration rate λ and emigration rate μ. A good solution enjoys a higher μ and lower λ and vice versa. The immigration and emigration rates are the functions of the number of species in the island and defined for the *k*th island as

$$\mu_k = E\left(\frac{k}{n}\right) \tag{41.3}$$

$$\lambda_k = I\left(1 - \frac{k}{n}\right) \tag{41.4}$$

when $E = I$, the immigration and emigration rates can be related as

$$\lambda_k + \mu_k = E \tag{41.5}$$

A population of candidate solutions is represented as a vector of islands similar to any other evolutionary algorithm. The features between the islands are shared through migration operation, which is probabilistically controlled through island modification probability, P^{mod}. If an island S_i in the population is selected for modification, then its λ is used to probabilistically decide whether or not to modify each *SIV* in that island. The μ of other solutions are thereafter used to select which of the solutions among the islands shall migrate randomly chosen *SIVs* to the selected solution S_i.

The cataclysmic events that drastically change the *SI* of an island is represented by mutation of *SIV* and the species count probabilities are used to determine mutation rates. The probability of each species count indicates the likelihood that it exists as a solution for a given problem. If the probability of a given solution is very low, then that solution is likely to mutate to some other solution. The solutions with very high *SI* and very low *SI* therefore eclipse to create more improved *SIV* in the later stage. The mutation scheme tends to increase diversity among the population, avoids the dominance of highly probable solutions and provides a chance of improving the low *SI* solutions.

As with the other population based strategies, some kind of elitism, which retains the best island with the highest *SI* without performing migration operation, is used to prevent the best solutions from being ruined by immigration.

41.3 Problem Formulation

41.3.1 Economic Load Dispatch

The ELD problem may be expressed by minimizing the fuel cost of generating units while satisfying several equality and inequality constraints as

$$\begin{aligned} Min \quad F(P_G) = \sum_{i=1}^{ng} a_i P_{Gi}^2 + b_i P_{Gi} + c_i \\ + \left| d_i \sin\left\{ e_i (P_{Gi}^{\min} - P_{Gi}) \right\} \right| \end{aligned} \tag{41.6}$$

Subject to

$$\sum_{i=1}^{ng} P_{Gi} - P_D - P_L = 0 \tag{41.7}$$

$$P_{Gi}^{\min} \le P_{Gi} \le P_{Gi}^{\max} \quad i = 1, 2, \ldots, ng \tag{41.8}$$

where

$$P_L = \sum_{i=1}^{ng} \sum_{j=1}^{ng} P_{Gi} B_{ij} P_{Gj} + \sum_{i=1}^{ng} B_{oi} P_{Gi} + B_{oo} \tag{41.9}$$

41.3.2 Minimum Emission Dispatch

The objective of MED is to minimize the emissions of all the generating units due to the burning of fuels for production of power to meet the load demand and expressed as

$$\begin{aligned} &Min \\ &E(P_G) = \sum_{i=1}^{ng} 10^{-2} \left(\alpha_i P_{Gi}^2 + \beta_i P_{Gi} + \gamma_i \right) + \xi_i \exp(\delta_i P_{Gi}) \end{aligned} \tag{41.10}$$

41.3.3 Economic Emission Dispatch

The EED problem is to determine optimal real power generations that minimize the two conflicting objectives of fuel cost and emissions, while satisfying several equality and inequality constraints. The bi-objective of EED problem can be mathematically formulated as

$$\text{Min} \quad \Phi = [F(P_G), E(P_G)] \tag{41.11}$$

The bi-objective EED problem is represented as a single objective optimization problem by assigning different weights for each objective.

$$\text{Min} \quad \Phi(P_G) = \sum_{i=1}^{ng} \{ w F_i(P_{Gi}) + (1-w)\, h E_i(P_{Gi}) \} \tag{41.12}$$

The value of w indicates relative significance between the two objectives. When w is 1, the problem becomes ELD that minimizes only the fuel cost. The fuel cost increases and the emission cost decreases when w is reduced in steps from 1 to 0. The problem becomes MED that minimizes only the emissions when w equals 0. The price penalty factor, h, called scaling factor, is multiplied with emission function to get an equivalent cost curve in \$/h. The procedure for computing h-parameter is outlined below:

- Evaluate the *ratio* between fuel cost and emissions corresponding to $P_{Gi}^{\max}$ for each generator-i.

$$ratio_i = \frac{F_i(P_{Gi}^{\max})}{E_i(P_{Gi}^{\max})} \quad \$/\text{ton}: \quad i = 1, 2, \ldots, ng$$

- Arrange the *ratios* in ascending order.
- Add the maximum capacity of each generator ($P_{Gi}^{\max}$) one at a time, starting from the smallest *ratio* until $\sum P_{Gi}^{\max} \geq P_D$.
- At this stage, $ratio_i$, associated with the last unit is the price penalty factor h for the given power demand P_D.

41.4 Proposed Method

The BBO based solution process involves representation of problem variables and formation of a fitness function. The *SIVs* of each island with P_G as the problem variables, can be represented as

$$Island = [P_{G1}, P_{G2}, P_{G3}, \ldots, P_{Gng}] \tag{41.13}$$

The BBO searches for optimal solution by maximizing *SI*, like fitness function in any other stochastic optimization techniques. The *SI* function can be obtained from the problem objective and constraint equations. The *SIVs* of each island during the solution process can be limited to satisfy the generation limit constraint of Eq. (41.8) but the power balance constraint of Eq. (41.7) is handled through penalty function approach. The penalty terms are incorporated in the *SI* function and are set to reduce the suitability of the island depending on the magnitude of the violation. The main advantage of penalty method compared to other approaches is that it does not disregard infeasible solutions; instead it uses these solutions in such a way as to aid the search process. Sometimes, these infeasible solutions may provide much more useful information about the optimum than the feasible solutions. The *SI* function can be obtained by transforming the objective function and power balance constraint into *SI* function as

$$Max \quad SI = \frac{1}{1+\Psi} \tag{41.14}$$

where Ψ is the augmented objective function, which can be formed by blending Eq. (41.12) with Eq. (41.7) through penalty factor approach as

$$Min \quad \Psi = K_1 \, \Phi(P_G) + K_2 \left\{ \sum_{i=1}^{ng} P_{Gi} - P_D - P_L \right\}^2 \tag{41.15}$$

The process of generating a new set of population from the initial set, which is generated randomly through migration and mutation, may be called an iteration. The genetic iterations may be continued by taking the population obtained in the previous iteration as the initial population for next iteration. In each and every genetic iteration, the island having the maximum *SI* is stored along with its objective function. The BBO iterative process of generating new population can be terminated either after a fixed number of iterations or if there is no further significant improvement in the global best solution. The flow of the Proposed Method (PM) is shown in Fig. 41.2.

41.5 Simulations

The PM is tested on three different test cases with varying degree of complexity for studying its performance. The first one is the standard IEEE 30-bus 6 generator system, the second system comprises 13 generators and the last one consists of 40 generators. The data for fuel cost, emissions and loss coefficients are available in Ref. [12].

The optimal generations, fuel cost and emissions for the test case-1 for a load demand of 2.834 per unit are given in Table 41.1. The fuel cost and emissions corresponding to ELD obtained through Eq. (41.12) with w as one, are 606.260

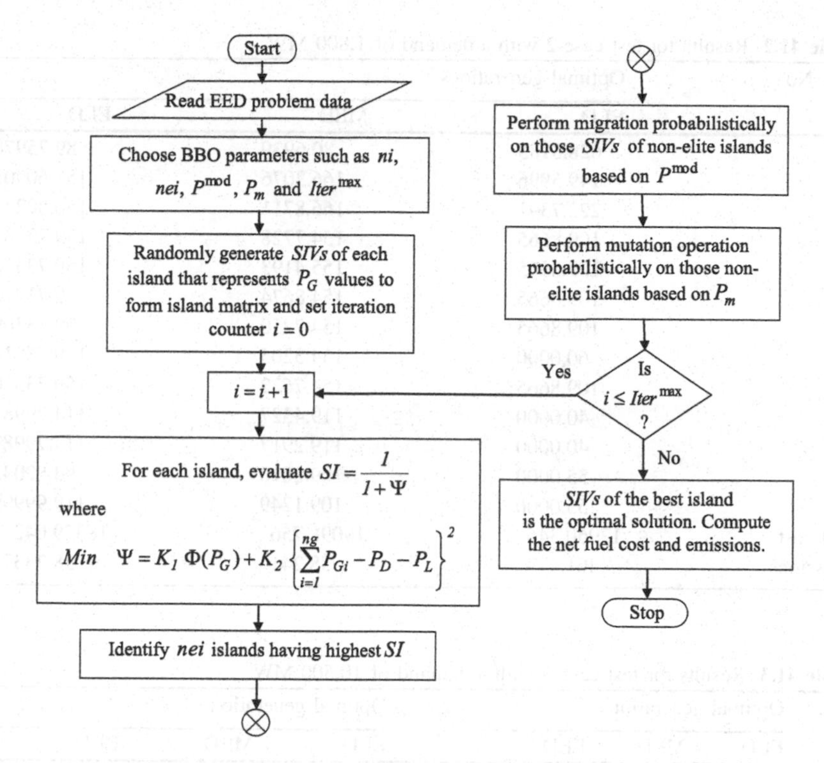

Fig. 41. 2 Flow chart of PM

Table 41.1 Results for test case-1

	ELD	MED	EED
P_{G1}	0.117966	0.408881	0.349500
P_{G2}	0.305320	0.461795	0.425763
P_{G3}	0.624894	0.541602	0.553406
P_{G4}	0.958791	0.387349	0.506009
P_{G5}	0.501513	0.541801	0.542032
P_{G6}	0.350323	0.513311	0.477033
Fuel cost	606.260	642.925	629.188
Emission	0.21875	0.19419	0.195204

\$/h and 0.21875 ton/h respectively. The algorithm enables the system to generate the required amount of power with lowest fuel cost. The solution of MED attained through Eq. (41.12) with w as zero results in lowest emission of 0.19419 ton/h and fuel cost of 642.925. The mechanism permits the system to offer the desired amount of power with smaller emissions. However, the ELD and MED cause higher emissions of 0.21875 ton/h and fuel cost of 642.925 \$/h respectively. The EED results available in the same table are acquired by assigning w equal to 0.5 in Eq. (41.12). The PM attempts to lower the fuel cost, minimizes emissions and extracts satisfactory results in relation to those obtained by ELD and MED.

Table 41.2 Results for test case-2 with a demand of 1,800 MW

Gen. No	Optimal generations		
	ELD	MED	EED
1	628.3185	80.6939	89.759789
2	149.5996	166.3076	151.609019
3	222.7391	166.8711	156.802426
4	109.8665	154.7728	159.733099
5	109.8665	155.4193	159.733100
6	109.8665	154.8674	159.733100
7	109.8665	154.7250	159.733099
8	60.0000	154.5205	159.733104
9	109.8665	154.7622	159.733100
10	40.0000	119.4327	114.799834
11	40.0000	119.2917	114.799872
12	55.0000	109.2010	93.820456
13	55.0000	109.1249	119.999999
Fuel cost	17960.346	19098.756	18379.042
Emissions	461.479	58.241	58.733729

Table 41.3 Results for test case-3 with a demand of 10,500 MW

Gen. No	Optimal generations			Gen. No	Optimal generations		
	ELD	MED	EED		ELD	MED	EED
1	110.8074	114.0000	114.000000	21	523.2794	413.1607	433.519572
2	110.8080	114.0000	113.999999	22	523.2794	413.1619	433.519443
3	97.4004	116.1198	120.000000	23	523.2796	413.1650	433.519585
4	179.7331	158.4352	166.600506	24	523.2794	413.1677	433.519580
5	87.8032	97.0000	97.000000	25	523.2794	413.1716	433.519582
6	140.0000	116.1145	120.410339	26	523.2794	413.1732	433.519598
7	259.6002	280.8856	290.056235	27	10.0000	150.0000	63.629783
8	284.6016	280.8548	286.794603	28	10.0000	150.0000	63.263622
9	284.6027	280.8445	286.762598	29	10.0000	150.0000	64.101222
10	130.0000	279.4388	281.252715	30	92.7131	97.0000	97.000000
11	168.7998	280.8624	289.779164	31	190.0000	158.4845	162.819706
12	168.7998	280.8568	289.651238	32	190.0000	158.4829	162.789955
13	214.7598	413.0698	416.009570	33	190.0000	158.4830	162.917010
14	394.2794	412.9083	417.016122	34	164.8024	200.0000	200.000000
15	304.5196	412.9209	417.071904	35	164.8044	200.0000	200.000000
16	394.2794	412.9194	417.129605	36	164.8023	200.0000	200.000000
17	489.2794	413.1602	421.427122	37	110.0000	93.8356	97.093992
18	489.2794	413.1625	421.303169	38	110.0000	93.8366	97.192969
19	511.2794	413.1607	421.519585	39	110.0000	93.8349	97.240706
20	511.2794	413.1587	421.519603	40	511.2794	413.1590	421.519592
Fuel Cost					121414.495	161366.113	133204.021
Emissions					285823.615	66654.838	81152.335

The ELD, MED and EED results in terms of real power generation, net fuel cost and emissions are presented for test cases 2 and 3 in Tables 41.2 and 41.3 respectively. It is clear from the table that the PM offers an EED solution that lies in between ELD and MED solutions.

It is observed from above discussions that the emissions are higher, when the fuel cost is lower and vice versa owing to the conflicting nature of the objectives in the problem. Thus, the PM offers the lowest fuel cost in economic dispatch, lowest emissions in the emission dispatch and strikes a compromise between fuel cost and emissions in EED.

41.6 Summary

A new strategy involving BBO for solving EED problem has been developed and studied on example problems. The ability of the PM to produce the global best solution that simultaneously minimizes the fuel cost and emissions has been projected. It has been chartered that the new approach fosters the continued use of BBO and will go a long way in serving as a useful tool in load dispatch centre.

Acknowledgments The authors gratefully acknowledge the authorities of Annamalai University for the facilities offered to carry out this work.

References

1. Chowdhury BH, Rahman S (1990) A review of recent advances in economic dispatch. IEEE Trans Power Syst 5(4):1248–1259
2. Lamont JW, Obessis EV (1995) Emission dispatch models and algorithms for the 1990s. IEEE Trans Power Syst 10(2):941–946
3. Abido M (2003) A niched Pareto genetic algorithm for multiobjective environmental/economic dispatch. Electric Power Energy Syst 25(2):97–105
4. Abido MA (2003) A novel multiobjective evolutionary algorithm for environmental/economic power dispatch. Electric Power Syst Res 65(1):71–81
5. Abido MA (2003) Environmental/economic power dispatch using multiobjective evolutionary algorithms. IEEE Trans Power Syst 18(4):1529–1537
6. Abido MA (2006) Multiobjective evolutionary algorithms for electric power dispatch problem. IEEE Trans Evol Comput 10(3):315–329
7. Hemamalini S, Sishaj PS (2008) Emission constrained economic dispatch with valve-point effect using particle swarm optimization. In: IEEE technical conference, 18–21 Nov 2008, TENCON-2008
8. Cai J, Ma X, Li Q, Li L, Peng H (2009) A multi-objective chaotic particle swarm optimization for environmental/economic dispatch. Energy Convers Manag 50(5):1318–1325
9. Sivasubramani S, Swarup KS (2011) Environmental/economic dispatch using multi-objective harmony search algorithm. Electric Power Syst Res 81(9):1778–1785
10. Simon D (2008) Biogeography-based optimization. IEEE Trans Evol Comput 12(6):702–713

11. Rarick R, Simon D, Villaseca F, Vyakaranam B (2009) Biogeography-based optimization and the solution of the power flow problem. In: IEEE conference on systems, man, and cybernetics, San Antonio, pp 1029–1034
12. Rajasomashekar S, Aravindhababu P (2012) Biogeography-based optimization technique for best compromise solution of economic emission dispatch. Swarm and Evolutionary Computations dx.doi.org/10.1016/j.swevo.2012.06.001

Chapter 42
BBO-Based TCSC Placement for Security Enhancement

K. Kavitha and R. Neela

Abstract Flexible AC Transmission System (FACTS) devices have brought in remarkable changes in the field of power system operation and control. Their development and application have made the traditional AC power system with an inherent time lag to respond quickly towards unprecedented changes. As these devices use power electronic switches unlike conventional controllers, they tend to be costlier and hence it is inevitable to consider not only the ratings but also the location for device placement to tap maximum benefit. In this work, the problem of optimal device placement is addressed taking into account the cost of the device also. Hence the objective function is formulated in such a way to take into account the cost of the device, line loadings and load voltage deviations. The problem is solved by applying BBO technique, simulating various load conditions on IEEE 14, 30 and 57 bus systems. The results of the proposed technique are compared with the results obtained through the application of PSO algorithm.

Keywords FACTS devices · Thyristor controlled series capacitor (TCSC) · Biogeography-based optimization (BBO) · Security enhancement

K. Kavitha (✉) · R. Neela
Department of Electrical Engineering, Annamalai University, Chidambaram, Tamil Nadu, India
e-mail: kavitha_au04@yahoo.com

R. Neela
e-mail: nishitha_30@rediffmail.com

R. Malathi and J. Krishnan (eds.), *Recent Advancements in System Modelling Applications*, Lecture Notes in Electrical Engineering 188,
DOI: 10.1007/978-81-322-1035-1_42, © Springer India 2013

42.1 Introduction

Modern power systems are highly complex interconnected systems, so it is essential to improve the electric power utilization while maintaining the reliability and security. Due to increase in load demand, the magnitude of the power flows in some of the transmission lines are well above their normal limits and in some other lines, it is below their normal. Its overall effect will detoriote the voltage profiles and decrease the security of the power system. To meet the increasing load demand and satisfy the stability and reliability criteria, either existing transmission or generation facilities must be utilized more efficiently or new facilities should be added to the power system. Due to the constraints such as lack of investment and difficulties in getting new transmission line right-of-ways, the later is often difficult but the former can be achieved by using Flexible Alternating Current Transmission Systems (FACTS) controllers [1].

FACTS is a concept introduced by Hingorani [2]. FACTS technology opens up new opportunities for controlling power flow and enhancing the usable capacity of present, as well as new and upgraded lines. FACTS devices can effectively control the load flow distribution, improve the usage of existing system by increasing transmission capability, compensate for reactive power, and improve stabilities of the power network. The possibility of controlling power flow in an electric power system without generation rescheduling or topological changes can improve the performance considerably [3]. However to achieve such benefits it is highly important to determine the suitable location and capacity of FACTS devices in the power system [4].

Thyristor Controlled Series Capacitor (TCSC) is a type of series compensator, that can provide many benefits for a power system including control of power flow in the line, damping power oscillations and mitigating sub synchronous resonance [5]. TCSC is a variable impedance type series compensator. It consists of a series compensating capacitor shunted by a thyristor controlled reactor. By controlling the firing angle of thyristor, TCSC can change the line reactance smoothly and rapidly. TCSC has one of the two possible characteristics either capacitive or inductive by increasing or decreasing the reactance of the line X_l [6]. Moreover to avoid the over compensation of the line, the maximum values of capacitance and inductance are fixed at $-0.8X_l$ and $0.2X_l$ [7]. World's first 3 phase, $2^{*}165$ MVAR, TCSC was installed in 1992 in Kayenta substation, Arizona. It raised the transmission capacity of transmission line by 30 % and effectively damped electromechanical power oscillations [8]. Optimal placement of TCSC is essential to tap the maximum benefits in terms of system performance and cost effectiveness.

A loss sensitivity index with respect to the control parameters of FACTS devices has been suggested and with the computed loss sensitivity index, the FACTS devices are placed on the most sensitive bus or line [9]. Fuzzy-based approach for the optimal placement of FACTS device for enhancing the system security under normal and network contingencies has been discussed [10]. The optimal location of a given number of FACTS devices is a problem of combinatorial analysis. To solve such kind of problems, heuristic methods can be used [11]. They permit to obtain acceptable solutions within a limited computation time. The application of Genetic Algorithm for the optimal location of multi type FACTS devices in order to maximize the system loadability is analysed in [12]. A Differential Evolution-based algorithm to decide the optimal location and device rating has been suggested in [13] with an objective of enhancing the system security under single line contingencies. The Particle Swarm Optimization (PSO) is applied for the optimal location of FACTS devices to achieve minimum cost of installation and to improve system loadability, by considering thermal limit for the lines and bus voltage limit for the load buses as constraints [14]. Sensitivity analysis approach for finding the optimal location and PSO for the optimal parameter setting of TCSC has been suggested in [15] so as to maximize the loadability. BBO, a population-based algorithm, which uses the immigration and emigration behavior of the species based on various natural factors is explained in [16]. Application of BBO to solve the economic dispatch problem is described in [17] where it has been proved that BBO gives a solution which is comparable with evolutionary programming and differential evolution techniques.

The objective function to be minimized comprises of cost of the device, line loadings and voltage deviations at the load buses.

42.2 Problem Formulation

42.2.1 Objective of the Optimization

As the cost of the FACTS devices, especially TCSC is high, in order to achieve the maximum benefit, the devices are to be installed at the optimal locations. Minimizing the cost of installation of the TCSC is chosen as the primary objective and the objective function is augmented with two indices one for the load voltage deviation and the other for line loading thereby making it a comprehensive one, whose minimization leads to a cost effective, security oriented solution.

The objective function is formulated as

$$MinF = W_1[C_{TCSC} * S] + W_2[LVD] + W_3[LL] \tag{42.1}$$

F is the objective function;
C_{TCSC} is the cost of TCSC device in US \$/KVar;
S is the operating range of TCSC;
LVD is the Load voltage deviation;
LL is the Line loading;
W_1, W_2 and W_3 are the weight factors.

(i) Cost (C_{TCSC})
The first term of the objective function C_{TCSC} presents the installation cost of TCSC device in the network, which is given by the following equation.

$$C_{TCSC} = 0.0015s^2 - 0.7130s + 153.75 \tag{42.2}$$

(ii) Load voltage deviation (LVD)
The excessive high or low voltage can lead to an unacceptable service quality and can create voltage instability problems. A FACTS device connected at the appropriate location plays a leading role in improving voltage profile thereby avoiding voltage collapse in the power system. The second term considered in this work to minimize the load voltage deviations so as to prevent the under or over voltages at network buses is given by

$$LVD = \sum_{m=1}^{nb} \left(\frac{V_{mref} - V_m}{V_{mref}} \right)^n \tag{42.3}$$

V_m is the voltage magnitude at bus m.
V_{mref} is the nominal voltage at bus m and is considered as 1.0 pu.
m refers to the load buses, where V_m is less than V_{mref}.

(iii) Line loading (LL)
TCSC is located in order to remove the overloads and to distribute the load flows uniformly. To achieve this, line loading is considered as the third term in the objective function.

$$LL = \sum_{l=1}^{nl} \left(\frac{S_l}{S_{l\max}} \right)^n \tag{42.4}$$

S_l is the apparent power in the line l.
$S_{l\max}$ is the apparent power rating of line l.

42.2.2 The Optimization Variables

The optimization variables considered in this work are

(a) The number of TCSC devices to be installed is taken as the first variable.
(b) TCSC location is considered as the second variable to be optimized. It is assumed that no TCSC is installed in the lines where the transformers exist.
(c) The reactance of the TCSC is considered as the third variable. The working range of TCSC is considered as follows.

$$-0.8X_l \leq X_{TCSC} \leq 0.2X_l \tag{42.5}$$

X_{TCSC} is the reactance added to the line by placing TCSC.
X_l is the reactance of the line where TCSC is located.

42.3 Overveiw of BBO Technique

Biogeography Optimization, an efficient optimization technique was introduced by Dan Simon [16]. BBO algorithm tries to solve the optimization problem through the simulation of immigration and emigration behaviour of the species in and out of a habitat. Species move in and out of the habitats depending upon various factors such as availability of food, temperature prevailing in that habitat, already existing species count in that area, diversity of vegetation, and species in that area etc. and the process strikes a balance when the rate of immigration is equal to the rate of migration. But these behaviours are probabilistic in nature. BBO algorithm exploits the search of the individuals to find them a suitable habitat to probe into the promising regions of the search space. A habitat is formed by a set of integers that form a feasible solution for the problem and an ecosystem consist of a number of such habitats.

A set of habitats are generated randomly, satisfying the constraints and their HSI is evaluated. In order to retain elitism, extremely good solutions are retained while modification operation is performed over the rest of the members, HSI recalculated for the modified ones thereafter mutation operation is carried out over the extremely good and bad solutions leaving aside the solutions in the middle range. Stopping criteria is similar to any other popular population based algorithm where the algorithm terminates after a pre defined number of trials or after the elapsing of the stipulated time or where there is no significant change in the solution after several successive trials.

42.4 Algorithm

The algorithm of the proposed work is explained below.

Step1: The system data and the load factor are initialized.

Step2: BBO parameters such as the size of the suitability index variable n, maximum number of iterations, limits of each variable in the habitat are initialized.

Step3: An initial set of solutions is randomly generated considering the variables to be optimized (the number of TCSCs, location of TCSC, parameter setting of TCSC).

Step4: The immigration rate λ and emigration rate μ are determined for each of the habitats.

Step5: Elite habitats are identified and they are exempted from modification procedure.

Step6: A habitat *Hi* is selected for modification proportional to its immigration rate λi and the source for this modification will be from the habitat *Hj* proportional to its emigration rate μ_j. This represents the migration phenomena of the species wherein the new habitats are formed through migration.

Step7: The probability of mutation P_i calculated from λ and μ is used to decide the habitat *Hi* for mutation and its *j*th SIV is replaced by a randomly generated SIV.

Step8: Already existing set of elite solutions along with those resulting from the migration and mutation operations result in a new ecosystem over which the Steps 4–6 are applied until any one of the stopping criteria is reached.

Step9: The same procedure is repeated for different load factors.

HSI Habitat Suitability Index.

SIV Suitability index variable.

λ Immigration rate.

μ Emigration rate.

P The probability of mutation.

42.5 Simulated Results

42.5.1 Case (I) 14 Bus System

See Tables 42.1, 42.2.

Table 42.1 Line loading for different load factors

Load factor	1	1.1	1.2	1.3	1.4	1.5
Without TCSC	17.5892	19.2093	20.9319	22.4464	24.0519	25.7766
With TCSC (PSO)	17.3907	19.0305	20.6862	22.3651	23.3228	25.0535
With TCSC (BBO)	17.3732	18.966	20.4223	22.143	23.011	24.9989

Table 42.2 Line voltage deviation for different load factors

Load factor	1	1.1	1.2	1.3	1.4	1.5
Without TCSC	0.3509	0.3696	0.4075	0.4875	0.5689	0.5922
With TCSC (PSO)	0.2884	0.3079	0.3275	0.3479	0.3801	0.4024
With TCSC (BBO)	0.2864	0.3062	0.3266	0.3445	0.37	0.4006

42.5.2 Case (II) 30 Bus System

See Tables 42.3, 42.4.

Table 42.3 Line loading for different load factors

Load factor	1	1.1	1.2	1.3	1.4	1.5
Without TCSC	14.5592	16.2116	17.9504	19.7258	21.5352	23.1761
With TCSC (PSO)	14.5522	16.1973	17.9397	19.7155	21.5319	23.1159
With TCSC (BBO)	14.5510	16.1071	17.9296	19.7017	21.5144	23.103

Table 42.4 Line voltage deviation for different load factors

Load factor	1	1.1	1.2	1.3	1.4	1.5
Without TCSC	0.6967	0.6974	0.7145	0.7342	0.7548	0.7834
With TCSC (PSO)	0.6965	0.6973	0.7128	0.7335	0.7529	0.7578
With TCSC (BBO)	0.6945	0.6953	0.7112	0.7228	0.7414	0.7492

42.5.3 Case (III) 57 Bus System

See Tables 42.5, 42.6.

The effectiveness of the proposed algorithm has been tested on IEEE 14, 30 and 57 bus systems. The objective function comprises three parts. First one is the quadratic cost function of the device for which data has been obtained from Siemen's data base. Second part of the objective function is the aggregate of line loadings expressed as a fraction of maximum loadability of the corresponding lines and the third part of the objective function takes care of line voltage deviations which are expressed as an aggregate of fractions of absolute values of the deviations at the load buses to the corresponding maximum values of the voltages. As line loadings and load voltage deviations form a part of the objective function, overload conditions are simulated by simultaneously varying the real and reactive powers on the load buses keeping the load power factor constant. Maximum number of devices is limited to 2 on a 14 bus system, 3 on a 30 bus system and 5 on a 57 bus system.

Table 42.5 Line loading for different load factors

Load factor	1	1.1	1.2	1.3	1.4	1.5
Without TCSC	53.33	61.29	70.01	79.86	92.01	105.16
With TCSC (PSO)	53.26	61.23	69.92	79.66	91.59	104.23
With TCSC (BBO)	53.22	61.21	69.81	79.52	91.49	104.14

Table 42.6 Line voltage deviation for different load factors

Load factor	1	1.1	1.2	1.3	1.4	1.5
Without TCSC	3.88	4.16	4.46	4.46	5.97	6.88
With TCSC (PSO)	3.86	4.16	4.45	4.46	5.65	6.78
With TCSC (BBO)	3.80	4.13	4.34	4.35	5.44	6.56

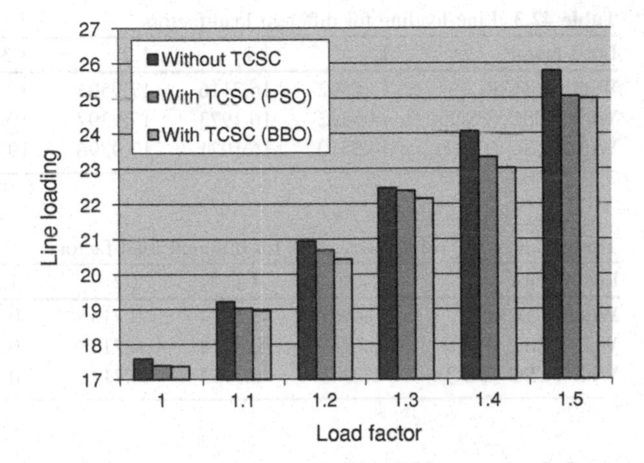

Fig. 42.1 Line loading versus percentage of load

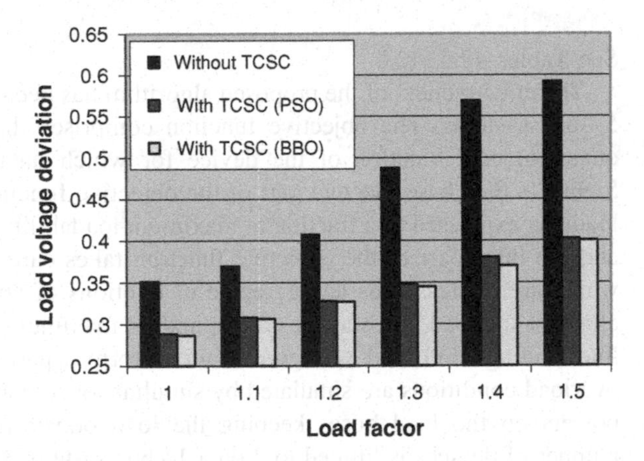

Fig. 42.2 Load voltage deviation versus percentage of load

Figures 42.1, 42.2, 42.3, 42.4, 42.5, and 42.6 represents the variation of line loadings and load voltage deviations for IEEE (14, 30 and 57) bus systems before and after the placement of the device. Device placement is carried out by fixing the location and ratings by applying BBO technique. It has been observed that there is a considerable reduction in line loadings and improvement in bus voltage profile after the placement of TCSC.

In order to validate the proposed method, PSO has been applied for solving the same objective function under the same simulated load conditions and from the bar

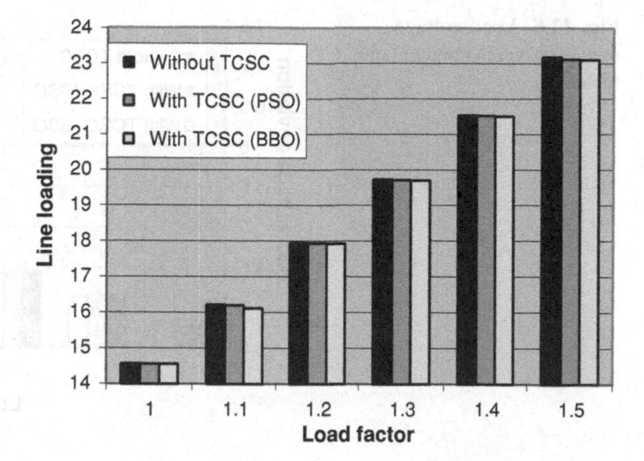

Fig. 42.3 Line loading versus percentage of load

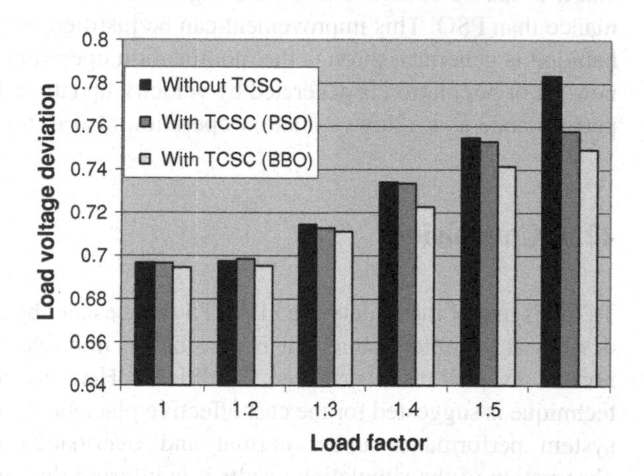

Fig. 42.4 Load voltage deviation versus percentage of load

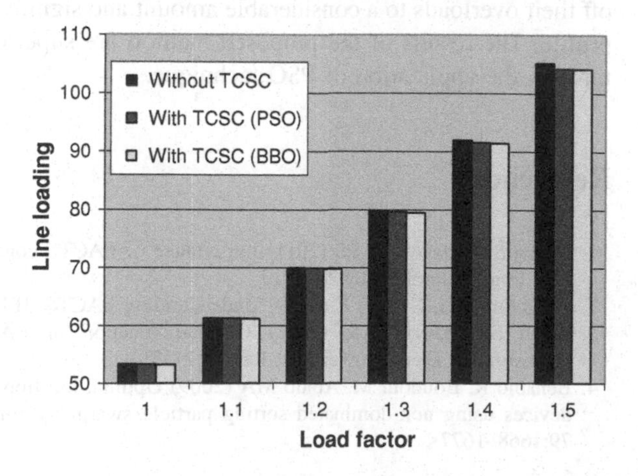

Fig. 42.5 Line loading versus percentage of load

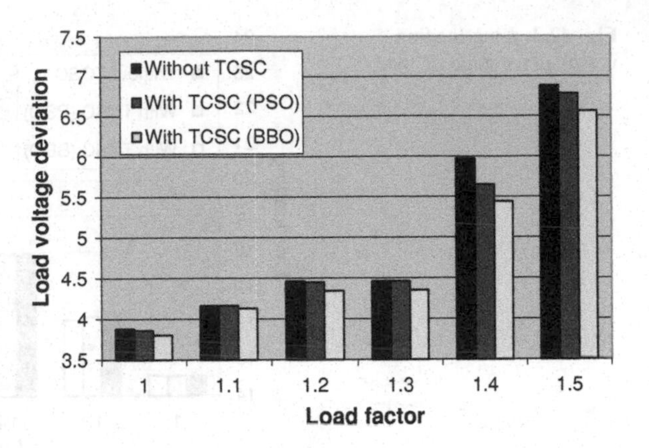

Fig. 42.6 Load voltage deviation versus percentage of load

chart, it can be observed that BBO algorithm shows comparatively better performance than PSO. This improvement can be justified by the fact that in BBO a new habitant is generated through the modification operation just like the PSO, where a new set of population is generated by velocity updation, BBO shows a cutting edge performance as it allows mutation operation also to take place.

42.6 Conclusions

TCSC is one of the widely used FACTS devices and by connecting optimally rated devices at suitable locations it is possible to improve the system performance at normal as well as overloaded conditions. Here an algorithm based on BBO technique is suggested for the cost effective placement of the device to improve the system performance under normal and overloaded conditions and from the observation of the simulation results it is inferred that the device relieves the lines off their overloads to a considerable amount and significantly improves the voltage profile. The results of the proposed method are superior than the result obtained through the application of PSO technique.

References

1. Manoj S, Puttaswamy PS (2011) Importance of FACTS controllers in power systems. Int J Adv Eng Technol II(III):207–212
2. Hingorani NG, Gyugyi L (1999) Understanding FACTS. IEEE Press, New York
3. Singh SN, David AK (2001) Optimal location of FACTS devices for congestion management. Electr Power Syst Res 58:71–79
4. Benabid R, Boudour M, Abido MA (2009) Optimal location and setting of SVC and TCSC devices using non-dominated sorting particle swarm optimization. Electr Power Syst Res 79:1668–1677

5. Ambriz-Perez H, Acha E, Fuerte-Esquivel CR (2006) TCSC-firing angle model for optimal power flow solutions using Newton's method. Electr Power Syst Res 28:77–85
6. Acharya N, Mithulanathan N (2007) Locating series FACTS devices for congestion management in deregulated electricity markets. Electr Power Syst Res 77:352–360
7. Rashed GI, Sun Y, Liu K-P (2011) Optimal placement of thyristor controlled series compensation in power system based on differential evolution algorithm. IEEE, Seventh Int Conf Nat Comput 4:2204–2210
8. Meikandasivam S, Nema RK, Jain SK (2008) Behavioral study of TCSC device—a matlab/simulink implementation. World Acad Sci, Eng Technol 45:694–699
9. Preedavichit P, Srivastava SC (1998) Optimal reactive power dispatch considering FACTS devices. Electr Power Syst Res 46:251–257
10. Visaka K, Thukaram D, Jenkins L (2003) Application of UPFC for system security improvement under normal and network contingencies. Electr Power Syst Res 70:46–55
11. Song S-H, Lim J-U, Moon S-II (2004) Installation and operation of FACTS devices for enhancing steady-state security. Electr Power Syst Res 70:7–15
12. Gerbex S, Cherkaoui R, Germond AJ (2001) Optimal location of multi-type FACTS devices in a power system by means of genetic algorithms. IEEE Trans Power Syst 16:537–544
13. Shaheen HI, Rashed GI, Cheng SJ (2010) Optimal location and parameter setting of UPFC for enhancing power system security based on differential evolution algorithm. Electr Power Syst Res 33:94–105
14. Saravanan M, Mary Raja Slochanal S, Venkatesh P, Prince Stephen Abraham J (2007) Application of particle swarm optimization technique for optimal location of FACTS devices considering cost of installation and system loadability. Electr Power Syst Res 77:276–283
15. Satyanarayana K, Prasad BKV, Devanand G, Siva Prasad N (2011) Optimal location of TCSC with minimum installation cost using PSO. Int J Comput Sci Technol 2:156–160
16. Simon D (2008) Biography-based optimization. IEEE Trans Evol Comput 12:702–713
17. Bhattacharya A, Chattopadhyay PK (2010) Solving complex economic dispatch problems using biogeography-based optimization. Expert Syst Appl 37:3605–3615

Chapter 43
Reliable Design of Embedded System with Minimal Resource Using SFT and Mode Algorithm

M. V. Raja and R. Srivatsan

Abstract In multi-process embedded system, optimizing the design of hardware and software reliability is fairly hard. In this paper, hardware replication and software re-execution techniques are combined to tolerate transient faults. It helps to reduce the usage of space and time in terms of embedded system design with minimal resource. The probability of information about embedded systems reliability is analyzed using System Fault Tree (SFT). SFT analyzer integrates SFT into an optimization process. An optimization algorithm which is used in this paper is known as Multi-Objective Differential Evolution (MODE), which makes effective design space exploration. It is also used to map the fault-tolerance policy information into chromosomes. The experimental result shows the achievement of maximum reliability in spatial redundancy using minimal resource and minimal fault tolerance.

Keywords Fault-tolerance · System fault tree · Transient fault · Multi-objective differential evolution · Reliability

43.1 Introduction

Now a day's transistor size plays major role in embedded systems. Every system evolves with new techniques and design, but it's critical to produce a reliable system with hardware replication and software re-execution to reduce fault

M. V. Raja (✉)
Periyar Maniyammai University, Thanjavur, India
e-mail: mvrajasastra@gmail.com

R. Srivatsan
M.A.R. College of Engineering and Technology, Pudukkottai, India
e-mail: shri2345@gmail.com

R. Malathi and J. Krishnan (eds.), *Recent Advancements in System Modelling Applications*, Lecture Notes in Electrical Engineering 188,
DOI: 10.1007/978-81-322-1035-1_43, © Springer India 2013

tolerance. Reduction in size of transistor makes more faults during design. Faults occurred due to defect in the device or transient. Transient faults keep on increasing than the permanent faults even in new technologies [1]. In evolving technologies, reliability of the system can be measured using transient faults (must be low).

Spatial redundancy of the system (also known as hardware redundancy) is a usual way to enhance reliability. Next most important part in embedded system which helps to increase reliability is software. Mostly in software's, execution time and functions are plays a major role. If execution time exceeds the deadline then it may leads to several faults in the system. Faults are detected in software by making check points and re-execution (in case of faults) [2, 3]. Scheduling of task in software system also kept intact in case of software redundancy [4–6].

Spatial and temporal redundancy is used to achieve optimal implementation in system design constrains [7, 8]. It also finds some solution for cost and scheduling. Not only large amount software redundancy helps to achieve reliability. Hardware safety is applied through Safety Integrity Level (SIL) 4 by reducing faults [9]. It is very critical to integrate and analyze several fault tolerance technique. A Multi-processor System-on-Chip (MPSoC) platform provides hardware resources and flexibility to well explore different fault tolerance techniques.

The objective of this paper focus on the reliable design of embedded system. Hardware replication and software re-execution techniques are used by us to tolerate transient faults. The work is based on fault and process models, e.g., models which is used [10, 11]. Multi-Objective Differential Evolution (MODE) has used to achieve optimal design of the system. MODE works on the principle of mapping the tasks to Processing Elements (PEs), the exact task and message schedule and the amount of the hardware/software redundancy is applied. The main aspects of this paper are: (1) A system fault tree analysis that computes the system-level failure probability in the presence of hardware/software redundancy. (2) An approach that integrates the proposed reliability analysis into a MODE-based optimization framework. (3) An inter-job slack sharing scheme for further reliability enhancement.

The rest of the paper is organized as follows. The system design used in this paper is introduced in Sect. 43.2. Sections 43.3 and 43.4 present the proposed reliability analysis and design optimization approaches. Experimental results are presented in Sect. 43.5. Section 43.6 concludes this paper.

43.2 System Design

Let us consider an application A as a set of independent cyclic jobs executing concurrently in the system. A job J ∈ A is a directed acyclic graph, whose vertices V represent a set of tasks to be executed and the edges E capture data dependencies between tasks. For each edge, a message is associated to characterize the data transfer. If jobs have originally different periods, they are first transformed into larger graphs representing a hyper-period (LCM of all periods) of the application. We use T to denote the set of all tasks in application A and T (n) to represent the set of tasks mapped to processor n.

Our target architecture is a MPSoC with time-triggered on-chip communication. The GENESYS platform is one example of such architecture. The set of available PEs is denoted using N. The communication bus is arbitrated using TDMA. If the two communicating tasks are mapped to the same PE, data transfer can be realized via local memory and no bus slot is needed. Otherwise, a dedicated time slot needs to be reserved. The message transfer is currently assumed to be fault-free. The consideration of message fault and optimization of the fault handling techniques are addressed in future work. Timing predictability is highly desirable for safety-related applications. In this paper, we target on synthesizing static time-triggered schedules. Such a schedule S is a set of sub-schedules, each describing the scheduling information for a specific PE. There must be exactly one sub-schedule Sn for each PE n ∈ N. A sub-schedule consists of a set of scheduling slots. A scheduling slot is a 3-tuples = (t_s, t_f, T), where t_s is the start time of the slot, t_f is the finish time and T is a set of tasks that may execute in the slot. A slot can be a normal task execution slot or a re-execution slot (also called slack slot). The later is meant to be shared by multiple tasks and used for re-execution of instances misbehaving due to transient faults. Figure 43.1a depicts an example schedule that has three slots for processor n1 and n2 each.

The actual utilization of scheduling slots depends on how the scheduler responds to the occurred faults. In this paper, we consider a static non-preemptive scheduling approach. In this case, the task t ∈ s.T2 that has the highest priority among all pending tasks acquires the slot. Figure 43.1 demonstrates two example execution scenarios. In Fig. 43.1b, the slot S2 is used to re-execute t1, since both S_O and S1 fail. In Fig. 43.1c, the same slot is used for t2, since t1 is already finished with S_O. We use the same assumption as in [6, 8, 12] that the transient faults are detected using sanity checks at the completion of a tasks' execution. The timing

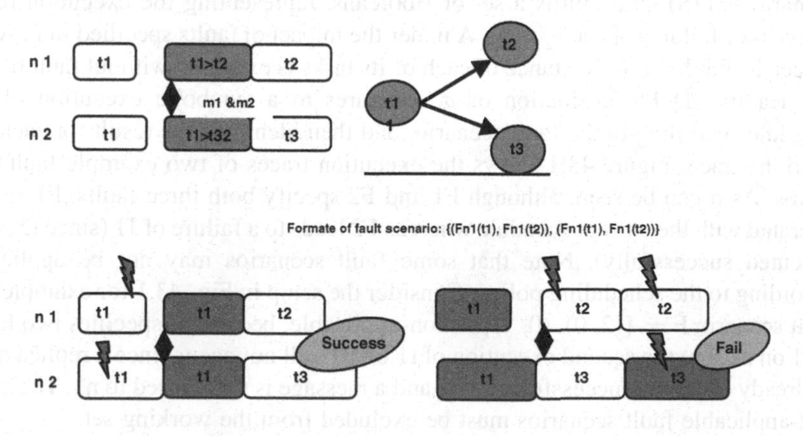

Fig. 43.1 Fault tolerance scenario 1

overhead of fault detection is assumed to be contained in the Worst- Case Execution Time (WCET) of tasks. The combination of faults that occur is described by a fault scenario F, which is a set of partial fault scenarios, one for each PE. A partial fault scenario Fn for PE n is a vector of integers of length |T(n)|, specifying the number of faults happening to each of the tasks mapped to n. A partial fault scenario with x faults in total compares to a selection of x tasks out of T(n), where each task can be selected multiple times and the order does not matter [8, 13]. For a specific PE n, the maximum number of tolerable faults depends on the amount of redundancy. We define the fault-tolerance capability max (S, n) as the number of re-execution slots scheduled on n plus the number of tasks which are replicated at least once in other PEs. Obviously, a fault-scenario is not expected to be tolerable if the amount of faults specified for any processor n is higher than max (S, n). To analyze the system-level reliability, we are interested in identifying the set of fault scenarios that can be tolerated using the current schedule. To do this it is safe to investigate only the finite set of fault-scenarios:

$$\widehat{F}(S) = \left\{ F\,\forall_n \in N: \sum_{t \in T(n)} F_n(t) \le \max(S, n)) \right\}$$

43.3 Design of System Fault Tree

System Fault Tree (SFT) analysis helps to compute the system failure probability (SFP) by component failure rate and the amount of software and hardware redundancy. In general, to compute the failure probability of a specific job J, we need to identify the complete set of tolerable fault scenarios that does not lead to a failure of J. Such a set is called the working set of J with schedule S, denoted by W (J, S). We define a function σ, which takes an application A, schedule S and a fault scenario F∊$\hat{F}$(S) and returns a set of Booleans, representing the execution result (success or failure) of each job J ∊ A under the impact of faults specified in F. A job succeeds if at least one instance of each of its tasks is executed without fault before the deadline D(T). Evaluation of σ compares to a symbolic execution of the schedule according to the fault scenario, and then identifies the result for each job from the trace. Figure 43.1 shows the execution traces of two example fault scenarios. As it can be seen, although F1 and F2 specify both three faults, F1 can be tolerated with the current schedule whereas F2 leads to a failure of J1 (since t2 is not executed successfully). Note that some fault scenarios may not be applicable according to the scheduling policy. Consider the setup in Fig. 43.1 for example, the fault scenario F = {(2, 0), (0, 1)} is non-applicable, because it specifies two faults in t1 on n1, but the second execution of t1 on n1 will not occur since a replica of p1 is already executed successfully on n2 and a message is transferred to n1. The set of non-applicable fault scenarios must be excluded from the working set.

Computing the working set can be done via fault tree analysis. Figure 43.2 illustrates the analysis for the schedule depicted in Fig. 43.1. From the none-fault case {(0, 0) (0, 0)}, we iteratively increase the number of faults and check if the new fault scenario is still tolerable. The procedure builds a tree structure representing the possible fault scenarios. The depth of the tree is restricted by the fault-tolerance capabilities max (S, n) of the PEs. The failure or non applicable nodes of the tree will not spawn further branches.

After obtaining the working set, the success probability of a job (denoted by Pr(J, S)) can be computed by summarizing the occurrence probability of the set of tolerable fault scenarios Pr(F):

$$\Pr(F) = \sum_{F \in W(J,S)} \Pr(F)$$

Let Pr(t, n) denote the success probability of task t on node n, the occurrence probability of fault scenario F is:

$$\Pr(F) = \prod_{n \in N} \Big(\prod_{t \in T(n)} (1 - \Pr(t,n))^{F_n(t)} \cdot \prod_{\substack{t \in Success \\ (F,S,n)}} \Pr(t,n) \Big)$$

where T(n) is the set of tasks mapped on n, Fn(t) is the number of faults on task t specified in Fn and Success (F, S, n) is the set of successfully finished tasks on n. Success(F,S,n) can be obtained from the trace of evaluating σ(A, S, F).

The number of nodes visited during the fault-tree analysis increases exponentially with the depth. Let

$$|F| = \sum_{n \in N} |F_n|$$

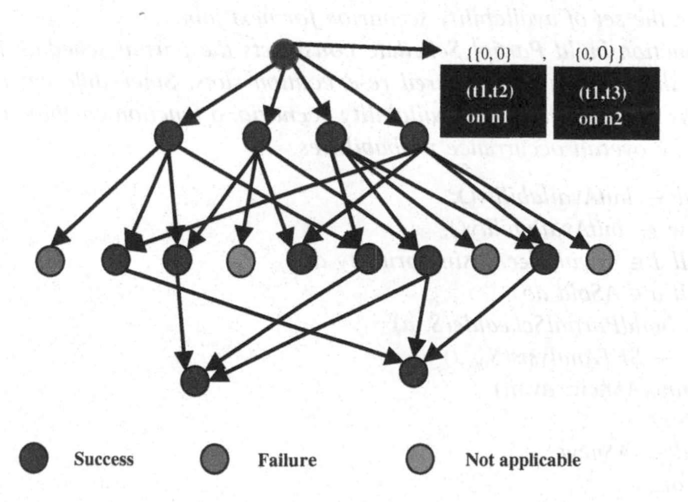

Fig. 43.2 Fault tolerance scenario 2

be the total number of entries in the fault scenario, we need to visit

$$\binom{|F| + d - 1}{d}$$

nodes at depth d in the worst case. This implies that the computational complexity also increases exponentially. Since the stand-alone failure rates of the tasks are typically very low, the nodes located deeper in the fault tree have much lower occurrence probability. Moreover, the portion of non tolerable fault-scenarios will also increase significantly as the depth increases. In Fig. 43.2 for example, only 2 out of 20 fault scenarios are tolerable at depth 3. Thus, in many circumstances, a safe underestimation of system reliability would be to consider only a bounded depth during analysis. This compares to assuming a maximum number of faults that may occur anywhere in one period of execution.

43.3.1 Inter-Job Slack Sharing

Inter-job slack sharing is motivated by the emerging needs to cope with mixed-criticality jobs, i.e. applications with highly distinct reliability requirements running in the same platform. For high criticality jobs, significant amount of software redundancy is needed to meet the high reliability requirements. However, the probability that the software slack is actually used is typically very low. In this case, reusing the slack time for low criticality jobs using the static-priority approach may lead to significant saving of hardware resources.

Algorithm 1 Iterative Tree Analysis(): iterative SFT for multiple jobs using the inter-job slack sharing scheme.

ASold: the set of availability scenarios from previous job.

ASnew: the set of availability scenarios for next job.

The function build Partial Schedule constructs the partial schedule for a job based on the availability of shared re-execution slots. Since different fault scenarios may result in the same availability scenario, a function combine is used to compute the overall occurrence probabilities.

1. *ASold* $\leftarrow$ *initAvailability();*
2. *ASnew* $\leftarrow$ *initAvailability();*
3. *for all* $J \in A$ *with decreasing priority do*
4. *for all* $a \in$ *ASold do*
5. *SO = buildPartialSchedule(S, a)*
6. *avail* $\leftarrow$ *SFTAnalysis(*S_0*, J)*
7. *combine(ASnew, avail)*
8. *end for*
9. *ASold* $\leftarrow$ *ASnew*
10. *end for*

Fig. 43.3 Work flow of DE

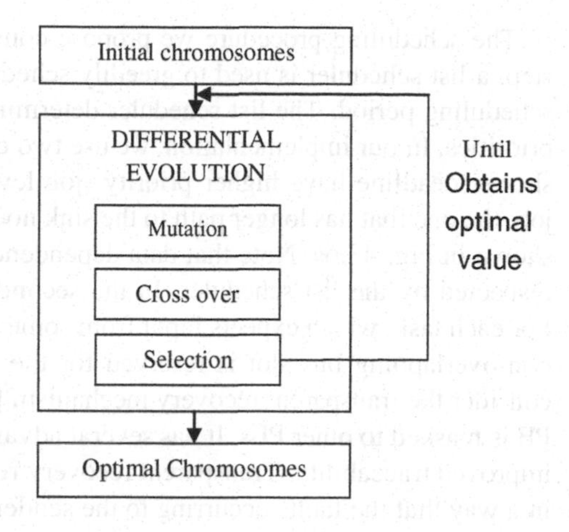

43.4 Optimization Process

Multi Objective Differential Evolution (MODE) is used as an optimization strategy. Differential Evolution is a population based algorithm. In DE, population represents the set of solutions. In each iteration, some set of solution are produced called Parent, mutation and Crossovers produce children. Fitness function find the optimal solution by replacing population set with new population set. Iteration takes place until optimal solution is produced.

The efficiency of MODE is heavily influenced by the length of the chromosome, since it determines the overall search space. However, a direct encoding of the schedule as discussed in Sect. 43.3 results in a very large chromosome. To cope with this problem, we utilize a two-step optimization process as shown in Fig. 43.3 inspired from [14]. The main idea is that, instead of encoding the entire schedule, we only put partial information, namely the mapping and fault-tolerance policy, into the chromosomes. A scheduler is integrated into the general variation-selection process to transform the chromosome to an optimized schedule. The resulting schedule is then used for fitness evaluation, e.g., the SFT analysis.

Using the above approach, we encode the mapping and the fault-tolerance policy information.The chromosome contains one gene for each task $t \in T$.

Each gene is a pair $g = (i, j)$, where i is the integer index of the task and j is a list of integer indicating the set of PEs onto which task i is mapped. Multiple mappings of the same task onto the same PE are interpreted as re-execution slots multiple mappings of the same task onto different PEs are interpreted as spatial replications Reconstruction of the schedule from the chromosome is the same as scheduling the task executions with known mapping and fault-tolerance policy. A notable advantage of the two-step approach is that the scheduling algorithm is orthogonal to our encoding scheme in the sense that any existing scheduling algorithms for this purpose can be used. Moreover, as long as the scheduler is implemented correctly, the variation of the chromosome will always produce valid solutions.

The scheduling procedure we propose consists of three main steps. In the first step, a list scheduler is used to greedily schedule the tasks to the beginning of the scheduling period. The list scheduler determines the order of tasks based on their priorities. In our implementation, we use two criteria: tasks belonging to a job with shorter deadline have higher priority (job-level EDF); for tasks within the same job, the one that has longer path to the sink node has higher priority. An example is shown in Fig. 43.6a. Note that data dependencies between tasks are automatically respected by the list scheduler. In the second step, bus scheduling is performed. For each task, which expects input from some predecessors mapped to other PEs, a non-overlapping bus slot is reserved for thc message transfer. In this paper, we consider the transparent recovery mechanism [15], where a fault happening on one PE is masked to other PEs. It has several advantages such as fault-containment and improved traceability. Transparent recovery requires that the message is scheduled in a way that the faults occurring to the sender are not visible to the receiver. Intra job and inter job slack sharing schemes can be integrated in this step.

For improving the performance of the DE, we have implemented some cross-over and mutation operators that add problem-specific knowledge to the optimization. We present those operators in the following:

Task Implementation Crossover: This operator randomly selects a set of tasks and swaps the entire implementation of these tasks between two chromosomes, including the amount of spatial/temporal redundancy and mapping. The rest of the chromosome remains unchanged.

Task Mapping Crossover: This operator performs crossover for the implementation of each task separately. Given two chromosomes, the mapping entries for a chosen task are randomly swapped. Figure 43.4b shows an example in which 3 mapping entries are selected for crossover in total. Increment Redundancy: This mutation operator inserts a new mapping entry for a randomly selected task. Insertion of the new mapping x to task t might result in: (1) a slack slot, if the chromosome already contains a mapping of t to x, or (2) a spatial replication, if t has not been mapped to x. Decrement Redundancy: The counterpart of Increment Redundancy, removes one mapping entry from a random task. There must be at least one mapping for each task. Re-mapping: This mutation

Re-Mapping: This mutation operator randomly changes selected mapping entries. The result might be: (1) re-mapping of the tasks to other PEs or (2) transformation of a slack slot to a spatial replication or vice verse.

43.5 Experim

In our design we utilize optimization in the design using MPEG decoder example [15]. The optimization algorithm and analysis put into operation in JAVA and run on a Windows machine with 6 GHz CPU and 6 GB memory. With a population more than 200 implementations and runs for 600 generations put together using MODE algorithm. We assume that the target platform consists of two types of PEs, namely a ARM processor and a DSP. The failure probability of each task on a certain PE is randomly generated between 1×10^{-5} and 1×10^{-6} (Fig. 43. 5).

The Pareto front results is shown in the figure after optimization. The horizontal axis shows the reliability and the vertical axis is the overhead of schedule length with respect to the deadline, i.e. overhead 0 implies meeting the deadline and a positive value means a deadline violation. It can be seen that, the Pareto fronts obtained with more PEs dominate in most cases the Pareto fronts obtained with less PEs, i.e. the application is finished with shorter time and higher reliability. This is due to the increased opportunity for spatial redundancy.

For each platform, we are interested in the solutions that achieve maximum reliability with deadline constraints fulfilled. These solutions are marked with 1–3 in Fig. 43.6. As it can be seen, the 2ARM+ 2DSP platform is the minimal one to achieve System Failure Probability (SFP) of 1×10^{-6} and the 2ARM+ 3DSP platform is necessary to achieve SFP of 1×10^{-9}. An important observation from Fig. 43. 5 is that, for the 2 ARM+ 1DSP platform, several solutions with SFP around 1×10^{-6} are very close to meeting the deadline. The same is observed for the platform 3ARM+ 2DSP, where several solutions are close to achieve SFP of 1×10^{-9}. This implies that, if some PEs can be replaced by faster ones, using 3 or

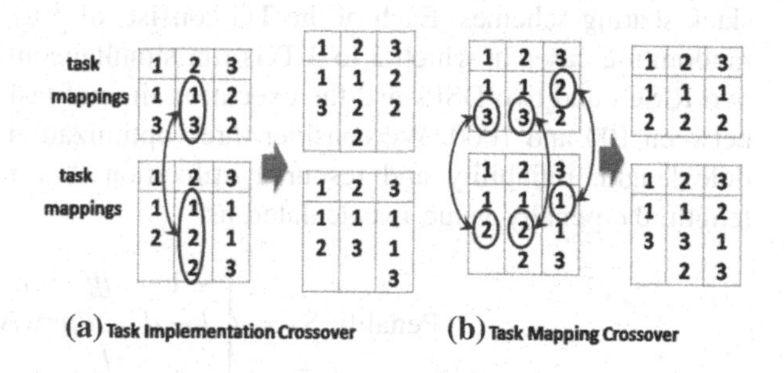

Fig. 43.4 Task implementation crossover

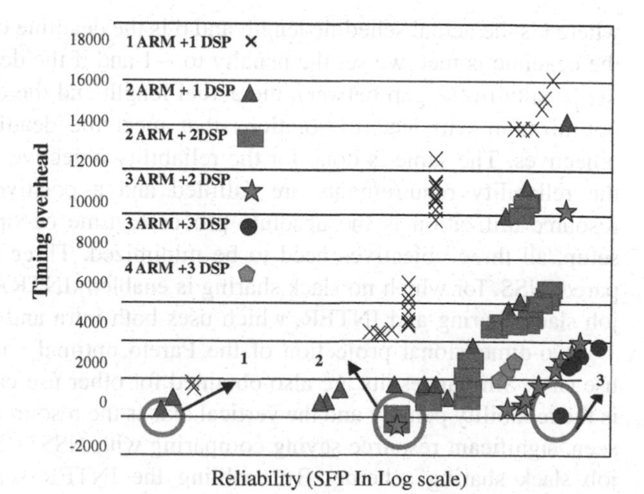

Fig. 43.5 Achievable reliability under different platform

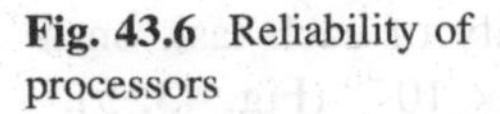

Fig. 43.6 Reliability of processors

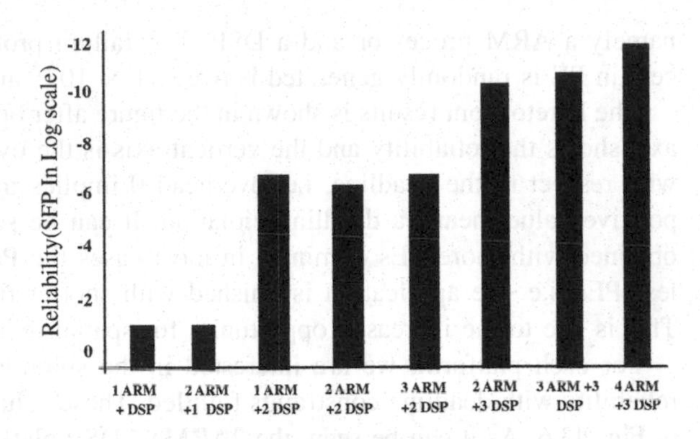

5 PEs might already be sufficient and become more cost-efficient design choices. We therefore tested two additional platforms with 1 ARM+ 2 DSP and 2 ARM+ 3 DSP (the DSP is faster for the mpeg 2 application). Figure 43.6 compares for each platform the maximum reliability achieved under deadline constraints. Clearly, the new platforms with 1 ARM+ 2 DSP and 2 ARM+ 3 DSP are the most cost-efficient solution to achieve SFP of 1×10^{-6} and 1×10^{-9} respectively.

For the second experiment we use a set of synthetic task graphs to evaluate the slack sharing schemes. Each of the TG consists of 5 to 15 tasks and we consider random use cases in which 2 to 3 TGs run simultaneously. The platform contains two RISCs and two DSPs and the execution time of tasks are generated randomly between 100 and 1000. We consider three optimization objectives, namely schedule length, reliability and resource utilization. For the objective of schedule length, the penalty value is calculated as:

$$\text{Penality(S)} = \begin{cases} -1 & \textit{iff } l \leq d \\ l-d & \textit{otherwise} \\ & l \end{cases}$$

where l is the actual schedule length and d is the deadline of the task. The idea is, if the deadline is met, we set the penalty to -1 and if the deadline is violated we set the penalty to the gap between the actual length and the deadline. In this way the optimization will lead to solutions that meet the deadline and optimize other objectives. The same is done for the reliability objective, i.e. the penalty is -1 if the reliability requirements are fulfilled and a positive value otherwise. The resource utilization is the absolute processor time occupation. Using the above setup, all three objectives need to be minimized. Three configurations are compared: NSS, for which no slack sharing is enabled; INTRA, which only uses intra-job slack sharing and INTER, which uses both intra and inter job slack sharing.

Two-dimensional projection of the Pareto optimal solutions for one example use case. Similar results are also obtained for other use cases. The horizontal axis is the reliability penalty and the vertical axis is the resource utilization. As it can be seen, significant resource saving comparing with NSS is achieved using the intra-job slack sharing scheme. By enabling the INTER scheme, further saving in resource consumption is observed. In the right part of the curve (reliability penalty larger than 4), the performance of INTRA and INTER is very close to each other. The reason is, for those solutions, the reliability is actually very low which suggests that the available re- execution slots are very limited, resulting in limited opportunity for further improvement using inter-job slack sharing. In contrast, in the left part of the curve, INTER shows notable benefit. In particular, considering the minimum resources that are needed to fulfill the reliability requirement, INTRA and INTER saves 14 and 20 % total resource consumption, respectively. Figure 43.6 compares the solution that has the minimum resource consumption and fulfills all deadline and reliability requirements. The resource consumption is normalized with respect to the NSS approach. On average, I N T RA and I N T ER saves 12 and 20 % resources, respectively.

43.6 Conclusion

From our work the reliability-aware Design Space Exploration (DSE) problem for real-time embedded systems. The main contribution is a SFT analysis that provides probabilistic information about system reliability and an approach that integrates SFT into an evolutionary algorithm based optimization process. We have proposed a two-step approach for efficient encoding of the chromosome and a set of problem-specific operators for manipulation of the chromosome. We have also proposed and evaluated an inter- job slack sharing scheme for further reliability enhancement. Next step will be the integration of DSE procedure into a model-based development framework. The optimization process will then take a set of input models and produce a set of transformed models that fulfill certain reliability requirements. Another direction of future work is to consider other non-functional properties besides reliability.

References

1. Sosnowski J (1994) Transient fault tolerance in digital systems. IEEE Micro 14(1):517–519
2. Pop P, Izosimov V, Eles P, Peng Z (2009) Design optimization of time- and cost-constrained fault-tolerant embedded systems with checkpointing and replication. IEEE Trans Very Large Scale Integr Syst 17(3):389–402
3. Zhang Y, Chakrabarty K (2006) A unified approach for fault tolerance and dynamic power management in fixed-priority real-time embedded systems. Comput Aided Des Integr Circuits Syst IEEE Trans 25(1):115–125
4. Liberato F, Melhem R, Mosse D (2000) Tolerance to multiple transient faults for aperiodic tasks in hard R-T systems. IEEE Trans Comput 49(9):906–914
5. Han C, Shin K, Wu J (2003) A fault-tolerant scheduling algorithm for real-time periodic tasks with possible software faults. IEEE Trans Comput 52(3):362–372
6. Pinello C, Carloni LP, Sangiovanni-Vincentelli AL (2004) Fault-tolerant deployment of embedded software for cost- sensitive real-time feedback-control applications. In: DATE, 2004

7. Izosimov V, Pop P, Eles P, Peng Z (2005) Design optimization of time-and cost-constrained fault-tolerant distributed ES. In: DATE, 2005
8. Izosimov V, Polian I, Pop P, Eles P, Peng Z (2009) Analysis and optimization of fault-tolerant embedded systems with hardened processorsIn. In: DATE, 2009
9. Zhu D, Aydin H (2009) Reliability-aware energy management for periodic real-time tasks. IEEE Trans Comput 58(10):1382–1397
10. Architectural Requirements IEC61508-2, chapter .4.3.1.1, Tab. 2 and 3
11. Birolini A (2004) Reliability engineering—theory and practice, 4th edn. Springer, Heidelberg
12. Reimann F, Glaß M, Lukasiewycz M, Haubelt C, Keinert J, Teich J (2008) Symbolic voter placement for dependability- aware system synthesis. In: CODE + ISSS
13. Bjorner A, Stanley RP A combinatorial miscellany
14. Kandasamy N, Hayes J, Murray B (2003) Transparent recovery from intermittent faults in time-triggered distributed systems. IEEE Trans Comput 52(2):113–125
15. Lukasiewycz M, Glaß M, Haubelt C, Teich J (2007) Sat-decoding in evolutionary algorithms for discrete constrained optimization problems. In: CEC, 2007

Chapter 44
ANN Based Word Sense Identifying Scheme for Question Answering Systems

C. Meenakshi and P. Thangaraj

Abstract The chapter develops an artificial neural network (ANN) based strategy to identify the sense of an ambiguous word as part of an information retrieval mechanism in a question answering system. The philosophy echoes to support the broad domain of natural language processing through its trained capabilities and endeavour to resolve the lexical ambiguity. The neural network is designed using the relations between the target word and associated words that appear in the sentence. It evolves a feed forward procedure to allow the weights to be adjusted and arrive at the precise contextual sense of the word. The inputs and weights are assigned in tune with the frequently appearing words in the English literature to train the model. The performance is investigated for a set of words that inherit a similar context with the words used in the training phase and the results claim the emergence of a promising word sense identifying tool.

Keywords Artificial neural network · Self organising map · Word sensing system · Word sense identification

44.1 Introduction

Word Sense Identification (WSI) is concerned with the identification of the correct sense of an ambiguous word in a preferred context. In spite of the fact that it is an independent task, still its importance is realized when it is applied to specific tasks

C. Meenakshi (✉) · P. Thangaraj
Research Scholar, Mother Terasa Womens University, Kodaikanal, India
e-mail: meenasi.c@gmail.com

P. Thangaraj
e-mail: ctptr@yahoo.co.in

R. Malathi and J. Krishnan (eds.), *Recent Advancements in System Modelling Applications*, Lecture Notes in Electrical Engineering 188,
DOI: 10.1007/978-81-322-1035-1_44, © Springer India 2013

such as page ranking, morphological analysis and transliteration. It continues to espouse interest in natural language processing issues and create impacts in tasks that relate to discourse, reference resolution, and textual inference [1].

The word sense ambiguity is ubiquitous in all natural languages, with a large number of the words in any given language carrying more than one meaning. For instance, the English noun *pen* can mean *writing material* or *fenced enclosure for animals*; similarly the French word *cours* can mean *lesson* or *compete*. The correct sense of an ambiguous word can be selected based on the context where it occurs, and correspondingly the problem of *word sense disambiguation* is defined as the task of automatically assigning the most appropriate meaning to a polysemous word within a given context [2]. If the ambiguity is in a sentence or clause, it is called structural (syntactic) ambiguity. If it is in a single word, it is called lexical ambiguity.

Lexical ambiguity can however refer to both homonymy and polysemy [3]. Homonyms are words that are written the same way, but are (historically or conceptually) really two different words with different meanings which seem unrelated [4, 5]. The examples are *lead* ("metal" and "start off in front") and *bear* ("animal" and "carry"). If a word's meanings are related, it is called a polyseme. The word *party* is polysemous because its senses can be generalized as "group of people", that is they are related.

44.2 Literature Review

A method that attempts to identify all the nouns, verbs, adjectives and adverbs in a text using their senses has been presented [6]. The reported results have been found to enjoy very good accuracy for the differently ranked senses. The semantic similarity between words and concepts has been measured through a lexical taxonomy structure with corpus statistical information. It has been found to involve edge based approach and enhanced by the node based paradigm of information content calculation [7].The performance of WSD scheme that incorporates lexical and structural semantic information has been evaluated using precise information from dictionary sentences and found to be effective [8]. A statistical method has been suggested for assigning senses to words through christening procedure about the context and the error rate in machine translation system found to decrease using statistical methods [9]. The results of word sense scheme generated from a concept map has been found to enhance understanding of the concept and sort the meaning of the word [10]. A neural network based algorithm suitable for very large corpora has been developed to address the needs of real time speech recogniser [11].

44.3 Problem Statement

The primary objective revolves around the creation of an artificial neural network model with a view to establish a relational type strategy among words that orient to a similar context. The scheme acquires the corpus from word net and evinces a choice of weights in accordance with the predictive feed forward corrective action to land at the correct target sense.

44.3.1 Proposed Methodology

The philosophy of WSD orients to determine the meaning of target words in a particular context which appears to be extraneously ambiguous in theoretical realities. Ambiguity is a crucial problem that invites a large focus on the part of computational linguists. The users in general are unaware of the ambiguities in the language they use in view of the fact that they are very good at resolving them using the context and their knowledge of the world. However, computer systems do not imbibe this knowledge and therefore may not perform a good task for the job on hand.

Various problems exist for which it is hard to arrive at exact algorithms as the solution. The traditional methods may not subscribe to the needs of the present day world in the sense identification is conceptual and pertains to a large number of domains [12, 13]. The emergence of intelligent methodologies foray their role through map based approaches and the training involved suits them to address reasoning related problems.

Neural network appears to be a viable tool that attracts a wide range of tasks such as classification, pattern recognition, approximation, character prediction, image clustering, function approximation, and many others. The artificial neural networks (ANNs) constituted of a large number of processors with many interconnections inherit parallel computing facility and can be used in areas of information retrieval.

The architecture seen in Fig. 44.1 is composed with three layers with input layer nodes being passive and serves to relay their single inputs to multiple outputs. The hidden layers and output layers are active and attempt to modify the signals in accordance with the desired output. The methodology avails the benefits of the appropriate choice of the weights in the hidden layers to arrive at the exact sense of the word. The flow diagram explaining the algorithm is depicted in Fig. 44.2.

The initial configuration involves the fixing up of map width and map height in order to gather the groups. The group size is roughly determined based on the probability of the words that may appear with the same context. The data in the

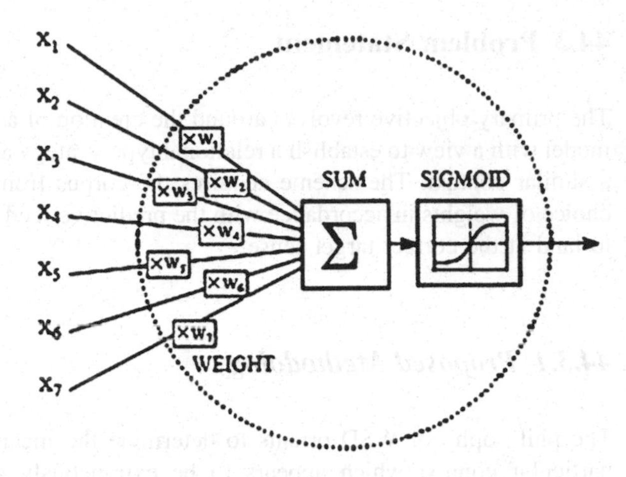

Fig. 44.1 Neural network architecture

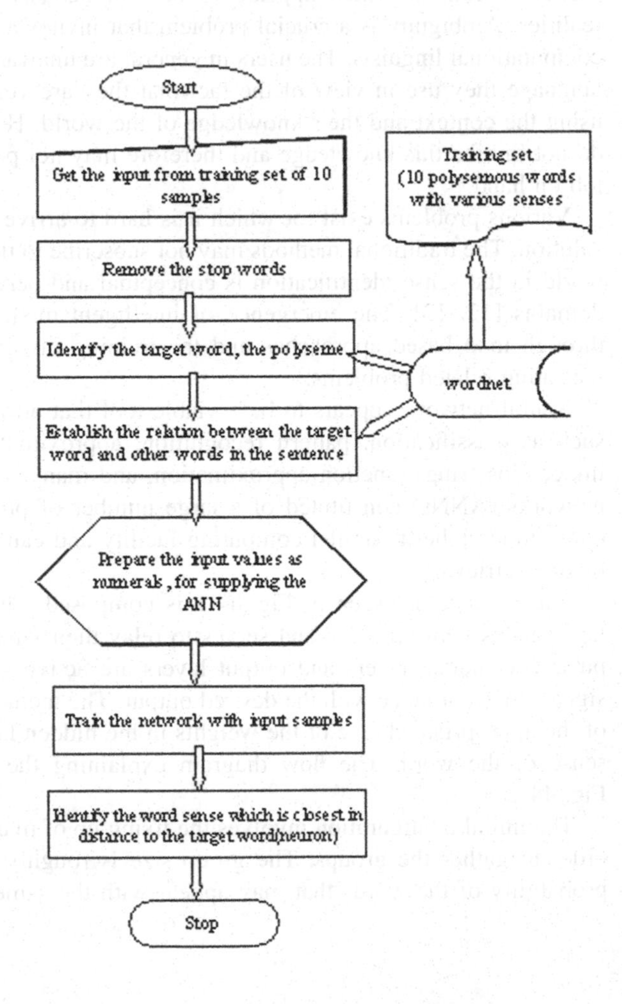

Fig. 44.2 Flow diagram

training set is fed one after another to find the distance between the available nodes i.e., the neurons in the map and the one with shortest distance is taken as the exact sense.

44.4 Simulation Results

The performance of the strategy is investigated with a data base obtained from word net [14] for different samples in training and testing phases and the results compared with that computed through an ANN based heuristic proposition. The procedure employs unsupervised learning to produce 2D representation of the input space of the training data set samples. The topology is initially defined and preserved by the neighbourhood function. The self-organizing map (SOM) shown in Fig. 44.3 projects the closeness in sense among the words. It depicts the plots for which exact senses are identified.

The denser cells relate to the closeness in the senses and orient to the context whereas the sparse ones illustrate the unidentical senses of the word with respect to the context.

The graphs seen in Figs. 44.4 and 44.5 represent the variation in mean square error accrued in the process of identifying the word sense for two different epochs 8 and 15 respectively.

The methodology extends to examine the performance for the same five words namely book, mark pen, stress and ruler through the use of a heuristic approach with a view to establish its merits and there from claim its applicability in the practical world. The relative comparison elaborates the efficacy of the scheme to rig out a superior correlative sense and intrigue out a much better accuracy as depicted in the bar diagram seen in Fig. 44.6. The line chart in Fig. 44.7 explains to the higher precision of the proposed scheme over the heuristic identifying capability and augurs its suitability in question answering systems.

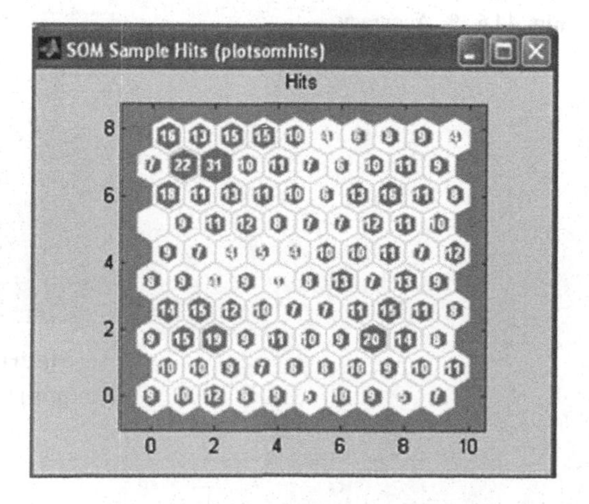

Fig. 44.3 Self organizing map

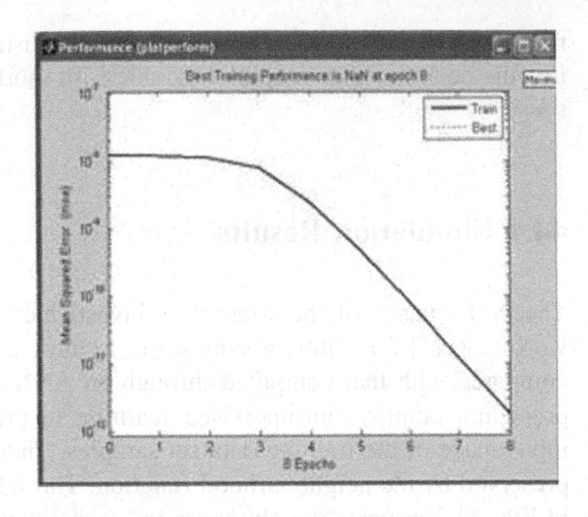

Fig. 44.4 Mean square error (8 epochs)

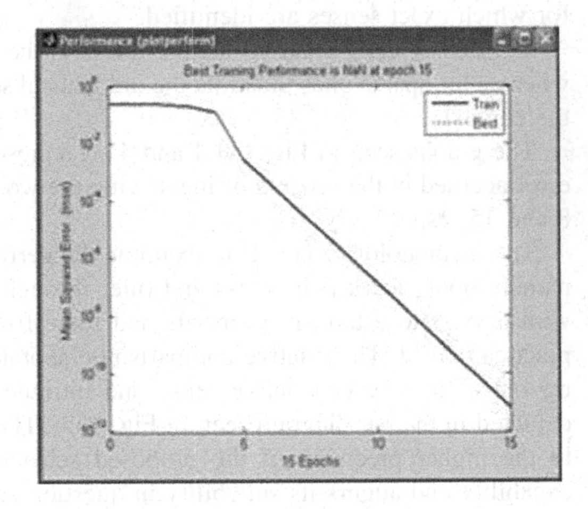

Fig. 44.5 Mean square error (15 epochs)

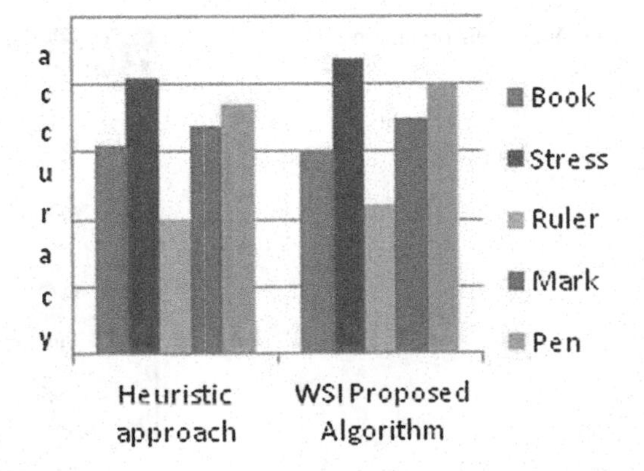

Fig. 44.6 % Accuracy

Fig. 44.7 % Precision

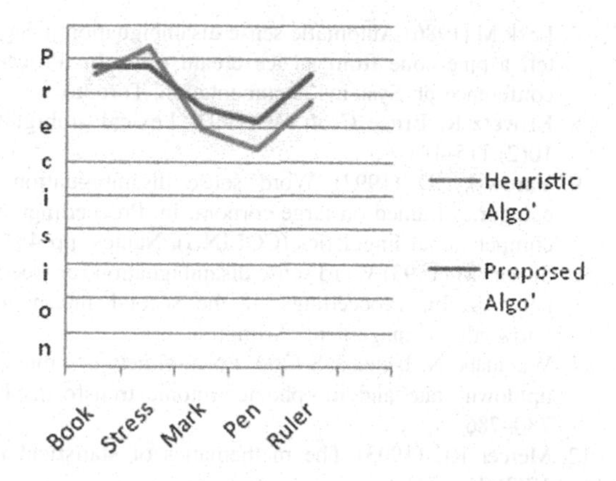

44.5 Conclusion

An ANN model has been constructed and trained suitably to extract the sense of a word in a question answering system. The strategy has been tailored to avail the data base from the word net and evolved to suit different varieties of words in an unambiguous domain. The results have been found to address the realistic requirements of a word sense identifying fortitude. The ability of the new formulation to churn out a better performance than a heuristic philosophy reinforces the use of neural network for disambiguation based interpretations. The significance of the proposition evinces a new direction in this field and propels to perpetuate its use in new related areas.

References

1. Amsler R, Walker D (1986) The use of machine readable dictionaries in sublanguage analysis. In: Grishman R, Kittredge R (eds) Analyzing language in restricted domains. LEA Press, Germany pp 69–83
2. Yarowsky D (1995) Unsupervised word sense disambiguation rivaling supervised methods. In: Proceedings of the 33rd annual meeting of the association for computational linguistics (ACL), Cambridge, pp 189–196
3. Resnik P (1995) Disambiguating noun groupings with respect to wordNet senses. In: Proceedings of the third workshop on very large corpora, MIT
4. Resnik P (1999) Semantic similarity in a taxonomy: an information-based measure and its application to problems of ambiguity in natural language. J Artif Intell Res 11:95–130
5. Wilks Y, Fass D, Guo C, McDonal J, Plate T, Slator B (1993) Providing machine tractablle dictionary tools. In: Pustejovsky J (ed) semantics and the lexicon, pp 341–401
6. Agirre E, Rigau G (1996) Word sense disambiguation using conceptual density. In: Proceedings of the 16th International Conference on Computational Linguistics (COLING), Copenhagen

7. Lesk M (1986) Automatic sense disambiguation using machine readable dictionaries: how to tell a pine cone from an ice cream cone. In: Proceedings of the 5th annual international conference on Systems documentation, Toronto
8. Krovetz R, Bruce Croft W (1992) Lexical ambiguity and information retrieval. Inf Syst 10(2):115–141
9. Yarowsky D (1992) Word sense disambiguation using statistical models of Roget's categories trained on large corpora. In: Proceedings of the 14th international conference on computational linguistics (COLING), Nantes, pp 454–460
10. Sussna M (1993) Word sense disambiguation for free-text indexing using a massive semantic network. In: Proceedings of the second international conference on information and knowledge management, Arlington
11. Watanabe N, Ishizaki S (2007) Neural network model for word sense disambiguation using up/down state and morphoelectrotonic transform. J Adv Comput Intell Intell Inf 11(7): 780–786
12. Mercer RL (1993) The mathematics of statistical machine translation. Comput Linguist 19(2):263–331
13. Tanaka T, Bond F, Baldwin T, Fujita S, Hashimoto C (2007) Word sense disambiguation incorporating lexical and structural semantic information. In: Proceedings of the 2007 joint conference on empirical methods in natural language processing and computational natural language learning, pp 477–485, Prague, (Association for Computational Linguistics) June 2007
14. Meenakshi C, Thangaraj P, Ramasamy M (2011) A novel scheme to identify the word sense in question answering systems. Int J Comput Sci Telecommun 2(9): 26–29

Author Index

R. Malathi and J. Krishnan (eds.), *Recent Advancements in System Modelling Applications*, Lecture Notes in Electrical Engineering 188,
DOI: 10.1007/978-81-322-1035-1, © Springer India 2013

Printed in the United States
By Bookmasters

Printed in the United States
By Bookmasters